Recent Advances of Novel Pharmaceutical Designs for Anti-cancer Therapies

Recent Advances of Novel Pharmaceutical Designs for Anti-cancer Therapies

Editor

Bernhard Biersack

MDPI • Basel • Beijing • Wuhan • Barcelona • Belgrade • Manchester • Tokyo • Cluj • Tianjin

Editor
Bernhard Biersack
Organic Chemistry
Laboratory
University of Bayreuth
Bayreuth
Germany

Editorial Office
MDPI
St. Alban-Anlage 66
4052 Basel, Switzerland

This is a reprint of articles from the Special Issue published online in the open access journal *International Journal of Molecular Sciences* (ISSN 1422-0067) (available at: www.mdpi.com/journal/ijms/special_issues/anticancer_pharmaceutical).

For citation purposes, cite each article independently as indicated on the article page online and as indicated below:

LastName, A.A.; LastName, B.B.; LastName, C.C. Article Title. *Journal Name* **Year**, *Volume Number*, Page Range.

ISBN 978-3-0365-8311-2 (Hbk)
ISBN 978-3-0365-8310-5 (PDF)

Contents

International Journal of
Molecular Sciences

Editorial

Editorial for the Special Issue—"Recent Advances of Novel Pharmaceutical Designs for Anti-Cancer Therapies"

Bernhard Biersack

Organische Chemie 1, Universität Bayreuth, Universitätsstrasse 30, 95440 Bayreuth, Germany; bernhard.biersack@uni-bayreuth.de or bernhard.biersack@yahoo.com

Cancer is one of the leading causes of death worldwide, despite the promising developments in terms of the curing and management of certain cancer types that have occurred over the last decades saving and prolonging the lives of numerous patients. Severe side effects, as well as intrinsic and acquired cancer drug resistance, pose constant problems, which lead to inefficient therapy outcomes and consequently to worse prognoses for affected patients. Although the development of a new anticancer drug is costly and takes much time, it is necessary to identify new drug candidates, which can overcome the drawbacks of the existing arsenal of cancer medicines, and which can become new treatment options for hitherto untreatable and/or highly aggressive cancer types.

The Special Issue "Recent Advances of Novel Pharmaceutical Designs for Anti-cancer Therapies" was launched in the autumn of 2021 in order to cover outstanding anticancer projects and to unveil new developments and breakthroughs in terms of pharmaceutical drug and target discovery. Eminent scientists working in the field contributed innovative methods, targets, and new interesting hit-and-lead candidates. At this point, eleven research and review articles were published in this Special Issue, and the contents of nine of them are briefly summarized in the following. For more detail, I strongly recommend reading the original full length open access articles published so far.

Zhou et al. described the synthesis and antitumor activities of new bis-cinnamoyl derivatives, which were prepared by simple methods from inexpensive starting materials [1]. The most promising compound I_{23} was active against three cancer cell lines in the low micromolar concentration range. Mechanistically, I_{23} induced apoptosis, stopped tumor cell migration, and stabilized microtubules.

The anticancer activity of the natural phenol pterostilbene was investigated by Wawszczyk et al. in melanotic and amelanotic melanoma models [2]. The described anti-melanoma effects of pterostilbene were mediated by p21, cyclin D1, and caspase-3. Interestingly, pterostilbene induced apoptosis in amelanotic melanoma cells in a caspase-dependent way, while apoptosis in melanotic melanoma cells remained unaffected by pterostilbene. These tumor-specific effects of pterostilbene deserve more profound investigation.

Curcumin is another prominent natural phenolic compound with promising anticancer properties, and Bhattacharyya et al. studied the effects of the fluorinated semi-synthetic curcumin derivative CDF and its new formulation with 2-hydroxypropyl-β-cyclodextrin (CDFHCD) on pancreatic ductal adenocarcinoma (PDAC) [3]. The CDFHCD formulation displayed increased stability and water solubility while conserving the anti-PDAC activity of CDF. Thus, the formulated CDF derivative CDFHCD appears to be a promising anticancer drug for clinical trials with PDAC patients in the future.

Adenosine deaminases acting on RNA (ADAR) are RNA-editing enzymes with relevance for various human diseases including cancer. ADAR2 was identified as a promising target for the design of inhibitors, and computer chemistry techniques (molecular docking and molecular dynamics simulation) were applied to identify potential ADAR2 inhibitors [4]. Various natural polyphenols, bis-naphthoquinones, and the well-known

Citation: Biersack, B. Editorial for the Special Issue—"Recent Advances of Novel Pharmaceutical Designs for Anti-Cancer Therapies". *Int. J. Mol. Sci.* **2023**, *24*, 8238. https://doi.org/10.3390/ijms24098238

Received: 28 April 2023
Accepted: 2 May 2023
Published: 5 May 2023

dye indigo were found to be promising ADAR2 inhibitor candidates, which should be considered for further studies on their anticancer properties as ADAR2 inhibitors.

Metal complexes are a promising class of anticancer compounds, and platinum-based drugs hold a salient position in the treatment of various solid tumors. Complexes of other metals were also found to be highly active against tumors. Schleser et al. prepared and evaluated a series of new iodidogold(I)-NHC (*N*-heterocyclic carbene) complexes in esophageal adenocarcinoma (EAC) cells [5]. The most active gold complexes of this study exhibited high antiproliferative activity, induced apoptosis, suppressed *c*-Myc and cyclin D1 expression, and inhibited EAC colony and spheroid formation.

The identification and validation of new cancer drug targets is of great importance for the development of new innovative anticancer drugs. Rotermund et al. studied the expression of carbonic anhydrases IX and XII in colorectal cancers, as well as the potential of these vital enzymes as possible drug targets [6]. Four consensus molecular subgroups were established based on their differing carbonic anhydrase expression states, and the response to treatment with the inhibitor SLC-0111 was clearly associated with target expression in colorectal cancer spheroids of these subgroups.

The specific targeting of the translation machinery of proliferating tumor cells is a promising strategy to fight cancer. Steinmann et al. identified the eukaryotic initiation factor 4A1 (eIF4A1) as a prognostic factor in hepatocellular carcinoma (HCC), whose molecular function can be suppressed by treatment with inhibitors [7]. These eIF4A1 inhibitors led to high antiproliferative and strong pro-apoptotic effects in HCC, and the anticancer activities of the eIF4A1 inhibitors were significantly augmented by the combination with a pan-mTOR inhibitor.

In addition to these original research manuscripts, three review articles were published in this Special Issue. One of them was provided by Zhang et al., which also deals with a promising target of the cellular translation machinery, the eukaryotic elongation factor 1A (eEF1A) [8]. Among the described eEF1A inhibitors, plitidepsin is already approved for multiple myeloma treatment, and another compound (metarrestin) is currently in clinical trials. In addition, further natural and synthetic eEF1A inhibitors with sound anticancer effects were described and discussed.

Finally, a comprehensive review on the application of the proteolysis-targeting chimera (PROTAC) strategy for the treatment of cancer was published in this Special Issue, which prompted a high interest among the readers of the International Journal of Molecular Sciences [9]. Bifunctional conjugates of small molecule target protein binders with E2 ligase binding scaffolds led to pronounced anticancer activities and remarkable curing effects. It is noteworthy that this strategy also holds considerable promise for the treatment of other human diseases.

All in all, this Special Issue covers a well-balanced collection of state-of-the-art topics in the prospering and competitive field of anticancer drug design and development, which can address and inspire oncologists, biologists, and medicinal chemists among the readership of this journal.

Conflicts of Interest: The authors declare no conflict of interest.

References

1. Zhou, X.; Fu, Y.-H.; Zou, Y.-Y.; Meng, J.; Ou-Yang, G.-P.; Ge, Q.-S.; Wang, Z.-C. Discovery of simple diacylhydrazine functionalized cinnamic acid derivatives as potential microtubule stabilizers. *Int. J. Mol. Sci.* **2022**, *23*, 12365. [CrossRef] [PubMed]
2. Wawszczyk, J.; Jesse, K.; Kapral, M. Pterostilbene-mediated inhibition of cell proliferation and cell death induction in amelanotic and melanotic melanoma. *Int. J. Mol. Sci.* **2023**, *24*, 1115. [CrossRef] [PubMed]
3. Bhattacharyya, S.; Ghosh, H.; Covarrubias-Zambrano, O.; Jain, K.; Swamy, K.V.; Kasi, A.; Hamza, A.; Anant, S.; VanSaun, M.; Weir, S.J.; et al. Anticancer activity of novel difluorinated curcumin analog and its inclusion complex with 2-hydroxypropyl-β-cyclodextrin against pancreatic cancer. *Int. J. Mol. Sci.* **2023**, *24*, 6336. [CrossRef] [PubMed]
4. Broni, E.; Striegel, A.; Ashley, C.; Sakyi, P.O.; Peracha, S.; Velazquez, M.; Bebla, K.; Sodhi, M.; Kwofie, S.K.; Ademokunwa, A.; et al. Molecular docking and dynamics simulation studies predict potential anti-ADAR2 inhibitors: Implications for the treatment of cancer, neurological, immunological and infectious diseases. *Int. J. Mol. Sci.* **2023**, *24*, 6795. [CrossRef] [PubMed]

5. Schleser, S.W.; Ghosh, H.; Hörner, G.; Seib, J.; Bhattacharyya, S.; Weber, B.; Schobert, R.; Dandawate, P.; Biersack, B. New 4,5-diarylimidazol-2-ylidene-iodidogold(I) complexes with high activity against esophageal adenocarcinoma cells. *Int. J. Mol. Sci.* **2023**, *24*, 5738. [CrossRef] [PubMed]
6. Rotermund, A.; Brandt, S.; Staege, M.S.; Luetzkendorf, J.; Mueller, L.P.; Mueller, T. Differential CMS-related expression of cell surface carbonic anhydrases IX and XII in colorectal cancer models–implications for therapy. *Int. J. Mol. Sci.* **2023**, *24*, 5797. [CrossRef] [PubMed]
7. Steinmann, S.M.; Sánchez-Martín, A.; Tanzer, E.; Cigliano, A.; Pes, G.M.; Simile, M.M.; Desaubry, L.; Marin, J.J.; Evert, M.; Calvisi, D.F. eIF4A1 is a prognostic marker and actionable target in human hepatocellular carcinoma. *Int. J. Mol. Sci.* **2023**, *24*, 2055. [CrossRef] [PubMed]
8. Zhang, H.; Cai, J.; Yu, S.; Sun, B.; Zhang, W. Anticancer small-molecule agents targeting eukaryotic elongation factor 1A: State of the art. *Int. J. Mol. Sci.* **2023**, *24*, 5184. [CrossRef] [PubMed]
9. Yao, T.; Xiao, H.; Wang, H.; Xu, X. Recent advances in PROTACs for drug targeted protein research. *Int. J. Mol. Sci.* **2022**, *23*, 10328. [CrossRef] [PubMed]

Molecular Sciences

Article

Molecular Docking and Dynamics Simulation Studies Predict Potential Anti-ADAR2 Inhibitors: Implications for the Treatment of Cancer, Neurological, Immunological and Infectious Diseases

Emmanuel Broni [1], Andrew Striegel [1,2], Carolyn Ashley [1], Patrick O. Sakyi [3,4], Saqib Peracha [1], Miriam Velazquez [1,5], Kristeen Bebla [5], Monsheel Sodhi [5], Samuel K. Kwofie [6,7], Adesanya Ademokunwa [1,8], Sufia Khan [1,9] and Whelton A. Miller III [1,5,*]

1 Department of Medicine, Loyola University Medical Center, Loyola University Chicago, Maywood, IL 60153, USA
2 Department of Chemical and Biochemistry, College of Science, University of Notre Dame, Notre Dame, IN 46556, USA
3 Department of Chemistry, School of Physical and Mathematical Sciences, College of Basic and Applied Sciences, University of Ghana, Legon, Accra P.O. Box LG 56, Ghana
4 Department of Chemical Sciences, School of Sciences, University of Energy and Natural Resources, Sunyani P.O. Box 214, Ghana
5 Department of Molecular Pharmacology & Neuroscience, Loyola University Medical Center, Loyola University Chicago, Maywood, IL 60153, USA
6 Department of Biomedical Engineering, School of Engineering Sciences, College of Basic & Applied Sciences, University of Ghana, Legon, Accra P.O. Box LG 77, Ghana
7 Department of Biochemistry, Cell and Molecular Biology, West African Centre for Cell Biology of Infectious Pathogens, College of Basic and Applied Sciences, University of Ghana, Accra P.O. Box LG 54, Ghana
8 Department of Cognitive and Behavioral Neuroscience, Loyola University Chicago, Chicago, IL 60660, USA
9 Department of Biology, Loyola University Chicago, Chicago, IL 60660, USA
* Correspondence: wmiller6@luc.edu

Citation: Broni, E.; Striegel, A.; Ashley, C.; Sakyi, P.O.; Peracha, S.; Velazquez, M.; Bebla, K.; Sodhi, M.; Kwofie, S.K.; Ademokunwa, A.; et al. Molecular Docking and Dynamics Simulation Studies Predict Potential Anti-ADAR2 Inhibitors: Implications for the Treatment of Cancer, Neurological, Immunological and Infectious Diseases. *Int. J. Mol. Sci.* **2023**, *24*, 6795. https://doi.org/10.3390/ijms24076795

Academic Editor: Bernhard Biersack

Received: 14 February 2023
Revised: 1 April 2023
Accepted: 4 April 2023
Published: 5 April 2023

Abstract: Altered RNA editing has been linked to several neurodevelopmental disorders, including autism spectrum disorder (ASD) and intellectual disability, in addition to depression, schizophrenia, some cancers, viral infections and autoimmune disorders. The human ADAR2 is a potential therapeutic target for managing these various disorders due to its crucial role in adenosine to inosine editing. This study applied consensus scoring to rank potential ADAR2 inhibitors after performing molecular docking with AutoDock Vina and Glide (Maestro), using a library of 35,161 compounds obtained from traditional Chinese medicine. A total of 47 compounds were predicted to be good binders of the human ADAR2 and had insignificant toxicity concerns. Molecular dynamics (MD) simulations, including the molecular mechanics Poisson–Boltzmann surface area (MM/PBSA) procedure, also emphasized the binding of the shortlisted compounds. The potential compounds had plausible binding free energies ranging from −81.304 to −1068.26 kJ/mol from the MM/PBSA calculations. ZINC000085511995, a naphthoquinone had more negative binding free energy (−1068.26 kJ/mol) than inositol hexakisphosphate (IHP) [−873.873 kJ/mol], an agonist and a strong binder of ADAR2. The potential displacement of IHP by ZINC000085511995 in the IHP binding site of ADAR2 could be explored for possible deactivation of ADAR2. Bayesian-based biological activity prediction corroborates the neuropharmacological, antineoplastic and antiviral activity of the potential lead compounds. All the potential lead compounds, except ZINC000014612330 and ZINC000013462928, were predicted to be inhibitors of various deaminases. The potential lead compounds also had probability of activity (Pa) > 0.442 and probability of inactivity (Pi) < 0.116 values for treating acute neurologic disorders, except for ZINC000085996580 and ZINC000013462928. Pursuing these compounds for their anti-ADAR2 activities holds a promising future, especially against neurological disorders, some cancers and viral infections caused by RNA viruses. Molecular interaction, hydrogen bond and per-residue decomposition analyses predicted Arg400, Arg401, Lys519, Trp687, Glu689, and Lys690 as hot-spot residues in the ADAR2 IHP binding site. Most of the top compounds were observed to have naphthoquinone, indole, furanocoumarin or benzofuran moieties. Serotonin and tryptophan, which are beneficial in digestive regulation, improving sleep cycle and mood, are indole derivatives. These chemical series may have the potential to treat neurological disorders, prion diseases, some cancers, specific viral infections, metabolic disorders and eating disorders through the disruption of ADAR2 pathways. A total of nine potential lead compounds were shortlisted as plausible modulators of ADAR2.

Keywords: adenosine deaminases acting on RNA (ADAR); anti-ADAR2; natural products; cancer; depression; anxiety disorders; autism spectrum disorder (ASD); molecular docking; molecular dynamics simulation

1. Introduction

RNA editing is a post-transcriptional process that alters RNA sequences with profound physiological consequences. The most common form of RNA editing in mammalian brain is catalyzed by the adenosine deaminases acting on RNA (ADAR) enzymes, ADAR1 and ADAR2, resulting in adenosine (A) to inosine (I) substitutions in mRNA and microRNAs, which is implicated in various diseases [1,2]. A-to-I editing is involved in regulating the immune system, RNA splicing, protein recoding, microRNA biogenesis, and in the formation of heterochromatin [3]. By catalyzing the conversion of A-to-I, these enzymes frequently alter the sequence of mRNA before translation, and potentially the primary structure of the resultant proteins [4]. Moreover, RNA editing of non-coding RNAs, e.g., microRNAs, can alter microRNA-mediated regulation of gene expression. RNA editing plays a critical role in the development of several organ systems including the brain, and has been associated with the molecular pathophysiology of a broad range of human diseases.

ADAR2-mediated RNA editing activity is crucial for the nervous system functions, thus altered RNA editing has been linked to neurological disorders [2,4,5]. A few studies linked increased RNA editing of the 5-Hydroxytryptamine (or serotonin) 2C receptor (5-HT2CR) to the neurodevelopmental disorder, Prader-Willi syndrome (PWS) [6]. Patients with PWS have a multisystem genetic disorder that includes cognitive disability and a behavioral phenotype with similarities to autism spectrum disorders [6,7]. Indeed, several studies of postmortem brain reveal higher levels of RNA editing in autism patients, specifically for the AMPA-type receptor subunits, kainate-type receptor subunits, NEAT1, CTSB and 5-HT2CR [8]. Other studies have shown that 5-HT2CR RNA editing is increased in the prefrontal cortex of individuals who die by suicide [9–16]. In rodents, increased 5-HT2CR RNA editing results in aggressive and anxiety-like behaviors, and other behaviors that may align with post-traumatic stress disorder [17–20]. Increased 5-HT2CR RNA editing decreases G-protein coupling of the receptor, thereby reducing its signal transduction and reducing the constitutive activity of the receptor [14,21].

It has been shown that a synthetic helix-threading peptide that binds 5-HT2CR editing sites was able to selectively inhibit ADAR2 editing of 5-HT2C RNA in vitro [22]. However, the bioavailability of such molecules is limited. Small molecule compounds that inhibit 5-HT2CR RNA editing have not been identified and these compounds are potential treatments for neurological illnesses that are triggered or exacerbated by psychological stress. Inhibition of ADAR-mediated RNA editing of the 5-HT2C receptor increases the activity of the 5-HT2C receptor, leading to reduced food intake [23–25]. This effect may be therapeutic in metabolic disorders, such as diabetes and in hyperphagia-induced obesity. This may also be therapeutic in binge-eating disorders, such as bulimia nervosa.

While there is an urgent need to identify inhibitors of ADAR2 that may have specificity for the RNA editing of 5-HT2CR, modifying the RNA editing of other ADAR targets may be beneficial. Epilepsy has been associated with increased RNA editing of glutamate receptors (kainate subtypes), the glutamate transporter EAAT2 (SLC1A2), CTSB, Rpa, Sparc, and OVCA2 (Figure 1). Therefore, inhibiting the RNA editing activity of ADAR2 may be beneficial for the treatment of some forms of epilepsy. In addition, one transcriptome-wide study showed increased RNA editing of more than 100 mRNAs in the frontal cortex of schizophrenia patients. Therefore, ADAR2 inhibition may be therapeutic in schizophrenia and other psychotic disorders [6].

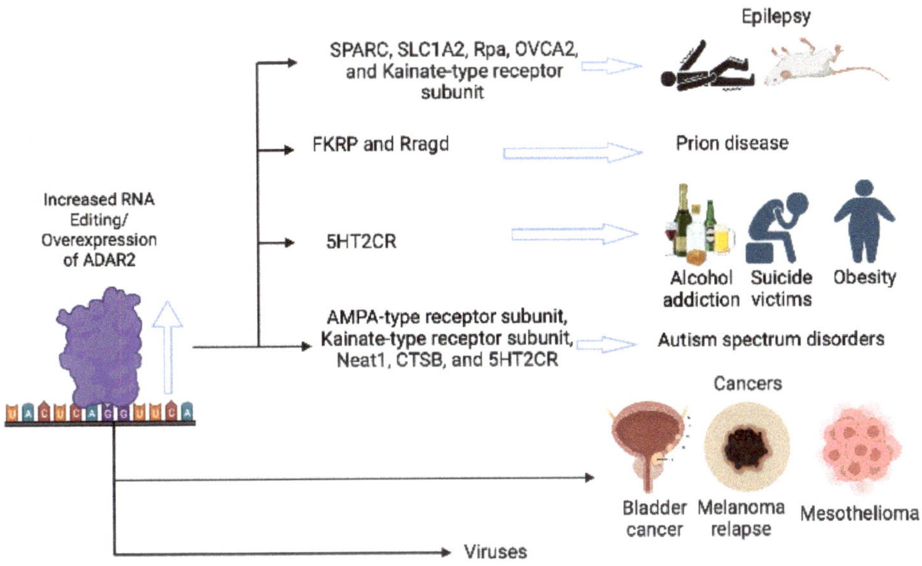

Figure 1. Schema representing the mechanism of ADAR2 overexpression (or increased RNA editing) and the associated diseases. The figure was created using BioRender (https://biorender.com/ accessed on 30 March 2023).

Prion diseases have been associated with increased RNA editing of FKRP and Rragd in mice, and may be improved by administration of ADAR inhibitor drugs. Moreover, elevated RNA editing activity has been reported as a major contributor to transcriptomic diversity in tumors [7], therefore an inhibitor of RNA editing may be therapeutic in some forms of cancer. ADAR enzymes modify the sequence of double-stranded RNA loop structures, which includes the nucleic acids within RNA viruses, often with proviral effects (Figure 1). Effects of editing have been documented only in a few groups of viruses, such as HIV, measles, Zika virus, SARS-CoV-2, mink enteritis virus (MEV), Ebola virus (EBOV), human parainfluenza viruses (HPIV), mumps virus (MuV), vesicular stomatitis virus (VSV), plant viruses and Sendai virus (SeV) [26]. Inhibition of ADAR enzymes may be therapeutic in the treatment of infections caused by some RNA viruses.

The availability of NMR and X-ray structures of the human ADAR2 (hADAR2) with the dsRNA of GluA2 offer an opportunity for the in silico screening of potential inhibitors of RNA editing [27,28]. This project therefore sought to employ structure-based drug design (SBDD) methods to identify natural products-derived compounds with potential inhibitory activity against the hADAR2 enzyme. SBDD approaches have become beneficial in identifying small molecules for treating various diseases [29] and natural products serve as an extensive reservoir of therapeutic candidates [30]. Natural products are structurally diverse and can be explored for the treatment of various disorders [31,32]. Since antiquity, traditional Chinese medicine has been used to improve neural regeneration and to repair neurological disorders [33–35]. Therefore, this study carefully predicted good binders of the ADAR2 with probable inhibitory activity by using structure-based virtual screening, molecular dynamics simulations and molecular mechanics Poisson–Boltzmann surface area calculations using compounds obtained from the Chinese flora and fauna. The study further predicted the biological activity and toxicity profiles of the identified biomolecules.

2. Results and Discussions

2.1. Prediction of Binding Sites

ADAR proteins share a common structure which consists of N-terminal dsRNA binding domains (dsRBDs) and a carboxy (C)-terminal catalytic deaminase domain [36,37]. The human ADAR2 (hADAR2) has been reported to contain two dsRBMs, which are separated by a 90-amino acid linker, and followed by the C-terminal catalytic domain [37]. The N terminus has been suggested to contain sequences that are involved in the auto-inhibition of the enzyme [37]. The IHP binds in the active site of the hADAR2, which is lined by residues Asn391, Ile397, Arg400, Arg401, Thr513, Lys519, Arg522, Gly530, Ser531, Lys629, Tyr658, Lys662, Tyr668, Lys672, Glu689, Lys690 and Asp695 [27,38]. A previous study has suggested that ADAR2 activation requires binding to an RNA which interacts with both double-stranded RNA binding motifs (dsRBMs) of the ADAR2 [37]. The IHP has also been suggested to activate ADAR2 upon binding [27,38].

Computed Atlas of Surface Topography of proteins (CASTp) predicted 52 potential binding sites, out of which only 3 were selected as plausible based on the area and volumes of the cavities (Figure 2 and Table 1). The other predicted sites either had no openings or were relatively small, such that no ligand could dock into [39,40]. Pocket 1 as predicted via CASTp (Figure 2 and Table 1) is consistent with the RNA binding loop of the ADAR2 [27,41], while pocket 2 is the IHP-binding site of the ADAR2 [27,38] (Figure 2 and Table 1).

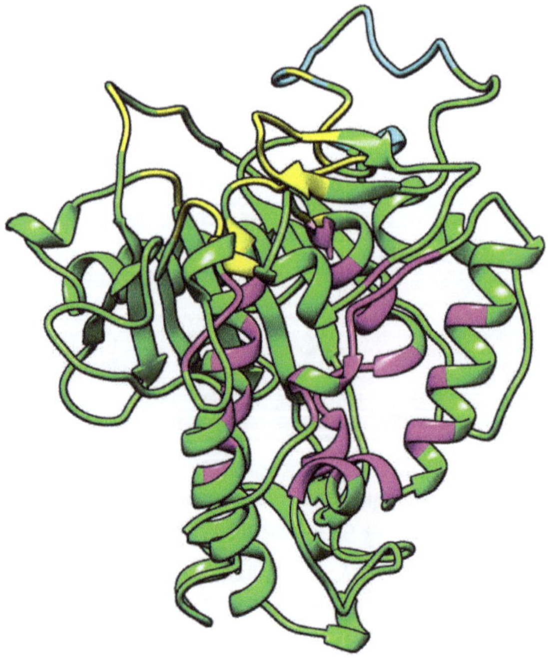

Figure 2. 3D structure of ADAR2 (PDB ID 5ED2) showing the predicted binding sites. Predicted binding sites are colored yellow, magenta and cyan for pockets 1, 2 and 3, respectively. Image was generated using UCSF Chimera version 1.16 [42].

Table 1. Predicted binding cavities via CASTp with their area, volumes and residues lining each pockets.

Pocket No.	Area (Å^2)	Volume (Å^3)	Residues Lining the Pocket
1	397.972	455.471	Lys350, Val351, Gly374, Thr375, Lys376, Cys377, Ile378, Asn379, His394, Ala395, Glu396, Ile446, Thr448, Ser449, Pro450, Cys451, Gly452, Arg455, Ile456, Pro459, Lys483, Ile484, Glu485, Ser486, Gly487, Gln488, Gly489, Thr490, Leu511, Thr513, Cys516, Arg590, Lys594, and Ala595.
2	541.281	342.976	Ala389, Leu390, Asn391, Asp392, Ile397, Arg400, Arg401, Leu404, Tyr408, Gln500, Leu512, Thr513, Met514, Lys519, Arg522, Trp523, Val526, Gly527, Ile528, Gln529, Gly530, Ser531, Leu532, Leu533, Lys629, Leu632, Tyr658, His659, Lys662, Leu663, Tyr668, Gln669, Lys672, Phe676, Trp687, Val688, Glu689, Lys690, Pro691, Thr692, Gln694, and Asp695.
3	120.812	129.477	Ser458, His460, Glu461, Pro462, Ile463, Glu466, Pro467, Ala468, Asp469, Arg470, His471, His552, Asp554, and His555.

2.2. *Molecular Docking*

2.2.1. Validation of Docking Protocols

The classical root mean square deviation (RMSD) was used to assess the ability of each docking tool to predict poses similar to that of the crystallographic structure. RMSD equal to or lower than 2 Å is considered good, 2 Å < RMSD < 3 Å is acceptable and RMSD > 3 Å is bad [43]. However, RMSD < 2 Å cut-off value is widely regarded as the most effective threshold value for validating correctly posed molecules [44,45]. The most negative IHP poses from AutoDock Vina and Glide had RMSD values of 1.6516 and 0.893 Å, respectively, when compared to the crystallographic or reference structure. The RMSD values obtained herein indicate the near accuracy pose prediction of AutoDock Vina and Glide for ADAR2 binders.

2.2.2. Molecular Docking via AutoDock Vina

A total of 25,131 compounds were successfully screened against the human ADAR2 protein. For each compound, AutoDock Vina generates up to nine conformers during docking. The pose or conformation with the most negative binding energy was selected as the best for each compound. The top 10% of the initial docking library (that is 2520 compounds) were shortlisted for analysis. The bound ligand, IHP had a binding energy of −8.6 kcal/mol when docked against the hADAR2. For the TCM natural products, ZINC000095913861 had the most negative binding energy when docked against the ADAR2 protein with a binding energy of −12.2 kcal/mol. ZINC000044350981 had a binding energy of −11.4 kcal/mol, while ZINC000070450936 and ZINC000070454365 both had a binding energy of −11.2 kcal/mol. Compounds ZINC000003203078, ZINC000085546044, ZINC000085593200, ZINC000085594057, and ZINC000103536976 had a binding energy of −11.1 kcal/mol.

The highest binding energy was observed to be −8.9 kcal/mol among the top 10%, which is lower than the −7.0 kcal/mol threshold specified for AutoDock Vina to differentiate between putative binders and non-binders of proteins [46]. The −7.0 threshold has been shown to filter ~95% of non-inhibitors while passing ~98% of known inhibitors [46]. The more negative binding energies of the shortlisted compounds (−8.9 kcal/mol and lesser) suggest that a greater number of the predicted ligands may be good ADAR2 binders. Additionally, the IHP had the least binding affinity (highest binding energy) to the hADAR2 as compared to the top 10%.

2.2.3. Molecular Docking via Maestro

All 37,398 TCM compounds generated via LigPrep were screened against the ADAR2 protein using Glide [47]. For compounds with two or more tautomers, the pose with the most negative binding energy was chosen. The control, IHP, had a binding energy of −7.97 kcal/mol. ZINC000103569281 had the most negative binding energy of −9.892 kcal/mol, followed by ZINC000013432666, ZINC000085488553, and ZINC000103584108 with docking scores of

−9.718, −9.606, and −9.546 kcal/mol, respectively. Among the top 10%, the compounds which had the highest binding energies (least affinity) with the ADAR2 protein had docking scores of −6.10 kcal/mol.

Most of the top compounds were observed to have indole, naphthoquinone, coumarin, furanocoumarin or benzofuran moieties in their structures. Compounds including ZINC000030726422, ZINC000070450892, ZINC000070454387, ZINC000070455595, and ZINC000085492822, among others were observed to have the indole moiety. Serotonin or 5-hydroxytryptamine (5-HT) and tryptophan also have the indole moiety. Tryptophan is the only precursor of serotonin production in humans [48]. Tryptophan has been shown to regulate mood and cognition [49]. Serotonin and tryptophan are known to contribute to well-being and happiness [50–53]. Serotonin reduces anxiety and promotes good mood and happiness by regulating brain function and cognition [54]. Serotonin also helps regulate the precursor to melatonin, a major control hormone of the sleep cycle [55–58] and digestive system regulation [49]. Selective serotonin reuptake inhibitors (SSRIs), including sertraline and fluoxetine, have been shown to have positive effects on patients with neurological or neurodevelopmental disorders including autism spectrum disorders, depression, anxiety, and obsessive-compulsive disorder [59–62].

Compounds ZINC000014690026, ZINC000070455383, ZINC000085488553, ZINC000085488571, and ZINC000085488602, among others had the benzofuran and/or coumarin moieties in their structure. Coumarins have recently been explored for treating neurological or brain disorders [63–67]. These chemical series need further exploration in order to identify more potent and effective ADAR2 inhibitors to increase our therapeutic arsenal against ADAR2-implicated disorders. De novo design and lead optimization via scaffold hopping, bioisosteric replacement, and other available methods will aid in identifying more effective and potent anti-ADAR2 compounds from these chemical series [68–71]. New pharmacologic agents from these series may demonstrate novel mechanisms of action with improved efficacy, tolerability and safety.

2.2.4. Shortlisting Compounds Based on Consensus Score

A total of 516 compounds were observed to be in the top 10% for both AutoDock Vina and Maestro. A consensus docking score was calculated by determining the averages of the AutoDock Vina binding energy and the Glide docking score of the top compounds. The 516 compounds were then ranked based on the consensus score. The known ADAR2 binder, IHP, had a consensus score of –8.26 kcal/mol (Table 2). ZINC000049888963 had the most negative consensus score of –9.58 kcal/mol followed by ZINC000044417732 (Table 2), ZINC000095909990, ZINC000095909350, and ZINC000085509666 with scores of –9.42, –9.41, –9.34, and –9.27 kcal/mol, respectively. Compounds ZINC000085505346, ZINC000085950180 (Table 2), and ZINC000095909587 also had consensus scores of –9.24, –9.19, and –9.13 kcal/mol, respectively. ZINC000085569204 had the least negative or highest consensus score of –7.50 kcal/mol among the 516 shortlisted compounds.

The study analyzed the effect of molecular weight on the binding scores. No compound with molecular weights 150 g/mol to 250 g/mol were found in the top 516 hits (Figure 3). It was also observed that the compounds with molecular weights between 250 g/mol and 350 g/mol had lower binding affinity (higher binding energy) than those with molecular weight above 350 g/mol (Figures 3 and S1A,B). All compounds with molecular weight <350 g/mol had consensus docking scores higher than −8.5 kcal/mol (Figure 3). Similar trends were observed for the AutoDock Vina and the Glide binding energy scores of the top 516 compounds. For AutoDock Vina and Glide, no compound with molecular weight <350 g/mol had binding energies lower than −10.4 (Figure S1A) and −7.7 kcal/mol (Figure S1B), respectively.

Table 2. Binding energies of IHP and some selected compounds with ADAR2. The interacting residues are also shown. The glide docking and consensus scores presented here are rounded to 2 decimal places.

Compound	Binding Energy (kcal/mol)			Interacting Residues		
	AutoDock Vina	Glide	Consensus Score	H-Bond	Pi-Cation	Salt Bridges
ZINC000044417732	−10.9	−7.95	−9.42	Arg401, Leu532, Lys662 and Glu689	Arg400 and Lys662	Arg400, Arg401, Lys629 and Lys662
ZINC000085950180	−10.5	−7.88	−9.19	Arg401, Ser531 (2), Lys629, Trp687 and Asp695	Lys662	-
ZINC000085511995	−10.9	−7.27	−9.08	Arg522, Lys629 and Trp687	Arg400 and Lys662	Arg401, Lys519, Lys629, Lys662 and Lys690
ZINC000085850673	−10.5	−7.24	−8.87	Arg401, Arg522, Lys629, Tyr658 and Lys690	Lys519 and Lys629.	Lys519 and Lys690
ZINC000085996580	−11	−6.62	−8.81	Arg401 (2), Lys519 and Lys690	Arg400, Arg522 and Lys662 (4)	-
ZINC000085734971	−10.6	−6.93	−8.76	Arg401, Ser531, Lys629, Lys662 (2) and Asp695	Lys629	-
ZINC000034517814	−9.4	−8.11	−8.75	Met514, Arg522, Ser531, Glu689 (2) and Lys690	-	-
ZINC000014613520	−9.9	−7.39	−8.64	Arg401, Ser531, Lys662 and Trp687	Arg522 and Lys662	Arg522
ZINC000095911588	−9.4	−7.86	−8.63	Arg401, Ser531, Arg522, Trp523 and Lys629	Lys662	Arg400
ZINC000085569519	−10.1	−6.74	−8.42	Tyr408, Lys629, Tyr658, Lys662 and Lys690	-	-
ZINC000008234342	−9.7	−7.08	−8.39	Leu532	Arg400 and Lys662	Arg400, Arg401 and Lys629
ZINC000085569292	−9.3	−7.48	−8.39	Arg400 (2), Arg401 and Lys662	-	Lys519 and Lys690
ZINC000095911414	−10.1	−6.65	−8.38	Arg401 (2), Ser531, Lys629 and Trp687	Arg400 and Lys662	-
ZINC000086050572	−9.8	−6.94	−8.37	Arg401, Ser531, Lys629, Lys662 (2) and Asp695	-	-
ZINC000014612330	−10.2	−6.51	−8.36	Arg401, Ser531 and Lys629	-	-
ZINC000014814624	−10.0	−6.57	−8.28	Arg522, Ser531 and Lys690	Arg400, Arg522 and Lys662	Arg400, Arg401 and Lys629
ZINC000004098700	−9.0	−7.50	−8.25	Arg401 (2), Arg522 and Ser531	Arg400 and Lys662	Lys519
ZINC000095912516	−9.8	−6.57	−8.18	Arg401, Arg522, Leu532, Lys629, Tyr658 and Glu689	-	-
ZINC000085488788	−9.5	−6.84	−8.17	Arg401, Ser531 and Lys629	Lys519	Arg401, Lys629 (2) and Lys662
ZINC000070454227	−9.0	−7.34	−8.17	Arg401, Arg522, Ser531 and Lys690	Lys662	Lys519 and Lys690
IHP	−8.6	−7.97	−8.26	Asn391, Arg400, Arg401 (2), Lys519, Ser531, Lys672, Trp687, Val688, Glu689 and Lys690	-	Arg401, Lys519, Arg522, Lys629, Lys662, Lys672 and Lys690

2.3. ADAR2-Ligand Interaction Profiling

The interaction profiles of the hADAR2 protein–ligand complexes for the shortlisted compounds were determined using Maestro. The best docking conformation from the Glide docking results were used to study the ADAR2-ligand interactions. IHP was predicted to form 12 hydrogen bonds with residues Asn391, Arg400, Arg401 (2 H-bonds), Lys519 (2 H-bonds), Ser531, Lys672, Trp687, Val688, Glu689, and Lys690 of the ADAR2 protein (Figure 4A and Table 2). It also formed salt bridges with Arg401, Lys519, Arg522, Lys629, Lys662, Lys672, and Lys690 (Figure 4A and Table 2). ZINC000049888963, which had the most negative consensus docking score, interacted via 7 H-bonds with the ADAR2 including residues Arg400, Lys519, Trp523, Gln529, Leu532, and Lys690 (2 H-bonds). It was also involved in 4 pi-cation interactions with Arg400, Arg522, and Lys629 (2 contacts). ZINC000044417732 formed hydrogen bonds with Arg401, Leu532, Lys662, and Glu689; pi-cation interactions with Arg400 and Lys662; and salt bridges with Arg400, Arg401, Lys629, and Lys662 (Figure 4B and Table 2).

ZINC000095909990 formed 8 hydrogen bonds with residues Asn391, Tyr408, Met514, Arg522, Lys629, Lys662, Val688, and Pro691. It also formed pi-cation interactions with Lys629 and Lys662, and four salt bridges with Arg400, Lys519 (2), and Lys690. ZINC000095909350 had four hydrogen bond interactions with residues Asp392, Ser531, His659, and Val688; five pi-cation interactions with Arg400, Arg401, Lys519, Lys662, and Lys672; and five salt bridges with residues Arg400, Arg401, Lys519, Lys629, and Lys662. ZINC000085509666 interacted with ADAR2 via 10 hydrogen bonds with residues Asn391, Leu532, Gln529, Lys629, Tyr658, Lys662, Glu689 (3 H-bonds), and Lys690; 2 salt bridges with Lys519 and Lys690; and formed a pi-cation with Arg400.

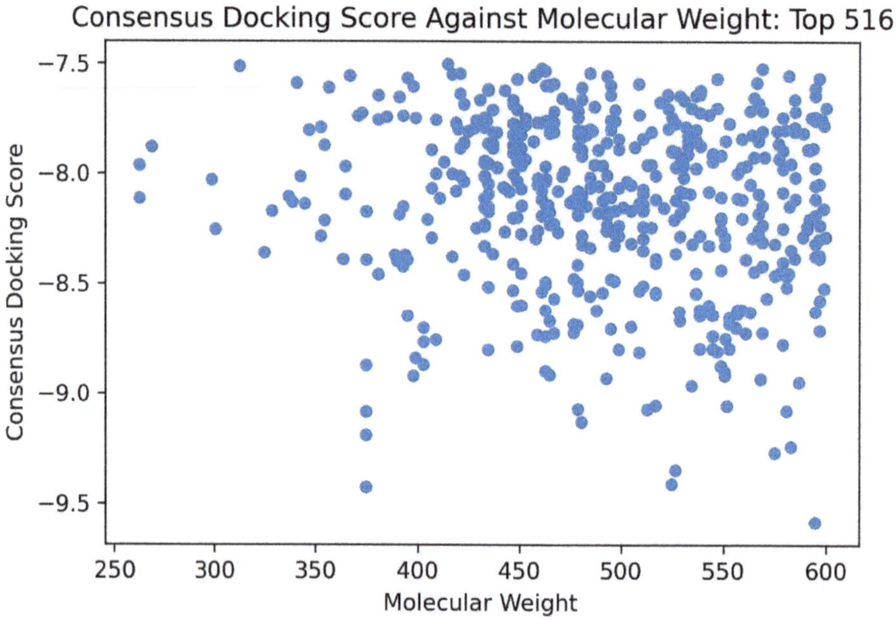

Figure 3. Consensus docking score against the molecular weight plot of the top 516 shortlisted compounds from molecular docking.

ZINC000085505346 interacted with ADAR2 via hydrogen bonds with Arg400, Arg401 (2), Trp523, Gln529, Leu532, Lys629, His659, and Trp687; pi-cation interaction with Arg522; and salt bridges with Arg522, Lys672, and Glu689. Compound ZINC000085950180 formed six hydrogen bonds with Arg401, Ser531 (2 H-bonds), Lys629, Trp687, and Asp695 and a pi-cation interaction with Lys662 (Table 2). ZINC000095909587 was involved in 6 hydrogen bonds with residues Arg522, Ser531, Lys629, Tyr668, Trp687, and Val688; and formed 2 salt bridges with Lys519 and Lys690. Multiple hydrogen bonds existing in protein–ligand complexes influence ligand binding affinity and contributes to ligand activity [72,73]. Residues Arg400, Arg401, Lys519, Ser531, Leu532, Trp687, and Lys690 were observed to interact with most of the ligands and may be crucial for ligand binding. The protein–ligand interaction profiles of the potential lead compounds are shown in Figures 4B and S2A–H.

2.4. ADMET Prediction

The top 516 compounds were subjected to ADMET testing via SwissADME to predict their pharmacokinetic profiles [74]. The compounds were assessed using Lipinski's and Veber's rules [75–79]. Lipinski's rule of five requires an orally active drug to have less than two violations of four criteria: less than 5 hydrogen bond donors; less than 10 hydrogen bond acceptors; molecular mass less than 500 Da; and an octanol-water partition coefficient (logP) less than 5 [75,79]. Veber's rule, on the other hand, requires compounds with good oral bioavailability to have not more than 10 rotatable bonds and topological polar surface area (TPSA) not more than 140 Å^2 [78]. Fluoxetine (antidepressant), nebularine (ADAR inhibitor) and doxorubicin (anti-cancer) were used as controls for the ADMET prediction. Fluoxetine and nebularine were predicted to not violate both Lipinski's and Veber's rules, while doxorubicin had 3 and 1 Lipinski's and Veber's violations, respectively (Table 3).

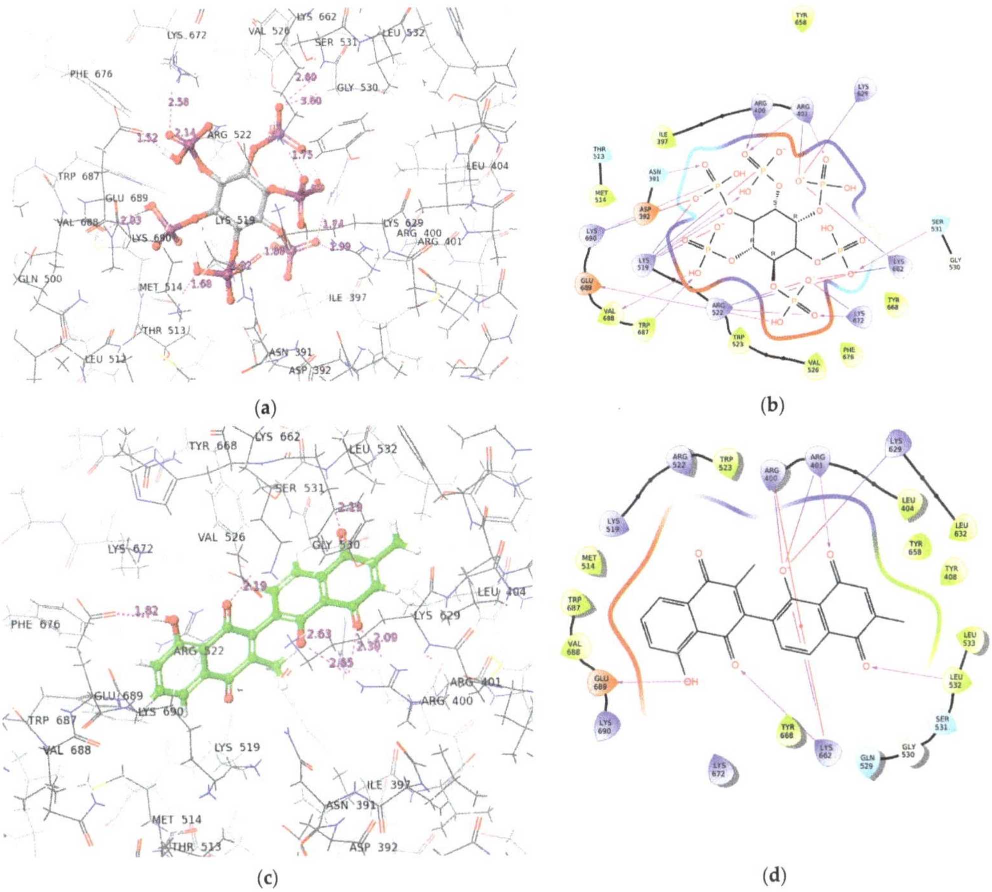

Figure 4. Conformation and protein–ligand interaction profiles of the binding pose with the most negative binding energy of IHP (**a**,**b**) and ZINC000044417732 (**c**,**d**). For the interaction profiles, purple arrows, red lines and combination of "red and blue" lines represent hydrogen bonds, pi-cation interactions and salt-brides, respectively.

Table 3. Pharmacokinetic evaluation of IHP, fluoxetine, nebularine, doxorubicin, and the 9 potential lead compounds. The consensus logP value (SwissADME) is reported in this table.

Compound	MW (g/mol)	logP o/w	TPSA (Å^2)	BBB Permeant	GI Absorption	ESOL Solubility Class	No of Lipinski's Rule Violations	No. of Veber's Rule Violations
IHP	660.04	-6.77	459.42	No	Low	High	3	2
ZINC000044417732	374.34	2.88	108.74	No	High	Moderate	0	0
ZINC000085950180	374.34	2.71	108.74	No	High	Moderate	0	0
ZINC000085511995	374.34	2.82	108.74	No	High	Moderate	0	0
ZINC000085850673	374.34	2.9	108.74	No	High	Moderate	0	0
ZINC000085996580	508.48	4.47	136.66	No	Low	Poor	1	0
ZINC000085734971	402.4	3.39	108.74	Yes	High	Moderate	0	0
ZINC000014612330	324.37	3.05	63.6	Yes	High	Moderate	0	0
ZINC000100513617	262.26	2.16	58.2	Yes	High	Moderate	0	0
ZINC000013462928	352.34	3.22	63.22	No	High	Moderate	0	0
Fluoxetine	309.33	4.32	21.26	Yes	High	Moderate	0	0
Nebularine	252.23	−1.16	113.52	No	High	Very	0	0
Doxorubicin	543.52	0.44	206.07	No	Low	Soluble	3	1

Compounds which violated two or more components of Lipinksi's rule of five and/or Veber's rule were eliminated. A total of 343 and 447 compounds failed Lipinski's and Veber's rules, respectively. Compound ZINC000049888963, which had the most negative consensus docking score, failed both Lipinski's and Veber's rules with 3 and 1 violations, respectively. ZINC000049888963 had a TPSA of 216.58 Å^2, molecular weight of 594.52 g/mol, 13 hydrogen bond acceptors and 7 hydrogen bond donors. ZINC000044417732, the second most negative compound with consensus score of −9.42 kcal/mol, passed both Lipinski's and Veber's rules, having a TPSA of 108.74 Å^2; 6 and 2 H-bond acceptors and donors, respectively; and a molecular weight of 374.34 g/mol. In total, 69 compounds passed both Lipinski's and Veber's rules and were thus considered for further analysis. The molecular weight of the 69 compounds ranged between 262 and 599 g/mol, while their TPSAs ranged between 58 and 140 Å^2.

Most neuroactive and psychoactive drugs bypass the blood-brain barrier (BBB) to be effective [80]. Thus the BBB permeability of the shortlisted compounds and controls were predicted. Fluoxetine was predicted as BBB permeant while nebularine and doxorubicin were predicted as non-permeant (Table 3). Only four compounds, comprising ZINC000013462928, ZINC000014612330, ZINC000100014196, and ZINC000100513617 were predicted to permeate the blood-brain barrier, making these compound interesting candidates to probe further. The BBB is a huge challenge for the development of neuroprotective drugs [81]. However, studies have shown that there are other effective ways of administering drugs bypassing the BBB [81–84]. The intranasal route, which is a noninvasive delivery route to bypass the barrier, can be exploited to administer compounds which cannot permeate the BBB [81,82]. Additionally, direct brain administration, which is an invasive approach to target the brain region with therapeutic molecules, can also be considered [82].

OSIRIS Datawarrior version 5.5.0 was also employed to predict toxicity properties including mutagenicity, tumorigenicity, irritancy and reproductive effects of the 69 shortlisted compounds [85] (Table S1). Datawarrior predicts the toxicity risk of compounds by classifying them as "none", "low" or "high" for each of the four toxicity properties. Compounds that were predicted to possess low or high toxicity for more than one property were eliminated. Moreover, compounds that were predicted to be either mutagenic or tumorigenic were also removed from the potential hit list. Mutagens and tumorigenic agents were eliminated because ADAR2 inhibition by death associated with protein 3 (DAP3) has been reported to be involved in certain cancer type development and progression [86]. Down-regulation of ADAR2 in lung cancers has been previously reported [87]. ADAR2 is a tumor suppressor and its inhibition may promote tumor growth [88–90]. Thus, ADAR2 inhibitors with mutagenic or tumorigenic properties may promote or speed up neoplasm due to ADAR2 inhibition.

A total of 51, 64, 52 and 52 compounds were predicted to be non-mutagenic, non-tumorigenic, non-irritant and exhibiting no reproductive effects, respectively (Table S1). However, only 29 compounds, including the top 3 comprising ZINC000044417732, ZINC000085950180 and ZINC000085511995, were predicted as non-mutagenic, non-tumorigenic, non-irritant and with no reproductive effects (Table S1). Of the 4 compounds that were predicted to be BBB permeants, 3 comprising ZINC000014612330, ZINC000100014196, and ZINC000100513617 were predicted to be non-tumorigenic, non-mutagenic, non-irritant and had no reproductive effects, while ZINC000013462928 was predicted to pose no toxicity risk except for its high reproductive effect (Table S1).

A total of 13 compounds were predicted to have two or more toxicity risks and were thus eliminated from the study. Moreover, a total of 9 compounds comprising ZINC000085569519, ZINC000085569501, ZINC000003918875, ZINC000085569417, ZINC000085569484, ZINC000085594490, ZINC000085569502, ZINC000085569474, and ZINC000059588402 were predicted to have high mutagenic effects while 9 others posed low mutagenic risks. The 9 compounds with low mutagenic risks include ZINC000014613520, ZINC000014814624, ZINC000004098700, ZINC000095912516, ZINC000015214955, ZINC000015254000, ZINC000095911983, ZINC000059587863, and ZINC000005714910. Furthermore, two (ZINC000059588402 and ZINC000085569204) and three compounds (ZINC000085594490, ZINC000103543244, and ZINC000059587863) were predicted to have high and low tumorigenic effects, respectively, and were also eliminated from the study

(Table S1). In all, 47 compounds passed the toxicity test and were considered as potential lead compounds with insignificant safety and toxicity concerns (Table S1).

2.5. Structural Similarity Search and Prediction of Biological Activity of Shortlisted Compounds

The biological activity of the identified compounds were predicted using the Prediction of Activity Spectra of Substances (PASS) by submitting the SMILES format of the compounds as inputs [91–93]. PASS predicts the activity of a molecule based on the similarity of the query with the structures of molecules available in its training dataset. The probability "to be active" (Pa) provides an estimation for the likelihood of the query compound to belong to sub-set of active compounds, whereas the probability "to be inactive" (Pi) provides an estimation for the query to be similar to the inactive subset of PASS training dataset. Compounds whose activities have Pa > Pi are considered likely to exhibit those activities.

Four known drugs namely quercetin, nebularine, lovastatin, and simvastatin were used as controls. Quercetin and nebularine, which are known adenosine deaminase (ADA) inhibitors [94–98], were used as positive controls while lovastatin and simvastatin were the negative controls. Lovastatin and simvastatin are both 3-hydroxy-3-methylglutaryl-CoA (HMG-CoA) reductase inhibitors which suppress cholesterol synthesis (IC_{50} < 5 μM) [99]. Both compounds did not inhibit ADA nor increase adenosine in hepatocytes, thus having no effect on autophagy [99]. Quercetin on the other hand demonstrated ADA inhibition with an IC_{50} of 170 μM (0.17 mM) in rats [94] while nebularine has been reported to inhibit ADA with an IC_{50} of 90 μM [100]. Nebularine is also a known ADAR inhibitor [98]. Both lovastatin and simvastatin showed no deaminase inhibition properties by the PASS predictions, supporting previous study that they may not be ADA inhibitors [99].

On the other hand, nebularine was predicted by PASS to be an inhibitor of AMP deaminase 2 and 3, and 11 other deaminases including adenine (Pa: 0.942 and Pi: 0.000), adenosine (Pa: 0.619 and Pi: 0.002), glucosamine-6-phosphate (Pa: 0.519 and Pi: 0.006), cytosine (Pa: 0.486 and Pi: 0.001), guanine (Pa: 0.404 and Pi: 0.003), deoxycytidine (Pa: 0.350 and Pi: 0.002), cytidine (Pa: 0.298 and Pi: 0.008), dCTP (Pa: 0.275 and Pi: 0.001), dCMP (Pa: 0.193 and Pi: 0.001), blasticidin-S (Pa: 0.205 and Pi: 0.066), and AMP (Pa: 0.118 and Pi: 0.001) deaminases. Nebularine was also predicted to be antineoplastic (Pa: 0.702 and Pi: 0.026) and antiviral (Pa: 0.886 Pi: 0.002). Quercetin was also predicted to be inhibitors of 8 deaminases including pterin (Pa: 0.352 and Pi: 0.104), blasticidin-S (Pa: 0.213 and Pi: 0.059), creatinine (Pa: 0.184 and Pi: 0.067), adenosine (Pa: 0.103 and Pi: 0.014), ornithine cyclodeaminase (Pa: 0.165 and Pi: 0.113), cytidine (Pa: 0.101 and Pi: 0.079), deoxycytidine (Pa: 0.093 and Pi: 0.070), and glucosamine-6-phosphate (Pa: 0.105 and Pi: 0.083) deaminases. Quercetin was also predicted as antineoplastic (Pa: 0.797 and Pi: 0.012) and antiviral (Pa: 0.498 and Pi: 0.005). The relatively higher Pa values (and lower Pi) for nebularine supports nebularine's better ADA inhibition than quercetin.

ZINC000044417732, ZINC000085950180, and ZINC000085511995 were predicted as inhibitors of various deaminases including ornithine cyclodeaminase, pterin, creatinine, glucosamine-6-phosphate, cytosine, deoxycytidine and ATP deaminases. The ADAR2 also belongs to the deaminase class of proteins, implying the potential of these compounds to inhibit the ADAR2. Furthermore, ZINC000014612330, ZINC000044417732, ZINC000085950180, and ZINC000085511995 were predicted to be useful in the treatment of acute neurologic disorders with Pa values of 0.627, 0.562, 0.562, and 0.442; and Pi values of 0.031, 0.051, 0.051, and 0.116, respectively. ZINC000085511995 and ZINC000100513617 were further predicted as antineurotic. Some of the compounds were predicted as neurotransmitter uptake inhibitors. Neurotransmitter uptake inhibitors, such as the SSRI class of drugs including sertraline and fluoxetine, have positive effects on depression, anxiety disorders, and certain types of obsessive-compulsive disorder [59–61]. Sertraline and fluoxetine are also beneficial for treating depression in epilepsy patients since they also lower the risk of triggering seizures [101].

ZINC000085511995 was predicted as an antidyskinetic, which might be relevant in treating dyskinesia in most neurological disorders. Most of the compounds were also predicted to be good for movement disorder treatment. Most neurological disorders including schizophrenia and autism spectrum disorder, are characterized by movement disorders [102–107]. The compounds were also predicted as kidney function stimulants with Pa values of 0.524, 0.526, and 0.711 and Pi values of 0.082, 0.080, and 0.008, respectively. Various studies have associated schizophrenia with high chronic kidney disease (CKD) risk [108,109], therefore these compounds may be helpful in treating renal complications associated with these disorders.

All the potential lead compounds, except ZINC000014612330 and ZINC000013462928, were predicted as inhibitors of various enzymes involved in phosphoinositide processes including inositol oxygenase, glycosylphosphatidylinositol phospholipase D, CDP-diacylglycerol-inositol 3-phosphatidyltransferase, glycosylphosphatidylinositol diacylglycerol-lyase, guanidinodeoxy-scyllo-inositol-4-phosphatase, inositol-3-phosphate synthase, phosphatidylinositol diacylglycerol-lyase and inositol 1,4,5-triphosphate 3-kinase. Since IHP is required for ADAR2 activation [27,38], inhibiting its production and transfer may help limit ADAR2 activity. ZINC000044417732, ZINC000085950180, and ZINC000085511995 were predicted as apoptosis agonists with Pa values of 0.837, 0.623 and 0.707; and Pi values of 0.005, 0.024 and 0.014, respectively. DAP3, an ADAR2 inhibitor has been previously reported to be involved in mitochondrial physiology and cell apoptosis [110,111]. Moreover, reduced expression of ADAR1 and ADAR2 have been reported to increase apoptosis in a region-specific manner in the hippocampus, parietal cortex and sub-ventricular zone of seizure-exposed brains in mouse [112].

Moreover, ZINC000044417732, ZINC000085950180, and ZINC000085511995 were predicted as Pin1 inhibitors with Pa values of 0.699, 0.702, and 0.807; and Pi values of 0.007, 0.007, and 0.004, respectively. Previous studies have shown that ADAR2-Pin1 interactions stabilizes ADAR2 in the nucleus [113,114]. Additionally, nuclear interactions between Pin1 and ADAR2 were observed to increase as neurons develop or mature [114]. Pin1 is also required for the editing of expressed GluA2 transcripts in cell lines [113] and to regulate ADAR2 protein levels as well as catalytic activity [113]. Pin1 is also known to facilitate multiple cancer-driving processes [115], supporting the prediction of these compounds as potential anti-cancer molecules. Aberrant A-to-I RNA editing has been shown to be implicated in cancers [86,116]. DAP3, which is an ADAR2 inhibitor, has been reported to affect editing targets involved in cancer-related signaling pathways and processes [86]. However, ZINC000044417732, ZINC000085950180, and ZINC000085511995 were predicted to be antineoplastics with Pa values of 0.839, 0.563 and 0.805; and Pi values of 0.008, 0.053 and 0.011, respectively. Furthermore, ZINC000044417732, ZINC000085950180, ZINC000085511995, and ZINC000085850673 were predicted as antimutagenic with Pa values of 0.793, 0.777, 0.845, and 0.795; and Pi of 0.004, 0.004, 0.003, and 0.004, respectively. They were also predicted to be beneficial in treating preneoplastic conditions.

ZINC000085511995 (mamegakinone) was also predicted to be similar to vitamin K with a Pa of 0.39 and a Pi of 0.002. Further structural similarity search via DrugBank revealed that ZINC000085950180 (isozeylanone) is similar to phylloquinone (vitamin K1), menatetrenone (vitamin K2), menaquinone-7, menaquinone-6 and menaquinone with similarity scores of 0.75, 0.748, 0.748, 0.748, and 0.748, respectively. ZINC000085734971 was also predicted to be structurally similar to these compounds. All the aforementioned compounds are naphthoquinones, a class of compounds widely known for their anti-cancer properties [117–120]. Naphthoquinones have demonstrated remarkable applications in medicinal chemistry owing to their good synthetic accessibility, making it possible to obtain many chemical substances [121]. Compounds ZINC000085950180 and ZINC000085511995 are composed of two naphthoquinones each in their structures. High levels of vitamin K has been shown to improve cognitive function and its deficiency may be associated with an increased risk of cognitive decline and dementia in older adults [122,123].

ZINC000014612330 is structurally similar to lapachone with a score of 0.636. β-lapachone has been reported to attenuate cognitive impairment and neuroinflammation in an Alzheimer's disease mouse model [124]. ZINC000034517814 is structurally similar to acteoside and echinacoside, both with a similarity score of 0.752. Both compounds have been shown to be beneficial in the treatment of neurological conditions including Parkinson's and Alzheimer's diseases [125–127]. Echinacoside has been shown to selectively reverse dopaminergic neuronal injury in rat models of Parkinson's disease induced by rotenone [126]. However, echinacoside and acteoside have significant pharmacokinetics concerns as they have very fast absorption and elimination rates in rats [128]. In humans, echinacoside was not identified in plasma samples after echinacea tablets were ingested [129]. Herein, SwissADME predicted a bioavailability score of 0.17 for both echinacoside and acteoside, which corroborates the poor bioavailability of both compounds [128,129]. However, ZINC000034517814 had a better score of 0.55, suggesting that ZINC000034517814 has the best oral bioavailability compared to echinacoside and acteoside [130].

ZINC000100014196 (indigo), ZINC000100513617 (indigo) and indirubin are isomers [131,132]. About 2.3 to 23.3 μg/kg of an indirubin derivative, and 7-bromoindirubin-3-oxime, has been recently shown to prevent β-amyloid (Aβ) oligomer-induced impairments of spatial cognition and recognition in mice and has been suggested for the treatment of Alzheimer's disease [131]. Its anti-cancer activity at an IC_{50}s less than 35 μM has also been reported [132,133]. ZINC000013462928 is structurally similar to arctigenin and matairesinol with scores of 0.632 and 0.626, respectively. Both arctigenin and matairesinol have been reported to have neuroprotective, anti-diabetic and anti-depressive activities [134–136]. The predicted biological activities of the shortlisted compounds corroborate their potential anti-ADAR2 activity.

2.6. Molecular Dynamics Simulations

This study investigated the structural conformation changes and atomic motions by performing 100 ns MD simulations on the unbound protein and ten ADAR2-ligand complexes. The protein–ligand complexes investigated include ADAR2 bound with IHP, ZINC000044417732, ZINC000085950180, ZINC000085511995, ZINC000085850673, ZINC000085996580, ZINC000085734971, ZINC000014612330, ZINC000100513617, and ZINC000013462928. The top six compounds based on the most negative consensus docking score and good toxicity profiles and three compounds that were predicted as BBB permeants (ZINC000085734971, ZINC000014612330, and ZINC000100513617) were selected for the MD simulations. ZINC000100014196 (indigo) and ZINC000100513617 are isomers and thus the latter was selected for the MD simulation due to its higher binding affinity. The structural stability, folding and conformational fluctuations of the proteins caused by ligand binding were investigated using the RMSD, radius of gyration (Rg) and the root mean square fluctuation (RMSF) after the 100 ns simulations. Hydrogen bond analysis throughout the 100 ns simulation period were performed for each complex. Snapshots were also generated to assess the position of the ligands with respect to the ADAR2 protein.

2.6.1. RMSD of ADAR2 and ADAR2-Ligand Complexes

The RMSD measures the deviation of the final conformation of the protein's backbone with regards to the protein's initial structural conformation [137]. RMSD provides insights into the stability of a protein—a protein with lower RMSD tends to be more stable than a protein with higher deviations in its backbone [137,138]. Herein, all structures reached equilibrium after ~15 ns (Figures 5 and S3A). The unbound protein experienced minor fluctuations until about 50 ns and then remained stable till the end of the 100 ns simulation period with an average RMSD of 0.23 nm (Figure 5). The ADAR2-IHP (Figure 5) and the ADAR2-ZINC000014612330 (Figure S3A) complexes were observed to be more stable than the unbound protein throughout the simulation period, maintaining RMSD averages of 0.21 and 0.22 nm, respectively. Although the ADAR2-IHP complex showed the greatest stability among all the structures, the ADAR2-ZINC000014612330 complex demonstrated comparable stability (Figure S3A). The stability of the ADAR2-IHP is not surprising since the IHP is a known binder of ADAR2 and is required for ADAR2 activation [27,38].

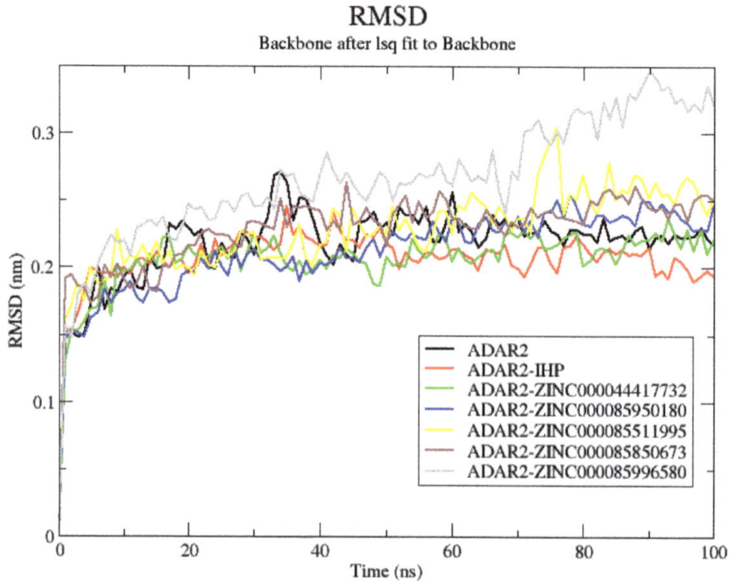

Figure 5. RMSD plot of the unbound ADAR2 and ADAR2-ligand complexes. The unbound protein, ADAR2-IHP, ZINC000044417732, ZINC000085950180, ZINC000085511995, ZINC000085850673, and ZINC000085996580 complexes are colored black, red, green, blue, yellow, brown, and grey, respectively.

The ADAR2-ZINC000085511995 complex was mostly stable throughout the simulation period. It reached relative equilibrium at around 10 ns (0.225 nm), maintained stability till 75 ns where the RMSD rose to ~0.3 ns and fell to 0.25 nm (around 80 ns) and remained stable till the end (Figure 5). The ADAR2-ZINC000100513617, ADAR-ZINC000085734971, and ADAR2-ZINC000085850673 complexes experienced similar RMSD trends (Figures 5 and S3A). The ADAR2-ZINC000100513617 complex rose to 0.227 nm at around 50 ns and remained stable till the end of the simulation time with an average RMSD of 0.23 nm (Figure S3A). The RMSD of ADAR2-ZINC000085850673 complex rose to 0.25 nm at around 35 ns and maintained stability till the end of the simulation with an average of 0.25 nm (Figure 5). The ADAR2-ZINC00008573491 complex rose to 0.25 nm at around 25 ns and remained stable till the end of the 100 ns simulation period (Figure S3A).

The ADAR2-ZINC000085950180, ADAR2-ZINC000013462928, and ADAR2-ZINC000085996580 complexes were the least stable among all the structures. The ADAR2-ZINC000085950180 complex rose to 0.2 nm at around 25 ns, then rose to 0.25 nm at 50 ns till about 80 ns and then remained stable till the end of the 100 ns period with an average RMSD of 0.24 nm (Figure 5). For the ADAR2-ZINC000085996580 complex, the RMSD experienced a gradual rise till about 35 ns, maintained an average of 0.26 nm till about 70 ns, then rose to about 0.3 nm (Figure 5). The RMSD of the ADAR2-ZINC000013462928 complex rose to 0.25 nm after 17 ns and remained stable till about 70 ns, where it experienced a rise to 0.3 nm (around 80 ns) and maintained it till 100 ns (Figure S3A).

2.6.2. Radius of Gyration of ADAR2 and ADAR2-Ligand Complexes

The radius of gyration provides information about the stability, folding and compactness of a protein, and can be further defined as the RMSD of atoms from the centroid of a protein [139,140]. The Rg plots of all the structures (unbound protein and complexes) revealed good stability and compactness of the ADAR2 protein. The Rg values of all the systems ranged between 2.04 nm and 2.1 nm (Figures 6 and S3B), signifying their stable folding [140]. The unbound protein maintained an average Rg of about 2.085 nm till about

27 ns, then fell to 2.055 nm at around 40 ns, then it rose and maintained an average of 2.07 ns until the end of the simulation (Figure 6). The IHP maintained an average Rg of 2.06 nm throughout the simulation (Figure 6). The ADAR2-ZINC000044417732 complex was relatively stable, maintaining an average of 2.075 nm throughout the simulation, although a high rise in Rg (to about 2.095 nm) was observed between 50 and 60 ns (Figure 6). The ADAR2-ZINC000085950180, ADAR2-ZINC000014612330 and ADAR2-ZINC000085850673 complexes demonstrated similar trends (Figures 6 and S3B). All three complexes were very stable throughout the simulation period, maintaining an average of 2.065 nm. The Rg of the ADAR2-ZINC000085734971 complex was also very stable until 80 ns with an average of ~2.05 nm, where a rise to 2.06 nm was observed (Figure S3B).

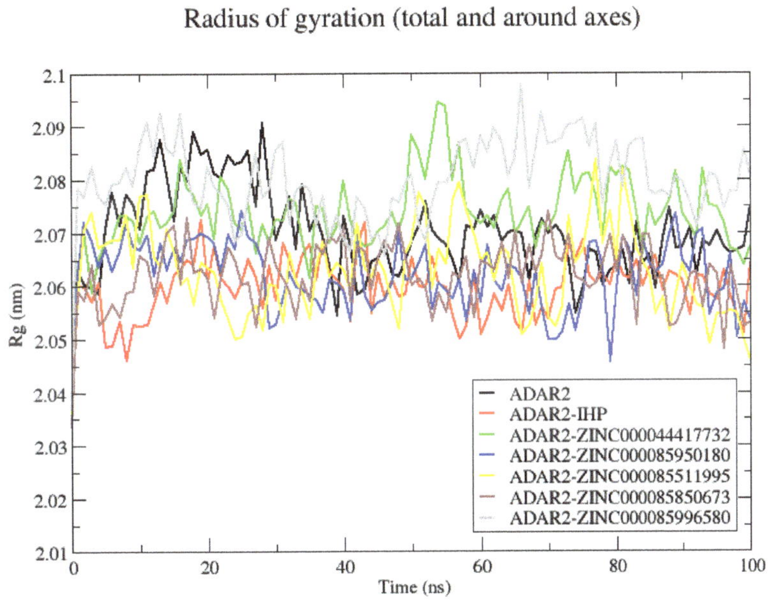

Figure 6. Radius of gyration plot of the unbound ADAR2 and ADAR2-ligand complexes. The unbound protein, ADAR2-IHP, ZINC000044417732, ZINC000085950180, ZINC000085511995, ZINC000085850673, and ZINC000085996580 complexes are colored black, red, green, blue, yellow, brown, and grey, respectively.

The Rg values of ADAR2-ZINC000085996580 complex fell until ~50 ns where a sharp rise was observed (Figure 6). The Rg rose to ~2.09 averagely until 90 ns and then fell again (Figure 6). The ADAR2-ZINC000100513617, ZINC000013462928, and ZINC000085511995 complexes also demonstrated fluctuations similar to that of the ADAR2-ZINC000085996580 complex. The ADAR2-ZINC000013462928 complex maintained quite a stable Rg with an average of 2.075 nm until ~80 ns and then declined to an average of 2.06 nm until the end of the 100 ns simulation time (Figure S3B). The Rg of the ADAR2-ZINC000100513617 complex, however, increased after 90 ns (Figure S3B). Collectively, all the complexes and the unbound protein had relatively low Rg values, indicating that they were stable and compact in their folding.

2.6.3. RMSF of ADAR2-Ligand Complexes

Herein, the RMSF of the various complexes were evaluated to determine the ADAR2 residues involved in the different mobility in the RMSD plots [139]. The RMSF in the binding cavity provides insights into the residues that make strong interactions with the ligand [139]. A high RMSF indicates the residues that do not make strong interaction, and thus have higher mobility [139].

For all the complexes, fluctuations were observed at similar residue indexes. Major fluctuations were observed at residue indexes 460–480, 490–515, and 582–600 (Figure S3C). Minor fluctuations were observed at residues 340–350, 360–370, 380–390, 415–430, 565–578, 620–630, and 650–665 (Figure S3C). ZINC000085996580 induced the highest fluctuations on the ADAR2 protein at residues Ala468 and Gln507 with RMSF values of 0.7276 and 0.7262 nm, respectively. For the ADAR2-ZINC000085996580 complex, residues Asp469, Arg470, and Leu506 were also observed to have high fluctuations with values of 0.6178, 0.647, and 0.6638 nm, respectively.

Residue indexes 390–415, 440–450, 515–560, 595–620, 635–650, and 660–700 experienced the least fluctuations, which is suggestive of their very strong interactions with the ligands in the active site [139]. For residue indexes 390–415, the interactions with Arg400 and Arg401 could be the reason for the low RMSF values, while Lys519, Ser531 and Leu532 account for the low fluctuations around the indexes 515–560. The interactions with residues Trp687, Glu689 and Lys690 account for the low RMSF observed in the residues 660–700 region.

2.6.4. Snapshots and Hydrogen Bond Analysis of Complexes

Since RMSD of the complex is calculated based on the protein backbone, snapshots were generated to verify the position of the ligands during the simulation [141]. Snapshots of each complex during the 100 ns MD simulation were generated at 25 ns intervals. For all the complexes, the ligands were tightly bound to the IHP binding site of the ADAR2. Additionally, aligning the resulting structures with the initial structure (time = 0 ns) as reference, further corroborated the stability of the complexes throughout the MD simulations. The RMSDs were determined for the structural alignment of each complex using the align module in PyMOL.

The RMSD values between the structures at times 25, 50, 75, and 100 ns when aligned to the initial structure (time = 0 ns) were 1.064, 1.161, 1.074, and 1.231 Å, respectively, for the ADAR2-IHP complex. For the ADAR2-ZINC000044417732 complex, RMSDs of 1.052, 1.264, 1.395, and 1.631 Å, respectively, were observed. For the ADAR2-ZINC000085950180 complex, 1.091, 1.064, 0.945 and 1.050 Å, respectively, were the observed RMSDs. Moreover, the ADAR2-ZINC000085511995 complex had RMSDs of 0.914, 0.932, 1.108, and 1.272 Å, respectively. The ADAR2-ZINC000085850673 structures at 25, 50, 75 and 100 ns also had RMSD values of 0.991, 1.204, 1.196, and 1.189 Å, respectively, when aligned to the initial structure. The ADAR2-ZINC000085734971 complex had RMSD values of 1.082, 1.149, 1.001, and 1.104, Å, respectively. The ADAR2-ZINC000100513617 complex also demonstrated stability with RMSD values of 1.359, 0.988, 0.996 and 1.292 Å, respectively. The relatively low RMSDs demonstrated by these complexes corroborate their good stability observed in the RMSD plots (Figures 5 and S3A).

The ADAR2-ZINC000014612330 complex also had RMSDs of 1.332, 1.322, 1.393, and 1.355 Å, when the initial structure was compared to the structures at times 25, 50, 75, and 100 ns, respectively. Although there were minor deviations from the starting structure, the ADAR2-ZINC000014612330 complex remained very stable after 50 ns (Figure S3A). Aligning the structures at 75 and 100 ns to the 50 ns structure revealed RMSDs of 0.890 and 0.816 Å, respectively, which are consistent with the RMSD plot (Figure S3A), implying that the ADAR2-ZINC000014612330 complex remained stable after 50 ns.

For the ADAR2-ZINC000085996580 complex, RMSDs of 1.414, 1.420, 1.330, and 1.360 Å were obtained when the structures at times 25, 50, 75, and 100 ns, respectively, were compared to the starting structure. Moreover, the ADAR2-ZINC000013462928 complex at times 25, 50, 75, and 100 ns demonstrated RMSDs of 1.291, 1.359, 1.415, and 1.233 Å, respectively, when aligned to the 0 ns structure. The relatively high RMSDs observed for ADAR2-ZINC000085996580 and ADAR2-ZINC000013462928 complexes (as compared to the other complexes herein) are consistent with the RMSD plot (Figures 5 and S3A) where the least stability (most deviation) was seen.

This study also analyzed the number of hydrogen bond interactions between the protein and each ligand during the MD simulations. Only hydrogen bond interactions within 0.35 nm and with an angle of 30° were reported using the GROMACS "gmx hbond" analysis (Figure 7). Snapshots of the complexes at 0 and 100 ns were visualized using Maestro and protein–ligand interaction profiles were also determined. The predicted number of hydrogen bonds via "gmx hbond" may be different from what is predicted via a molecular interaction visualization tool due to the cut-off.

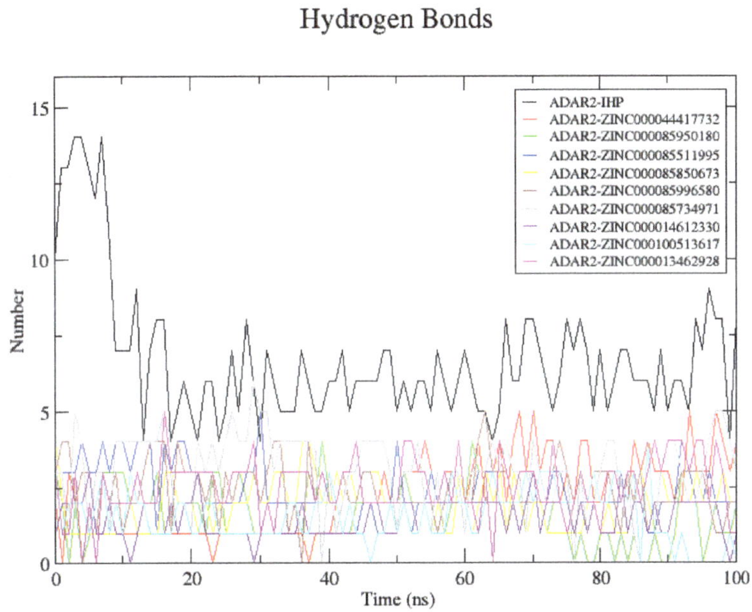

Figure 7. Hydrogen bond analyses of each ADAR2-ligand complex throughout the 100 ns MD simulation.

For the ADAR2-IHP complex, 10 hydrogen bonds were within the 0.35 nm distance and 30° angle threshold as predicted via GROMACS although 12 were identified from the protein–ligand interaction profile. The 12 hydrogen bonds were formed with residues Arg400 (2 H-bonds), Arg401 (2 H-bonds), Lys519 (2 H-bonds), Arg522, Tyr658, Trp687, Val688, Glu689, and Lys690. At 4 ns, the hydrogen bond interactions increased to 14, after which a decline in the number of hydrogen bonds was observed. At the end of the 100 ns simulation, 8 hydrogen bonds with Thr513, Lys519, Tyr658, Lys662, Glu689 (2 H-bonds), Lys690, and Gln694 were observed.

For the ADAR2-ZINC000013462928 complex, 3 hydrogen bonds were predicted by GROMACS. However, only 2 hydrogen bonds with Met514 and Arg400 were seen using Maestro at 0 ns. At 100 ns, 3 hydrogen bonds with Tyr408, Ser531, and Leu532 were observed via Maestro while GROMACS predicted only one hydrogen bond. The hydrogen bond interactions with Arg400 and Met514 were lost after the 100 ns period. For the ADAR2-ZINC000014612330 complex, there was only one hydrogen bond with Ser531, prior to the MD simulation. After the 100 ns simulations, the hydrogen bond with Ser531 was lost and 3 new hydrogen bonds were formed with Arg401 and Trp523 (2 H-bonds) although only one passed the 0.35 nm and 30° angle cut-off.

At the start of the simulation, ADAR2-ZINC000044417732 had 4 hydrogen bonds with Arg401 and Lys662 (2 each) and at the end of the simulation, the complex maintained the 4 hydrogen bonds. However, the contacts with Lys662 were lost and 2 new hydrogen bonds with Asn391 and Tyr658 were formed. For ADAR2-ZINC000085511995, 4 hydrogen bond contacts with Arg401, Lys519, Arg522, and Trp687 existed before the MD simulation.

After the simulation, only 3 were observed: 1 with Asn391 and 2 with Lys519. The ADAR2-ZINC000085734971 complex was predicted via "gmx hbond" to have 3 hydrogen bonds at both the initial and final states. However, after visualizing with Maestro, 4 hydrogen bonds with Ser531, Leu532, and Lys662 (2 bonds) were observed initially and 5 with residues Tyr408, Ser531 (2 bonds), Leu532, and Gln669 were observed at the end of the simulation. Compound ZINC000085850673 formed 3 hydrogen bonds with Met514, Lys519, and Tyr658 although 4 were predicted via GROMACS. After the simulation, 2 hydrogen bonds with Lys519 and Arg522 were observed. Hydrogen bond interactions with Met514 and Tyr658 were lost at the end of the simulation.

The ADAR2-ZINC000085950180 complex had 3 hydrogen bonds with Ser531, Leu532, and Asp695 initially and 3 hydrogen bonds with Arg401, Tyr408, and Asp695 after the simulation (2 H-bonds were reported via "gmx hbond"). ZINC000085996580 was predicted to form 3 hydrogen bonds with ADAR2 at initial time, although only 1 bond with Ser532 was observed via Maestro. After the 100 ns MD simulation, 2 hydrogen bonds with Arg400 and Gln669 were observed. ZINC000100513617 formed 2 hydrogen bonds with Arg401 before the simulation (although 3 were predicted by "gmx hbond") which were lost after the simulation. Multiple hydrogen bonds between a protein and a ligand have been shown to improve ligand binding and influence ligand activity [72,73]. However, the hydrogen bonds existing between both the donor and receptor must be either weaker or stronger than that between hydrogen and oxygen atoms in water, in order to influence ligand binding affinity [73]. Apart from ZINC000100513617, all the potential leads maintained hydrogen bonds with the ADAR2 throughout the simulation.

Residues Arg400, Arg401, Lys519, Ser531, and Leu532 were observed to be involved in hydrogen bond interactions with most of the potential lead compounds throughout the MD simulations, which is consistent with the interaction profiles from the molecular docking study.

2.7. Evaluating Potential Leads via MM/PBSA Calculation

The MM/PBSA method has become a more efficient and reliable approach to model protein–ligand interactions. MM/PBSA provides reasonable approximations for free energy calculations and is more reliable than the conventional molecular docking process, yielding higher enrichment factors than docking [142]. MM/PBSA helps to prioritize compounds for experimental testing [143]. The Gibbs free energy of binding ($\Delta G_{(bind)}$) can be determined using Equation (1) [144,145].

$$\Delta G_{(bind)} = \Delta G_{(complex)} - [\Delta G_{(receptor)} + \Delta G_{(ligand)}], \quad (1)$$

where $\Delta G_{(complex)}$, $\Delta G_{(protein)}$ and $\Delta G_{(ligand)}$ are the total free energies of the protein–ligand complex, protein, and ligand, respectively.

Herein, the MM/PBSA approach was employed to determine the binding free energies as well as the energy contributions per-residue of the ADAR2-ligand complexes. Ligand structures of the top nine potential lead compounds and IHP are presented in Figure 8, while those of the other compounds listed in Table 2 are shown in Figure S4. The other contributing energy terms, including van der Waals (vdW), electrostatic, polar solvation and solvent accessible surface area (SASA) energies were also computed (Table 4). The vdW energy of the complexes ranged between −138 and −177.93 with the ADAR2-IHP having the highest vdW energy (−138.816 kJ/mol) [Table 4]. ZINC000085734971 demonstrated the most negative vdW energy with −177.923 kJ/mol, followed by ZINC000013462928 (−177.25 kJ/mol) and ZINC000085996580 (−174.655 kJ/mol) [Table 4]. For SASA energy, the values ranged from −14.949 to −22.88 kJ/mol with ZINC000085996580 and ZINC000100513617 demonstrating the most and least negative energy values, respectively (Table 4). SASA is linearly related to the non-polar solvation energy [144,146,147]. The SASA term is usually small and has little variations among similar ligands [144].

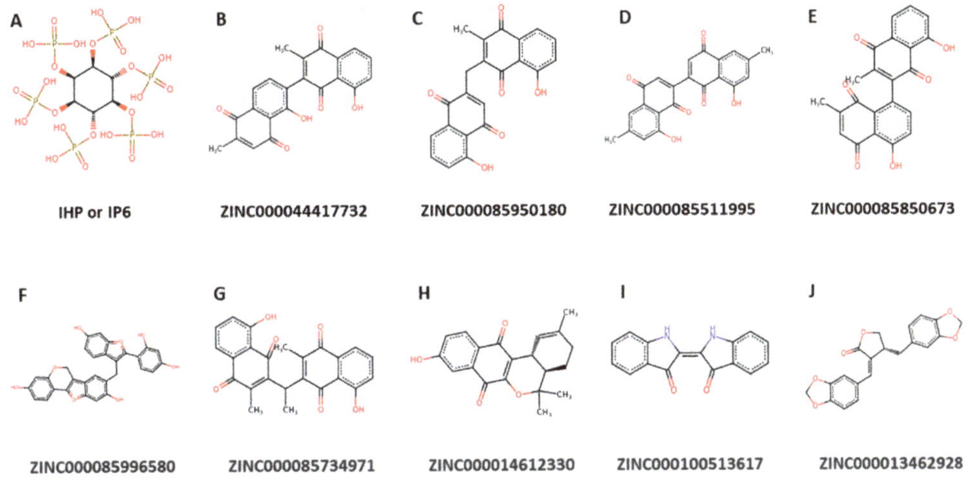

Figure 8. Chemical structures of IHP and the nine potential lead compounds.

Table 4. Contributing energy terms for the protein–ligand complexes determined via MM/PBSA calculations. The energy values are presented as "energy ± standard deviation". All energy values are in kJ/mol.

Compound	vdW	Electrostatic Energy	Polar Solvation Energy	SASA Energy	Binding Energy
IHP	−138.816 ± 2.243	−1597.111 ± 7.065	883.140 ± 5.474	−20.866 ± 0.095	−873.873 ± 6.225
ZINC000044417732	−164.45 ± 1.303	−798.403 ± 4.671	340.231 ± 3.618	−19.937 ± 0.089	−642.856 ± 3.746
ZINC000085950180	−172.457 ± 1.278	−66.957 ± 2.459	123.492 ± 2.987	−19.151 ± 0.101	−135.075 ± 3.285
ZINC000085511995	−157.3 ± 1.578	−1550.773 ± 5.372	659.676 ± 5.323	−19.816 ± 0.108	−1068.26 ± 4.122
ZINC000085850673	−170.78 ± 1.79	−818.946 ± 5.622	357.645 ± 5.284	−18.914 ± 0.09	−650.863 ± 4.925
ZINC000085996580	−174.655 ± 1.248	−87.798 ± 1.888	203.977 ± 2.763	−22.88 ± 0.144	−81.304 ± 2.269
ZINC000085734971	−177.923 ± 1.177	−37.521 ± 2.675	141.105 ± 3.714	−20.808 ± 0.093	−95.133 ± 4.263
ZINC000014612330	−167.284 ± 1.304	−774.68 ± 2.258	310.493 ± 1.672	−17.154 ± 0.074	−648.56 ± 2.801
ZINC000100513617	−114.531 ± 1.602	−764.552 ± 3.591	326.488 ± 4.91	−14.949 ± 0.093	−567.619 ± 4.003
ZINC000013462928	−177.25 ± 1.238	−19.523 ± 1.104	56.374 ± 1.273	−18.245 ± 0.108	−158.661 ± 1.669

The ADAR2-IHP complex had a binding energy of −873.873 kJ/mol (Table 4). This is expected since IHP is a strong binder of ADAR2. Previous studies have shown that RNA binding neither alters the hydrogen bond network nor the binding of IHP in the active site of ADAR2 [27,38]. Interestingly, ZINC000085511995 demonstrated a more negative binding energy to the ADAR2 than IHP, with an energy value of −1068.26 kJ/mol (Table 4). ZINC000085511995 may function as an ADAR2 antagonist by first displacing IHP (an ADAR2 agonist) due to its more negative binding energy, which is suggestive of a stronger attraction. Naloxone, a competitive antagonist of opiate receptors, functions in a similar manner by selectively displacing agonists, such as morphine, thereby reversing their actions [148,149]. ZINC000044417732, ZINC000085850673, ZINC000014612330, and ZINC000100513617 also had comparable binding free energies of −642.856, −650.863, −648.56, and −567.619 kJ/mol, respectively (Table 4). The relatively good binding energy values of the potential lead compounds warrant the development and implementation of a low throughput functional assay complemented by biochemical assays for GluA2 and 5-HT2CR, among others, in a neuronal culture system to ascertain their potency for treating various neurological disorders.

Per-Residue Energy Decomposition

The energetic contribution of each residue to the interaction with the shortlisted compounds was investigated using the per-residue energy decomposition via g_mmpbsa [146]. Understanding the binding modes of proteins with other molecules and elucidating the hot-spot residues (residues directly involved in the interactions) could help in designing drugs to disrupt proteins [150,151]. Hotspot residues have been suggested to contribute more than 1 or 2 kcal/mol in protein-protein complexes [152]. However, for protein—ligand complexes, hot-spot residues have been suggested to contribute energies greater than 5 kJ/mol or less than −5 kJ/mol [153].

From the molecular interaction profiles, Arg400, Arg401, Lys519, Ser531, Leu532, Trp687, Glu689, and Lys690 were predicted to be critical in ligand binding. Therefore, we sought to investigate their importance in ligand binding via the energy decomposition method. For the ADAR2-IHP complex, several residues contributed more than the ±5 kJ/mol threshold. A total of 53 residues contributed favorably (less than −5 kJ/mol) to IHP binding. Arg400, Arg401, Lys519, Ser531, Leu532, Trp687, Glu689, and Lys690 contributed −51.4229, −73.4334, −74.3903, −1.9345, −2.1645, −3.9056, 113.0625, and −83.6689 kJ/mol, respectively. Other residues worth mentioning are Arg522, Lys629, Lys662, Lys672 and Asp695 which contributed −50.0431, −88.4847, −63.9655, −56.4605 and 91.1192 kJ/mol, respectively, to IHP binding. This result corroborates previous studies which report that IHP strongly and stably binds in the active site of ADAR2 and is not influenced by RNA binding to the ADAR2 [27,38]. The extremely basic nature of the ADAR2 active site favors IHP binding which in turn stabilizes multiple lysine and arginine residues [36,38].

The ADAR2-ZINC000085511995 complex, which had a binding free energy of −1068.26 kJ/mol, experienced a lot of residues contributing energies above the ±5 kJ/mol threshold, comparable to the ADAR2-IHP complex. A total of 54 residues contributed favorably to its binding in the active site of the ADAR2. The predicted critical residues comprising Arg400, Arg401, Lys519, Ser531, Leu532, Trp687, Glu689, and Lys690 had energies of −49.6330, −52.8445, −63.4549, −3.3215, −3.9087, −6.4576, 71.2105, and −83.5814 kJ/mol, respectively. The multiple residues and their contribution energies involved in ZINC000085511995 binding make ZINC000085511995 an interesting drug candidate to consider for ADAR2.

The ADAR2-ZINC000044417732 complex also had several residues contributing above the threshold. The critical residues, Arg400, Arg401, Lys519, Ser531, Leu532, Trp687, Glu689, and Lys690 were observed to contribute energies of −35.9893, −50.7197, −38.7647, −1.5284, −7.1948, −2.1630, 51.3166, and −34.4276 kJ/mol, respectively. For the ADAR2-ZINC000085950180 complex, only residues Arg400, Arg401, Leu404, Glu509, Lys519, Arg522, Leu532, Lys629, Leu632, Arg635, Lys662, Lys672, Glu689, Lys690, and Asp695 contributed above the threshold with energies of 6.9611, −8.1196, −6.0607, 8.0841, −5.3107, −7.2416, −6.6838, −26.3468, −9.2015, −6.0062, −10.7480, −8.1484, 28.8698, −9.8404, and 30.7624 kJ/mol, respectively. From the per-residue energy decomposition, residues Arg400, Arg401, Lys519, Trp687, Glu689 and Lys690 seem to be very crucial in ligand binding in the IHP binding site of ADAR2 than Ser531 and Leu532. Future ADAR2 drug initiatives may consider developing drugs with higher specificity to these residues.

2.8. Re-Docking Predicted Hits against 5-HT2C Receptor

Ritanserin had the most negative binding energy of −12.7 kcal/mol followed by ZINC000085996580 and ZINC000085850673 with binding energies of −11.5 and −10.9 kcal/mol, respectively (Table S2). Ritanserin is a serotonin receptor antagonist, an antidepressant and has anxiolytic properties. Ritanserin is also an antiparkinsonian agent and has been shown to improve sleep [154–156]. Ritanserin improves motor deficits, including akinesia in mouse and humans [157]. Significant success was recorded when ritanserin (average dose of 13.5 mg/day) was administered to 10 patients with neuroleptic-induced Parkinsonism for 2 to 4 days. Out of the 10 patients with akathisia, 8 responded positively, observing a drop in Hillside Akathisia Scale (HAS) baseline ratings from 16.4 (±6) to 7.4 (±5.2) after 3 days of treatment [158]. However, ritanserin has not been approved for medical use due to safety and toxicity concerns, although it is widely used for research purposes [159–161]. Thus, there is the need to develop more potent and less toxic biomolecules that target serotonin receptors.

ZINC000085734971 had a binding energy of −10.6 kcal/mol, while both ZINC000085511995 and ZINC000085950180 had −10.5 kcal/mol. Both ZINC000044417732 and ZINC000014612330 also had binding energies of −10.3 kcal/mol while ZINC000013462928 and ZINC000100513617 had binding energies of −9.9 and −9.6 kcal/mol, respectively (Table S2). A previous study docked compounds with the ethyl 2-(*p*-tolyloxy)acetate skeleton against the 5-HT2CR (PDB ID: 6BQH) using AutoDock Vina and the compound with the best affinity had a binding energy of −6.2 kcal/mol [162]. The strong binding affinities of the predicted leads could make them potential 5-HT2CR modulators. The PASS predictions and safety profiles of the identified molecules warrant further experimental studies to corroborate or validate their anti-ADAR2 and anti-schizophrenia properties. Further functional assays, both in a reduced system and in cultured neurons, are required to determine with greater resolution the relative efficacy of the predicted compounds on ADAR2, 5-HT2CR and GluA2.

2.9. Origin and Source of the Potential Lead Compounds

The natural sources of the nine potential lead compounds were investigated from various databases, including ZINC15 [163], PubChem [164,165], ChEMBL [166–168], LOTUS [169], KNApSAcK [170,171], Natural Product Activity and Species Source (NPASS) [172], and Indian Medicinal Plants, Phytochemistry And Therapeutics (IMPPAT) [173] as previously conducted [174]. The pharmacological activities of the plant sources were also probed from the existing literature.

ZINC000044417732 (chitranone) has been isolated from the roots of *Plumbago zeylanica* [175,176] and *P. capensis* [177], and stem bark and fruits of *Diospyros maritima* [178–180] and *D. kaki* [180]. Chitranone has been previously reported to be cytotoxic against four cancer cell lines including human oral epidermoid carcinoma (KB), human lung cancer (Lu1), hormone-dependent human prostate cancer (LNCaP), and human umbilical vein endothelial cells (HUVEC) at EC_{50} values of 0.3, 1.1, 2.2, and 2.1 μg/mL, respectively [178]. The antimicrobial activity of chitranone has also been reported previously [178]. ZINC000085950180 (isozeylanone) is also found in *D. maritima* [179,180], *P. europaea* [181], and *P. zeylanica* [182]. Organic extracts of *P. zeylanica* were shown to have strong bactericidal activity against *Helicobacter pylori*, a type of bacteria associated with gastric cancer and peptic ulceration [183].

ZINC000085511995 (mamegakinone) has also been isolated from various *Diospyros* spp. [184], includng *D. batocana* [185], *D. obliquifolia* [186], *D. mollis* [184], *D. chamaethamnus* [184], *D. kaki* [187], *D. lotus* [188,189], *D. lycioides* [186], *D. montana* [189], *D. usambarensis* [190], and *D. zombensis* [184]. Additionally, ZINC000085850673 (3,8′-Bi[2-methyl-5-hydroxy-1,4-naphthoquinone] or 3,8′-biplumbagin) can be found in the fruit extract of *D. maritima* [179,180]. ZINC000085734971 (3,3′-Ethylidenebis(2-methyl-5-hydroxy-1,4-naphthoquinone) or ethylidene-3,3′-biplumbagin) can be found in extracts of *D. maritima* [179,180]. *D. kaki* has been used to treat patients with mental and physical complaints caused by trauma due to war, disaster, and burning [191].

ZINC000085996580 (lespedezol B2 or 8-[[2-(2,4-dihydroxyphenyl)-6-hydroxy-1-benzofuran-3-yl]methyl]-6H-[1]benzofuro[3,2-c]chromene-3,9-diol) has been extracted from the stem of *Lespedeza homoloba* [192]. Compounds derived from *L. homoloba* were found to exhibit strong antioxidative activity in the rat brain homogenate test against lipid peroxidation [192]. ZINC000014612330 ((4aR,12bR)-9-hydroxy-2,5,5-trimethyl-3,4,4a,12b-tetrahydronaphtho[3,2-c]isochromene-7,12-dione) and pyranokunthone A are isomers. Pyranokunthone A, obtained from the root bark of *Stereospermum kunthianum*, demonstrated moderate activity against two strains of *Plasmodium falciparum* comprising chloroquine-sensitive strain (IC_{50} of 11.7 μg/mL) and chloroquine-resistant clone (IC_{50} > 25 μg/mL) while having insignificant toxicity on the endothelial cell line ECV-304 (IC_{50} > 200 μg/mL) [193].

ZINC000100513617 (indigo) is found in the leaves and seeds of both *Isatis tinctoria* [194] and *I. indigotica* [195], fruits of *Couroupita guianensis* [196], and leaves of *Eupatorium laeve* [197]. Indigo and indirubin are widely known for their anti-inflammatory properties as they have been used in Central Europe since ancient times [194,198]. Bisindigotin, a derivative of indigo and indirubin, demonstrated a dose-dependent (0.1–5 μM) inhibition of 2,3,7,8-tetrachlorodibenzo-p-dioxin (TCDD)-induced ethoxyresorufin O-deethylase (EROD) activity in human HepG2 hepatoma cells with an IC_{50} of 0.8 μM [195]. In addition to its antipyretic, antiviral, anti-inflammatory, and anti-endotoxin properties, *I. indigotica* leaf extracts inhibit human hepatoma cell growth [199]. Moreover, several compounds isolated from *I. indgotica* and *tinctoria* have shown neuroprotective activities [200–204], thereby warranting the experimental testing of indigo.

ZINC000013462928 (isohibalactone or BDBM512896 or 3′,4′-bis(methylenedioxy)-lign-7(E)-en-9,9′-olide) can be obtained from *Hypoestes purpurea* [205], *Linum corymbulosum* [206], and leaves of *Juniperus chinensis* [207]. 3′,4′-bis(methylenedioxy)-lign-7(E)-en-9,9′-olide has a Z isomer [206] which is similar to cubebin. The anti-cancer, trypanocidal, anti-inflammatory, analgesic, anti-proliferative and leishmanicidal activities of cubebin have been highlighted in the literature [208,209]. Moreover, cubebin has been reported to have neuroprotective properties, as pretreatment of mice with cubebin (25 and 50 mg/kg) prevented scopolamine-induced learning and memory impairments [210]. Also, cubebin inhibited acetylcholinesterase with an IC_{50} of 992 μM [210], making ZINC000013462928 an interesting candidate to investigate for its neuroprotective and anti-ADAR2 property.

Some of the topmost compounds, including ZINC000085511995, ZINC000085950180, ZINC000044417732, ZINC000085850673, ZINC000085996580, and ZINC000085734971 are polyphenols. Polyphenols are known for their antioxidant [211,212], anti-inflammatory [211,213], and anti-cancer properties [214–216]. They also possess neuroprotective effects and help improve cognitive function [217–219]. However, similar to all bioactive compounds, polyphenols also have off-target interactions. One major off-target interaction of polyphenols is their ability to bind and inhibit the activity of certain drug metabolizing enzymes [220,221]. Some polyphenols have also been reported to modulate estrogen receptors and may exhibit estrogen-like effects [222–225]. Polyphenols also have the ability to interfere with nutrient absorption and metabolism in the body, especially, chelating iron and zinc, thereby inhibiting their absorption [226,227]. While this can be beneficial in preventing iron overload in patients with hemochromatosis, it can also lead to iron deficiency anemia in individuals with poor iron status.

Furthermore, polyphenols can also interact with the gut microbiota [228,229]. Some polyphenols have been found to promote the growth of beneficial gut bacteria, thereby improving gut health and immune function [230,231]. However, some polyphenols can also inhibit the growth of certain pathogenic bacteria, which can alter the gut microbiota composition and lead to negative effects [232,233]. Notwithstanding this, metal-phenolic networks (MPNs) can help with precise drug delivery, improve the efficacy, and limit the off-target toxicity of polyphenols as MPNs have negligible cytotoxicity [234].

3. Materials and Methods

Structure-based virtual screening (SBVS) was employed to identify potentially good ADAR2 binders with ADAR inhibitory activity (Figure 9). The study used IHP as a control, which is a known binder and agonist of the ADAR2 [27,38]. Two widely used molecular docking programs, AutoDock Vina and Maestro, were used for the molecular docking runs [47,235–237]. The top compounds found in both screening outputs were selected for further analysis. The molecular interactions between the ADAR2 and the novel potential inhibitors were determined (Figure 9). Furthermore, pharmacokinetics and toxicity profiling of the shortlisted compounds were performed to assess their drug-likeness. The biological activities of the compounds were then predicted using the Bayesian approach. The top potential lead compounds and the known binders were then subjected to molecular dynamics simulations including the MM/PBSA computations (Figure 9).

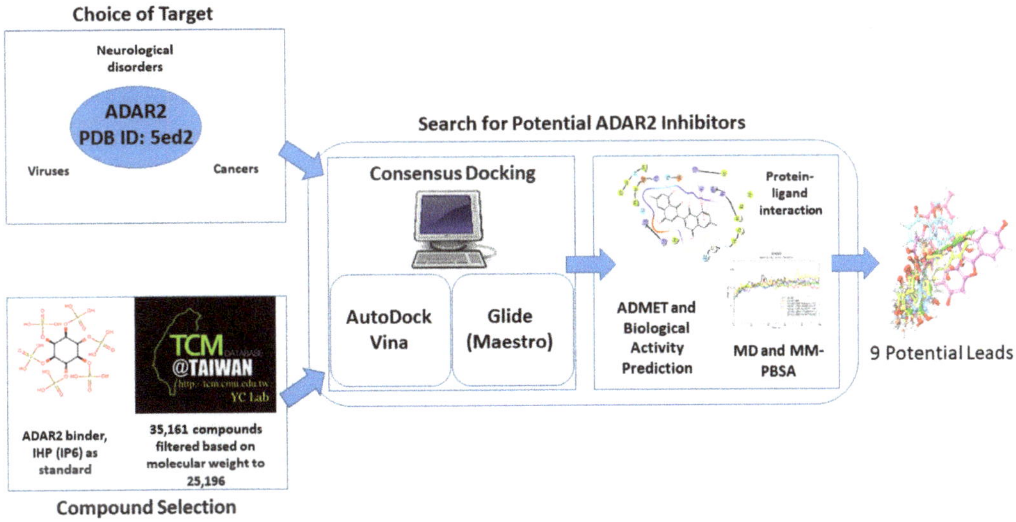

Figure 9. Schema detailing the step-by-step approach employed in this study to identify novel potential ADAR2 inhibitors.

3.1. Obtaining and Preparing Protein and Ligand Structures

The hADAR2 structure was retrieved from the Research Collaboratory for Structural Bioinformatics Protein Data Bank (RCSB PDB) with corresponding PDB ID of 5ed2 [238,239]. The 5ed2 comprises a human ADAR2 protein structure complex with inositol hexakisphosphate (also known as phytic acid, myo-inositol hexakisphosphate, InsP6, IP_6 or IHP), bound to a double-stranded ribonucleic acid (dsRNA) [27]. The dsRNA, ligand (IHP) and the Zinc atoms which were bound to the hADAR2 protein were removed using PyMOL (version 2.3.0). The resulting structure was processed using the Protein Preparation Wizard in Maestro (Schrödinger, LLC, New York, NY, USA). Additionally, the OPLS4 force field was employed to optimize protein energies and to remove any steric hindrance [240].

A total of 35,161 natural products from the Traditional Chinese Medicine (TCM) database were obtained from TCM@Taiwan, the world's largest non-commercial TCM database, which is a catalog of ZINC15 database [163,241]. The compounds were pre-filtered based on molecular weight as previously conducted [242]. Compounds with molecular weights below 150 g/mol and above 600 g/mol were eliminated. A total of 25,196 compounds had molecular weights within the threshold and were used for the study. The ADAR2-bound ligand (IHP) was extracted from the 5ed2 structure and used in the study as a standard or control.

Ligand structures were prepared using the Ligand Preparation Wizard (LigPrep) in Maestro. For the TCM library, a total of 37,398 compounds (some being conformers of the 25,196 pre-filtered library) were generated using LigPrep. Various ionic states, tautomeric states and stereo chemistries were generated from each input ligand molecule at a pH of 7.0 ± 2.0 using Epik [243]. LigPrep optionally expands tautomeric and ionization states, ring conformations, and stereoisomers to generate chemical diversity from the input structures, accounting for the increase in number of ligands. LigPrep also optimizes the output structures for molecular docking in Maestro. For the bound ligand, IHP, 3 conformers were generated.

3.2. Determining Binding Sites

A literature search was conducted to identify previously reported active sites of the human ADAR2. Moreover, CASTp 3.0 (http://sts.bioe.uic.edu/castp/calculation.html

accessed on 14 July 2022) [244] was employed to predict potential binding cavities of the hADAR2. Binding site predictions which had no openings or were relatively small for ligands to dock into were not considered.

3.3. Molecular Docking

Two molecular docking applications comprising Xglide (Glide Cross Docking) module in Maestro [47,235] and AutoDock Vina [237] were employed to screen natural products-derived compounds against the ADAR2 protein. Molecular docking tools tend to differ in their search and scoring algorithms, therefore, combining multiple docking tools compensates for individual limitations [245–248]. A total of 25,196 pre-filtered TCM compounds and IHP as control were used as the screening library.

For each docking tool, the top 10% (2520 compounds) of the total number of compounds were shortlisted after the docking process based on the docking scores. Traditional consensus scoring method was used to shortlist suitable compounds. In molecular docking, choosing compounds based on consensus scoring from a variety of docking programs produces a far higher predictive performance than using the docking scores from a single docking program [44]. Thus, compounds which were found in the top 10% for both docking tools were selected for further analysis.

3.3.1. Validation of Molecular Docking Protocols

Inhibitor identification has been greatly aided by molecular docking simulations, which have allowed the prediction and investigation of potential protein–ligand interactions [249]. However, the pose quality prediction of the available docking tools must be assessed in order to attain higher performance. Thus, this study validated the ability of the two docking programs to accurately predict conformations similar to the experimental scenario as previously conducted [45,250–252]. The IHP was docked against the ADAR2 protein using both docking tools and the pose with the most negative binding energy from each tool was structurally aligned with the conformation in the crystallographic structure using the rigid module of LS-Align [253]. The RMSD values of the alignments were then determined.

3.3.2. Molecular Docking via AutoDock Vina

The first step of the molecular docking run was performed using the AutoDock Vina module in PyRx (version 0.9.2) [236,237]. The energy of the 25,196 pre-filtered TCM compounds were minimized using the UFF force field, out of which 25,189 were successfully converted to AutoDock's compatible format, Protein Data Bank, Partial Charge (Q), and Atom Type (T) (PDBQT). The docking process was performed with exhaustiveness set to 8, using grid box dimensions of 64.306 × 53.098 × 51.413 Å^3 and the protein centered at 30.697 Å, 27.166 Å and 82.548 Å.

3.3.3. Molecular Docking via XGlide (Maestro)

For the second phase of the molecular docking process, Glide Cross Docking (XGlide) was employed [47,235]. For the grid generation, SiteMap [254] was used to locate active sites which were used to determine the grid box. Default inner (10 × 10 × 10 Å^3) and outer box sizes (26 × 26 × 26 Å^3) were used for the virtual screening. Ligand vdW scale factor and cut-off for a 'good' RMSD were set to default (0.80 and 2.0 Å, respectively). The Standard Precision (SP) algorithm was used for the molecular docking run [235]. For compounds with more than one tautomer, the docking pose with the most negative binding energy was selected after the molecular docking process.

3.3.4. Shortlisting Compounds using Consensus Scoring

To facilitate compound selection, the consensus scoring approach was used. A previous study had success rates of ~82% when multiple docking tools, including AutoDock [255], AutoDock Vina [237] and DOCK6 [256], were combined to identify inhibitors [44]. The accuracies of the individual docking tools were 55%, 64% and 58%, respectively [44], implying a far greater success rate when multiple docking tools are used. The top 10% (2520) of the initial pre-filtered library were shortlisted based on binding energy for further studies. Traditional consensus scoring was employed to rank the top compounds [257]—the consensus score for each compound was obtained by averaging the docking scores for both AutoDock Vina and Maestro (Glide).

3.4. Determining the Interactions between the ADAR2-Ligand Complexes

Protein–ligand interactions help us to understand the mechanisms of ligand binding, which is crucial in drug design and development [258]. Ligand binding via specific molecular recognition can trigger protein activation or inactivation and affect protein function [259]. Thus, elucidating the mechanisms of binding and the residues involved in these interactions is relevant for drug design. This study therefore visualized the molecular interactions between the hADAR2 protein and docked ligands (selected compounds) using Maestro.

3.5. Determining ADMET Properties

The pharmacokinetic profiles comprising absorption, distribution, metabolism, excretion (ADME) of the shortlisted compounds were determined using SwissADME [74]. OSIRIS Datawarrior 5.5.0 was also used to determine the toxicity risks of the compounds by predicting their mutagenicity, tumorigenicity, irritancy and reproductive effects [85,260]. Compounds which were predicted to have more than one toxicity risk were ignored in this study. Additionally, compounds that were predicted to be either tumorigenic or mutagenic were eliminated.

3.6. Structural Similarity Search and Prediction of Biological Activity of Compounds

Prediction of Activity Spectra of Substances (PASS) was employed to predict the biological activity of the identified compounds [91–93]. PASS, with average prediction accuracy of ~95% based on leave-one-out cross validation (LOO CV) estimation, uses a Bayesian approach to predict the possible biological activity of a molecule based on its structure [92]. Furthermore, similarity search of the shortlisted compounds was performed via DrugBank [261,262] in order to identify structural analogs with relevant biological activity.

3.7. Molecular Dynamics Simulations

The unbound protein and selected ADAR2-ligand complexes were subjected to 100 ns molecular dynamics (MD) simulations using the GROningen MAchine for Chemical Simulations (GROMACS) version 5.1.5. Ligand topologies of the compounds for OPLS force field were generated using the LigParGen [263]. Each system was solvated in a cubic box using the "TIP4P" water model and the OPLS/AA force field [264,265]. Either sodium or chlorine ions were added to each system to neutralize the charges. The systems were then subjected to constant-number, constant-volume and constant-temperature (NVT) and isothermal-isobaric or constant-number, constant-pressure and constant-temperature (NPT) ensembles, prior to the MD simulation. The RMSD, Rg and RMSF of each system were analyzed after the MD simulations. Moreover, the number of hydrogen bonds during the simulation was assessed for each system. Snapshots at 25 ns intervals (time step = 0, 25, 50, 75 and 100 ns) were generated for each complex.

3.8. Molecular Mechanics Poisson–Boltzmann Surface Area (MM/PBSA) Computation of Potential Leads

The binding free energies and the other energy components including vdW, electrostatic, polar solvation and solvent accessible surface area (SASA) energies of each complex were determined using the MM/PBSA method via the g_mmpbsa tool [146]. The energetic contribution per-residue to ligand binding in the IHP binding site was also evaluated [146].

3.9. Re-Docking Potential Leads against the 5-HT2CR

The identified potential leads were finally screened against the 5-Hydroxytryptamine (or serotonin) 2C receptor (5-HT2CR) to evaluate their binding to the receptor using AutoDock Vina. The 3D structure of 5-HT2CR in complex with ritanserin was retrieved from RCSB PDB with corresponding ID 6BQH [266]. Ritanserin, which is a known antagonist and a strong binder of 5-HT2CR, was extracted from the complex and used as the control for the molecular docking. Molecular docking was performed using the ritanserin binding site with docking grid box dimensions of 25.0 × 25.0 × 43.0801246543 $Å^3$ and the protein centered at 40.5731 Å, 33.1458 Å and 52.5793299511 Å.

4. Conclusions

This study carefully predicted potential ADAR2 inhibitors which can be beneficial in the treatment of various diseases including some cancers, viral infections and neurological disorders. We employed two widely used molecular docking tools, namely AutoDock Vina and Glide (Maestro, Schrodinger Suite) to predict the binding affinities of natural compounds from the Chinese flora and fauna. The top nine shortlisted compounds had favorable binding to the human ADAR2 and 5-HT2C receptor and were predicted to have insignificant toxicities. They also had favorable results when subjected to MD simulations and MM/PBSA calculations. One of the potential leads, ZINC000085511995, had higher binding free energy (−1068.26 kJ/mol) to the ADAR2 than the known binder, IHP (−873.873 kJ/mol), after MM/PBSA computations. The biological activity predictions and structural similarity search also corroborated the activity of the predicted compounds. The identified biomolecules can help accelerate the pace of ADAR2 research since they can be potential tools for understanding the biology of RNA editing and as a starting point for the development of potential therapeutic agents. Further experimental studies on the predicted molecules are required to determine their anti-ADAR2 activity. This study also observed the presence of naphthoquinone, indole, furanocoumarin and benzofuran moieties in most of the top compounds. Serotonin and tryptophan, which are beneficial in digestive regulation, sleep cycle and improved mood, also have the indole moiety [48,49,53,55]. Moreover, vitamin K, a naphthoquinone derivative, has been reported to improve cognitive function [122,123]. The identified chemical series are worthy of further experimental testing to ascertain their potency for extensive medicinal chemistry and biological characterization. These series may give rise to new putative ADAR2 inhibitors which can be designed using de novo/rational design, virtual focused combinatorial libraries (vFCL) generation, and in silico screening of the vFCL.

Supplementary Materials: The following supporting information can be downloaded at: https://www.mdpi.com/article/10.3390/ijms24076795/s1.

Author Contributions: W.A.M.III, E.B. and M.S. conceptualized the project. E.B. and W.A.M.III predominantly undertook all the computational analysis with inputs from A.S., C.A., P.O.S., S.P., M.V., K.B., S.K.K., A.A. and S.K. W.A.M.III and E.B. co-wrote the first draft. The final manuscript was written through contributions of all authors. All authors have read and agreed to the published version of the manuscript.

Funding: This research received no external funding.

Institutional Review Board Statement: Not applicable.

Informed Consent Statement: Not applicable.

Data Availability Statement: Not applicable.

Conflicts of Interest: The authors declare no conflict of interest.

References

1. Nishikura, K. A-to-I editing of coding and non-coding RNAs by ADARs. *Nat. Rev. Mol. Cell Biol.* **2016**, *17*, 83–96. [CrossRef] [PubMed]
2. Yang, Y.; Okada, S.; Sakurai, M. Adenosine-to-inosine RNA editing in neurological development and disease. *RNA Biol.* **2021**, *18*, 999–1013. [CrossRef] [PubMed]
3. Shevchenko, G.; Morris, K.V. All I's on the RADAR: Role of ADAR in gene regulation. *FEBS Lett.* **2018**, *592*, 2860–2873. [CrossRef] [PubMed]
4. Werry, T.D.; Loiacono, R.; Sexton, P.M.; Christopoulos, A. RNA editing of the serotonin 5HT2C receptor and its effects on cell signalling, pharmacology and brain function. *Pharmacol. Ther.* **2008**, *119*, 7–23. [CrossRef] [PubMed]
5. Slotkin, W.; Nishikura, K. Adenosine-to-inosine RNA editing and human disease. *Genome Med.* **2013**, *5*, 105. [CrossRef]
6. Morabito, M.V.; Abbas, A.I.; Hood, J.L.; Kesterson, R.A.; Jacobs, M.M.; Kump, D.S.; Hachey, D.L.; Roth, B.L.; Emeson, R.B. Mice with altered serotonin 2C receptor RNA editing display characteristics of Prader–Willi syndrome. *Neurobiol. Dis.* **2010**, *39*, 169–180. [CrossRef]
7. Paz-Yaacov, N.; Bazak, L.; Buchumenski, I.; Porath, H.T.; Danan-Gotthold, M.; Knisbacher, B.A.; Eisenberg, E.; Levanon, E.Y. Elevated RNA Editing Activity Is a Major Contributor to Transcriptomic Diversity in Tumors. *Cell Rep.* **2015**, *13*, 267–276. [CrossRef]
8. Karagianni, K.; Pettas, S.; Christoforidou, G.; Kanata, E.; Bekas, N.; Xanthopoulos, K.; Dafou, D.; Sklaviadis, T. A Systematic Review of Common and Brain-Disease-Specific RNA Editing Alterations Providing Novel Insights into Neurological and Neurodegenerative Disease Manifestations. *Biomolecules* **2022**, *12*, 465. [CrossRef]
9. Di Narzo, A.F.; Kozlenkov, A.; Roussos, P.; Hao, K.; Hurd, Y.; Lewis, D.A.; Sibille, E.; Siever, L.J.; Koonin, E.; Dracheva, S. A unique gene expression signature associated with serotonin 2C receptor RNA editing in the prefrontal cortex and altered in suicide. *Hum. Mol. Genet.* **2014**, *23*, 4801–4813. [CrossRef]
10. Dracheva, S.; Patel, N.; Woo, D.A.; Marcus, S.M.; Siever, L.J.; Haroutunian, V. Increased serotonin 2C receptor mRNA editing: A possible risk factor for suicide. *Mol. Psychiatry* **2008**, *13*, 1001–1010. [CrossRef]
11. Gurevich, I.; Tamir, H.; Arango, V.; Dwork, A.J.; Mann, J.J.; Schmauss, C. Altered Editing of Serotonin 2C Receptor Pre-mRNA in the Prefrontal Cortex of Depressed Suicide Victims. *Neuron* **2002**, *34*, 349–356. [CrossRef]
12. Karanović, J.; Šviković, S.; Pantović, M.; Durica, S.; Brajušković, G.; Damjanović, A.; Jovanović, V.; Ivković, M.; Romac, S.; Savić Pavićević, D. Joint effect of ADARB1 gene, HTR2C gene and stressful life events on suicide attempt risk in patients with major psychiatric disorders. *World J. Biol. Psychiatry* **2015**, *16*, 261–271. [CrossRef]
13. Lyddon, R.; Dwork, A.J.; Keddache, M.; Siever, L.J.; Dracheva, S. Serotonin 2c receptor RNA editing in major depression and suicide. *World J. Biol. Psychiatry* **2013**, *14*, 590–601. [CrossRef]
14. Niswender, C. RNA Editing of the Human Serotonin 5-HT2C Receptor Alterations in Suicide and Implications for Serotonergic Pharmacotherapy. *Neuropsychopharmacology* **2001**, *24*, 478–491. [CrossRef]
15. Pandey, G.N.; Dwivedi, Y.; Ren, X.; Rizavi, H.S.; Faludi, G.; Sarosi, A.; Palkovits, M. Regional Distribution and Relative Abundance of Serotonin2c Receptors in Human Brain: Effect of Suicide. *Neurochem. Res.* **2006**, *31*, 167–176. [CrossRef]
16. Weissmann, D.; van der Laan, S.; Underwood, M.D.; Salvetat, N.; Cavarec, L.; Vincent, L.; Molina, F.; Mann, J.J.; Arango, V.; Pujol, J.F. Region-specific alterations of A-to-I RNA editing of serotonin 2c receptor in the cortex of suicides with major depression. *Transl. Psychiatry* **2016**, *6*, e878. [CrossRef]
17. Bombail, V.; Qing, W.; Chapman, K.E.; Holmes, M.C. Prevention of 5-hydroxytryptamine 2C receptor RNA editing and alternate splicing in C57BL/6 mice activates the hypothalamic-pituitary-adrenal axis and alters mood. *Eur. J. Neurosci.* **2014**, *40*, 3663–3673. [CrossRef]
18. Martin, C.B.P.; Ramond, F.; Farrington, D.T.; Aguiar, A.S.; Chevarin, C.; Berthiau, A.-S.; Caussanel, S.; Lanfumey, L.; Herrick-Davis, K.; Hamon, M.; et al. RNA splicing and editing modulation of 5-HT2C receptor function: Relevance to anxiety and aggression in VGV mice. *Mol. Psychiatry* **2013**, *18*, 656–665. [CrossRef]
19. Règue, M.; Poilbout, C.; Martin, V.; Franc, B.; Lanfumey, L.; Mongeau, R. Increased 5-HT2C receptor editing predisposes to PTSD-like behaviors and alters BDNF and cytokines signaling. *Transl. Psychiatry* **2019**, *9*, 100. [CrossRef]
20. Warhaftig, G.; Sokolik, C.M.; Khermesh, K.; Lichtenstein, Y.; Barak, M.; Bareli, T.; Levanon, E.Y.; Yadid, G. RNA editing of the 5-HT2C receptor in the central nucleus of the amygdala is involved in resilience behavior. *Transl. Psychiatry* **2021**, *11*, 137. [CrossRef]
21. Sodhi, M.S.; Burnet, P.W.J.; Makoff, A.J.; Kerwin, R.W.; Harrison, P.J. RNA editing of the 5-HT2C receptor is reduced in schizophrenia. *Mol. Psychiatry* **2001**, *6*, 373–379. [CrossRef] [PubMed]
22. Schirle, N.T.; Goodman, R.A.; Krishnamurthy, M.; Beal, P.A. Selective inhibition of ADAR2-catalyzed editing of the serotonin 2c receptor pre-mRNA by a helix-threading peptide. *Org. Biomol. Chem.* **2010**, *8*, 4898. [CrossRef] [PubMed]
23. Zhang, Z.; Shen, M.; Gresch, P.J.; Ghamari-Langroudi, M.; Rabchevsky, A.G.; Emeson, R.B.; Stamm, S. Oligonucleotide-induced alternative splicing of serotonin 2C receptor reduces food intake. *EMBO Mol. Med.* **2016**, *8*, 878–894. [CrossRef] [PubMed]
24. Heisler, L.K.; Chu, H.-M.; Tecott, L.H. Epilepsy and Obesity in Serotonin 5-HT2C Receptor Mutant Mice. *Ann. N. Y. Acad. Sci.* **1998**, *861*, 74–78. [CrossRef] [PubMed]
25. Tecott, L.H.; Sun, L.M.; Akana, S.F.; Strack, A.M.; Lowenstein, D.H.; Dallman, M.F.; Julius, D. Eating disorder and epilepsy in mice lacking 5-HT2C serotonin receptors. *Nature* **1995**, *374*, 542–546. [CrossRef]

26. Piontkivska, H.; Wales-McGrath, B.; Miyamoto, M.; Wayne, M.L. ADAR Editing in Viruses: An Evolutionary Force to Reckon with. *Genome Biol. Evol.* **2021**, *13*, evab240. [CrossRef]
27. Matthews, M.M.; Thomas, J.M.; Zheng, Y.; Tran, K.; Phelps, K.J.; Scott, A.I.; Havel, J.; Fisher, A.J.; Beal, P.A. Structures of human ADAR2 bound to dsRNA reveal base-flipping mechanism and basis for site selectivity. *Nat. Struct. Mol. Biol.* **2016**, *23*, 426–433. [CrossRef]
28. Stefl, R.; Oberstrass, F.C.; Hood, J.L.; Jourdan, M.; Zimmermann, M.; Skrisovska, L.; Maris, C.; Peng, L.; Hofr, C.; Emeson, R.B.; et al. The Solution Structure of the ADAR2 dsRBM-RNA Complex Reveals a Sequence-Specific Readout of the Minor Groove. *Cell* **2010**, *143*, 225–237. [CrossRef]
29. Maia, E.H.B.; Assis, L.C.; de Oliveira, T.A.; da Silva, A.M.; Taranto, A.G. Structure-Based Virtual Screening: From Classical to Artificial Intelligence. *Front. Chem.* **2020**, *8*, 343. [CrossRef]
30. Mushtaq, S.; Abbasi, B.H.; Uzair, B.; Abbasi, R. Natural products as reservoirs of novel therapeutic agents. *EXCLI J.* **2018**, *17*, 420–451. [CrossRef]
31. Li, G.; Lou, H.-X. Strategies to diversify natural products for drug discovery. *Med. Res. Rev.* **2018**, *38*, 1255–1294. [CrossRef]
32. Kumar, G.B.; Nair, B.G.; Perry, J.J.P.; Martin, D.B.C. Recent insights into natural product inhibitors of matrix metalloproteinases. *Medchemcomm* **2019**, *10*, 2024–2037. [CrossRef]
33. Ren, Z.-L.; Zuo, P.-P. Neural Regeneration: Role of Traditional Chinese Medicine in Neurological Diseases Treatment. *J. Pharmacol. Sci.* **2012**, *120*, 139–145. [CrossRef]
34. Zhang, X.; Li, L.; Chen, T.; Sun, Z.; Tang, W.; Wang, S.; Wang, T.; Wang, Y.; Zhang, H. Research Progress in the Effect of Traditional Chinese Medicine for Invigoration on Neurotransmitter Related Diseases. *Evid.-Based Complement. Altern. Med.* **2018**, *2018*, 4642018. [CrossRef]
35. Liu, Y.; Wang, S.; Kan, J.; Zhang, J.; Zhou, L.; Huang, Y.; Zhang, Y. Chinese Herbal Medicine Interventions in Neurological Disorder Therapeutics by Regulating Glutamate Signaling. *Curr. Neuropharmacol.* **2020**, *18*, 260–276. [CrossRef]
36. Savva, Y.A.; Rieder, L.E.; Reenan, R.A. The ADAR protein family. *Genome Biol.* **2012**, *13*, 252. [CrossRef]
37. Macbeth, M.R.; Lingam, A.T.; Bass, B.L. Evidence for auto-inhibition by the N terminus of hADAR2 and activation by dsRNA binding. *RNA* **2004**, *10*, 1563–1571. [CrossRef]
38. Macbeth, M.R.; Schubert, H.L.; VanDemark, A.P.; Lingam, A.T.; Hill, C.P.; Bass, B.L. Inositol Hexakisphosphate Is Bound in the ADAR2 Core and Required for RNA Editing. *Science* **2005**, *309*, 1534–1539. [CrossRef]
39. Broni, E.; Kwofie, S.K.; Asiedu, S.O.; Miller, W.A.; Wilson, M.D. A Molecular Modeling Approach to Identify Potential Antileishmanial Compounds Against the Cell Division Cycle (cdc)-2-Related Kinase 12 (CRK12) Receptor of Leishmania donovani. *Biomolecules* **2021**, *11*, 458. [CrossRef]
40. Kwofie, S.K.; Broni, E.; Asiedu, S.O.; Kwarko, G.B.; Dankwa, B.; Enninful, K.S.; Tiburu, E.K.; Wilson, M.D. Cheminformatics-Based Identification of Potential Novel Anti-SARS-CoV-2 Natural Compounds of African Origin. *Molecules* **2021**, *26*, 406. [CrossRef]
41. Wang, Y.; Beal, P.A. Probing RNA recognition by human ADAR2 using a high-throughput mutagenesis method. *Nucleic Acids Res.* **2016**, *44*, 9872–9880. [CrossRef] [PubMed]
42. Pettersen, E.F.; Goddard, T.D.; Huang, C.C.; Couch, G.S.; Greenblatt, D.M.; Meng, E.C.; Ferrin, T.E. UCSF Chimera—A visualization system for exploratory research and analysis. *J. Comput. Chem.* **2004**, *25*, 1605–1612. [CrossRef] [PubMed]
43. Ramírez, D.; Caballero, J. Is It Reliable to Take the Molecular Docking Top Scoring Position as the Best Solution without Considering Available Structural Data? *Molecules* **2018**, *23*, 1038. [CrossRef] [PubMed]
44. Houston, D.R.; Walkinshaw, M.D. Consensus docking: Improving the reliability of docking in a virtual screening context. *J. Chem. Inf. Model.* **2013**, *53*, 384–390. [CrossRef] [PubMed]
45. Alves, M.; Froufe, H.; Costa, A.; Santos, A.; Oliveira, L.; Osório, S.; Abreu, R.; Pintado, M.; Ferreira, I. Docking Studies in Target Proteins Involved in Antibacterial Action Mechanisms: Extending the Knowledge on Standard Antibiotics to Antimicrobial Mushroom Compounds. *Molecules* **2014**, *19*, 1672–1684. [CrossRef]
46. Chang, M.W.; Lindstrom, W.; Olson, A.J.; Belew, R.K. Analysis of HIV wild-type and mutant structures via in silico docking against diverse ligand libraries. *J. Chem. Inf. Model.* **2007**, *47*, 1258–1262. [CrossRef]
47. Friesner, R.A.; Banks, J.L.; Murphy, R.B.; Halgren, T.A.; Klicic, J.J.; Mainz, D.T.; Repasky, M.P.; Knoll, E.H.; Shelley, M.; Perry, J.K.; et al. Glide: A New Approach for Rapid, Accurate Docking and Scoring. Method and Assessment of Docking Accuracy. *J. Med. Chem.* **2004**, *47*, 1739–1749. [CrossRef]
48. Richard, D.M.; Dawes, M.A.; Mathias, C.W.; Acheson, A.; Hill-Kapturczak, N.; Dougherty, D.M. L -Tryptophan: Basic Metabolic Functions, Behavioral Research and Therapeutic Indications. *Int. J. Tryptophan Res.* **2009**, *2*, IJTR.S2129. [CrossRef]
49. Jenkins, T.; Nguyen, J.; Polglaze, K.; Bertrand, P. Influence of Tryptophan and Serotonin on Mood and Cognition with a Possible Role of the Gut-Brain Axis. *Nutrients* **2016**, *8*, 56. [CrossRef]
50. Baixauli, E. Happiness: Role of Dopamine and Serotonin on Mood and Negative Emotions. *Emerg. Med. Open Access* **2017**, *7*, 33–51. [CrossRef]
51. Yousef, I. Serotonin-Happiness and Satisfaction. *Biomed. J. Sci. Tech. Res.* **2021**, *33*, 25870–25871. [CrossRef]
52. Alenina, N.; Klempin, F. The role of serotonin in adult hippocampal neurogenesis. *Behav. Brain Res.* **2015**, *277*, 49–57. [CrossRef]
53. Voracek, M.; Tran, U.S. Dietary tryptophan intake and suicide rate in industrialized nations. *J. Affect. Disord.* **2007**, *98*, 259–262. [CrossRef]

54. Munir, S.; Shahid, A.; Aslam, B.; Ashfaq, U.A.; Akash, M.S.H.; Ali, M.A.; Almatroudi, A.; Allemailem, K.S.; Rajoka, M.S.R.; Khurshid, M. The Therapeutic Prospects of Naturally Occurring and Synthetic Indole Alkaloids for Depression and Anxiety Disorders. *Evid.-Based Complement. Altern. Med.* **2020**, *2020*, 8836983. [CrossRef]
55. Melancon, M.O.; Lorrain, D.; Dionne, I.J. Exercise and sleep in aging: Emphasis on serotonin. *Pathol. Biol.* **2014**, *62*, 276–283. [CrossRef]
56. Kohyama, J. Sleep, Serotonin, and Suicide in Japan. *J. Physiol. Anthropol.* **2011**, *30*, 1–8. [CrossRef]
57. Tricklebank, M.D. Serotonin and sleep. In *The Serotonin System*; Elsevier: Amsterdam, The Netherlands, 2019; pp. 181–192. ISBN 9780128133231.
58. Ursin, R. Serotonin and sleep. *Sleep Med. Rev.* **2002**, *6*, 55–67. [CrossRef]
59. Fava, M.; Rosenbaum, J.F.; Hoog, S.L.; Tepner, R.G.; Kopp, J.B.; Nilsson, M.E. Fluoxetine versus sertraline and paroxetine in major depression: Tolerability and efficacy in anxious depression. *J. Affect. Disord.* **2000**, *59*, 119–126. [CrossRef]
60. Cipriani, A.; La Ferla, T.; Furukawa, T.A.; Signoretti, A.; Nakagawa, A.; Churchill, R.; McGuire, H.; Barbui, C. Sertraline versus other antidepressive agents for depression. *Cochrane Database Syst. Rev.* **2010**. [CrossRef]
61. Bergeron, R.; Ravindran, A.V.; Chaput, Y.; Goldner, E.; Swinson, R.; van Ameringen, M.A.; Austin, C.; Hadrava, V. Sertraline and Fluoxetine Treatment of Obsessive-Compulsive Disorder: Results of a Double-Blind, 6-Month Treatment Study. *J. Clin. Psychopharmacol.* **2002**, *22*, 148–154. [CrossRef]
62. Soorya, L.; Kiarashi, J.; Hollander, E. Psychopharmacologic Interventions for Repetitive Behaviors in Autism Spectrum Disorders. *Child Adolesc. Psychiatr. Clin. N. Am.* **2008**, *17*, 753–771. [CrossRef] [PubMed]
63. Wang, J.; Zhang, W.; Li, M.; Li, X. The new coumarin compound Bis 3 ameliorates cognitive disorder and suppresses brain-intestine-liver systematic oxidative stress in high-fat diet mice. *Biomed. Pharmacother.* **2021**, *137*, 111293. [CrossRef] [PubMed]
64. Yan, Y.; Kong, L.; Xia, Y.; Liang, W.; Wang, L.; Song, J.; Yao, Y.; Lin, Y.; Yang, J. Osthole promotes endogenous neural stem cell proliferation and improved neurological function through Notch signaling pathway in mice acute mechanical brain injury. *Brain. Behav. Immun.* **2018**, *67*, 118–129. [CrossRef] [PubMed]
65. Kong, L.; Yao, Y.; Xia, Y.; Liang, X.; Ni, Y.; Yang, J. Osthole alleviates inflammation by down-regulating NF-κB signaling pathway in traumatic brain injury. *Immunopharmacol. Immunotoxicol.* **2019**, *41*, 349–360. [CrossRef]
66. Du, G.; Tu, H.; Li, X.; Pei, A.; Chen, J.; Miao, Z.; Li, J.; Wang, C.; Xie, H.; Xu, X.; et al. Daphnetin, a Natural Coumarin Derivative, Provides the Neuroprotection Against Glutamate-Induced Toxicity in HT22 Cells and Ischemic Brain Injury. *Neurochem. Res.* **2014**, *39*, 269–275. [CrossRef]
67. Jin, X.; Wang, Y.; Li, X.; Tan, X.; Miao, Z.; Chen, Y.; Hamdy, R.C.; Chua, B.H.L.; Kong, J.; Zhao, H.; et al. 7,8-Dihydroxy-4-methylcoumarin Provides Neuroprotection by Increasing Hippocalcin Expression. *Neurotox. Res.* **2015**, *27*, 268–274. [CrossRef]
68. Langdon, S.R.; Ertl, P.; Brown, N. Bioisosteric Replacement and Scaffold Hopping in Lead Generation and Optimization. *Mol. Inform.* **2010**, *29*, 366–385. [CrossRef]
69. Wassermann, A.M.; Bajorath, J. Identification of target family directed bioisosteric replacements. *Medchemcomm* **2011**, *2*, 601–606. [CrossRef]
70. Lazzara, P.R.; Moore, T.W. Scaffold-hopping as a strategy to address metabolic liabilities of aromatic compounds. *RSC Med. Chem.* **2020**, *11*, 18–29. [CrossRef]
71. Sakyi, P.O.; Broni, E.; Amewu, R.K.; Miller, W.A.I.; Wilson, M.D.; Kwofie, S.K. Homology Modeling, de Novo Design of Ligands, and Molecular Docking Identify Potential Inhibitors of Leishmania donovani 24-Sterol Methyltransferase. *Front. Cell. Infect. Microbiol.* **2022**, *12*, 657. [CrossRef]
72. Wen, C.C.; Kuo, Y.H.; Jan, J.T.; Liang, P.H.; Wang, S.Y.; Liu, H.G.; Lee, C.K.; Chang, S.T.; Kuo, C.J.; Lee, S.S.; et al. Specific plant terpenoids and lignoids possess potent antiviral activities against severe acute respiratory syndrome coronavirus. *J. Med. Chem.* **2007**, *50*, 4087–4095. [CrossRef]
73. Chen, D.; Oezguen, N.; Urvil, P.; Ferguson, C.; Dann, S.M.; Savidge, T.C. Regulation of protein-ligand binding affinity by hydrogen bond pairing. *Sci. Adv.* **2016**, *2*, e1501240. [CrossRef]
74. Daina, A.; Michielin, O.; Zoete, V. SwissADME: A free web tool to evaluate pharmacokinetics, drug-likeness and medicinal chemistry friendliness of small molecules. *Sci. Rep.* **2017**, *7*, 42717. [CrossRef]
75. Barret, R. Lipinski's Rule of Five. In *Therapeutical Chemistry*; Elsevier: Amsterdam, The Netherlands, 2018.
76. Lipinski, C.A. Rule of five in 2015 and beyond: Target and ligand structural limitations, ligand chemistry structure and drug discovery project decisions. *Adv. Drug Deliv. Rev.* **2016**, *101*, 34–41. [CrossRef]
77. Lipinski, C.A. Lead- and drug-like compounds: The rule-of-five revolution. *Drug Discov. Today Technol.* **2004**, *1*, 337–341. [CrossRef]
78. Veber, D.F.; Johnson, S.R.; Cheng, H.; Smith, B.R.; Ward, K.W.; Kopple, K.D. Molecular Properties That Influence the Oral Bioavailability of Drug Candidates. *J. Med. Chem.* **2002**, *45*, 2615–2623. [CrossRef]
79. Lipinski, C.A.; Lombardo, F.; Dominy, B.W.; Feeney, P.J. Experimental and computational approaches to estimate solubility and permeability in drug discovery and development settings. *Adv. Drug Deliv. Rev.* **2001**, *46*, 3–26. [CrossRef]
80. Majerova, P.; Olesova, D.; Golisova, G.; Buralova, M.; Michalicova, A.; Vegh, J.; Piestansky, J.; Bhide, M.; Hanes, J.; Kovac, A. Analog of kynurenic acid decreases tau pathology by modulating astrogliosis in rat model for tauopathy. *Biomed. Pharmacother.* **2022**, *152*, 113257. [CrossRef]

81. Crowe, T.P.; Greenlee, M.H.W.; Kanthasamy, A.G.; Hsu, W.H. Mechanism of intranasal drug delivery directly to the brain. *Life Sci.* **2018**, *195*, 44–52. [CrossRef]
82. Bors, L.; Erdő, F. Overcoming the Blood–Brain Barrier. Challenges and Tricks for CNS Drug Delivery. *Sci. Pharm.* **2019**, *87*, 6. [CrossRef]
83. Hanson, L.R.; Frey, W.H. Intranasal delivery bypasses the blood-brain barrier to target therapeutic agents to the central nervous system and treat neurodegenerative disease. *BMC Neurosci.* **2008**, *9*, S5. [CrossRef] [PubMed]
84. Han, L.; Jiang, C. Evolution of blood–brain barrier in brain diseases and related systemic nanoscale brain-targeting drug delivery strategies. *Acta Pharm. Sin. B* **2021**, *11*, 2306–2325. [CrossRef] [PubMed]
85. Sander, T.; Freyss, J.; Von Korff, M.; Rufener, C. DataWarrior: An open-source program for chemistry aware data visualization and analysis. *J. Chem. Inf. Model.* **2015**, *55*, 460–473. [CrossRef] [PubMed]
86. Han, J.; An, O.; Hong, H.Q.; Chan, T.H.M.; Song, Y.; Shen, H.; Tang, S.J.; Lin, J.S.; Ng, V.H.E.; Tay, D.J.T.; et al. Suppression of adenosine-to-inosine (A-to-I) RNA editome by death associated protein 3 (DAP3) promotes cancer progression. *Sci. Adv.* **2020**, *6*, eaba5136. [CrossRef] [PubMed]
87. Valles, I.; Pajares, M.J.; Segura, V.; Guruceaga, E.; Gomez-Roman, J.; Blanco, D.; Tamura, A.; Montuenga, L.M.; Pio, R. Identification of Novel Deregulated RNA Metabolism-Related Genes in Non-Small Cell Lung Cancer. *PLoS ONE* **2012**, *7*, e42086. [CrossRef]
88. Chen, Y.-B.; Liao, X.-Y.; Zhang, J.-B.; Wang, F.; Qin, H.-D.; Zhang, L.; Shugart, Y.Y.; Zeng, Y.-X.; Jia, W.-H. ADAR2 functions as a tumor suppressor via editing IGFBP7 in esophageal squamous cell carcinoma. *Int. J. Oncol.* **2017**, *50*, 622–630. [CrossRef]
89. Tomaselli, S.; Galeano, F.; Alon, S.; Raho, S.; Galardi, S.; Polito, V.A.; Presutti, C.; Vincenti, S.; Eisenberg, E.; Locatelli, F.; et al. Modulation of microRNA editing, expression and processing by ADAR2 deaminase in glioblastoma. *Genome Biol.* **2015**, *16*, 5. [CrossRef]
90. Sakata, K.-I.; Maeda, K.; Sakurai, N.; Liang, S.; Nakazawa, S.; Yanagihara, K.; Kubo, T.; Yoshiyama, H.; Kitagawa, Y.; Hamada, J.-I.; et al. ADAR2 Regulates Malignant Behaviour of Mesothelioma Cells Independent of RNA-editing Activity. *Anticancer Res.* **2020**, *40*, 1307–1314. [CrossRef]
91. Filimonov, D.A.; Lagunin, A.A.; Gloriozova, T.A.; Rudik, A.V.; Druzhilovskii, D.S.; Pogodin, P.V.; Poroikov, V.V. Prediction of the Biological Activity Spectra of Organic Compounds Using the Pass Online Web Resource. *Chem. Heterocycl. Compd.* **2014**, *50*, 444–457. [CrossRef]
92. Parasuraman, S. Prediction of activity spectra for substances. *J. Pharmacol. Pharmacother.* **2011**, *2*, 52–53. [CrossRef]
93. Lagunin, A.; Stepanchikova, A.; Filimonov, D.; Poroikov, V. PASS: Prediction of activity spectra for biologically active substances. *Bioinformatics* **2000**, *16*, 747–748. [CrossRef]
94. El-Said, K.S.; Atta, A.; Mobasher, M.A.; Germoush, M.O.; Mohamed, T.M.; Salem, M.M. Quercetin mitigates rheumatoid arthritis by inhibiting adenosine deaminase in rats. *Mol. Med.* **2022**, *28*, 24. [CrossRef]
95. Li, G.; Nakagome, I.; Hirono, S.; Itoh, T.; Fujiwara, R. Inhibition of adenosine deaminase (ADA)-mediated metabolism of cordycepin by natural substances. *Pharmacol. Res. Perspect.* **2015**, *3*, e00121. [CrossRef]
96. Lindell, S.D.; Maechling, S.; Klein, R.; Freigang, J.; Laber, B.; Blanazs, L.; Leonhardt, M.; Haupt, S.; Petry, T.; Sabina, R.L. Mechanism and structure based design of inhibitors of AMP and adenosine deaminase. *Bioorganic Med. Chem.* **2021**, *43*, 116272. [CrossRef]
97. Lougiakis, N.; Marakos, P.; Pouli, N.; Fragopoulou, E.; Tenta, R. Synthesis of New Nebularine Analogues and Their Inhibitory Activity against Adenosine Deaminase. *Chem. Pharm. Bull.* **2015**, *63*, 134–142. [CrossRef]
98. Véliz, E.A.; Easterwood, L.M.; Beal, P.A. Substrate Analogues for an RNA-Editing Adenosine Deaminase: Mechanistic Investigation and Inhibitor Design. *J. Am. Chem. Soc.* **2003**, *125*, 10867–10876. [CrossRef]
99. Samari, H.R.; Seglen, P.O. Inhibition of Hepatocytic Autophagy by Adenosine, Aminoimidazole-4-carboxamide Riboside, and N 6-Mercaptopurine Riboside. *J. Biol. Chem.* **1998**, *273*, 23758–23763. [CrossRef]
100. Bojack, G.; Earnshaw, C.G.; Klein, R.; Lindell, S.D.; Lowinski, C.; Preuss, R. Design and Synthesis of Inhibitors of Adenosine and AMP Deaminases. *Org. Lett.* **2001**, *3*, 839–842. [CrossRef]
101. Górska, N.; Słupski, J.; Cubała, W.J.; Wiglusz, M.S.; Gałuszko-Węgielnik, M. Antidepressants in epilepsy. *Neurol. Neurochir. Pol.* **2018**, *52*, 657–661. [CrossRef]
102. Abboud, R.; Noronha, C.; Diwadkar, V.A. Motor system dysfunction in the schizophrenia diathesis: Neural systems to neurotransmitters. *Eur. Psychiatry* **2017**, *44*, 125–133. [CrossRef]
103. Walther, S.; Strik, W. Motor Symptoms and Schizophrenia. *Neuropsychobiology* **2012**, *66*, 77–92. [CrossRef] [PubMed]
104. Lefebvre, S.; Pavlidou, A.; Walther, S. What is the potential of neurostimulation in the treatment of motor symptoms in schizophrenia? *Expert Rev. Neurother.* **2020**, *20*, 697–706. [CrossRef] [PubMed]
105. Varlet, M.; Marin, L.; Raffard, S.; Schmidt, R.C.; Capdevielle, D.; Boulenger, J.P.; Del-Monte, J.; Bardy, B.G. Impairments of social motor coordination in schizophrenia. *PLoS ONE* **2012**, *7*, e29772. [CrossRef] [PubMed]
106. Posar, A.; Visconti, P. Early Motor Signs in Autism Spectrum Disorder. *Children* **2022**, *9*, 294. [CrossRef] [PubMed]
107. Mohd Nordin, A.; Ismail, J.; Kamal Nor, N. Motor Development in Children With Autism Spectrum Disorder. *Front. Pediatr.* **2021**, *9*, 598276. [CrossRef]
108. Tzeng, N.-S.; Hsu, Y.-H.; Ho, S.-Y.; Kuo, Y.-C.; Lee, H.-C.; Yin, Y.-J.; Chen, H.-A.; Chen, W.-L.; Chu, W.C.-C.; Huang, H.-L. Is schizophrenia associated with an increased risk of chronic kidney disease? A nationwide matched-cohort study. *BMJ Open* **2015**, *5*, e006777. [CrossRef]

109. Tzur Bitan, D.; Krieger, I.; Berkovitch, A.; Comaneshter, D.; Cohen, A. Chronic kidney disease in adults with schizophrenia: A nationwide population-based study. *Gen. Hosp. Psychiatry* **2019**, *58*, 1–6. [CrossRef]
110. Kissil, J.L.; Cohen, O.; Raveh, T.; Kimchi, A. Structure-function analysis of an evolutionary conserved protein, DAP3, which mediates TNF-α- and Fas-induced cell death. *EMBO J.* **1999**, *18*, 353–362. [CrossRef]
111. Kissil, J.L.; Deiss, L.P.; Bayewitch, M.; Raveh, T.; Khaspekov, G.; Kimchi, A. Isolation of DAP3, a Novel Mediator of Interferon-γ-induced Cell Death. *J. Biol. Chem.* **1995**, *270*, 27932–27936. [CrossRef]
112. Jung, S.; Ballheimer, Y.E.; Brackmann, F.; Zoglauer, D.; Geppert, C.-I.; Hartmann, A.; Trollmann, R. Seizure-induced neuronal apoptosis is related to dysregulation of the RNA-edited GluR2 subunit in the developing mouse brain. *Brain Res.* **2020**, *1735*, 146760. [CrossRef]
113. Marcucci, R.; Brindle, J.; Paro, S.; Casadio, A.; Hempel, S.; Morrice, N.; Bisso, A.; Keegan, L.P.; Del Sal, G.; O'Connell, M.A. Pin1 and WWP2 regulate GluR2 Q/R site RNA editing by ADAR2 with opposing effects. *EMBO J.* **2011**, *30*, 4211–4222. [CrossRef]
114. Behm, M.; Wahlstedt, H.; Widmark, A.; Eriksson, M.; Öhman, M. Accumulation of nuclear ADAR2 regulates A-to-I RNA editing during neuronal development. *J. Cell Sci.* **2017**, *130*, 745–753. [CrossRef]
115. Chen, Y.; Wu, Y.; Yang, H.; Li, X.; Jie, M.; Hu, C.; Wu, Y.; Yang, S.; Yang, Y. Prolyl isomerase Pin1: A promoter of cancer and a target for therapy. *Cell Death Dis.* **2018**, *9*, 883. [CrossRef]
116. Tay, D.J.T.; Song, Y.; Peng, B.; Toh, T.B.; Hooi, L.; Toh, D.-F.K.; Hong, H.; Tang, S.J.; Han, J.; Gan, W.L.; et al. Targeting RNA editing of antizyme inhibitor 1: A potential oligonucleotide-based antisense therapy for cancer. *Mol. Ther.* **2021**, *29*, 3258–3273. [CrossRef]
117. Shen, X.; Wang, Y.; Han, X.; Sheng, L.; Wu, F.; Liu, X. Design, synthesis and anticancer activity of naphthoquinone derivatives. *J. Enzyme Inhib. Med. Chem.* **2020**, *35*, 773–785. [CrossRef]
118. Wang, H.; Luo, Y.; Shen, G.; Piao, X.; Xu, W.; Zhang, Y.; Wang, J.; Feng, Y.; Li, J.; Zhang, Y.; et al. Two novel 1,4-naphthoquinone derivatives induce human gastric cancer cell apoptosis and cell cycle arrest by regulating reactive oxygen species-mediated MAPK/Akt/STAT3 signaling pathways. *Mol. Med. Rep.* **2019**, *20*, 2571–2582. [CrossRef]
119. Kayashima, T.; Mori, M.; Yoshida, H.; Mizushina, Y.; Matsubara, K. 1,4-Naphthoquinone is a potent inhibitor of human cancer cell growth and angiogenesis. *Cancer Lett.* **2009**, *278*, 34–40. [CrossRef]
120. Byrne, F.L.; Olzomer, E.M.; Marriott, G.R.; Quek, L.E.; Katen, A.; Su, J.; Nelson, M.E.; Hart-Smith, G.; Larance, M.; Sebesfi, V.F.; et al. Phenotypic screen for oxygen consumption rate identifies an anti-cancer naphthoquinone that induces mitochondrial oxidative stress. *Redox Biol.* **2020**, *28*, 101374. [CrossRef]
121. Kumar, S.; Malachowski, W.P.; DuHadaway, J.B.; LaLonde, J.M.; Carroll, P.J.; Jaller, D.; Metz, R.; Prendergast, G.C.; Muller, A.J. Indoleamine 2,3-Dioxygenase Is the Anticancer Target for a Novel Series of Potent Naphthoquinone-Based Inhibitors. *J. Med. Chem.* **2008**, *51*, 1706–1718. [CrossRef]
122. Presse, N.; Belleville, S.; Gaudreau, P.; Greenwood, C.E.; Kergoat, M.-J.; Morais, J.A.; Payette, H.; Shatenstein, B.; Ferland, G. Vitamin K status and cognitive function in healthy older adults. *Neurobiol. Aging* **2013**, *34*, 2777–2783. [CrossRef]
123. Kiely, A.; Ferland, G.; Ouliass, B.; O'Toole, P.W.; Purtill, H.; O'Connor, E.M. Vitamin K status and inflammation are associated with cognition in older Irish adults. *Nutr. Neurosci.* **2020**, *23*, 591–599. [CrossRef] [PubMed]
124. Mokarizadeh, N.; Karimi, P.; Erfani, M.; Sadigh-Eteghad, S.; Fathi Maroufi, N.; Rashtchizadeh, N. β-Lapachone attenuates cognitive impairment and neuroinflammation in beta-amyloid induced mouse model of Alzheimer's disease. *Int. Immunopharmacol.* **2020**, *81*, 106300. [CrossRef] [PubMed]
125. Chen, Y.; Li, Y.-Q.; Fang, J.-Y.; Li, P.; Li, F. Establishment of the concurrent experimental model of osteoporosis combined with Alzheimer's disease in rat and the dual-effects of echinacoside and acteoside from Cistanche tubulosa. *J. Ethnopharmacol.* **2020**, *257*, 112834. [CrossRef] [PubMed]
126. Feng, X. Selective protection of nigral dopaminergic neurons by echinacoside in a rat model of Parkinson disease induced by rotenone. *J. Chin. Integr. Med.* **2012**, *10*, 777–783. [CrossRef] [PubMed]
127. Dai, Y.; Han, G.; Xu, S.; Yuan, Y.; Zhao, C.; Ma, T. Echinacoside Suppresses Amyloidogenesis and Modulates F-actin Remodeling by Targeting the ER Stress Sensor PERK in a Mouse Model of Alzheimer's Disease. *Front. Cell Dev. Biol.* **2020**, *8*, 593659. [CrossRef]
128. Jia, C.; Shi, H.; Wu, X.; Li, Y.; Chen, J.; Tu, P. Determination of echinacoside in rat serum by reversed-phase high-performance liquid chromatography with ultraviolet detection and its application to pharmacokinetics and bioavailability. *J. Chromatogr. B* **2006**, *844*, 308–313. [CrossRef]
129. Matthias, A.; Addison, R.S.; Penman, K.G.; Dickinson, R.G.; Bone, K.M.; Lehmann, R.P. Echinacea alkamide disposition and pharmacokinetics in humans after tablet ingestion. *Life Sci.* **2005**, *77*, 2018–2029. [CrossRef]
130. Martin, Y.C. A bioavailability score. *J. Med. Chem.* **2005**, *48*, 3164–3170. [CrossRef]
131. Chen, L.; Huang, C.; Shentu, J.; Wang, M.; Yan, S.; Zhou, F.; Zhang, Z.; Wang, C.; Han, Y.; Wang, Q.; et al. Indirubin Derivative 7-Bromoindirubin-3-Oxime (7Bio) Attenuates Aβ Oligomer-Induced Cognitive Impairments in Mice. *Front. Mol. Neurosci.* **2017**, *10*, 393. [CrossRef]
132. Ribas, J.; Bettayeb, K.; Ferandin, Y.; Knockaert, M.; Garrofé-Ochoa, X.; Totzke, F.; Schächtele, C.; Mester, J.; Polychronopoulos, P.; Magiatis, P.; et al. 7-Bromoindirubin-3′-oxime induces caspase-independent cell death. *Oncogene* **2006**, *25*, 6304–6318. [CrossRef]
133. Ribas, J.; Yuste, V.J.; Garrofé-Ochoa, X.; Meijer, L.; Esquerda, J.E.; Boix, J. 7-Bromoindirubin-3′-oxime uncovers a serine protease-mediated paradigm of necrotic cell death. *Biochem. Pharmacol.* **2008**, *76*, 39–52. [CrossRef]

134. Xu, X.; Piao, H.; Aosai, F.; Zeng, X.; Cheng, J.; Cui, Y.; Li, J.; Ma, J.; Piao, H.; Jin, X.; et al. Arctigenin protects against depression by inhibiting microglial activation and neuroinflammation via HMGB1/TLR4/NF-κB and TNF-α/TNFR1/NF-κB pathways. *Br. J. Pharmacol.* **2020**, *177*, 5224–5245. [CrossRef]
135. Wei, L.; Xue, Z.; Lan, B.; Yuan, S.; Li, Y.; Guo, C.; Zhang, R.; Ding, R.; Shen, H. Arctigenin Exerts Neuroprotective Effect by Ameliorating Cortical Activities in Experimental Autoimmune Encephalomyelitis In Vivo. *Front. Immunol.* **2021**, *12*, 733101. [CrossRef]
136. Xu, P.; Huang, M.-W.; Xiao, C.-X.; Long, F.; Wang, Y.; Liu, S.-Y.; Jia, W.-W.; Wu, W.-J.; Yang, D.; Hu, J.-F.; et al. Matairesinol Suppresses Neuroinflammation and Migration Associated with Src and ERK1/2-NF-κB Pathway in Activating BV2 Microglia. *Neurochem. Res.* **2017**, *42*, 2850–2860. [CrossRef]
137. Mangat, H.K.; Rani, M.; Pathak, R.K.; Yadav, I.S.; Utreja, D.; Chhuneja, P.K.; Chhuneja, P. Virtual screening, molecular dynamics and binding energy-MM-PBSA studies of natural compounds to identify potential EcR inhibitors against Bemisia tabaci Gennadius. *PLoS ONE* **2022**, *17*, e0261545. [CrossRef]
138. Sharma, J.; Kumar Bhardwaj, V.; Singh, R.; Rajendran, V.; Purohit, R.; Kumar, S. An in-silico evaluation of different bioactive molecules of tea for their inhibition potency against non structural protein-15 of SARS-CoV-2. *Food Chem.* **2021**, *346*, 128933. [CrossRef]
139. De Vita, S.; Chini, M.G.; Bifulco, G.; Lauro, G. Insights into the Ligand Binding to Bromodomain-Containing Protein 9 (BRD9): A Guide to the Selection of Potential Binders by Computational Methods. *Molecules* **2021**, *26*, 7192. [CrossRef]
140. Lobanov, M.Y.; Bogatyreva, N.S.; Galzitskaya, O.V. Radius of gyration as an indicator of protein structure compactness. *Mol. Biol.* **2008**, *42*, 623–628. [CrossRef]
141. Asiedu, S.O.; Kwofie, S.K.; Broni, E.; Wilson, M.D. Computational Identification of Potential Anti-Inflammatory Natural Compounds Targeting the p38 Mitogen-Activated Protein Kinase (MAPK): Implications for COVID-19-Induced Cytokine Storm. *Biomolecules* **2021**, *11*, 653. [CrossRef]
142. Sgobba, M.; Caporuscio, F.; Anighoro, A.; Portioli, C.; Rastelli, G. Application of a post-docking procedure based on MM-PBSA and MM-GBSA on single and multiple protein conformations. *Eur. J. Med. Chem.* **2012**, *58*, 431–440. [CrossRef]
143. Rifai, E.A.; Van Dijk, M.; Vermeulen, N.P.E.; Yanuar, A.; Geerke, D.P. A Comparative Linear Interaction Energy and MM/PBSA Study on SIRT1−Ligand Binding Free Energy Calculation. *J. Chem. Inf. Model.* **2019**, *59*, 4018–4033. [CrossRef] [PubMed]
144. Genheden, S.; Ryde, U. The MM/PBSA and MM/GBSA methods to estimate ligand-binding affinities. *Expert Opin. Drug Discov.* **2015**, *10*, 449–461. [CrossRef] [PubMed]
145. Borkotoky, S.; Meena, C.K.; Murali, A. Interaction Analysis of T7 RNA Polymerase with Heparin and Its Low Molecular Weight Derivatives—An in Silico Approach. *Bioinform. Biol. Insights* **2016**, *10*, BBI.S40427. [CrossRef] [PubMed]
146. Kumari, R.; Kumar, R.; Lynn, A. g_mmpbsa—A GROMACS Tool for High-Throughput MM-PBSA Calculations. *J. Chem. Inf. Model.* **2014**, *54*, 1951–1962. [CrossRef] [PubMed]
147. Kollman, P.A.; Massova, I.; Reyes, C.; Kuhn, B.; Huo, S.; Chong, L.; Lee, M.; Lee, T.; Duan, Y.; Wang, W.; et al. Calculating Structures and Free Energies of Complex Molecules: Combining Molecular Mechanics and Continuum Models. *Acc. Chem. Res.* **2000**, *33*, 889–897. [CrossRef]
148. Pergolizzi, J.V.; Dahan, A.; Ann LeQuang, J.; Raffa, R.B. Overdoses due to fentanyl and its analogues (F/FAs) push naloxone to the limit. *J. Clin. Pharm. Ther.* **2021**, *46*, 1501–1504. [CrossRef]
149. Hesselbarth, S.; Löwenstein, O.; Cegla, T. Effects of prolonged-release oxycodone/naloxone on pain control, bowel function and quality of life: A prospective observational study. *Scand. J. Pain* **2014**, *5*, 75–81. [CrossRef]
150. DeLano, W.L. Unraveling hot spots in binding interfaces: Progress and challenges. *Curr. Opin. Struct. Biol.* **2002**, *12*, 14–20. [CrossRef]
151. Zhong, S.; Macias, A.; MacKerell, A. Computational Identification of Inhibitors of Protein-Protein Interactions. *Curr. Top. Med. Chem.* **2007**, *7*, 63–82. [CrossRef]
152. Grosdidier, S.; Fernández-Recio, J. Identification of hot-spot residues in protein-protein interactions by computational docking. *BMC Bioinform.* **2008**, *9*, 447. [CrossRef]
153. Kwofie, S.; Dankwa, B.; Enninful, K.; Adobor, C.; Broni, E.; Ntiamoah, A.; Wilson, M. Molecular Docking and Dynamics Simulation Studies Predict Munc18b as a Target of Mycolactone: A Plausible Mechanism for Granule Exocytosis Impairment in Buruli Ulcer Pathogenesis. *Toxins* **2019**, *11*, 181. [CrossRef]
154. Silhol, S.; Glin, L.; Gottesmann, C. Study of the 5-HT2 antagonist ritanserin on sleep-waking cycle in the rat. *Pharmacol. Biochem. Behav.* **1992**, *41*, 241–243. [CrossRef]
155. Mayer, G. Ritanserin Improves Sleep Quality in Narcolepsy. *Pharmacopsychiatry* **2003**, *36*, 150–155. [CrossRef] [PubMed]
156. Viola, A.U.; Brandenberger, G.; Toussaint, M.; Bouhours, P.; Paul Macher, J.; Luthringer, R. Ritanserin, a serotonin-2 receptor antagonist, improves ultradian sleep rhythmicity in young poor sleepers. *Clin. Neurophysiol.* **2002**, *113*, 429–434. [CrossRef]
157. Henderson, J.; Yiannikas, C.; Graham, J.S. Effect of ritanserin, a highly selective 5-HT2 receptor antagonist, on Parkinson's disease. *Clin. Exp. Neurol.* **1992**, *29*, 277–282.
158. Poyurovsky, M.; Shardorodsky, M.; Fuchs, C.; Schneidman, M.; Weizman, A. Treatment of neuroleptic-induced akathisia with the 5-HT 2 antagonist mianserin. *Br. J. Psychiatry* **1999**, *174*, 238–242. [CrossRef] [PubMed]
159. Barone, J.A.; Bierman, R.H.; Cornish, J.W.; Hsuan, A.; Drake, N.D.; Colaizzi, J.L. Safety Evaluation of Ritanserin—An Investigational Serotonin Antagonist. *Drug Intell. Clin. Pharm.* **1986**, *20*, 770–775. [CrossRef]

160. Campbell, S.T.; Franks, C.E.; Borne, A.L.; Shin, M.; Zhang, L.; Hsu, K.-L. Chemoproteomic Discovery of a Ritanserin-Targeted Kinase Network Mediating Apoptotic Cell Death of Lung Tumor Cells. *Mol. Pharmacol.* **2018**, *94*, 1246–1255. [CrossRef]
161. Wesensten, N.J. Role of Pharmacological Interventions for Sleep Deprivation. *Encycl. Sleep* **2013**, 366–370. [CrossRef]
162. Daoui, S.; Kansiz, S.; Aktas, F.A.; Dege, N.; Saif, E.; Benchat, N.; Karrouchi, K. Crystal structure and molecular docking study of diethyl 2,2′-({[(1 E, 1′ E)-(hydrazine-1,2-diylidene)bis(methanylylidene)]bis(4,1-phenylene)}bis(oxy))diacetate. *Acta Crystallogr. Sect. E Crystallogr. Commun.* **2022**, *78*, 88–91. [CrossRef]
163. Sterling, T.; Irwin, J.J. ZINC 15—Ligand Discovery for Everyone. *J. Chem. Inf. Model.* **2015**, *55*, 2324–2337. [CrossRef] [PubMed]
164. Kim, S.; Thiessen, P.A.; Bolton, E.E.; Chen, J.; Fu, G.; Gindulyte, A.; Han, L.; He, J.; He, S.; Shoemaker, B.A.; et al. PubChem Substance and Compound databases. *Nucleic Acids Res.* **2016**, *44*, D1202–D1213. [CrossRef] [PubMed]
165. Kim, S.; Chen, J.; Cheng, T.; Gindulyte, A.; He, J.; He, S.; Li, Q.; Shoemaker, B.A.; Thiessen, P.A.; Yu, B.; et al. PubChem in 2021: New data content and improved web interfaces. *Nucleic Acids Res.* **2021**, *49*, D1388–D1395. [CrossRef] [PubMed]
166. Mendez, D.; Gaulton, A.; Bento, A.P.; Chambers, J.; De Veij, M.; Félix, E.; Magariños, M.P.; Mosquera, J.F.; Mutowo, P.; Nowotka, M.; et al. ChEMBL: Towards direct deposition of bioassay data. *Nucleic Acids Res.* **2019**, *47*, D930–D940. [CrossRef] [PubMed]
167. Gaulton, A.; Bellis, L.J.; Bento, A.P.; Chambers, J.; Davies, M.; Hersey, A.; Light, Y.; McGlinchey, S.; Michalovich, D.; Al-Lazikani, B.; et al. ChEMBL: A large-scale bioactivity database for drug discovery. *Nucleic Acids Res.* **2012**, *40*, D1100–D1107. [CrossRef]
168. Bento, A.P.; Gaulton, A.; Hersey, A.; Bellis, L.J.; Chambers, J.; Davies, M.; Krüger, F.A.; Light, Y.; Mak, L.; McGlinchey, S.; et al. The ChEMBL bioactivity database: An update. *Nucleic Acids Res.* **2014**, *42*, D1083–D1090. [CrossRef]
169. Rutz, A.; Sorokina, M.; Galgonek, J.; Mietchen, D.; Willighagen, E.; Gaudry, A.; Graham, J.G.; Stephan, R.; Page, R.; Vondrášek, J.; et al. The LOTUS initiative for knowledge sharing in Natural Products research. In Proceedings of the GA—69th Annual Meeting 2021, Virtual, 5–8 September 2021.
170. Nakamura, Y.; Mochamad Afendi, F.; Kawsar Parvin, A.; Ono, N.; Tanaka, K.; Hirai Morita, A.; Sato, T.; Sugiura, T.; Altaf-Ul-Amin, M.; Kanaya, S. KNApSAcK Metabolite Activity Database for Retrieving the Relationships Between Metabolites and Biological Activities. *Plant Cell Physiol.* **2014**, *55*, e7. [CrossRef]
171. Afendi, F.M.; Okada, T.; Yamazaki, M.; Hirai-Morita, A.; Nakamura, Y.; Nakamura, K.; Ikeda, S.; Takahashi, H.; Altaf-Ul-Amin, M.; Darusman, L.K.; et al. KNApSAcK family databases: Integrated metabolite-plant species databases for multifaceted plant research. *Plant Cell Physiol.* **2012**, *53*, e1. [CrossRef]
172. Zeng, X.; Zhang, P.; He, W.; Qin, C.; Chen, S.; Tao, L.; Wang, Y.; Tan, Y.; Gao, D.; Wang, B.; et al. NPASS: Natural product activity and species source database for natural product research, discovery and tool development. *Nucleic Acids Res.* **2018**, *46*, D1217–D1222. [CrossRef]
173. Mohanraj, K.; Karthikeyan, B.S.; Vivek-Ananth, R.P.; Chand, R.P.B.; Aparna, S.R.; Mangalapandi, P.; Samal, A. IMPPAT: A curated database of Indian Medicinal Plants, Phytochemistry And Therapeutics. *Sci. Rep.* **2018**, *8*, 4329. [CrossRef]
174. Broni, E.; Ashley, C.; Adams, J.; Manu, H.; Aikins, E.; Okom, M.; Miller, W.A.; Wilson, M.D.; Kwofie, S.K. Cheminformatics-Based Study Identifies Potential Ebola VP40 Inhibitors. *Int. J. Mol. Sci.* **2023**, *24*, 6298. [CrossRef]
175. Baskaran, M.; Shanmugam, L.; Raman, V. Effect of Plumbago zeylanica administration on brain neurotransmitter level in Wistar albino rats. *J. Appl. Pharm. Sci.* **2015**, *5*, 53–57. [CrossRef]
176. Lin, L.-C.; Yang, L.-L.; Chou, C.-J. Cytotoxic naphthoquinones and plumbagic acid glucosides from Plumbago zeylanica. *Phytochemistry* **2003**, *62*, 619–622. [CrossRef]
177. Sreelatha, T.; Hymavathi, A.; Murthy, J.M.; Rani, P.U.; Rao, J.M.; Babu, K.S. Bioactivity-guided isolation of mosquitocidal constituents from the rhizomes of Plumbago capensis Thunb. *Bioorg. Med. Chem. Lett.* **2010**, *20*, 2974–2977. [CrossRef]
178. Gu, J.-Q.; Graf, T.N.; Lee, D.; Chai, H.-B.; Mi, Q.; Kardono, L.B.S.; Setyowati, F.M.; Ismail, R.; Riswan, S.; Farnsworth, N.R.; et al. Cytotoxic and Antimicrobial Constituents of the Bark of Diospyros ma ritima Collected in Two Geographical Locations in Indonesia. *J. Nat. Prod.* **2004**, *67*, 1156–1161. [CrossRef]
179. Higa, M.; Takashima, Y.; Yokaryo, H.; Harie, Y.; Suzuka, T.; Ogihara, K. Naphthoquinone Derivatives from *Diospyros maritima*. *Chem. Pharm. Bull.* **2017**, *65*, 739–745. [CrossRef]
180. Higa, M.; Noha, N.; Yokaryo, H.; Ogihara, K.; Yogi, S. Three New Naphthoquinone Derivatives from Diospyros maritima Blume. *Chem. Pharm. Bull.* **2002**, *50*, 590–593. [CrossRef]
181. Sobhani, M.; Abbas-Mohammadi, M.; Ebrahimi, S.N.; Aliahmadi, A. Tracking leading anti-candida compounds in plant samples; Plumbago europaea. *Iran. J. Microbiol.* **2018**, *10*, 187–193.
182. Sankaram, A.V.B.; Rao, A.S.; Shoolery, J.N. Zeylanone and Isozeylanone, two novel quinones from *Plumbago Zeylanica*. *Tetrahedron* **1979**, *35*, 1777–1782. [CrossRef]
183. Wang, Y.C.; Huang, T.L. Anti-Helicobacter pylori activity of *Plumbago zeylanica* L. *FEMS Immunol. Med. Microbiol.* **2005**, *43*, 407–412. [CrossRef]
184. Mallavadhani, U.V.; Panda, A.K.; Rao, Y.R. Review article number 134 pharmacology and chemotaxonomy of diospyros. *Phytochemistry* **1998**, *49*, 901–951. [CrossRef] [PubMed]
185. Alves, A.C.; Costa, M.A.C.; Paul, M.I. Naphthaquinones of Diospyros batocana. *Planta Med.* **1983**, *47*, 121–124. [CrossRef] [PubMed]

186. Musgrave, O.C.; Skoyles, D. Ebenaceae extractives. Part IV. Diosindigo A, a blue pigment from several Diospyros species. *J. Chem. Soc. Perkin Trans.* **1974**, *1*, 1128. [CrossRef]
187. Tezuka, M.; Kuroyanagi, M.; Yoshihira, K.; Natori, S. Naphthoquinone Derivatives from the Ebenaceae. IV. Naphthoquinone Derivatives from Diospyros kaki THUNB. and D. kaki THUNB. var. sylvestris MAKINO. *Chem. Pharm. Bull.* **1972**, *20*, 2029–2035. [CrossRef]
188. Yoshihira, K.; Tezuka, M.; Natori, S. Naphthoquinone derivatives from SPP.: Bisisodiospyrin, a tetrameric naphthoquinone. *Tetrahedron Lett.* **1970**, *11*, 7–10. [CrossRef]
189. YOSHIHIRA, K.; TEZUKA, M.; NATORI, S. Naphthoquinone Derivatives from the Ebenaceae. II. Isodiospyrin, Bisisodiospyrin, and Mamegakinone from Diospyros lotus L. and D. morrisiana HANCE. *Chem. Pharm. Bull.* **1971**, *19*, 2308–2313. [CrossRef]
190. Marston, A.; Msonthi, J.; Hostettmann, K. Naphthoquinones of Diospyros usambarensis; their Molluscicidal and Fungicidal Activities. *Planta Med.* **1984**, *50*, 279–280. [CrossRef]
191. Creveld, M. Diospyros kaki—Der Weltbaum—Ein neues homöopathisches Mittel. *Allg. Homöopathische Zeitung* **2004**, *249*, 246–250. [CrossRef]
192. Miyase, T. Antioxidants from *Lespedeza homoloba*. (I). *Phytochemistry* **1999**, *52*, 303–310. [CrossRef]
193. Onegi, B.; Kraft, C.; Köhler, I.; Freund, M.; Jenett-Siems, K.; Siems, K.; Beyer, G.; Melzig, M.F.; Bienzle, U.; Eich, E. Antiplasmodial activity of naphthoquinones and one anthraquinone from Stereospermum kunthianum. *Phytochemistry* **2002**, *60*, 39–44. [CrossRef]
194. Mohn, T.; Plitzko, I.; Hamburger, M. A comprehensive metabolite profiling of Isatis tinctoria leaf extracts. *Phytochemistry* **2009**, *70*, 924–934. [CrossRef] [PubMed]
195. Wei, X.Y.; Leung, C.Y.; Wong, C.K.C.; Shen, X.L.; Wong, R.N.S.; Cai, Z.W.; Mak, N.K. Bisindigotin, a TCDD Antagonist from the Chinese Medicinal Herb Isatis indigotica. *J. Nat. Prod.* **2005**, *68*, 427–429. [CrossRef] [PubMed]
196. Bergman, J.; Lindström, J.-O.; Tilstam, U. The structure and properties of some indolic constituents in Couroupita guianensis aubl. *Tetrahedron* **1985**, *41*, 2879–2881. [CrossRef]
197. Hirschmann, G.S.; Ferro, E. Indigo from Eupatorium laeve. *J. Ethnopharmacol.* **1989**, *26*, 93–94. [CrossRef]
198. Hamburger, M. Isatis tinctoria—From the rediscovery of an ancient medicinal plant towards a novel anti-inflammatory phytopharmaceutical. *Phytochem. Rev.* **2002**, *1*, 333–344. [CrossRef]
199. Chung, Y.-C.; Tang, F.-Y.; Liao, J.-W.; Chung, C.-H.; Jong, T.-T.; Chen, S.-S.; Tsai, C.-H.; Chiang, E.-P. Isatis indigotica Induces Hepatocellular Cancer Cell Death via Caspase-Independent Apoptosis-Inducing Factor Translocation Apoptotic Pathway In Vitro and In Vivo. *Integr. Cancer Ther.* **2011**, *10*, 201–214. [CrossRef]
200. Xi, Y.-F.; Lou, L.-L.; Han, F.-Y.; Liu, S.-F.; Yao, G.-D.; Lin, B.; Huang, X.-X.; Wang, X.-B.; Song, S.-J. Four pairs of alkaloid enantiomers from Isatis indigotica Fortune Ex Land with neuroprotective effects against H_2O_2-induced SH-SY5Y cell injury. *Bioorg. Chem.* **2020**, *96*, 103650. [CrossRef]
201. Liu, S.F.; Zhang, Y.Y.; Zhou, L.; Lin, B.; Huang, X.X.; Wang, X.B.; Song, S.J. Alkaloids with neuroprotective effects from the leaves of Isatis indigotica collected in the Anhui Province, China. *Phytochemistry* **2018**, *149*, 132–139. [CrossRef]
202. Xi, Y.-F.; Lou, L.-L.; Han, F.-Y.; Wang, X.-B.; Huang, X.-X.; Yao, G.-D.; Song, S.-J. Discovery of alkaloids from the leaves of Isatis indigotica Fortune with neuroprotective activity. *Chin. J. Nat. Med.* **2021**, *19*, 680–685. [CrossRef]
203. Xi, Y.-F.; Liu, S.-F.; Hong, W.; Song, X.-Y.; Lou, L.-L.; Zhou, L.; Yao, G.-D.; Lin, B.; Wang, X.-B.; Huang, X.-X.; et al. Discovery of cycloneolignan enantiomers from Isatis indigotica Fortune with neuroprotective effects against MPP+-induced SH-SY5Y cell injury. *Bioorg. Chem.* **2019**, *88*, 102926. [CrossRef]
204. Xi, Y.-F.; Lou, L.-L.; Xu, Z.-Y.; Hou, Z.-L.; Wang, X.-B.; Huang, X.-X.; Song, S.-J. Alkaloid Enantiomers from Isatis tinctoria with Neuroprotective Effects against H2O2-Induced SH-SY5Y Cell Injury. *Planta Med.* **2019**, *85*, 1374–1382. [CrossRef]
205. Shen, C.-C.; Ni, C.-L.; Huang, Y.-L.; Huang, R.-L.; Chen, C.-C. Furanolabdane Diterpenes from Hypoestes p urpurea. *J. Nat. Prod.* **2004**, *67*, 1947–1949. [CrossRef]
206. Mohagheghzadeh, A.; Schmidt, T.; Bayindir, Ü.; Fuss, E.; Mehregan, I.; Alfermann, A. Diarylbutyrolactone Lignans from Linum corymbulosum in vitro Cultures. *Planta Med.* **2006**, *72*, 1165–1167. [CrossRef]
207. Jim-Min, F.; Ching-Kuo, L.; Yu-Shia, C. Lignans from leaves of Juniperus chinensis. *Phytochemistry* **1992**, *31*, 3659–3661. [CrossRef]
208. Niwa, A.M.; Marcarini, J.C.; Sartori, D.; Maistro, E.L.; Mantovani, M.S. Effects of (−)-cubebin (Piper cubeba) on cytotoxicity, mutagenicity and expression of p38 MAP kinase and GSTa2 in a hepatoma cell line. *J. Food Compos. Anal.* **2013**, *30*, 1–5. [CrossRef]
209. Rajalekshmi, D.S.; Kabeer, F.A.; Madhusoodhanan, A.R.; Bahulayan, A.K.; Prathapan, R.; Prakasan, N.; Varughese, S.; Nair, M.S. Anticancer activity studies of cubebin isolated from Piper cubeba and its synthetic derivatives. *Bioorg. Med. Chem. Lett.* **2016**, *26*, 1767–1771. [CrossRef]
210. Somani, G.S.; Nahire, M.S.; Parikh, A.D.; Mulik, M.B.; Ghumatkar, P.J.; Laddha, K.S.; Sathaye, S. Neuroprotective effect of Cubebin: A dibenzylbutyrolactone lignan on scopolamine-induced amnesia in mice. *Indian J. Med. Res.* **2017**, *146*, 255–259. [CrossRef]
211. Bucciantini, M.; Leri, M.; Nardiello, P.; Casamenti, F.; Stefani, M. Olive Polyphenols: Antioxidant and Anti-Inflammatory Properties. *Antioxidants* **2021**, *10*, 1044. [CrossRef]
212. Batista, Â.G.; Ferrari, A.S.; Da Cunha, D.C.; Da Silva, J.K.; Cazarin, C.B.B.; Correa, L.C.; Prado, M.A.; De Carvalho-Silva, L.B.; Esteves, E.A.; Maróstica Júnior, M.R. Polyphenols, antioxidants, and antimutagenic effects of Copaifera langsdorffii fruit. *Food Chem.* **2016**, *197*, 1153–1159. [CrossRef]

213. Chen, G.-L.; Munyao Mutie, F.; Xu, Y.-B.; Saleri, F.D.; Hu, G.-W.; Guo, M.-Q. Antioxidant, Anti-inflammatory Activities and Polyphenol Profile of Rhamnus prinoides. *Pharmaceuticals* **2020**, *13*, 55. [CrossRef]
214. Hu, Z.; Li, M.; Cao, Y.; Akan, O.D.; Guo, T.; Luo, F. Targeting AMPK Signaling by Dietary Polyphenols in Cancer Prevention. *Mol. Nutr. Food Res.* **2022**, *66*, 2100732. [CrossRef] [PubMed]
215. Prakash, M.D.; Stojanovska, L.; Feehan, J.; Nurgali, K.; Donald, E.L.; Plebanski, M.; Flavel, M.; Kitchen, B.; Apostolopoulos, V. Anti-cancer effects of polyphenol-rich sugarcane extract. *PLoS ONE* **2021**, *16*, e0247492. [CrossRef] [PubMed]
216. Miyata, Y.; Shida, Y.; Hakariya, T.; Sakai, H. Anti-Cancer Effects of Green Tea Polyphenols Against Prostate Cancer. *Molecules* **2019**, *24*, 193. [CrossRef] [PubMed]
217. Sakagami, H.; Shi, H.; Bandow, K.; Tomomura, M.; Tomomura, A.; Horiuchi, M.; Fujisawa, T.; Oizumi, T. Search of neuroprotective polyphenols using the "overlay" isolation method. *Molecules* **2018**, *23*, 1840. [CrossRef]
218. Zhang, Z.; Zhang, Y.; Li, J.; Fu, C.; Zhang, X. The Neuroprotective Effect of Tea Polyphenols on the Regulation of Intestinal Flora. *Molecules* **2021**, *26*, 3692. [CrossRef]
219. Spagnuolo, C.; Napolitano, M.; Tedesco, I.; Moccia, S.; Milito, A.; Luigi Russo, G. Neuroprotective Role of Natural Polyphenols. *Curr. Top. Med. Chem.* **2016**, *16*, 1943–1950. [CrossRef]
220. Galli, F. Interactions of Polyphenolic Compounds with Drug Disposition and Metabolism. *Curr. Drug Metab.* **2007**, *8*, 830–838. [CrossRef]
221. Joshua, D.L.; Sang, S.; Anthony, Y.; Lu, H.; Chung, S. Yang Metabolism of Dietary Polyphenols and Possible Interactions with Drugs. *Curr. Drug Metab.* **2007**, *8*, 499–507. [CrossRef]
222. Chalopin, M.; Tesse, A.; Martínez, M.C.; Rognan, D.; Arnal, J.F.; Andriantsitohaina, R. Estrogen receptor alpha as a key target of red wine polyphenols action on the endothelium. *PLoS ONE* **2010**, *5*, e8554. [CrossRef]
223. Ganguly, M.; Hazarika, J.; Sarma, S.; Bhuyan, P.; Mahanta, R. Estrogen receptor modulation of some polyphenols extracted from Daucus carota as a probable mechanism for antifertility effect: An in silico study. *J. Theor. Comput. Chem.* **2020**, *19*, 2041004. [CrossRef]
224. Peng, N.; Clark, J.T.; Prasain, J.; Kim, H.; White, C.R.; Wyss, J.M. Antihypertensive and cognitive effects of grape polyphenols in estrogen-depleted, female, spontaneously hypertensive rats. *Am. J. Physiol. Integr. Comp. Physiol.* **2005**, *289*, R771–R775. [CrossRef]
225. Cipolletti, M.; Solar Fernandez, V.; Montalesi, E.; Marino, M.; Fiocchetti, M. Beyond the Antioxidant Activity of Dietary Polyphenols in Cancer: The Modulation of Estrogen Receptors (ERs) Signaling. *Int. J. Mol. Sci.* **2018**, *19*, 2624. [CrossRef]
226. Layrisse, M.; García-Casal, M.N.; Solano, L.; Barón, M.A.; Arguello, F.; Llovera, D.; Ramírez, J.; Leets, I.; Tropper, E. Iron Bioavailability in Humans from Breakfasts Enriched with Iron Bis-Glycine Chelate, Phytates and Polyphenols. *J. Nutr.* **2000**, *130*, 2195–2199. [CrossRef]
227. Dabbagh-Bazarbachi, H.; Clergeaud, G.; Quesada, I.M.; Ortiz, M.; O'Sullivan, C.K.; Fernández-Larrea, J.B. Zinc ionophore activity of quercetin and epigallocatechin-gallate: From hepa 1-6 cells to a liposome model. *J. Agric. Food Chem.* **2014**, *62*, 8085–8093. [CrossRef]
228. Ozdal, T.; Sela, D.A.; Xiao, J.; Boyacioglu, D.; Chen, F.; Capanoglu, E. The reciprocal interactions between polyphenols and gut microbiota and effects on bioaccessibility. *Nutrients* **2016**, *8*, 78. [CrossRef]
229. Mithul Aravind, S.; Wichienchot, S.; Tsao, R.; Ramakrishnan, S.; Chakkaravarthi, S. Role of dietary polyphenols on gut microbiota, their metabolites and health benefits. *Food Res. Int.* **2021**, *142*, 110189. [CrossRef]
230. Zhao, Y.; Jiang, Q. Roles of the Polyphenol–Gut Microbiota Interaction in Alleviating Colitis and Preventing Colitis-Associated Colorectal Cancer. *Adv. Nutr.* **2021**, *12*, 546–565. [CrossRef]
231. Cardona, F.; Andrés-Lacueva, C.; Tulipani, S.; Tinahones, F.J.; Queipo-Ortuño, M.I. Benefits of polyphenols on gut microbiota and implications in human health. *J. Nutr. Biochem.* **2013**, *24*, 1415–1422. [CrossRef]
232. Kanhere, H.S.; Rahangdale, Y.U.; Bodele, A.S.; Wadhwani, D.I.; Ghoshewar, A.R.; Karande, S.P. Karande Neurological disorders associated with impaired gut microbiota. *GSC Biol. Pharm. Sci.* **2021**, *15*, 29–39. [CrossRef]
233. Tomás-Barberán, F. Interaction of polyphenols with gut microbiota: Role in human health. *Planta Med.* **2014**, *80*, PL7. [CrossRef]
234. Zhang, Z.; Xie, L.; Ju, Y.; Dai, Y. Recent Advances in Metal-Phenolic Networks for Cancer Theranostics. *Small* **2021**, *17*, 2100314. [CrossRef] [PubMed]
235. Repasky, M.P.; Shelley, M.; Friesner, R.A. Flexible Ligand Docking with Glide. In *Current Protocols in Bioinformatics*; John Wiley & Sons, Inc.: Hoboken, NJ, USA, 2007.
236. Dallakyan, S.; Olson, A.J. Small-Molecule Library Screening by Docking with PyRx. In *Methods in Molecular Biology*; Humana Press: Clifton, NJ, USA, 2015; Volume 1263, pp. 243–250. ISBN 9781118435762. [CrossRef]
237. Trott, O.; Olson, A.J. AutoDock Vina: Improving the Speed and Accuracy of Docking with a New Scoring Function, EfficientOptimization, and Multithreading. *J. Comput. Chem.* **2010**, *31*, 455–461. [CrossRef] [PubMed]
238. Rose, P.W.; Prlić, A.; Altunkaya, A.; Bi, C.; Bradley, A.R.; Christie, C.H.; Di Costanzo, L.; Duarte, J.M.; Dutta, S.; Feng, Z.; et al. The RCSB protein data bank: Integrative view of protein, gene and 3D structural information. *Nucleic Acids Res.* **2017**, *45*, D271–D281. [CrossRef] [PubMed]
239. Burley, S.K.; Bhikadiya, C.; Bi, C.; Bittrich, S.; Chen, L.; Crichlow, G.V.; Christie, C.H.; Dalenberg, K.; Di Costanzo, L.; Duarte, J.M.; et al. RCSB Protein Data Bank: Powerful new tools for exploring 3D structures of biological macromolecules for basic and applied research and education in fundamental biology, biomedicine, biotechnology, bioengineering and energy sciences. *Nucleic Acids Res.* **2021**, *49*, D437–D451. [CrossRef]

240. Lu, C.; Wu, C.; Ghoreishi, D.; Chen, W.; Wang, L.; Damm, W.; Ross, G.A.; Dahlgren, M.K.; Russell, E.; Von Bargen, C.D.; et al. OPLS4: Improving Force Field Accuracy on Challenging Regimes of Chemical Space. *J. Chem. Theory Comput.* **2021**, *17*, 4291–4300. [CrossRef]
241. Chen, C.Y.-C. TCM Database@Taiwan: The world's largest traditional Chinese medicine database for drug screening in silico. *PLoS ONE* **2011**, *6*, e15939. [CrossRef]
242. Kwofie, S.; Broni, E.; Yunus, F.; Nsoh, J.; Adoboe, D.; Miller, W.; Wilson, M. Molecular Docking Simulation Studies Identifies Potential Natural Product Derived-Antiwolbachial Compounds as Filaricides against Onchocerciasis. *Biomedicines* **2021**, *9*, 1682. [CrossRef]
243. Shelley, J.C.; Cholleti, A.; Frye, L.L.; Greenwood, J.R.; Timlin, M.R.; Uchimaya, M. Epik: A software program for pK a prediction and protonation state generation for drug-like molecules. *J. Comput. Aided. Mol. Des.* **2007**, *21*, 681–691. [CrossRef]
244. Tian, W.; Chen, C.; Lei, X.; Zhao, J.; Liang, J. CASTp 3.0: Computed atlas of surface topography of proteins. *Nucleic Acids Res.* **2018**, *46*, W363–W367. [CrossRef]
245. Chaput, L.; Mouawad, L. Efficient conformational sampling and weak scoring in docking programs? Strategy of the wisdom of crowds. *J. Cheminform.* **2017**, *9*, 1–18. [CrossRef]
246. Gimeno, A.; Mestres-Truyol, J.; Ojeda-Montes, M.J.; Macip, G.; Saldivar-Espinoza, B.; Cereto-Massagué, A.; Pujadas, G.; Garcia-Vallvé, S. Prediction of novel inhibitors of the main protease (M-pro) of SARS-CoV-2 through consensus docking and drug reposition. *Int. J. Mol. Sci.* **2020**, *21*, 3793. [CrossRef]
247. Li, J.; Fu, A.; Zhang, L. An Overview of Scoring Functions Used for Protein–Ligand Interactions in Molecular Docking. *Interdiscip. Sci. Comput. Life Sci.* **2019**, *11*, 320–328. [CrossRef]
248. Dankwa, B.; Broni, E.; Enninful, K.S.; Kwofie, S.K.; Wilson, M.D. Consensus docking and MM-PBSA computations identify putative furin protease inhibitors for developing potential therapeutics against COVID-19. *Struct. Chem.* **2022**, *33*, 2221–2241. [CrossRef]
249. Thomas, B.N.; Parrill, A.L.; Baker, D.L. Self-docking and cross-docking simulations of G protein-coupled receptor-ligand complexes: Impact of ligand type and receptor activation state. *J. Mol. Graph. Model.* **2022**, *112*, 108119. [CrossRef]
250. Kwofie, S.K.; Dankwa, B.; Odame, E.A.; Agamah, F.E.; Doe, L.P.; Teye, J.; Agyapong, O.; Miller, W.; Mosi, L.; Wilson, M.D. In Silico Screening of Isocitrate Lyase for Novel Anti-Buruli Ulcer Natural Products Originating from Africa. *Molecules* **2018**, *23*, 1550. [CrossRef]
251. Jaundoo, R.; Bohmann, J.; Gutierrez, G.; Klimas, N.; Broderick, G.; Craddock, T. Using a Consensus Docking Approach to Predict Adverse Drug Reactions in Combination Drug Therapies for Gulf War Illness. *Int. J. Mol. Sci.* **2018**, *19*, 3355. [CrossRef]
252. Kapale, S.S.; Mali, S.N.; Chaudhari, H.K. Molecular modelling studies for 4-oxo-1,4-dihydroquinoline-3-carboxamide derivatives as anticancer agents. *Med. Drug Discov.* **2019**, *2*, 100008. [CrossRef]
253. Hu, J.; Liu, Z.; Yu, D.J.; Zhang, Y. LS-align: An atom-level, flexible ligand structural alignment algorithm for high-throughput virtual screening. *Bioinformatics* **2018**, *34*, 2209–2218. [CrossRef]
254. Halgren, T.A. Identifying and Characterizing Binding Sites and Assessing Druggability. *J. Chem. Inf. Model.* **2009**, *49*, 377–389. [CrossRef]
255. Seeliger, D.; De Groot, B.L. Ligand docking and binding site analysis with PyMOL and Autodock/Vina. *J. Comput. Aided. Mol. Des.* **2010**, *24*, 417–422. [CrossRef]
256. Lang, P.T.; Brozell, S.R.; Mukherjee, S.; Pettersen, E.F.; Meng, E.C.; Thomas, V.; Rizzo, R.C.; Case, D.A.; James, T.L.; Kuntz, I.D. DOCK 6: Combining techniques to model RNA-small molecule complexes. *RNA* **2009**, *15*, 1219–1230. [CrossRef] [PubMed]
257. Charifson, P.S.; Corkery, J.J.; Murcko, M.A.; Walters, W.P. Consensus Scoring: A Method for Obtaining Improved Hit Rates from Docking Databases of Three-Dimensional Structures into Proteins. *J. Med. Chem.* **1999**, *42*, 5100–5109. [CrossRef] [PubMed]
258. Du, X.; Li, Y.; Xia, Y.-L.; Ai, S.-M.; Liang, J.; Sang, P.; Ji, X.-L.; Liu, S.-Q. Insights into Protein–Ligand Interactions: Mechanisms, Models, and Methods. *Int. J. Mol. Sci.* **2016**, *17*, 144. [CrossRef] [PubMed]
259. Shortridge, M.D.; Bokemper, M.; Copeland, J.C.; Stark, J.L.; Powers, R. Correlation between Protein Function and Ligand Binding Profiles. *J. Proteome Res.* **2011**, *10*, 2538–2545. [CrossRef]
260. Thorat, B.R.; Mali, S.N.; Rani, D.; Yamgar, R.S. Synthesis, In silico and In vitro Analysis of Hydrazones as Potential Antituberculosis Agents. *Curr. Comput. Aided. Drug Des.* **2021**, *17*, 294–306. [CrossRef]
261. Wishart, D.S.; Feunang, Y.D.; Guo, A.C.; Lo, E.J.; Marcu, A.; Grant, J.R.; Sajed, T.; Johnson, D.; Li, C.; Sayeeda, Z.; et al. DrugBank 5.0: A major update to the DrugBank database for 2018. *Nucleic Acids Res.* **2018**, *46*, D1074–D1082. [CrossRef]
262. Wishart, D.S.; Knox, C.; Guo, A.C.; Cheng, D.; Shrivastava, S.; Tzur, D.; Gautam, B.; Hassanali, M. DrugBank: A knowledgebase for drugs, drug actions and drug targets. *Nucleic Acids Res.* **2008**, *36*, D901–D906. [CrossRef]
263. Dodda, L.S.; Cabeza de Vaca, I.; Tirado-Rives, J.; Jorgensen, W.L. LigParGen web server: An automatic OPLS-AA parameter generator for organic ligands. *Nucleic Acids Res.* **2017**, *45*, W331–W336. [CrossRef]
264. Nguyen, T.T.; Viet, M.H.; Li, M.S. Effects of water models on binding affinity: Evidence from all-atom simulation of binding of tamiflu to A/H5N1 neuraminidase. *Sci. World J.* **2014**, *2014*, 536084. [CrossRef]
265. Cao, Z.; Liu, L.; Zhao, L.; Wang, J. Effects of different force fields and temperatures on the structural character of abeta (12-28) peptide in aqueous solution. *Int. J. Mol. Sci.* **2011**, *12*, 8259–8274. [CrossRef]
266. Peng, Y.; McCorvy, J.D.; Harpsøe, K.; Lansu, K.; Yuan, S.; Popov, P.; Qu, L.; Pu, M.; Che, T.; Nikolajsen, L.F.; et al. 5-HT2C Receptor Structures Reveal the Structural Basis of GPCR Polypharmacology. *Cell* **2018**, *172*, 719–730.e14. [CrossRef]

International Journal of *Molecular Sciences*

MDPI

Article

Anticancer Activity of Novel Difluorinated Curcumin Analog and Its Inclusion Complex with 2-Hydroxypropyl-β-Cyclodextrin against Pancreatic Cancer

Sangita Bhattacharyya [1,†], Hindole Ghosh [1,†], Obdulia Covarrubias-Zambrano [1], Krishan Jain [1], K. Venkateswara Swamy [2], Anup Kasi [3], Ameer Hamza [4], Shrikant Anant [1], Michael VanSaun [1], Scott J. Weir [1,3,5], Stefan H. Bossmann [1], Subhash B. Padhye [1,6] and Prasad Dandawate [1,*]

1 Department of Cancer Biology, University of Kansas Medical Center, Kansas City, KS 66103, USA
2 MIT School of Bioengineering, Sciences & Research, MIT Art, Design and Technology University, Pune 412201, India
3 Division of Medical Oncology, University of Kansas, Kansas City, KS 66160, USA
4 Pathology and Laboratory Medicine, University of Kansas, Kansas City, KS 66160, USA
5 Institute for Advancing Medical Innovation, University of Kansas Medical Center, Kansas City, KS 66160, USA
6 Interdisciplinary Science & Technology Research Academy (ISTRA), Azam Campus, University of Pune, Pune 411001, India
* Correspondence: pdandawate@kumc.edu; Tel.: +1-913-945-6336
† These authors contributed equally to this work.

Citation: Bhattacharyya, S.; Ghosh, H.; Covarrubias-Zambrano, O.; Jain, K.; Swamy, K.V.; Kasi, A.; Hamza, A.; Anant, S.; VanSaun, M.; Weir, S.J.; et al. Anticancer Activity of Novel Difluorinated Curcumin Analog and Its Inclusion Complex with 2-Hydroxypropyl-β-Cyclodextrin against Pancreatic Cancer. *Int. J. Mol. Sci.* **2023**, *24*, 6336. https://doi.org/10.3390/ijms24076336

Academic Editor: Mauro Coluccia

Received: 27 December 2022
Revised: 14 March 2023
Accepted: 21 March 2023
Published: 28 March 2023

Abstract: Pancreatic ductal adenocarcinoma (PDAC) is the primary reason for cancer-related deaths in the US. Genetic mutations, drug resistance, the involvement of multiple signaling pathways, cancer stem cells (CSCs), and desmoplastic stroma, which hinders drug penetrance, contribute to poor chemotherapeutic efficacy. Hence, there is a need to identify novel drugs with improved delivery to improve treatment outcomes. Curcumin is one such compound that can inhibit multiple signaling pathways and CSCs. However, curcumin's clinical applicability for treating PDAC is limited because of its poor solubility in water and metabolic instability. Hence, we developed a difluorinated curcumin (CDF) analog that accumulates selectively in the pancreas and inhibits PDAC growth in vitro and in vivo. In the present work, we developed its 2-hydroxy-propyl-β-cyclodextrin (HCD) inclusion complex to increase its water solubility and hydrolytic stability. The CDFHCD inclusion complex was characterized by spectroscopic, thermal, and microscopic techniques. The inclusion complex exhibited increased aqueous solubility, hydrolytic stability, and antiproliferative activity compared to parent CDF. Moreover, CDF and CDFHCD inhibited colony and spheroid formation, and induced cell cycle and apoptosis in PDAC cell lines. Hence, CDFHCD self-assembly is an efficient approach to increase water solubility and anticancer therapeutic efficacy, which now warrants advancement towards a clinical proof of concept in PDAC patients.

Keywords: curcumin; difluorinated curcumin; CDF; pancreatic cancer; PDAC; 2-hydroxypropyl-β-cyclodextrin; cyclodextrin

1. Introduction

Pancreatic ductal adenocarcinoma (PDAC) is the 4th highest cause of cancer-related deaths in the US, with a poor five-year survival 11% [1,2]. PDAC is characterized by rapid progression, invasiveness, and profound drug resistance, resulting in poor treatment outcomes [1,2]. PDAC is expected to be the second leading reason for deaths associated with cancer by the end of this decade [3]. Despite the extensive research conducted in diagnostic and therapeutic PDAC treatment, it remains a significant problem worldwide [4,5]. Moreover, the existence of germline and acquired genetic mutations such as KRAS, TP53, CDK2NA, and SMAD4/DPC4 are associated with an even poorer prognosis [6]. The existence of cancer stem cells (CSCs) [7], the occurrence of tumor-associated immune cells in the tumor microenvironment, the activation of multiple signaling pathways, and desmoplastic stroma can result in drug resistance or cancer recurrence [8,9]. The current chemotherapy for PDAC consists of gemcitabine combined with nab paclitaxel [10] or 5-fluorouracil/leucovorin with irinotecan and oxaliplatin (FOLFIRINOX) [11,12], which only improves overall survival between 8.5 and 11 months with a response rate of 23–31% [10]. Hence, it is essential to identify novel targets that are more effective treatments and that improve survival for PDAC patients.

Nature is a rich source of compounds called phytochemicals, which inhibit multiple signaling pathways, multiple cell types, and CSCs and their signaling pathways [13,14]. Phytochemicals have gained wide attention in the last 20 years because of their excellent safety profiles and ability to target several signaling pathways in cancer cells [13,14]. We have synthesized and studied analogs of phytochemicals for their anti-cancer activities, including curcumin [15], resveratrol [16–18], honokiol [19], plumbagin [20–22], cucurbitacin B and I [23], celastrol [24], triptolide [24], chalcones [25], mangostin [26], and quinomycin A [27]. We have summarized the biological activities of these phytochemicals and their analogs against various cancers in recent review articles [13,14,28–35].

We selected curcumin from the plant *Curcuma longa*. Curcumin has been found to be effective against various cancers; however, its poor bioavailability and poor water solubility have limited its clinical utility [13,14,28–30]. To overcome these limitations, we developed a new chemical analog of curcumin, 3,4-difluoro-benzo-curcumin, commonly known as CDF. CDF showed superior bioavailability, effective delivery to and uptake into pancreatic tissues, and the inhibition of the PDAC cell growth [36,37].

Cancer studies investigating CDF have demonstrated its effects on multiple cancer-associated pathways and phenotypes. CDF significantly inhibited the sphere-forming ability (pancospheres) of PDAC cells by down-regulating the cancer stem cell (CSC) markers EPCAM and CD44. CDF treatment inhibited tumor growth and the expression of cyclooxygenase-2 and miR-21 in a mouse xenograft model, while increasing both PTEN and miR-200. The up-regulation of miR-200 in tumors remained [38,39], and the reduction in miR-21 resulted in the induction of PTEN [40]. The CDF compound inhibited VEGF and IL-6 production and, when in hypoxic conditions, further reduced Nanog, Oct4, and EZH2 expression and miR-210 and miR-21 levels in PDAC cells [41]. CDF inhibited cell growth [42] in prostate cancer as well as chemo-resistant colon cancer cells by eradicating CSCs [43]. It inhibits the growth of 5-Fluorouracil and oxaliplatin-resistant colon cancer cells by down-regulating miR-21 levels and restoring PTEN levels with decreased p-Akt levels [44–46]. Recent reports have also shown that CDF inhibits MMP9 expression and activity in 549 and H1299 NSCLC cells [47]. Basak and coworkers [48] reported the anti-cancer activity of CDF delivered via a liposomal formulation in cisplatin-resistant head and neck squamous cell carcinoma CSCs. Recently, CDF-folic acid-conjugated polymeric micelles have also been shown to have inhibitory effects on ovarian and cervical cancer cells by inhibiting NF-kB and causing significant apoptosis [49].

Although CDF showed remarkable anticancer activity compared to curcumin in vitro, poor water solubility and metabolic instability present significant barriers to advancing this compound to clinical proof of concept. Cyclodextrin inclusion complexes have been used to enhance the water solubility, improve the bioavailability and enhance the biological activity of several phytochemicals and their synthetic analogs [15,22,50]. Yallapu and coworkers [51] have recently reported on the enhanced therapeutic activity of the β-cyclodextrin–curcumin inclusion complex compared to free curcumin against prostate cancer. Previously, pharmacokinetic studies conducted in our laboratory showed that CDF-β-cyclodextrin conjugate increases the systemic bioavailability of CDF from 6 ng/mL to 110 ng/mL, as well as tissue uptake from 300 ng/mL to 410 ng/mL in the serum and the pancreas, respectively. Moreover, we could not detect CDF following four hours of administration. In contrast, we noticed the CDF- β-cyclodextrin inclusion complex at 35 ng/mL and 280 ng/mL in serum and the pancreas, respectively. These data indicated that β-cyclodextrin conjugation increased the solubility and stability of CDF in the blood and pancreas [15]. Although β-cyclodextrin conjugates have been extensively used to improve the solubility of the drugs, toxicities are associated with clinical use [52,53]. However, the 2-hydroxypropyl derivative of β-cyclodextrin (HCD) is well tolerated when dosed orally and utilized in several approved drug products [54,55]. Hence, we prepared and assessed the anticancer properties of the CDF 2-hydroxypropyl-β-cyclodextrin (HCD) inclusion complex against PDAC cells.

2. Results and Discussion

2.1. Synthesis of CDF-2-Hydroxypropyl-Cyclodextrin (HCD) Inclusion Complex

CDF was synthesized and spectroscopically characterized per our previously reported method [37]. Based on phase solubility studies, we described previously that CDF forms 1:2 complexes with β-cyclodextrin derivatives [15]. Hence, we prepared the CDFHCD inclusion complex in a 1:2 ratio using the kneading method and used it for further spectroscopic characterization and biological assays.

2.2. Infra-Red Studies

Fourier Transform Infrared (FTIR) spectroscopy was used to study the interaction of guest molecule CDF and HCD within the inclusion complex [56]. The CDF spectrum (Figure 1A,B) showed absorption bands in the region 3439.19–3302.24 cm^{-1}, suggestive of phenolic –OH stretching, while other bands were observed at 2596.97 (C-H, OCH_3), 1629.90–1593.25 (C=O, C=C), 1429.30 (C-H, olefinic), 1274.99 (C-F), 1184.33–1163.11 (Ar C–O), and 1037.74–823.63 (C–O–C), respectively. HCD spectrum showed major peaks at 3410.26–3329.25 and 2968.55–2926.11 of O-H and C-H bands, whereas 1410.01 and 1371.43 cm^{-1} bands indicated the C-H stretches from CH_2 and CH_3. Other peaks at 1330.93, 1151.54–1082.10, 1037.74, and 947.08 cm^{-1} indicated the existence of skeletal vibrations containing α-1,4 linkages of glucose and cyclodextrin. The FTIR spectra of the CDFHCD inclusion complex showed lower or higher wavenumber shifts in the absorption frequencies of major bands for HCD and CDF (Figure 1A,B), while the decreased sharpness or disappearance of the peaks were indicative of the typical characteristics of inclusion complex formation [57]. Within the CDFHCD complex itself, the peak of the phenolic hydroxyl group appeared at 3439.19–3302.24 compared to HCD alone, which exhibited peaks at 3410.26–3329.25 cm^{-1}. Similarly, the C-H and C–C–H, C–O, and C–C peaks at 2596.97 and 854.49 were shifted to 2968.55–2926.11 cm^{-1} and 844.85 cm^{-1}, respectively, compared to HCD. The minor shifts in the FTIR peaks of HCD and CDF in the CDFHCD complex indicated the successful formation of the CDFHCD inclusion complex [50,58].

2.3. Differential Scanning Calorimetric (DSC) Studies

Thermochemical analysis techniques are successfully used to examine the physical state of the drug in polymer or inclusion complexes [59–61]. We used DSC to study the thermal behavior of CDF and HCD in the CDFHCD inclusion complex (Figure 1C). HCD and CDF showed a sharp endothermic peak at 81.44 °C (beginning at 33.56 and ending at 123.96 °C) and 219.46 °C (beginning at 217.76 °C and ending at 222.06 °C), respectively, because of their melting temperature. Within the CDFHCD complex, however, the prominent melting peak of CDF at 219.46 °C disappeared, and a small peak appeared at 199.78 (with onset at 198.38 °C and end at 213.62 °C). Moreover, the peak of HCD was shifted from 81.44 °C to 74.75 °C (beginning at 27.94 °C and ending at 109.21 °C). The shift in/disappearance of the endothermic peaks of CDF and HCD indicated the successful formation of the CDFHCD complex [62,63] and demonstrated stronger solid-state interactions [51].

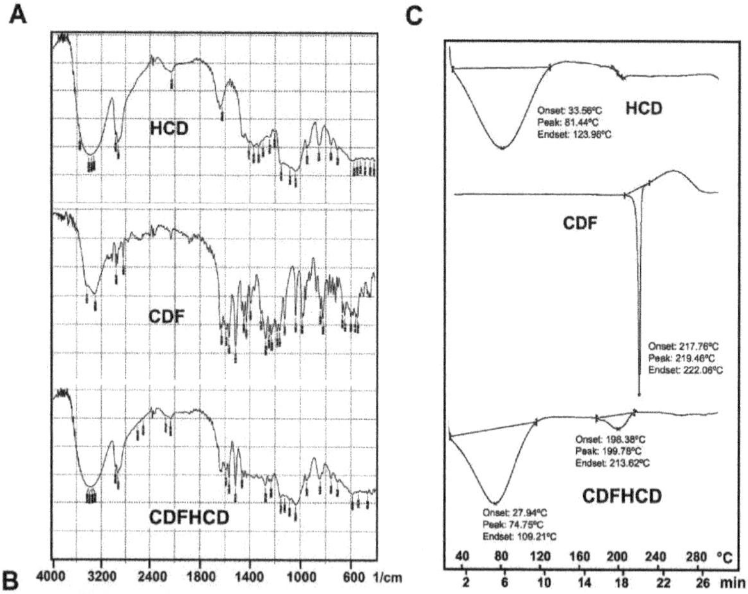

HCD	CDF	CDFHCD
3410.26-3329.25: (O-H)	3439.19-3302.24: (O-H)	3410.26-3335.03: (O-H)
2968.55-2926.11: (C-H)	2596.97: (C-H, OCH_3)	2966.62 and 2928.04: (C-H)
1410.01 and 1371.43: (C-H) from CH_2 and CH_3	1629.90-1593.25: (C=C, C=O)	1630 and 1595.18: (C=C, C=O)
1330.93: coupled (C–C–H), (C–O–H), (H–C–H))	1429.30: (olefinic C-H)	1430: (olefinic C-H)
1151.54 and 1082.10: (C–O), (C–C), (C–O–C)	1274.99: (C-F)	1415 and 1370: (C-H) from CH_2 and CH_3
1037.74: (O–C–H), (C–C–H), (C–C–O)	1184.33-1163.11: (aromatic C–O)	1274.99: (C-F)
947.08: skeletal vibration involving α-1,4 linkage		1153.47 and 1084.03: (C–O), (C–C), (C–O–C)
854.49: (C–C–H), (C–O), (C–C) from anomeric vibration	1037.74-823.63: (C–O–C)	949.01: skeletal vibration involving α-1,4 linkage
		844.85: (C–C–H), (C–O), (C–C) from anomeric vibration

Figure 1. Spectroscopic and thermal Characterization of CDFHCD inclusion complex. (**A**). FTIR spectra of 2-hydroxypropyl-β-cyclodextrin (HCD), CDF, and CDFHCD inclusion complex. (**B**). The FTIR spectral assignments are summarized. (**C**). Differential scanning calorimetry (DSC) spectra of HCD, CDF, and CDFHCD inclusion complex.

2.4. Nuclear Magnetic Resonance Spectroscopy

NMR spectroscopy aids the study of the interactions between the drug and guest molecule within the cyclodextrin cavity since the electronic and chemical surroundings of the protons of both the guest and host molecules are affected during complex formation and are reflected through the shifts in the δ values of the protons [64]. In the present study, the HCD protons undergo a considerable change to downfield (higher ppm) in the CDFHCD inclusion complex, suggesting a weak interaction (Van der Waals forces and hydrogen bonding) between HCD and CDF (Figure 2A) in the inner side of the HCD cavity. The upfield shift of the protons located within the CDF and HCD cavities indicate a major hydrophobic interaction [65]. An H^1-NMR analysis of HCD exhibited extensive resonances overlap in the 3.20 to 3.80 ppm spectral region. It therefore only allowed the identification of clusters of signals due to the different types of glycosidic or 2-hydroxypropyl protons, which made it difficult to appropriately assign the protons; meanwhile, the appearance of the methoxy protons of CDF in the CDFHCD spectra further complicated the assignments. These peaks in the methoxy groups of the CDF were shifted from 3.798 to 3.788 ppm (−0.010), and from 3.859 to 3.849 ppm (−0.010) after the formation of the inclusion complex, respectively, suggesting the involvement of aromatic rings in the complexation of CDF with HCD. The aromatic ring protons of CDF appeared at 6.77–7.64 ppm, which showed an upfield shift in the region of 6.75–7.63 ppm in the CDFHCD spectra, suggesting their involvement in the HCD complex formation. Nonetheless, the peaks of CDF between 6 and 8 ppm in the CDFHCD spectra were not clear enough to assign the shifts; this was due to the merging and reduced multiplicity of the peaks, which is characteristic of an inclusion complex [51,66,67]. These data indicated the successful formation of a complex between CDF and HCD, while it can be proposed that both aromatic rings of CDF participated in the inclusion of the HCD cavity.

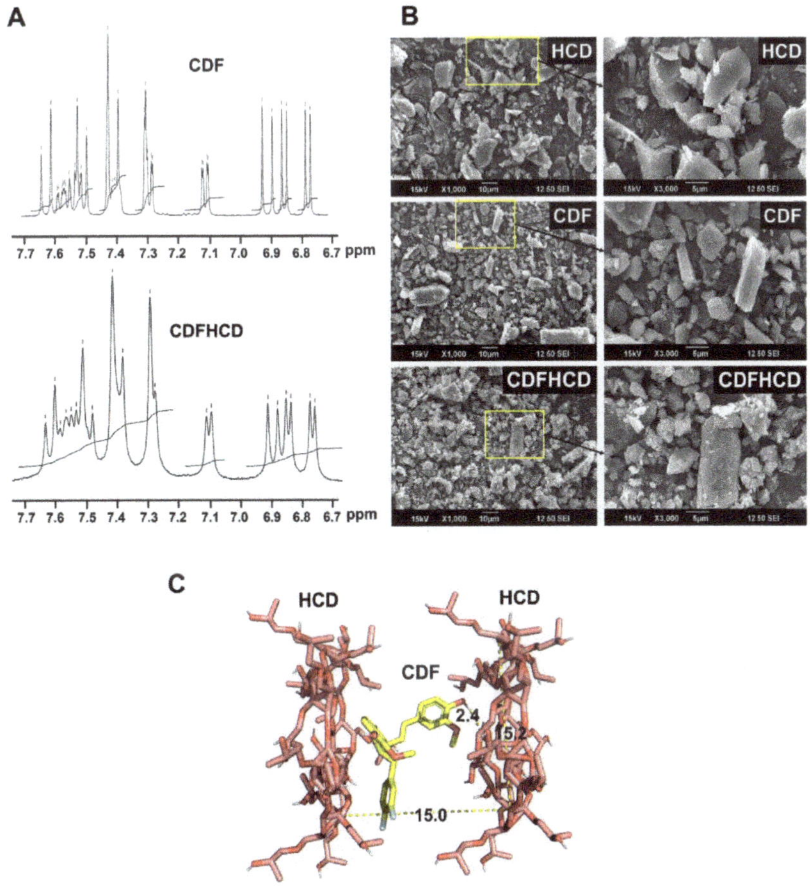

Figure 2. Morphological characterization of CDFHCD complex and the interactions of the CDF within the cyclodextrin cavity. (**A**). NMR spectra of 2- HCD, CDF, and CDFHCD inclusion complex.(**B**). SEM images of 2- HCD, CDF, and CDFHCD inclusion complex. (**C**). The binding mode of CDF in the cavity of HCD was proposed using molecular docking.

2.5. Scanning Electron Microscopic (SEM) Studies

SEM was utilized to analyze the bulk surface morphology of the CDF, HCD, and CDFHCD inclusion complexes (Figure 2B). The analysis indicated that HCD exhibited amorphous, 'shrunken' spheres and flakes, while CDF showed rod-like-shaped or hexagonal crystalline structures. The CDFHCD inclusion complex did not show flakes or rod-shaped crystals but presented irregularly shaped aggregates or clump formations. This modification in the morphology of the CDFHCD inclusion complex has been ascribed mainly to the inclusion of CDF in the HCD cavity. These data suggest the formation of the CDFHCD inclusion complex [68].

2.6. Molecular Docking Studies

We used the molecular docking technique to further study the CDF's binding mode with the HCD cavity in the inclusion complex [15,58,69]. We found that CDF can interact with HCD in a 1:2 ratio. The possible stable structure of the CDFHCD (1:2) interaction is shown in Figure 2C. CDF was predicted to bind with HCD (1:2) with the binding energy of −6.3 Kcal/mol, while the methoxy and hydroxyl groups on the aromatic ring were found to interact with the HCD cavity via two hydrogen bonds (2.4 Å) (Figure 2C). These observations agree with the NMR study predictions described in previous reports [15].

2.7. Anticancer Activity

2.7.1. CDF and CDFHCD Inhibit the Proliferation of PDAC Cells

First, we studied the effects of CDF and CDFHCD (dose range 0–5 μM) on the proliferation of PDAC cell lines based on their mutation, origin, and metastatic potential. Both CDF and CDFHCD inhibited (time- and dose-dependent effects) the proliferation of MiaPaCa-2 ($KRAS^{G12C}$, $p53^{R248W}$), Panc-1 ($KRAS^{G12D}$, $p53^{R273H}$), Panc 01728 (patient-derived cell line), and S2-007 (Metastatic) cells (Figure 3A). The IC_{50} values of CDF and CDFHCD against PDAC cell lines at different time points are summarized (Figure 3B). We observed the significant increase in the antiproliferative activity of CDF after HCD inclusion complexation (Figure 3B, $p < 0.05$), which can be attributed to a higher water solubility and a higher uptake of CDF into pancreatic cancer cells through the formation of the cyclodextrin inclusion complex [15]. We then used the clonogenic assay to study the long-lasting effects of CDF and CDFHCD on PDAC cell lines. MiaPaCa-2 and Panc-1 cells were treated with CDF and CDFHCD at IC_{50} and $\frac{1}{2}IC_{50}$ concentrations for 24 h and 48 h. After 48 h, the cell culture media containing CDF or CDFHCD was replaced with fresh drug-free media and grown for 10–14 days. Both CDF and CDFHCD treatment significantly ($p < 0.01$) reduced the colony formation (both size and number) in MiaPaCa-2 and Panc-1 cells ($p < 0.01$) (Figure 4A–D), suggesting that the anticancer effects of CDF and CDFHCD are non-reversible. Further, we evaluated the effect of CDF and CDFHCD on HPNE cells (immortalized epithelial pancreatic ductal cell line) to characterize the cytotoxicity potential. While CDF produced cytotoxicity in HPNE cells at a 1 μM concentration, CDFHCD did not induce any toxicity up to a 5 μM concentration (Supplementary Figure S1). This suggests the potentially lower cytotoxic effects on immortalized cells. However, more detailed studies with different normal cell lines are needed in the future to evaluate this. Given the greater efficacy of the CDFHCD inclusion complex, we used IC_{50} concentrations of CDF and CDFHCD for further studies in order to understand whether there are differences in their mechanism(s) regarding anticancer activity.

A

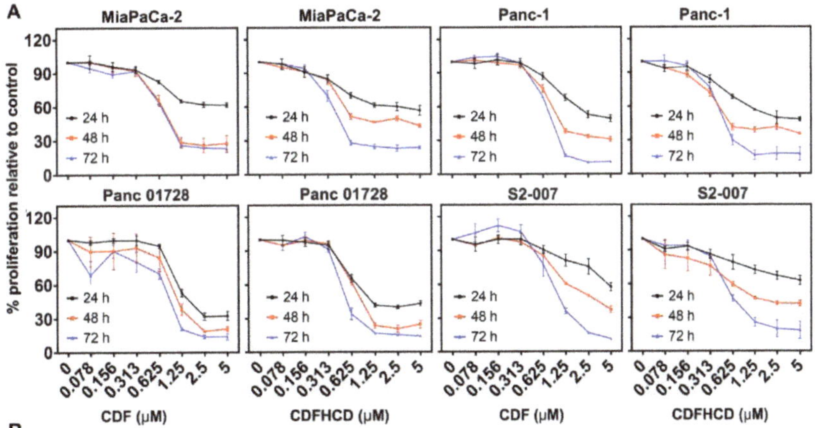

B

IC50 values	CDF (µM)			CDFHCD (µM)		
MiaPaCa-2	>2.5	0.81 ± 0.04	0.78 ± 0.03	>2.5	0.63 ± 0.07	0.47 ± 0.06
Panc-1	>2.5	1.00 ± 0.09	0.77 ± 0.02	>2.5	0.48 ± 0.03	0.43 ± 0.01
Panc01728	1.33 ± 0.11	1.01 ± 0.11	0.81 ± 0.03	0.94 ± 0.06	0.74 ± 0.02	0.48 ± 0.02
S2-007	>2.5	2.47 ± 0.15	0.95 ± 0.08	>2.5	1.01 ± 0.09	0.56 ± 0.04

Figure 3. CDF and CDFHCD inhibit the proliferation of PDAC cell lines. (**A**). PDAC cells (MiaPaCa-2, Panc-1, Panc01728, and S2-007) were treated with increasing concentrations of CDF and CDFHCD (0–5 µM) for up to 72 h and studied using hexosaminidase assay. (**B**). The IC_{50} values of the antiproliferative activity are summarized in the tabular format as mean ± SD.

2.7.2. CDF and CDFHCD Induce Cell Cycle Arrest

We studied the effects of CDF and CDFHCD on cell cycle progression using flow cytometry. The treatment of CDF and its inclusion of complex CDFHCD caused a significant ($p < 0.01$) increase in the number of cells in the sub-G0 phase at 24 h and 48 h in PDAC cell lines (Figures 5A,B and S2A,B). Moreover, the concentration of CDF that is equivalent to CDFHCD produced less potent effects than CDFHCD. The overexpression of Cyclin D1 plays a role in cancer progression and development [70] by controlling the cell cycle progression. Hence, we performed the western blot analysis and observed the significant ($p < 0.01$) downregulation of cyclin D1 after CDF and CDFHCD treatment (Figure 5C,D). These data suggested that CDF and CDFHCD induce cell cycle arrest in PDAC cells.

2.7.3. CDF and CDFHCD Induce Apoptosis

The accumulation of cancer cells in a sub-G0 stage after CDF and CDFHCD treatment can result from DNA fragmentation, suggesting probable cytotoxic effects. To study this, we utilized the Annexin V/PI assay with flow cytometry. We observed an increased cell number over 48 h in the late apoptosis stage after CDF and CDFHCD treatment compared to the control cells (Figures 6A and S3A). We observed CDF autofluorescence interfering in early apoptosis readings differentially; hence, we cannot decisively determine the role that CDF and CDFHCD play in early apoptosis. We further confirmed apoptosis using caspase 3/7 assays to examine the effector activity of caspase. CDF and CDFHCD treatment increased the caspase 3/7 activity (Figure 6B, $p < 0.01$) in PDAC cells. Mechanistically, the western blot analysis showed the enhanced expression of cleaved PARP protein ($p < 0.05$) in CDF and CDFHCD-treated cells (Figures 6C and S3B). The CDF and CDFHCD treatment reduced the levels of the anti-apoptotic markers Mcl1 ($p < 0.05$) and Bcl2 ($p < 0.01$), while proapoptotic protein Bax expression was not significantly changed (Figure 6C). A decrease in Bcl-2 levels is established as a mechanism of apoptosis in cancer cells [71]. These data suggest the utility of CDF and CDFHCD in combination with the first-line chemotherapeutic drugs [72,73]. These data indicate that the CDF and its inclusion in the CDFHCD complex can induce apoptosis in PDAC cells.

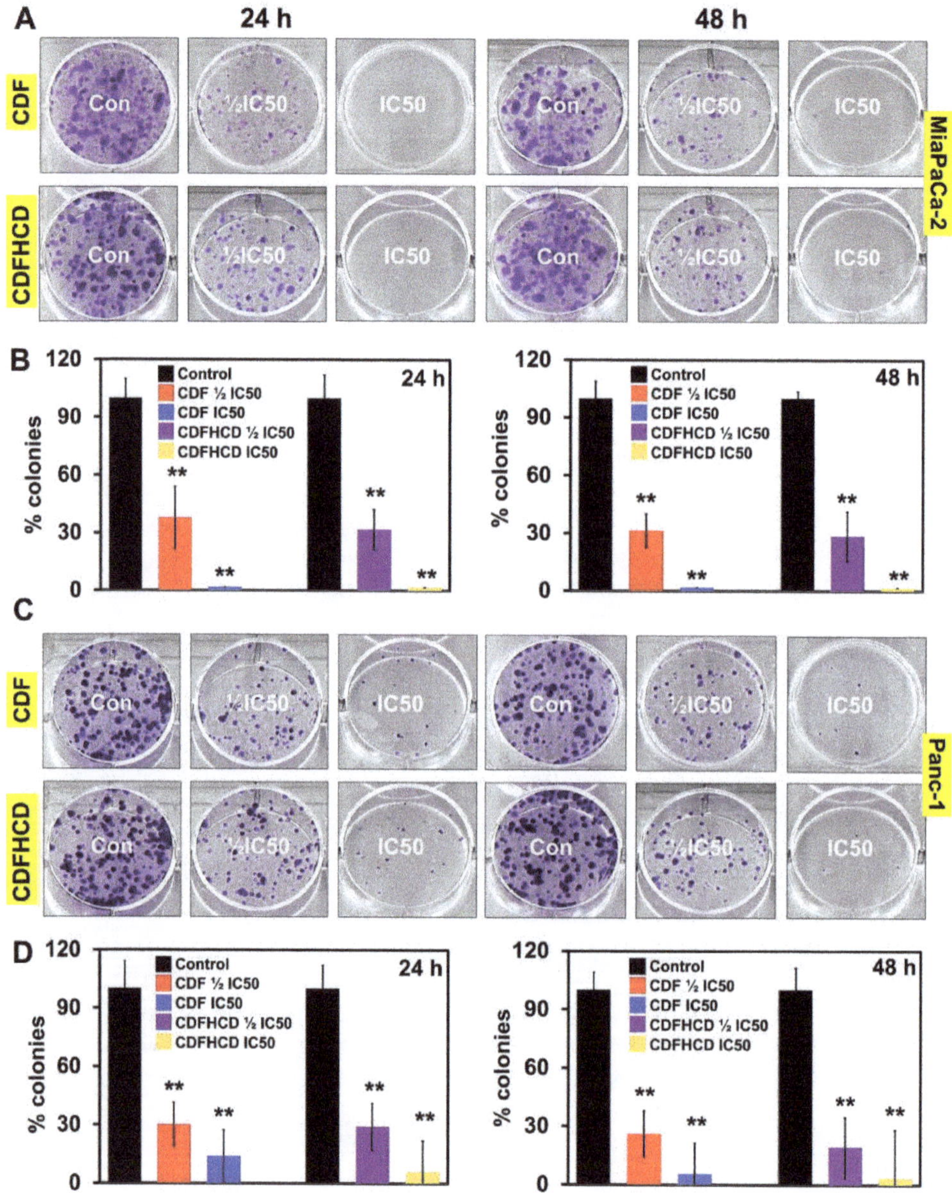

Figure 4. CDF and CDFHCD inhibit the colony formation of PDAC cell lines. (**A**). MiaPaCa-2 and (**C**). Panc-1 cells were incubated with $\frac{1}{2}$ IC50 and IC50 concentrations of CDF and CDFHCD for 48 h and grown into colonies. Treatment with CDF and CDFHCD inhibited the number of colonies in (**B**) MiaPaCa-2 and (**D**) Panc-1 cells (** $p < 0.01$).

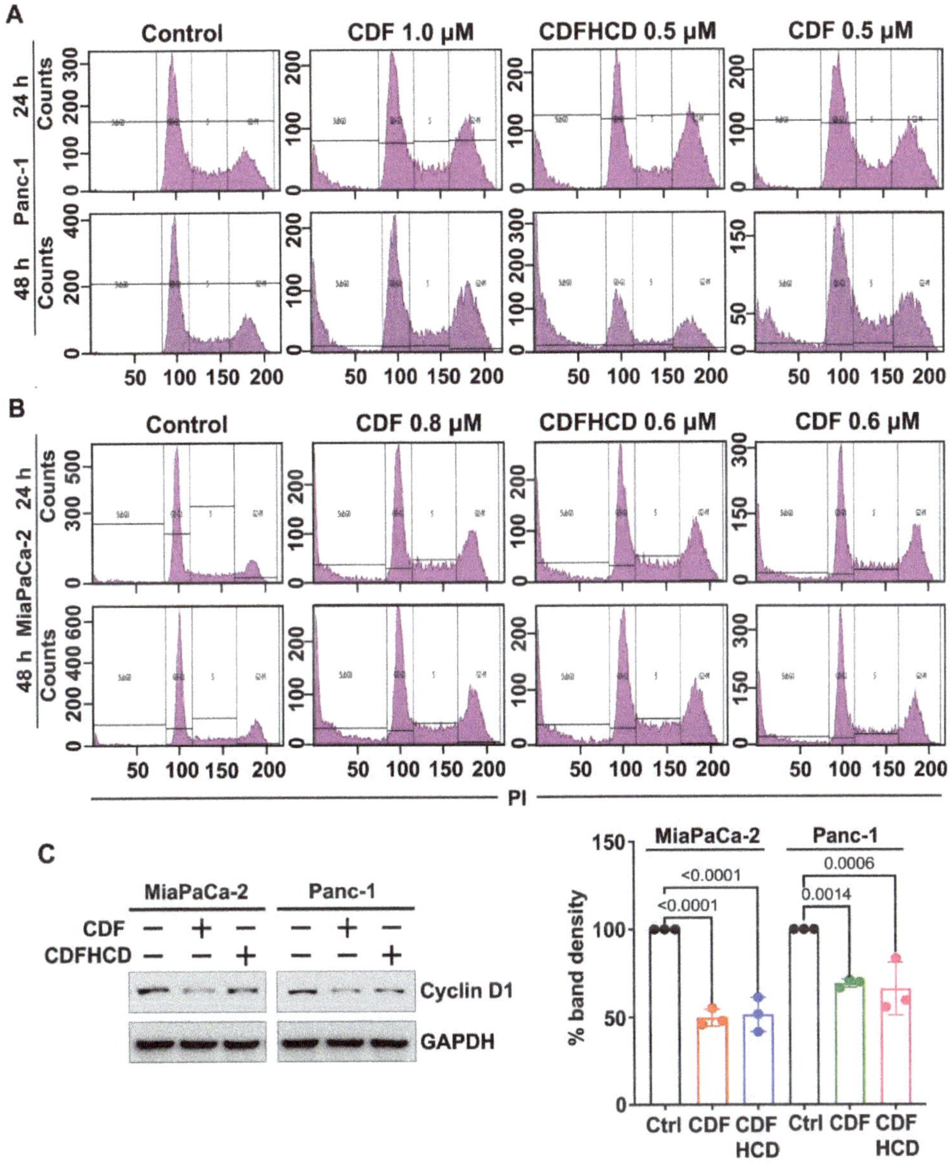

Figure 5. Cell cycle analysis of CDF and CDFHCD in PDAC cells. (**A**) MiaPaCa-2 and (**B**) Panc-1 cells were incubated with CDF and CDFHCD for 24 h and 48 h and analyzed by flow cytometry using FxCycle^{TM} PI/RNase staining solution. (**C**) Cell lysates from CDF and CDFHCD-treated MiaPaCa-2 and Panc-1 were analyzed using western blot to study the changes in cyclin D1 expression.

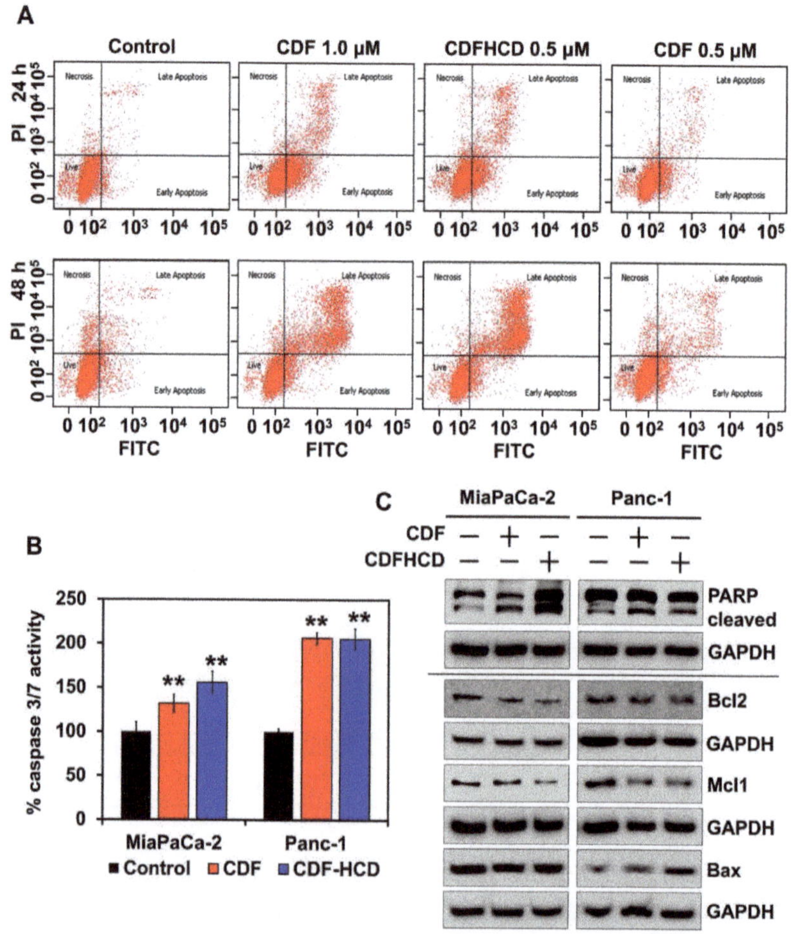

Figure 6. CDF and CDFHCD induce apoptosis. (**A**). Panc-1 cells were incubated with CDF and CDFHCD, stained with Annexin V-FITC and PI, and examined by flow cytometry. (**B**). Caspase 3/7 assay showed increased caspase activity in MiaPaCa-2 and Panc-1 cells at 48 h after treatment with CDF and CDFHCD (** $p < 0.01$). (**C**). PDAC cell lysates treated with CDF and CDFHCD were analyzed by western blot to study the changes in the expression of apoptotic marker proteins, PARP, Mcl1, Bax, and Bcl2.

2.7.4. CDF and CDFHCD Inhibit the Spheroid Formation

Increasing evidence suggests that CSCs are involved in causing drug resistance, aggressiveness, and recurrence in PDAC [74]. Several surface markers that mark CSCs, including CD44, have been identified in PDAC [75]. Moreover, the overexpression of CD44 induces the activation of cMyc, which in turn activates the MEK and ERK pathway to inhibit apoptosis in cancer cells [76]. cMyc is also overexpressed in CSCs in order to maintain the pluripotency [77], and its overexpression is involved in PDAC progression and drug resistance [78]. Recent studies have also shown that doublecortin-like kinase 1 (DCLK1) is a novel PDAC CSC marker [79]. Hence, to completely eradicate cancer, it is essential to understand CSCs' growth and develop novel therapeutic agents that target those mechanisms. It is reported that CSCs form spheroids in ultra-low attachment plates in suspension culture. Hence, we used a spheroid formation assay to analyze the effect of CDF and CDFHCD on CSCs. CDF and CDFHCD treatment reduced the spheroid formation (size and number) in MiaPaCa-2 and Panc-1 cells (Figure 7A,B). Moreover, CDF and CDFHCD treatment downregulated the expression of CD44 ($p < 0.01$), DCLK1 ($p < 0.01$) and cMYC ($p < 0.01$) in MiaPaCa-2 and Panc-1 cells (Figures 7C and S4). These datasets showed that CDF and CDFHCD inhibit spheroid growth and CSC marker protein expression.

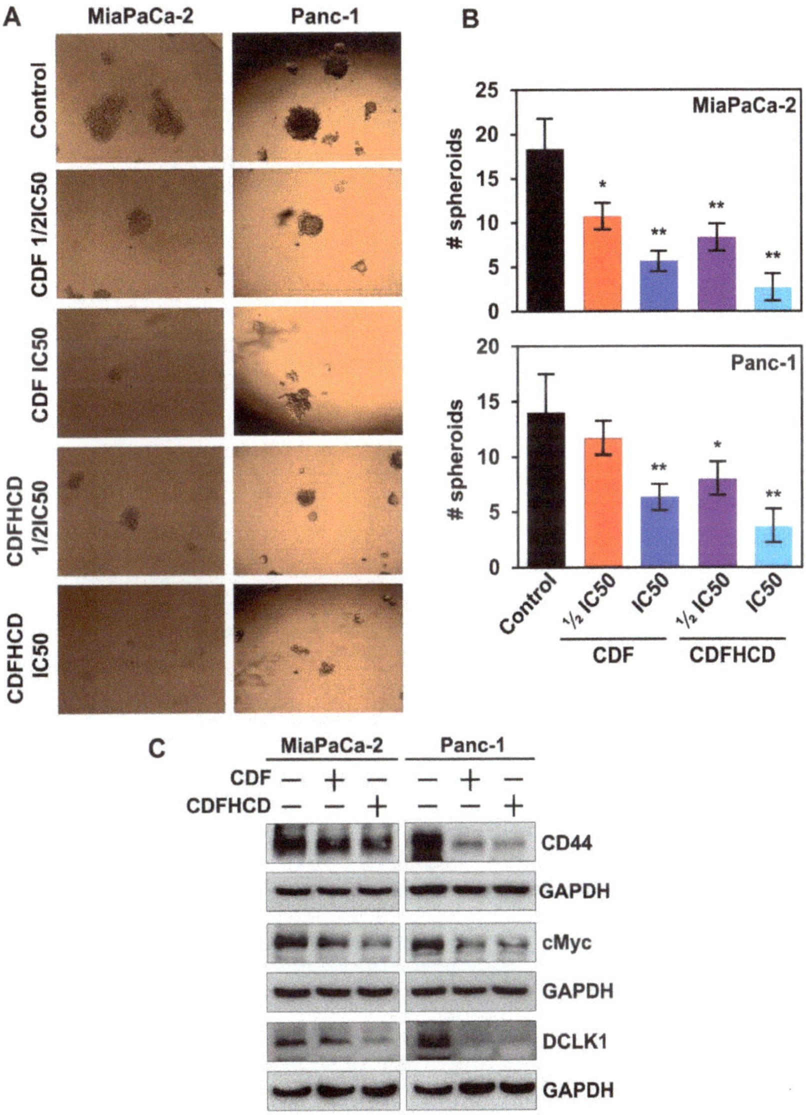

Figure 7. CDF and CDFHCD inhibit spheroid formation. (**A**). PDAC cells were grown in spheroid media in ultra-low attachment plates and treated with $\frac{1}{2}$IC50 and IC_{50} concentrations of CDF and CDF-HCD. After five days, the pancospheres were counted and imaged (10× magnification). (**B**). CDF and CDFHCD inhibited the number of pancospheres (* $p < 0.05$, ** $p < 0.01$). (**C**). Cell lysates from MiaPaCa-2 and Panc-1 cells treated with CDF and CDFHCD were examined by western blot to study the changes in CEC marker proteins, CD44, cMyc, and DCLK1 expression.

2.8. Hydrolytic Stability Study

Curcumin is prone to hydrolysis and is hydrolyzed to ferulic acid, feruloyl methane, and vanillin [80]. Cyclodextrin drug inclusion complexes, however, have been reported to improve the hydrolytic stability of guest drug molecules [81]. Hence, we performed the hydrolytic stability study of CDF with and without HCD in PBS solution [82] to study the hydrolysis of CDF. The UV-visible spectra of CDF showed hydrolysis from 15 min to 1 h in PBS, as demonstrated by a reduction in the intensity of the absorption bands. In comparison, we observed complete hydrolysis within 1 h (Figure 8A). CDF showed resistance to hydrolysis up to 5 h in PBS with 10% HCD (Figure 8B). The binding constant (K_D) of CDF in 10% HCD solution was calculated to be $4.50 \pm 0.27 \times 10^{-10}$ M^{-1}, which suggests the excellent stability of the complex for potential in vivo application. The study indicated that HCD complexation improved the hydrolytic stability of CDF.

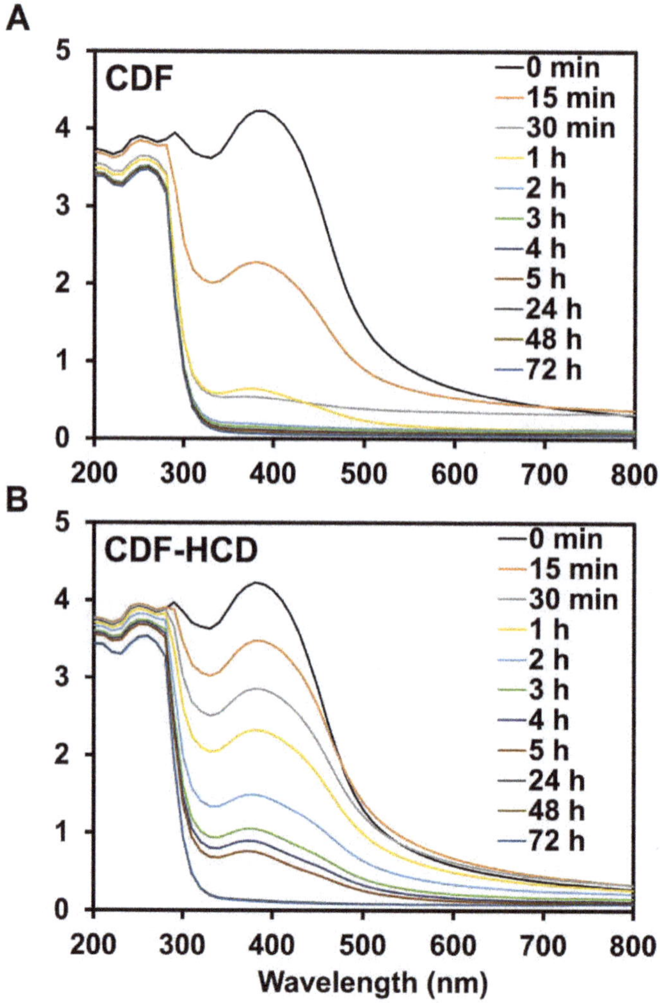

Figure 8. Hydrolytic stability study of CDF in PBS without (**A**) and with (**B**) 10% HCD solution over 0–72 h.

3. Materials and Methods

3.1. Synthesis of CDF

The difluorinated curcumin analog CDF was synthesized and characterized using our previously published method [37]. The curcumin was separated from the mixture of curcuminoids by column chromatography using dichloromethane and methanol in a 9:1 proportion (v/v). The first fraction was found to be curcumin and was used for further reactions. The purified curcumin (1 mmol) was dissolved in methanol, and 3,4-difluorobenzaldehyde (1 mmol) was added dropwise to this reaction with constant stirring in the presence of the catalytic amount of piperidine. The reaction was further stirred for 48 h and was monitored by TLC. The purified fraction by column chromatography was dried under a vacuum.

3.2. Preparation of Inclusion Complexes

We have previously shown that CDF forms a 1:2 complex with β-cyclodextrin by using phase solubility studies. We used the kneading method for making CDFHCD inclusion complexes. CDF and HCD were mixed in the proportion of 1:2 molar concentrations in a mortar for one hour by adding a mixture of methanol and deionized water (1:1) to get a slurry-like uniformity using a pestle. The slurry was further dried in an oven. The dried complex was used for further studies [15].

3.3. Infra-Red Studies

CDF, HCD, and CDFHCD were subjected to infra-red studies using the potassium bromide disk method and using the Shimadzu Fourier Transform Infrared (FTIR)-8700 spectrophotometer. A transparent disk of CDF, HCD and CDFHCD was made by mixing potassium bromide using high pressure that was applied by dyes. The resultant disk was positioned in an IR spectrophotometer, and the spectrum was recorded from 4000 to 400 cm^{-1}.

3.4. Differential Scanning Calorimetric (DSC)

CDF, HCD, and CDFHCD were studied using a DSC (Mettler Toledo, Switzerland). Then, 2.5–5 $\pm$ 0.5 mg of sample was positioned in sealed aluminum pans in the presence of liquid nitrogen as a cooling agent. The thermograms were recorded by scanning from 20 to 300 °C at ten °C/min intervals.

3.5. Nuclear Magnetic Resonance Spectroscopy

The proton NMR spectra were recorded in DMSO-d6 by the BRUKER AV III (^{1}H-NMR 500 MHz) spectrophotometer. The chemical shifts (δ) were expressed in parts per million (ppm) using residual solvent DMSO as a reference (2.50 ppm). The spectra were generated using Bruker's TOPSPIN-2.1 software. NMR experiments were conducted by setting the scan numbers at 64, the relaxation delay at 1.0 s, and the pulse degree at 25 °C.

3.6. Molecular Docking Studies

The 3D structure of CDF was prepared using CORINA [83] software from smiles that were further energy minimized in the PRODRG server [84]. The β-CD was extracted from the 3D structure of alpha-amylase (PDB id: 1JL8.pdb) and converted to HCD by adding 2-hydroxypropyl groups. The HCD was then prepared for docking by adding polar hydrogen and energy minimized. For studying CDFHCD interaction in a 1:2 ratio, the two HCD molecules were prepared and energy minimized in VMD software 1.9.3 [85] and the PRODRG server, respectively. Molecular docking was carried out to obtain the possible binding modes for the CD-HCD complex using AutoDock Vina software version 1.2.3 [86]. The grid was created around the X, Y, and Z-axis ($40 \times 40 \times 44$) using the Lamarckian Genetic Algorithm (LGA). The stable CDFHCD conformation was selected by considering the lowermost binding energy and hydrogen bonds formed between CDF and HCD. The CDFHCD complex was analyzed and visualized using Pymol software version 1.7.4 [87].

3.7. Cell Culture

The PDAC cells MiaPaCa-2, Panc-1 (obtained from American Type Culture Collection), Panc 01728 (a patient-derived cell line, a gift from Dr. Shrikant Anant's lab), HPNE and S2-007 (a gift from Dr. Animesh Dhar laboratory) were cultured in Dulbecco's Modified Eagle Medium (DMEM), supplemented with L-glutamine, 4.5 g/L of glucose, and sodium pyruvate (Corning, Tewksbury, MA, USA). Finally, heat-inactivated fetal bovine serum (10% concentration, FBS) (Sigma-Aldrich, St. Louis, MO, USA) and 1% antibiotic-antimycotic solution (Corning, Tewksbury, MA, USA) were added to make complete media. PDAC cells were grown in a 5% CO_2-humidified at 37 °C. PDAC cell lines were within 20 passages and validated by STR analysis. All methods followed the standard guidelines, regulations and the manufacturer's instructions.

3.8. Proliferation and Colony Formation Assays

For the proliferation and colony formation assays, 5×10^3 cells/well PDAC cells (MiaPaCa-2, Panc-1, S2-007, P01728) were plated in 96-well plates and grown in complete media. Cells were treated with increasing doses of CDF, CDFHCD, and the respective controls (DMSO and HCD). A hexosaminidase assay field recorded the cell viability at different times (24 h, 48 h, and 72 h) [88]. The percent of growth inhibition was calculated by comparing cell viability with the controls, while IC_{50} values were calculated by plotting a graph of % viability vs. concentration. The concentration that showed 50% cell viability was considered an IC_{50} value. In total, 500 cells/well of PDAC cells were plated in complete DMEM media for the colony formation assay. PDAC cells were treated with different doses of CDF and CDF–HCD. Media was changed following 24 h and 48 h of CDF and CDFHCD treatment to remove drug exposure. Further, these cells were grown and they formed colonies for 10–12 days. The resulting colonies were washed and fixed using formalin (10% solution). Further, formalin was aspirated, washed with PBS, and stained using a staining solution (1% crystal violet solution in 10% ethanol). The plates containing colonies were dried at room temperature, counted, and compared to untreated cells [23].

3.9. Cell Cycle Analysis by Flow Cytometry

For the cell cycle analysis, 5×10^5 PDAC cells/well (MiaPaCa-2 and Panc-1) were plated in a 10 cm cell culture dish and treated with CDF and CDFHCD at indicated time points. Cells were trypsinized, washed, resuspended in PBS, fixed using 70% ethanol in PBS, and stored at 4 °C overnight after the indicated time points. These cells were further permeabilized and stained with FxCycle^{TM} PI/RNase staining solution (Invitrogen, Eugene, OR, USA) at room temperature and subjected to flow cytometry. Flow cytometric analysis was performed with a FACS Calibur analyzer (Becton Dickinson, Mountain, View, CA, USA) and studied using BD FACSDiva 8.0.1 software (Verity Software House, Topsham, ME, USA)

3.10. Apoptosis Assays

For the first assay, the Apo-one Homogeneous Caspase-3/7 Assay kit (Promega Corporation, Madison, WI, USA) was used following the manufacturer's instructions in order to estimate caspase 3/7 activity in PDAC cells (MiaPaCa-2 and Panc-1) after CDF and CDFHCD treatment. In the second assay, Annexin V/PI staining was performed and studied using flow cytometry. Briefly, 1×10^5 PDAC cells were plated in 6-well plates. After 24 h of plating, PDAC cells (MiaPaCa-2 and Panc-1) were treated with vehicle control, CDF, and CDFHCD at IC_{50} doses. At the end of the treatment, cells were washed and stained with Annexin V-FITC antibody and propidium iodide (PI), as per the manufacturer's protocol, and were studied by flow cytometry.

3.11. Spheroid Formation Assay

In total, 500 PCAC cells were plated in each well of the ultralow attachment plates (Corning, Lowell, MA, USA). The spheroid media was composed of DMEM medium added to FGF (20 ng/mL), EGF (20 ng/mL), heparin (4 ug/mL), pen/strep (1%) (Invitrogen), and B27 (10 mL in 500 mL of 50×), which was used for growing spheroids. Spheroids were treated with vehicle, CDF, and CDFHCD at IC50 and $\frac{1}{2}$IC50 concentrations. Spheroids were counted and imaged after five days.

3.12. Western Blot Analysis

For the western blot analysis, 5×10^5 cells PDAC cells were plated in 10 cm dishes. After 24 h of plating, cells were treated with vehicle, CDF, and CDFHCD for 48 h at IC50 concentrations. Cell lysis was then performed in a protein lysis buffer and protease inhibitor (ThermoScientific, Rockford, IL, USA) solution and sonicated to achieve complete lysis. The resultant lysates were centrifugated at 6000 rpm for 10 mins at a cold temperature. The protein estimation was performed using a BCA reagent (Pierce™ BCA Protein Assay Kit) (ThermoScientific, Rockford, IL, USA), and 50 µg of protein for each group was used to load into the gels. Protein lysates were separated using polyacrylamide gel electrophoresis and transferred onto Immobilion membranes (PVDF, Millipore, Bedford, MA, USA). Post transfer, the membranes were removed from the transfer assembly, blocked using 5% milk for 1 h, washed, and probed with primary antibodies. The individual proteins were identified using the chemiluminescence system (GE Health Care, Piscataway, NJ, USA). The Bio-Rad ChemiDoc-XRS+ instrument quantified the data using image lab software and recorded the protein levels. All antibodies (Cyclin D1 (CST#2922), Bax (CST#2772), Bcl2 (CST#4223), Bcl-XL (CST#2762), Mcl-1 (CST#4572), PARP (CST#9542), CD44 (CST#3570), and cMyc (CST#9402) were purchased from Cell Signaling Technology (Beverly, MA, USA), and GAPDH (G-9) was purchased from Santa Cruz Biotech, Inc. (Santa Cruz, CA, USA). The DCLK1 antibody was purchased from Sigma-Aldrich (Sigma#SAB4200186). The antibodies were diluted per the manufacturer's instructions (1:1000 dilution in 5% BSA in TBST).

3.13. Hydrolytic Stability Study

The hydrolytic stability study used the previously reported method [50]. The stock solution of CDF was made in DMSO. Then, 100 µL of the stock solution was added to 10 mL of the 10% (w/v) HCD in PBS and PBS alone, respectively. The absorption spectra were recorded by taking aliquots at indicated intervals up to 72 h.

3.14. Statistical Analysis

All experiments were repeated three times, and the experiment values are shown as the mean ± SD. The experimental datasets were examined using an unpaired two-tailed t-test comparing it to the control group. The western blot quantifications were analyzed using a one-way ANOVA test, which was compared to the control group using GraphPad's Prism-9 (Boston, MA, USA). A $p < 0.05$ was chosen to be statistically significant.

4. Conclusions

In the present study, we successfully prepared and characterized the inclusion complex of CDF with 2-hydroxypropyl-β-cyclodextrin, CDF–HCD. CDFHCD treatment demonstrated greater anti-proliferative effects against PDAC cell lines compared to CDF treatment alone. CDF and CDFHCD inhibited colony and spheroid formation, producing cell cycle arrest and apoptosis in PDAC cell lines. Moreover, HCD improved the hydrolytic stability of CDF. Future studies will be directed toward characterizing the ability of the CDFHCD inclusion complex to deliver sufficient CDF into PDAC tumors, the engagement of CDF with the target(s) within PDAC cells, the effects on tumor development and progression, as well as survival in validated mouse models of PDAC.

Supplementary Materials: The following supporting information can be downloaded at: https://www.mdpi.com/article/10.3390/ijms24076336/s1.

Author Contributions: Conception and design: P.D.; Experiments performed: P.D., S.B., H.G., O.C.-Z. and K.J.; Molecular docking: K.V.S.; Acquisition of data: S.B., H.G., O.C.-Z. and P.D.; Analysis and interpretation of data: S.B., H.G., O.C.-Z., K.J., M.V., S.B.P., S.A. and P.D.; Writing, review, and revision: A.K., A.H., S.J.W., S.H.B., S.A., M.V., S.B.P. and P.D.; Administrative, technical, or material support: P.D.; Study supervision: P.D. All authors have read and agreed to the published version of the manuscript.

Funding: PD is supported by grants from the Department of Defense (CA201135) and the American Gastroenterological Association Research Scholar Award (2021RSA-12522). The Flow Cytometry Core Laboratory is sponsored, in part, by the NIH COBRE program of the NCRR P20 RR016443 and The University of Kansas Cancer Center P30CA168524–01 grants.

Institutional Review Board Statement: Not applicable.

Informed Consent Statement: Not applicable.

Acknowledgments: P.D. acknowledges CSIR, New Delhi, India, for providing a Senior Research Fellowship. We acknowledge SAIF IIT-Madras, STIC Cochin, University of Pune, and SID-IISc Bangalore for providing NMR, PXRD, SEM, FTIR, and DSC analysis facilities, respectively.

Conflicts of Interest: The authors declare no conflict of interest.

References

1. Gungor, C.; Hofmann, B.T.; Wolters-Eisfeld, G.; Bockhorn, M. Pancreatic cancer. *Br. J. Pharmacol.* **2014**, *171*, 849–858. [CrossRef]
2. Siegel, R.L.; Miller, K.D.; Jemal, A. Cancer statistics, 2020. *CA Cancer J. Clin.* **2020**, *70*, 7–30. [CrossRef]
3. Rahib, L.; Smith, B.D.; Aizenberg, R.; Rosenzweig, A.B.; Fleshman, J.M.; Matrisian, L.M. Projecting cancer incidence and deaths to 2030: The unexpected burden of thyroid, liver, and pancreas cancers in the United States. *Cancer Res.* **2014**, *74*, 2913–2921. [CrossRef]
4. Burris, H.A., 3rd; Moore, M.J.; Andersen, J.; Green, M.R.; Rothenberg, M.L.; Modiano, M.R.; Cripps, M.C.; Portenoy, R.K.; Storniolo, A.M.; Tarassoff, P.; et al. Improvements in survival and clinical benefit with gemcitabine as first-line therapy for patients with advanced pancreas cancer: A randomized trial. *J. Clin. Oncol.* **1997**, *15*, 2403–2413. [CrossRef]
5. Adamska, A.; Domenichini, A.; Falasca, M. Pancreatic Ductal Adenocarcinoma: Current and Evolving Therapies. *Int. J. Mol. Sci.* **2017**, *18*, 1338. [CrossRef]
6. Chiorean, E.G.; Coveler, A.L. Pancreatic cancer: Optimizing treatment options, new, and emerging targeted therapies. *Drug Des. Devel. Ther.* **2015**, *9*, 3529–3545. [CrossRef]
7. Patil, K.; Khan, F.B.; Akhtar, S.; Ahmad, A.; Uddin, S. The plasticity of pancreatic cancer stem cells: Implications in therapeutic resistance. *Cancer Metastasis Rev.* **2021**, *40*, 691–720. [CrossRef]
8. Ho, W.J.; Jaffee, E.M.; Zheng, L. The tumour microenvironment in pancreatic cancer—clinical challenges and opportunities. *Nat. Rev. Clin. Oncol.* **2020**, *17*, 527–540. [CrossRef]
9. Carvalho, T.M.A.; Di Molfetta, D.; Greco, M.R.; Koltai, T.; Alfarouk, K.O.; Reshkin, S.J.; Cardone, R.A. Tumor Microenvironment Features and Chemoresistance in Pancreatic Ductal Adenocarcinoma: Insights into Targeting Physicochemical Barriers and Metabolism as Therapeutic Approaches. *Cancers* **2021**, *13*, 6135. [CrossRef]
10. Von Hoff, D.D.; Ervin, T.; Arena, F.P.; Chiorean, E.G.; Infante, J.; Moore, M.; Seay, T.; Tjulandin, S.A.; Ma, W.W.; Saleh, M.N.; et al. Increased survival in pancreatic cancer with nab-paclitaxel plus gemcitabine. *N. Engl. J. Med.* **2013**, *369*, 1691–1703. [CrossRef]
11. Conroy, T.; Desseigne, F.; Ychou, M.; Bouche, O.; Guimbaud, R.; Becouarn, Y.; Adenis, A.; Raoul, J.L.; Gourgou-Bourgade, S.; de la Fouchardiere, C.; et al. FOLFIRINOX versus gemcitabine for metastatic pancreatic cancer. *N. Engl. J. Med.* **2011**, *364*, 1817–1825. [CrossRef]
12. Conroy, T.; Gavoille, C.; Adenis, A. Metastatic pancreatic cancer: Old drugs, new paradigms. *Curr. Opin. Oncol.* **2011**, *23*, 390–395. [CrossRef]
13. Dandawate, P.; Padhye, S.; Ahmad, A.; Sarkar, F.H. Novel strategies targeting cancer stem cells through phytochemicals and their analogs. *Drug Deliv. Transl. Res.* **2013**, *3*, 165–182. [CrossRef]
14. Dandawate, P.R.; Subramaniam, D.; Jensen, R.A.; Anant, S. Targeting cancer stem cells and signaling pathways by phytochemicals: Novel approach for breast cancer therapy. *Semin. Cancer Biol.* **2016**, *40*, 192–208. [CrossRef]
15. Dandawate, P.R.; Vyas, A.; Ahmad, A.; Banerjee, S.; Deshpande, J.; Swamy, K.V.; Jamadar, A.; Dumhe-Klaire, A.C.; Padhye, S.; Sarkar, F.H. Inclusion complex of novel curcumin analogue CDF and beta-cyclodextrin (1:2) and its enhanced in vivo anticancer activity against pancreatic cancer. *Pharm. Res.* **2012**, *29*, 1775–1786. [CrossRef]

16. Ronghe, A.; Chatterjee, A.; Singh, B.; Dandawate, P.; Abdalla, F.; Bhat, N.K.; Padhye, S.; Bhat, H.K. 4-(E)-{(p-tolylimino)-methylbenzene-1,2-diol}, 1 a novel resveratrol analog, differentially regulates estrogen receptors alpha and beta in breast cancer cells. *Toxicol. Appl. Pharmacol.* **2016**, *301*, 1–13. [CrossRef]
17. Ronghe, A.; Chatterjee, A.; Singh, B.; Dandawate, P.; Murphy, L.; Bhat, N.K.; Padhye, S.; Bhat, H.K. Differential regulation of estrogen receptors alpha and beta by 4-(E)-{(4-hydroxyphenylimino)-methylbenzene,1,2-diol}, a novel resveratrol analog. *J. Steroid Biochem. Mol. Biol.* **2014**, *144 Pt B*, 500–512. [CrossRef]
18. Siddiqui, A.; Dandawate, P.; Rub, R.; Padhye, S.; Aphale, S.; Moghe, A.; Jagyasi, A.; Venkateswara Swamy, K.; Singh, B.; Chatterjee, A.; et al. Novel Aza-resveratrol analogs: Synthesis, characterization and anticancer activity against breast cancer cell lines. *Bioorganic Med. Chem. Lett.* **2013**, *23*, 635–640. [CrossRef]
19. Subramaniam, D.; Ponnurangam, S.; Ramalingam, S.; Kwatra, D.; Dandawate, P.; Weir, S.J.; Umar, S.; Jensen, R.A.; Anant, S. Honokiol Affects Stem Cell Viability by Suppressing Oncogenic YAP1 Function to Inhibit Colon Tumorigenesis. *Cells* **2021**, *10*, 1607. [CrossRef]
20. Dandawate, P.; Ahmad, A.; Deshpande, J.; Swamy, K.V.; Khan, E.M.; Khetmalas, M.; Padhye, S.; Sarkar, F. Anticancer phytochemical analogs 37: Synthesis, characterization, molecular docking and cytotoxicity of novel plumbagin hydrazones against breast cancer cells. *Bioorganic Med. Chem. Lett.* **2014**, *24*, 2900–2904. [CrossRef]
21. Dandawate, P.; Khan, E.; Padhye, S.; Gaba, H.; Sinha, S.; Deshpande, J.; Venkateswara Swamy, K.; Khetmalas, M.; Ahmad, A.; Sarkar, F.H. Synthesis, characterization, molecular docking and cytotoxic activity of novel plumbagin hydrazones against breast cancer cells. *Bioorganic Med. Chem. Lett.* **2012**, *22*, 3104–3108. [CrossRef] [PubMed]
22. Dandawate, P.; Vemuri, K.; Venkateswara Swamy, K.; Khan, E.M.; Sritharan, M.; Padhye, S. Synthesis, characterization, molecular docking and anti-tubercular activity of Plumbagin-Isoniazid Analog and its beta-cyclodextrin conjugate. *Bioorganic Med. Chem. Lett.* **2014**, *24*, 5070–5075. [CrossRef] [PubMed]
23. Dandawate, P.; Subramaniam, D.; Panovich, P.; Standing, D.; Krishnamachary, B.; Kaushik, G.; Thomas, S.M.; Dhar, A.; Weir, S.J.; Jensen, R.A.; et al. Cucurbitacin B and I inhibits colon cancer growth by targeting the Notch signaling pathway. *Sci. Rep.* **2020**, *10*, 1290. [CrossRef] [PubMed]
24. Ramamoorthy, P.; Dandawate, P.; Jensen, R.A.; Anant, S. Celastrol and Triptolide Suppress Stemness in Triple Negative Breast Cancer: Notch as a Therapeutic Target for Stem Cells. *Biomedicines* **2021**, *9*, 482. [CrossRef] [PubMed]
25. Padhye, S.; Ahmad, A.; Oswal, N.; Dandawate, P.; Rub, R.A.; Deshpande, J.; Swamy, K.V.; Sarkar, F.H. Fluorinated 2′-hydroxychalcones as garcinol analogs with enhanced antioxidant and anticancer activities. *Bioorganic Med. Chem. Lett.* **2010**, *20*, 5818–5821. [CrossRef]
26. Krishnamachary, B.; Subramaniam, D.; Dandawate, P.; Ponnurangam, S.; Srinivasan, P.; Ramamoorthy, P.; Umar, S.; Thomas, S.M.; Dhar, A.; Septer, S.; et al. Targeting transcription factor TCF4 by gamma-Mangostin, a natural xanthone. *Oncotarget* **2019**, *10*, 5576–5591. [CrossRef] [PubMed]
27. Ponnurangam, S.; Dandawate, P.R.; Dhar, A.; Tawfik, O.W.; Parab, R.R.; Mishra, P.D.; Ranadive, P.; Sharma, R.; Mahajan, G.; Umar, S.; et al. Quinomycin A targets Notch signaling pathway in pancreatic cancer stem cells. *Oncotarget* **2016**, *7*, 3217–3232. [CrossRef]
28. Padhye, S.; Dandawate, P.; Yusufi, M.; Ahmad, A.; Sarkar, F.H. Perspectives on medicinal properties of plumbagin and its analogs. *Med. Res. Rev.* **2012**, *32*, 1131–1158. [CrossRef]
29. Pradhan, R.; Dandawate, P.; Vyas, A.; Padhye, S.; Biersack, B.; Schobert, R.; Ahmad, A.; Sarkar, F.H. From body art to anticancer activities: Perspectives on medicinal properties of henna. *Curr. Drug Targets* **2012**, *13*, 1777–1798. [CrossRef]
30. Vyas, A.; Dandawate, P.; Padhye, S.; Ahmad, A.; Sarkar, F. Perspectives on new synthetic curcumin analogs and their potential anticancer properties. *Curr. Pharm. Des.* **2013**, *19*, 2047–2069.
31. Angulo, P.; Kaushik, G.; Subramaniam, D.; Dandawate, P.; Neville, K.; Chastain, K.; Anant, S. Natural compounds targeting major cell signaling pathways: A novel paradigm for osteosarcoma therapy. *J. Hematol. Oncol.* **2017**, *10*, 10. [CrossRef]
32. Chestnut, C.; Subramaniam, D.; Dandawate, P.; Padhye, S.; Taylor, J., 3rd; Weir, S.; Anant, S. Targeting Major Signaling Pathways of Bladder Cancer with Phytochemicals: A Review. *Nutr. Cancer* **2021**, *73*, 2249–2271. [CrossRef]
33. Dandawate, P.; Ahmed, K.; Padhye, S.; Ahmad, A.; Biersack, B. Anticancer Active Heterocyclic Chalcones: Recent Developments. *Anticancer Agents Med. Chem.* **2021**, *21*, 558–566. [CrossRef]
34. Dandawate, P.R.; Subramaniam, D.; Padhye, S.B.; Anant, S. Bitter melon: A panacea for inflammation and cancer. *Chin. J. Nat. Med.* **2016**, *14*, 81–100. [CrossRef]
35. Tageja, N.; Padheye, S.; Dandawate, P.; Al-Katib, A.; Mohammad, R.M. New targets for the treatment of follicular lymphoma. *J. Hematol. Oncol.* **2009**, *2*, 50. [CrossRef]
36. Padhye, S.; Banerjee, S.; Chavan, D.; Pandye, S.; Swamy, K.V.; Ali, S.; Li, J.; Dou, Q.P.; Sarkar, F.H. Fluorocurcumins as cyclooxygenase-2 inhibitor: Molecular docking, pharmacokinetics and tissue distribution in mice. *Pharm. Res.* **2009**, *26*, 2438–2445. [CrossRef]
37. Padhye, S.; Yang, H.; Jamadar, A.; Cui, Q.C.; Chavan, D.; Dominiak, K.; McKinney, J.; Banerjee, S.; Dou, Q.P.; Sarkar, F.H. New difluoro Knoevenagel condensates of curcumin, their Schiff bases and copper complexes as proteasome inhibitors and apoptosis inducers in cancer cells. *Pharm. Res.* **2009**, *26*, 1874–1880. [CrossRef]
38. Bao, B.; Ali, S.; Kong, D.; Sarkar, S.H.; Wang, Z.; Banerjee, S.; Aboukameel, A.; Padhye, S.; Philip, P.A.; Sarkar, F.H. Anti-tumor activity of a novel compound-CDF is mediated by regulating miR-21, miR-200, and PTEN in pancreatic cancer. *PLoS ONE* **2011**, *6*, e17850. [CrossRef]

39. Bao, B.; Ali, S.; Banerjee, S.; Wang, Z.; Logna, F.; Azmi, A.S.; Kong, D.; Ahmad, A.; Li, Y.; Padhye, S.; et al. Curcumin analogue CDF inhibits pancreatic tumor growth by switching on suppressor microRNAs and attenuating EZH2 expression. *Cancer Res.* **2012**, *72*, 335–345. [CrossRef]
40. Ali, S.; Ahmad, A.; Banerjee, S.; Padhye, S.; Dominiak, K.; Schaffert, J.M.; Wang, Z.; Philip, P.A.; Sarkar, F.H. Gemcitabine sensitivity can be induced in pancreatic cancer cells through modulation of miR-200 and miR-21 expression by curcumin or its analogue CDF. *Cancer Res.* **2010**, *70*, 3606–3617. [CrossRef]
41. Bao, B.; Ali, S.; Ahmad, A.; Azmi, A.S.; Li, Y.; Banerjee, S.; Kong, D.; Sethi, S.; Aboukameel, A.; Padhye, S.B.; et al. Hypoxia-induced aggressiveness of pancreatic cancer cells is due to increased expression of VEGF, IL-6 and miR-21, which can be attenuated by CDF treatment. *PLoS ONE* **2012**, *7*, e50165. [CrossRef]
42. Li, Y.; Kong, D.; Wang, Z.; Ahmad, A.; Bao, B.; Padhye, S.; Sarkar, F.H. Inactivation of AR/TMPRSS2-ERG/Wnt signaling networks attenuates the aggressive behavior of prostate cancer cells. *Cancer Prev. Res.* **2011**, *4*, 1495–1506. [CrossRef] [PubMed]
43. Kanwar, S.S.; Yu, Y.; Nautiyal, J.; Patel, B.B.; Padhye, S.; Sarkar, F.H.; Majumdar, A.P. Difluorinated-curcumin (CDF): A novel curcumin analog is a potent inhibitor of colon cancer stem-like cells. *Pharm. Res.* **2011**, *28*, 827–838. [CrossRef] [PubMed]
44. Roy, S.; Levi, E.; Majumdar, A.P.; Sarkar, F.H. Expression of miR-34 is lost in colon cancer which can be re-expressed by a novel agent CDF. *J. Hematol. Oncol.* **2012**, *5*, 58. [CrossRef]
45. Roy, S.; Yu, Y.; Padhye, S.B.; Sarkar, F.H.; Majumdar, A.P. Difluorinated-curcumin (CDF) restores PTEN expression in colon cancer cells by down-regulating miR-21. *PLoS ONE* **2013**, *8*, e68543. [CrossRef]
46. Yu, Y.; Sarkar, F.H.; Majumdar, A.P. Down-regulation of miR-21 Induces Differentiation of Chemoresistant Colon Cancer Cells and Enhances Susceptibility to Therapeutic Regimens. *Transl. Oncol.* **2013**, *6*, 180–186. [CrossRef]
47. Ahmad, A.; Sayed, A.; Ginnebaugh, K.R.; Sharma, V.; Suri, A.; Saraph, A.; Padhye, S.; Sarkar, F.H. Molecular docking and inhibition of matrix metalloproteinase-2 by novel difluorinatedbenzylidene curcumin analog. *Am. J. Transl. Res.* **2015**, *7*, 298–308.
48. Basak, S.K.; Zinabadi, A.; Wu, A.W.; Venkatesan, N.; Duarte, V.M.; Kang, J.J.; Dalgard, C.L.; Srivastava, M.; Sarkar, F.H.; Wang, M.B.; et al. Liposome encapsulated curcumin-difluorinated (CDF) inhibits the growth of cisplatin resistant head and neck cancer stem cells. *Oncotarget* **2015**, *6*, 18504–18517. [CrossRef]
49. Luong, D.; Kesharwani, P.; Alsaab, H.O.; Sau, S.; Padhye, S.; Sarkar, F.H.; Iyer, A.K. Folic acid conjugated polymeric micelles loaded with a curcumin difluorinated analog for targeting cervical and ovarian cancers. *Colloids Surf. B Biointerfaces* **2017**, *157*, 490–502. [CrossRef]
50. Dandawate, P.; Vemuri, K.; Khan, E.M.; Sritharan, M.; Padhye, S. Synthesis, characterization and anti-tubercular activity of ferrocenyl hydrazones and their beta-cyclodextrin conjugates. *Carbohydr. Polym.* **2014**, *108*, 135–144. [CrossRef]
51. Yallapu, M.M.; Jaggi, M.; Chauhan, S.C. Poly(beta-cyclodextrin)/curcumin self-assembly: A novel approach to improve curcumin delivery and its therapeutic efficacy in prostate cancer cells. *Macromol. Biosci.* **2010**, *10*, 1141–1151. [CrossRef]
52. Bellringer, M.E.; Smith, T.G.; Read, R.; Gopinath, C.; Olivier, P. beta-Cyclodextrin: 52-week toxicity studies in the rat and dog. *Food Chem. Toxicol.* **1995**, *33*, 367–376. [CrossRef]
53. Olivier, P.; Verwaerde, F.; Hedges, A.R. Subchronic Toxicity of Orally Administered Beta-Cyclodextrin in Rats. *J. Am. Coll. Toxicol.* **1991**, *10*, 407–419. [CrossRef]
54. Gould, S.; Scott, R.C. 2-Hydroxypropyl-beta-cyclodextrin (HP-beta-CD): A toxicology review. *Food Chem. Toxicol.* **2005**, *43*, 1451–1459. [CrossRef]
55. Yokoo, M.; Kubota, Y.; Motoyama, K.; Higashi, T.; Taniyoshi, M.; Tokumaru, H.; Nishiyama, R.; Tabe, Y.; Mochinaga, S.; Sato, A.; et al. 2-Hydroxypropyl-beta-Cyclodextrin Acts as a Novel Anticancer Agent. *PLoS ONE* **2015**, *10*, e0141946. [CrossRef]
56. Rajendiran, N.; Siva, S.; Saravanan, J. Inclusion complexation of sulfapyridine with α- and β-cyclodextrins: Spectral and molecular modeling study. *J. Mol. Struct.* **2013**, *1054–1055*, 215–222. [CrossRef]
57. Wu, H.; Liang, H.; Yuan, Q.; Wang, T.; Yan, X. Preparation and stability investigation of the inclusion complex of sulforaphane with hydroxypropyl-β-cyclodextrin. *Carbohydr. Polym.* **2010**, *82*, 613–617. [CrossRef]
58. Bulani, V.D.; Kothavade, P.S.; Kundaikar, H.S.; Gawali, N.B.; Chowdhury, A.A.; Degani, M.S.; Juvekar, A.R. Inclusion complex of ellagic acid with β-cyclodextrin: Characterization and in vitro anti-inflammatory evaluation. *J. Mol. Struct.* **2016**, *1105*, 308–315. [CrossRef]
59. Craparo, E.F.; Cavallaro, G.; Bondi, M.L.; Mandracchia, D.; Giammona, G. PEGylated Nanoparticles based on a polyaspartamide. preparation, physico-chemical characterization, and intracellular uptake. *Biomacromolecules* **2006**, *7*, 3083–3092. [CrossRef]
60. Lemarchand, C.; Gref, R.; Lesieur, S.; Hommel, H.; Vacher, B.; Besheer, A.; Maeder, K.; Couvreur, P. Physico-chemical characterization of polysaccharide-coated nanoparticles. *J. Control. Release* **2005**, *108*, 97–111. [CrossRef]
61. Uner, M. Preparation, characterization and physico-chemical properties of solid lipid nanoparticles (SLN) and nanostructured lipid carriers (NLC): Their benefits as colloidal drug carrier systems. *Die Pharm.* **2006**, *61*, 375–386.
62. Aigner, Z.; Hassan, H.B.; Berkesi, O.; Kata, M.; Erős, I. Thermoanalytical, FTIR and X-ray studies of gemfibrozil-cyclodextrin complexes. *J. Therm. Anal. Calorim.* **2005**, *81*, 267–272. [CrossRef]
63. Dua, K.; Pabreja, K.; Ramana, M.V.; Lather, V. Dissolution behavior of β-cyclodextrin molecular inclusion complexes of aceclofenac. *J. Pharm. Bioallied Sci.* **2011**, *3*, 417–425. [CrossRef] [PubMed]
64. Garnero, C.; Zoppi, A.; Genovese, D.; Longhi, M. Studies on trimethoprim:hydroxypropyl-beta-cyclodextrin: Aggregate and complex formation. *Carbohydr. Res.* **2010**, *345*, 2550–2556. [CrossRef]

65. Fathy, M.; Sheha, M. In vitro and in vivo evaluation of an amylobarbitone/hydroxypropyl-beta-cyclodextrin complex prepared by a freeze-drying method. *Die Pharm.* **2000**, *55*, 513–517.
66. Marcolino, V.A.; Zanin, G.M.; Durrant, L.R.; Benassi Mde, T.; Matioli, G. Interaction of curcumin and bixin with beta-cyclodextrin: Complexation methods, stability, and applications in food. *J. Agric. Food Chem.* **2011**, *59*, 3348–3357. [CrossRef]
67. Yallapu, M.M.; Jaggi, M.; Chauhan, S.C. beta-Cyclodextrin-curcumin self-assembly enhances curcumin delivery in prostate cancer cells. *Colloids Surf. B Biointerfaces* **2010**, *79*, 113–125. [CrossRef]
68. Veiga, M.a.D.; Ahsan, F. Influence of surfactants (present in the dissolution media) on the release behaviour of tolbutamide from its inclusion complex with β-cyclodextrin. *Eur. J. Pharm. Sci.* **2000**, *9*, 291–299. [CrossRef]
69. Faucci, M.T.; Melani, F.; Mura, P. Computer-aided molecular modeling techniques for predicting the stability of drug–cyclodextrin inclusion complexes in aqueous solutions. *Chem. Phys. Lett.* **2002**, *358*, 383–390. [CrossRef]
70. Alao, J.P. The regulation of cyclin D1 degradation: Roles in cancer development and the potential for therapeutic invention. *Mol. Cancer* **2007**, *6*, 24. [CrossRef]
71. Niu, Y.; Sun, W.; Lu, J.J.; Ma, D.L.; Leung, C.H.; Pei, L.; Chen, X. PTEN Activation by DNA Damage Induces Protective Autophagy in Response to Cucurbitacin B in Hepatocellular Carcinoma Cells. *Oxidative Med. Cell. Longev.* **2016**, *2016*, 4313204. [CrossRef] [PubMed]
72. Choudhary, G.S.; Al-Harbi, S.; Mazumder, S.; Hill, B.T.; Smith, M.R.; Bodo, J.; Hsi, E.D.; Almasan, A. MCL-1 and BCL-xL-dependent resistance to the BCL-2 inhibitor ABT-199 can be overcome by preventing PI3K/AKT/mTOR activation in lymphoid malignancies. *Cell Death Dis.* **2015**, *6*, e1593. [CrossRef] [PubMed]
73. Lee, E.F.; Harris, T.J.; Tran, S.; Evangelista, M.; Arulananda, S.; John, T.; Ramnac, C.; Hobbs, C.; Zhu, H.; Gunasingh, G.; et al. BCL-XL and MCL-1 are the key BCL-2 family proteins in melanoma cell survival. *Cell Death Dis.* **2019**, *10*, 342. [CrossRef]
74. Subramaniam, D.; Kaushik, G.; Dandawate, P.; Anant, S. Targeting Cancer Stem Cells for Chemoprevention of Pancreatic Cancer. *Curr. Med. Chem.* **2018**, *25*, 2585–2594. [CrossRef] [PubMed]
75. Zhao, S.; Chen, C.; Chang, K.; Karnad, A.; Jagirdar, J.; Kumar, A.P.; Freeman, J.W. CD44 Expression Level and Isoform Contributes to Pancreatic Cancer Cell Plasticity, Invasiveness, and Response to Therapy. *Clin. Cancer Res.* **2016**, *22*, 5592–5604. [CrossRef]
76. Gzil, A.; Zarebska, I.; Bursiewicz, W.; Antosik, P.; Grzanka, D.; Szylberg, L. Markers of pancreatic cancer stem cells and their clinical and therapeutic implications. *Mol. Biol. Rep.* **2019**, *46*, 6629–6645. [CrossRef]
77. Shankar, S.; Nall, D.; Tang, S.N.; Meeker, D.; Passarini, J.; Sharma, J.; Srivastava, R.K. Resveratrol inhibits pancreatic cancer stem cell characteristics in human and KrasG12D transgenic mice by inhibiting pluripotency maintaining factors and epithelial-mesenchymal transition. *PLoS ONE* **2011**, *6*, e16530. [CrossRef]
78. Ala, M. Target c-Myc to treat pancreatic cancer. *Cancer Biol. Ther.* **2022**, *23*, 34–50. [CrossRef]
79. Ito, H.; Tanaka, S.; Akiyama, Y.; Shimada, S.; Adikrisna, R.; Matsumura, S.; Aihara, A.; Mitsunori, Y.; Ban, D.; Ochiai, T.; et al. Dominant Expression of DCLK1 in Human Pancreatic Cancer Stem Cells Accelerates Tumor Invasion and Metastasis. *PLoS ONE* **2016**, *11*, e0146564. [CrossRef]
80. Tonnesen, H.H.; Karlsen, J. Studies on curcumin and curcuminoids. V. Alkaline degradation of curcumin. *Z. Für Lebensm. Unters. Und Forsch.* **1985**, *180*, 132–134. [CrossRef]
81. Tomren, M.A.; Másson, M.; Loftsson, T.; Tønnesen, H.H. Studies on curcumin and curcuminoids: XXXI. Symmetric and asymmetric curcuminoids: Stability, activity and complexation with cyclodextrin. *Int. J. Pharm.* **2007**, *338*, 27–34. [CrossRef] [PubMed]
82. Buss, J.L.; Ponka, P. Hydrolysis of pyridoxal isonicotinoyl hydrazone and its analogs. *Biochim. Et Biophys. Acta (BBA) Gen. Subj.* **2003**, *1619*, 177–186. [CrossRef]
83. Renner, S.; Schwab, C.H.; Gasteiger, J.; Schneider, G. Impact of conformational flexibility on three-dimensional similarity searching using correlation vectors. *J. Chem. Inf. Model.* **2006**, *46*, 2324–2332. [CrossRef]
84. Schuttelkopf, A.W.; van Aalten, D.M. PRODRG: A tool for high-throughput crystallography of protein-ligand complexes. *cta Crystallogr. Sect. D Biol. Crystallogr.* **2004**, *60*, 1355–1363. [CrossRef]
85. Humphrey, W.; Dalke, A.; Schulten, K. VMD: Visual molecular dynamics. *J. Mol. Graph.* **1996**, *14*, 33–38. [CrossRef] [PubMed]
86. Trott, O.; Olson, A.J. AutoDock Vina: Improving the speed and accuracy of docking with a new scoring function, efficient optimization, and multithreading. *J. Comput. Chem.* **2010**, *31*, 455–461. [CrossRef] [PubMed]
87. Schrödinger, L.; DeLano, W. PyMOL. Available online: http://www.pymol.org/pymol (accessed on 11 December 2022).
88. Landegren, U. Measurement of cell numbers by means of the endogenous enzyme hexosaminidase. Applications to detection of lymphokines and cell surface antigens. *J. Immunol. Methods* **1984**, *67*, 379–388. [CrossRef]

International Journal of
Molecular Sciences

Article

Differential CMS-Related Expression of Cell Surface Carbonic Anhydrases IX and XII in Colorectal Cancer Models—Implications for Therapy

Arne Rotermund [1], Sarah Brandt [1], Martin S. Staege [2], Jana Luetzkendorf [1], Lutz P. Mueller [1] and Thomas Mueller [1,*]

1 Department of Internal Medicine IV (Hematology/Oncology), Medical Faculty, Martin Luther University Halle-Wittenberg, 06120 Halle, Germany
2 Department of Surgical and Conservative Pediatrics and Adolescent Medicine, Medical Faculty, Martin Luther University Halle-Wittenberg, 06120 Halle, Germany
* Correspondence: thomas.mueller@medizin.uni-halle.de; Tel.: +49-0345-5577211

Abstract: Tumor-associated carbonic anhydrases IX (CAIX) and XII (CAXII) have long been in the spotlight as potential new targets for anti-cancer therapy. Recently, CAIX/CAXII specific inhibitor SLC-0111 has passed clinical phase I study and showed differential response among patients with colorectal cancer (CRC). CRC can be classified into four different consensus molecular subgroups (CMS) showing unique expression patterns and molecular traits. We questioned whether there is a CMS-related CAIX/CAXII expression pattern in CRC predicting response. As such, we analyzed transcriptomic data of tumor samples for CA9/CA12 expression using Cancertool. Protein expression pattern was examined in preclinical models comprising cell lines, spheroids and xenograft tumors representing the CMS groups. Impact of CAIX/CAXII knockdown and SLC-0111 treatment was investigated in 2D and 3D cell culture. The transcriptomic data revealed a characteristic CMS-related CA9/CA12 expression pattern with pronounced co-expression of both CAs as a typical feature of CMS3 tumors. Protein expression in spheroid- and xenograft tumor tissue clearly differed, ranging from close to none (CMS1) to strong CAIX/CAXII co-expression in CMS3 models (HT29, LS174T). Accordingly, response to SLC-0111 analyzed in the spheroid model ranged from no (CMS1) to clear (CMS3), with moderate in CMS2 and mixed in CMS4. Furthermore, SLC-0111 positively affected impact of single and combined chemotherapeutic treatment of CMS3 spheroids. In addition, combined CAIX/CAXII knockdown and more effective treatment with SLC-0111 reduced clonogenic survival of CMS3 modelling single cells. In conclusion, the preclinical data support the clinical approach of targeted CAIX/CAXII inhibition by showing linkage of expression with response and suggest that patients with CMS3-classified tumors would most benefit from such treatment.

Keywords: carbonic anhydrases; colorectal cancer; consensus molecular subtypes; CMS; SLC-0111; targeted inhibition of carbonic anhydrases; chemotherapy

Citation: Rotermund, A.; Brandt, S.; Staege, M.S.; Luetzkendorf, J.; Mueller, L.P.; Mueller, T. Differential CMS-Related Expression of Cell Surface Carbonic Anhydrases IX and XII in Colorectal Cancer Models—Implications for Therapy. *Int. J. Mol. Sci.* **2023**, *24*, 5797. https://doi.org/10.3390/ijms24065797

Academic Editor: Nam Deuk Kim

Received: 22 February 2023
Revised: 14 March 2023
Accepted: 16 March 2023
Published: 18 March 2023

1. Introduction

Hypoxia in solid tumors has long been established as an essential factor for tumor progression and tumor malignancy and, therefore, presents a promising strategy in tumor therapy [1]. As intratumoral hypoxia leads to severe metabolic reprogramming, e.g., in pathways such as the Krebs cycle, fatty acid synthesis and the respiratory chain, a metabolic shift towards anaerobic glycolysis and excessive production of acidic metabolites such as lactate and protons (H^+) is commonly observed in many different tumors in order to meet the energy demands of the fast-growing tumor tissue, even in low-oxygen environments [2]. This metabolic reprogramming is partly induced via the hypoxia-inducible factor 1 alpha (HIF1α) pathway, which supports the shift towards glycolysis but also plays a role in several other key factors of tumor progression, e.g., invasion and metastasis, genomic instability and pH homeostasis [3]. Due to this metabolic shift, physiological homeostasis between intracellular and extracellular space changes drastically inside the tumor as the extracellular pH (pHe) is acidified towards a pH of below 7 while the intracellular pH (pHi) slightly alkalinizes [4]. Alteration in pH, especially in the tumor microenvironment (TME), is further supported by loss of tumor suppressor genes and overexpression of oncogenes, inhibiting the physiological mechanism of pH homeostasis even further [2]. This disruption in pH homeostasis is not exclusively found in highly hypoxic tissue as tumor cells in close proximity to blood vessels exposed to only moderate levels of hypoxia also show strong imbalance in pH homeostasis, therefore solidifying acidosis as a hypoxia-independent factor driving tumor progression [5]. Tumor cells greatly benefit from this pH alteration as an acidified pHe inhibits immune functions [6], supports selection of more malignant tumor cells [7], supports degradation of the extracellular matrix [8], supports angiogenesis [9] and inhibits efficacy of anti-tumor therapy as many drugs have pH-titrable groups [10–12]. Due to the permanently slightly alkaline pHi, intracellular energy production remains largely unaffected by excessive production of acidic metabolites, which aids tumor growth and cell survival [13].

The dysregulated balance between pHi and pHe is maintained by complex machinery consisting of several membrane transporters, e.g., Na^+/H^+-exchangers (NHEs) or sodium bicarbonate cotransporters (NBCs), as well as membrane-associated carbonic anhydrases IX and XII (CAIX/CAXII) and intracellular carbonic anhydrase II. Both the membrane transporters and CAIX/CAXII are upregulated via HIF1α and form a functional complex, a so called "Metabolon", which facilitates extrusion of acidic metabolites, such as lactate, in highly hypoxic and CO_2 in moderately hypoxic tumor tissue and ensures uptake of bicarbonate (HCO_3^-), which is used to buffer intracellular protons [14–16]. As both hypoxia and acidosis are essential factors driving tumor progression, both have already been discussed as potential targets for anti-tumor therapy [9,17]. HIF1α especially has long been a gene of interest as it is induced by both hypoxia and acidosis [18] and plays a role in many aspects of tumor development. However, due to its physiological functions, direct inhibition of HIF1α is impossible. Therefore, inhibition of HIF1α target genes seems to be a better approach for anti-tumor therapy.

Two of those genes are CA9 [19,20] and CA12 [21,22], both zinc metalloenzymes catalyzing reversible hydration of CO_2 to H^+ and HCO_3^-. In comparison to other carbonic anhydrases, especially, expression of CAIX is low in healthy tissue but high in several carcinomas, including carcinomas of the colon [23,24], breast [25], head and neck [26], kidney [27] and bladder [28], making them an interesting target for anti-tumor therapy [29]. Aside from their described role in pH-regulation, CAIX and CAXII also play an important role in chemotherapeutic resistance [30,31], tumor cell migration [32], tumorigenesis [33], cell adhesion [34,35] as well as tumor growth and survival [36,37]. Furthermore, CAIX and CAXII are prognostic markers in several carcinomas [38–45]. Due to their many functions in cancer, several studies on inhibition of CAIX and CAXII have been conducted so far, including development of monoclonal antibodies [46] and small-molecule inhibitors [47], some of which have already entered clinical trials. One small-molecule inhibitor, SLC-0111, an ureido-substituted benzenesulfonamide, has recently passed clinical phase I dose escalation study and showed a promising safety and tolerability profile in patients with previously treated advanced solid tumors [48]. Interestingly, one among four patients with colorectal cancer exhibited stable disease for prolonged time.

Colorectal cancer (CRC) represents a heterogeneous disease both from a molecular and clinical perspective. With the establishment of consensus molecular subtypes (CMS) by an international consortium, CRC was classified into four different molecular subgroups and an unclassified/mixed group showing distinguishing features: CMS1 (MSI immune, 14%), hypermutated, microsatellite unstable and strong immune activation; CMS2 (canonical, 37%), epithelial, marked WNT and MYC signaling activation; CMS3 (metabolic, 13%), epithelial and evident metabolic dysregulation; and CMS4 (mesenchymal, 23%), prominent TGF-beta activation, stromal invasion and angiogenesis [49]. As the different subgroups are very much distinct from each other, several studies were conducted examining the differences between those subgroups regarding different pathophysiological traits, including chemotherapeutic efficacy [50], tumor location [51], local anti-tumor immune response [52], tumor budding [53] and regarding their clinical predictive and prognostic value [54,55]. Notably, the CMS classification of CRC is recapitulated in preclinical model systems, enabling investigations to uncover new targets and test new therapy approaches [56–58]. Accordingly, we used our preclinical CRC models comprising cell lines, spheroids and xenograft tumors, which were classified into the four different CMS groups by applying the CMScaller [59].

The aim of this study was to analyze expression of CAIX and CAXII with special respect to CMS classification of CRC and investigate their specific role regarding differential tumor biological and therapeutic aspects.

2. Results

2.1. Differential Molecular-Subtype-Associated Expression of CA9 and CA12 in Colorectal Cancer

Based on the data of the phase 1 study showing differential response to CAIX/XII specific inhibitor SLC-0111 among patients with colorectal cancer [48], we questioned whether there is a characteristic differential expression pattern of both CAs in colorectal tumors, which could explain the differential response. To this end, we used Cancertool [60] to investigate gene expression of CA9 and CA12 in patient tumor samples. Cancertool is an online bioinformatics platform performing expression-, correlation- and gene-enrichment analyses based on seven different colorectal cancer transcriptomic datasets. Three of the seven datasets also contain normal tissue. The summarized data established with Cancertool are depicted in Figure 1. Compared to normal or normal-adjacent tissue, CA9 expression is increased in tumors, whereas CA12 is decreased, which were significant in all three datasets (Figure 1a). In addition, CA9 expression positively correlated with HIF1α expression in all seven datasets, although reaching statistical significance in only one dataset (Figure 1b). Correlation analysis of CA12 with HIF1α showed a more varying pattern among datasets yet with an overall positive correlation and with statistically significant correlations in two datasets. This suggests, that not only CA9-, but also CA12 expression is associated with hypoxia in colorectal tumors. Analyzing the relationship of both CAs directly revealed an overall positive correlation, reaching statistical significance in three datasets (Figure 1b). Thus, CA9 and CA12 seem to have no complementary roles in colorectal tumors but rather can also occur in a co-expressed manner.

To investigate CA9/CA12 expression depending on the molecular subtype, we employed the same CRC datasets and applied the CMScaller [59] to classify tumor samples according to the CMS classification system. This resulted in a characteristic pattern of expression, which was observed in all seven CRC datasets, exemplarily represented by two datasets in Figure 1c (see Figure S1 in Supplementary Materials for the other five datasets). CA9 was more expressed in CMS1/CMS3 tumors compared to CMS2/CMS4 tumors. CA12 showed the highest expression in CMS3 tumors, followed by CMS1 tumors. Thus, the transcriptomic data revealed pronounced co-expression of both CAs as a typical feature of CMS3- and CMS1 tumors. In addition, the characteristic relation between CA expression in tumors versus normal tissue was reproduced (Figure 1c).

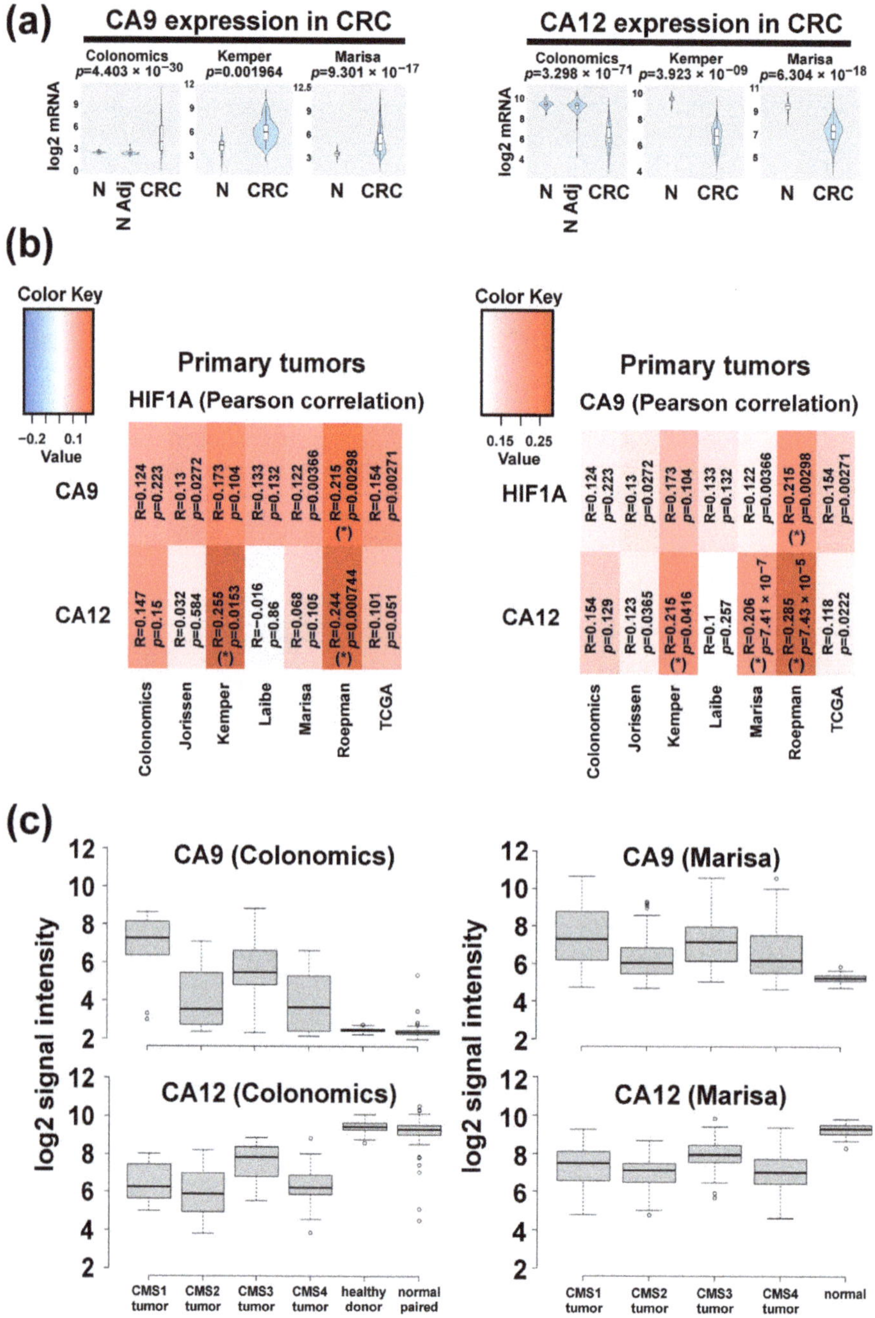

Figure 1. Analyses of expression and correlation of CA9, CA12 and HIF1α on transcriptomic levels in patient tumor samples using Cancertool [60]. (**a**): Expression of CA9 and CA12 in relation to normal or normal-adjacent tissue in 3 out of 7 CRC datasets. Other 4 datasets did not contain normal tissue and could, therefore, not be included for this analysis. (N = normal tissue; N Adj = normal tissue

adjacent to tumor; CRC = colorectal carcinoma) (**b**): Heat maps of correlation analyses based on HIF1α (left) and CA9 (right) in 7 datasets showing overall positive correlation of both CAs to HIF1α and both CAs to each other with different stringency depending on datasets. (**c**): CMS-related expression of CA9 and CA12. Depicted are 2 representative datasets (Colonomics, GSE44076; Marisa, GSE39582). The other 5 datasets can be found in Figure S1 (Supplementary Materials). Tumor samples of each dataset were classified using the CMScaller [59]. Schemes of (**a**,**b**) were created by Cancertool and modified to improve the lettering. Heat maps: in every cell, the corresponding R and p-values of the analysis are shown. The color of each cell represents its correlation R value, being red towards 1 and blue towards -1. Correlations with p-value ≤ 0.05 and $|R| \geq 0.2$ are indicated with (*).

2.2. Differential Expression of CAIX and CAXII in Preclinical Models Representing CMSs

To investigate CAIX/XII expression in tumor tissues on a protein level, we employed our established preclinical CRC model comprising luciferase-expressing cell-line-derived nude mouse xenograft tumors and spheroids representing the four molecular subtypes according to the CMS classification system. Detailed characterization of the model, including generation of the luciferase-expressing variants of cell lines, transcriptomic data and CMS classification using the CMScaller [59], will be published in a separate paper. The results of CMS classification were in accordance with wild-type CRC cell lines, which were previously characterized in the study by Berg et al. [56]. Immunohistochemical analyses of both CAs and hypoxia (pimonidazole) showed a characteristic expression pattern among xenograft tumors of different CMSs. Figure 2 shows examples featuring each xenograft tumor type (see Figure S2 in Supplementary Materials for original pictures with higher quality, including additional images and selected images with higher magnification). In tumors derived from cell line LOVO representing CMS1, expression of CAIX was low in all examined tumors, and, if positive cells were found, they were associated with hypoxic areas, especially in perinecrotic tissue. CAXII was found more frequently; however, in contrast to CAIX, CAXII expression also occurred outside of hypoxic tissue, even adjacent to vessels. In SW48 tumors (CMS1), CAIX expression was found in even fewer cells than in LOVO tumors in three out of four tumors, while, in one of the four tumors, much stronger expression of CA9 was found. If found, CAIX expression was associated with hypoxic tissue. CAXII expression was not observed in any of the SW48 tumors. Overall, CMS1 tumors showed an undifferentiated phenotype and large portions of the tumors were hypoxic, yet expression of CAIX was rare. CAXII expression occurred in one CMS1 tumor but was not necessarily bound to hypoxic tissue. The rare CAIX protein expression was in contrast to the transcriptomic data obtained from tumor samples (Figure 1c), whereas a differential CAXII protein expression pattern among CMS1 tumors could be expected.

CMS2 tumors represented by xenografts from cell lines SW1463 and LS1034 frequently showed a colonic-epithelium-like differentiation pattern and were overall less hypoxic compared to the other tumor types, with hypoxic areas observed in some distance to the blood vessels. CAIX staining in SW1463 tumors revealed nearly ubiquitous expression as even tissue adjacent to vessels showed positive staining. Therefore, expression of CAIX was not necessarily bound to hypoxic areas but was also found in non-hypoxic tissue. CAXII-stained cells could be frequently found; however, the signal was confined to the cytoplasmic compartment and was typically observed in colonic epithelial cells' building ducts. In LS1034 tumors, CAIX expression was frequent but rarer when compared to SW1463 as association of CAIX expression to hypoxia was stronger in LS1034 tumors. For CAXII, we observed the same phenomenon as described for SW1463. Together, CMS2 tumors showed very frequent expression of CAIX, with expression being more spread in SW1463 than in LS1034 tumors. Expression of CAIX can be associated with hypoxia, but both tumors also expressed CAIX in non-hypoxic tissue. Regarding CAXII, the clear cytoplasmic signal exclusively observed in distinct cell types in CMS2 tumors might be due to unspecific staining but could also be a specific feature of those cells. Otherwise, cytoplasmic expression of CAXII is probably less relevant regarding the therapeutic impact of CAIX/XII specific inhibitors. Strikingly, the overall strong CAIX protein expression in CMS2 xenograft tumors, in particular when compared to CMS1 tumors, was in clear contrast to the transcriptomic data of tumor samples (Figure 1c).

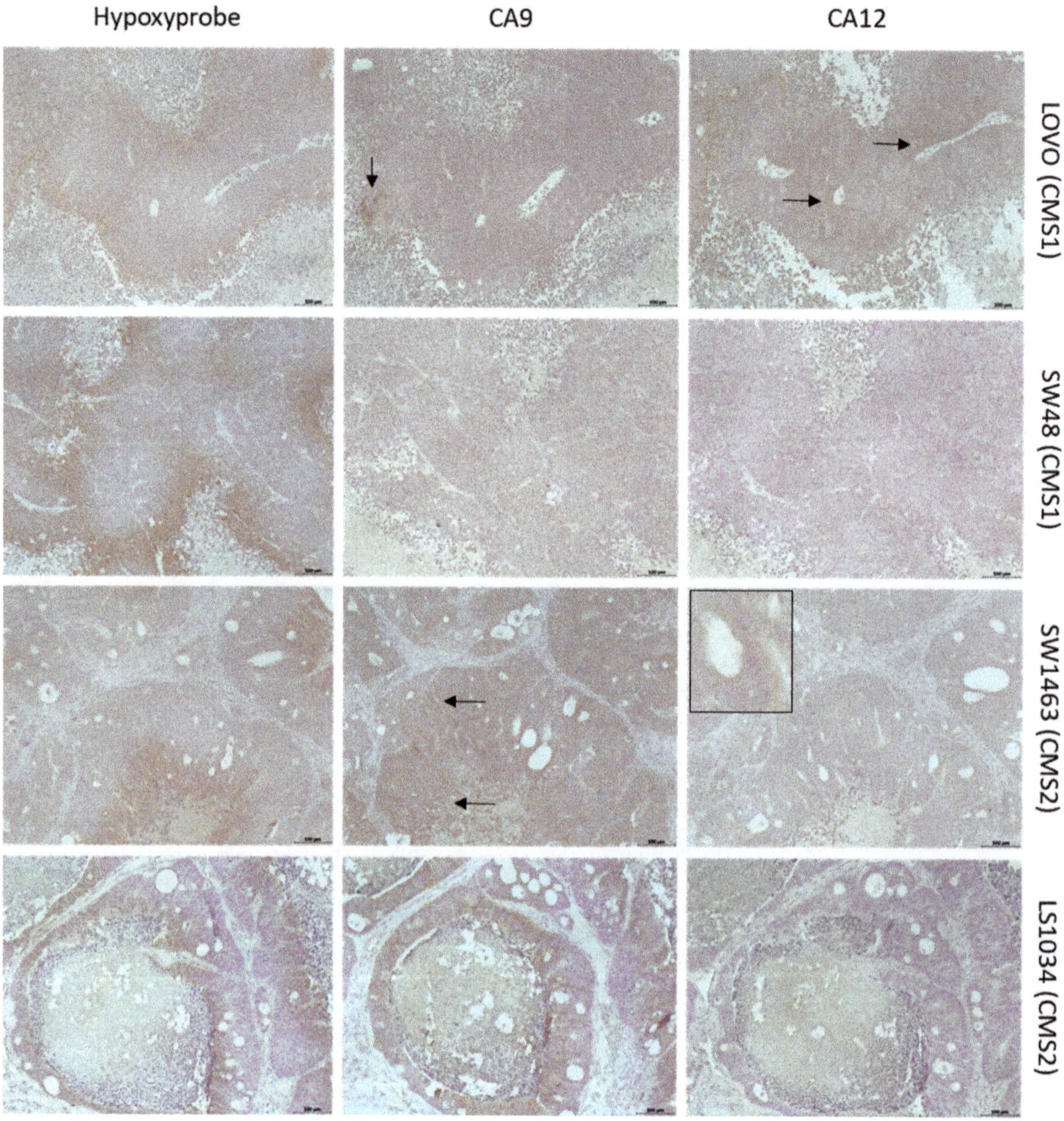

Figure 2. *Cont.*

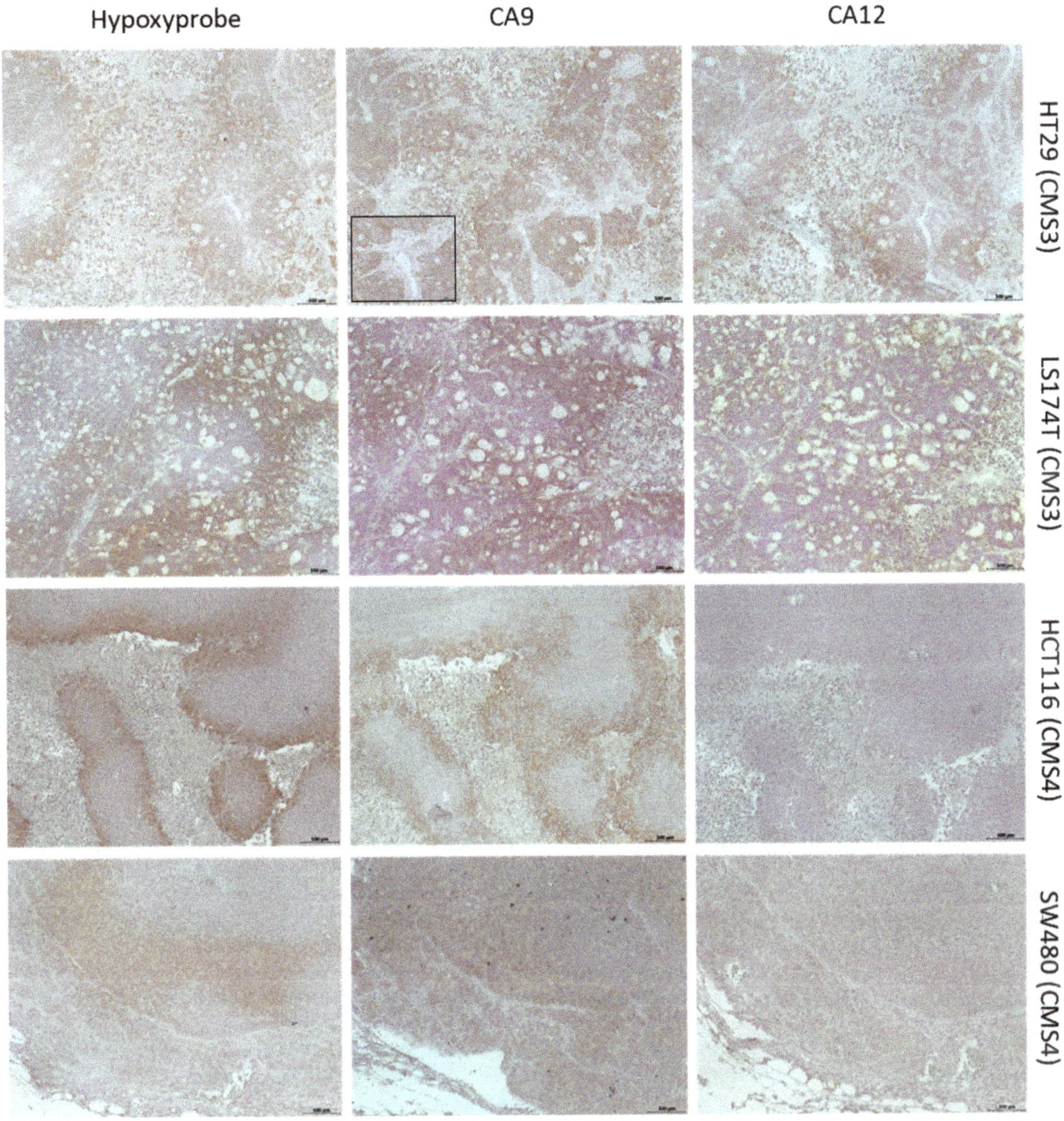

Figure 2. Immunohistochemical analyses of expression of CAIX and CAXII and presence of hypoxia (Hypoxyprobe, pimonidazole) in xenograft tumors representing different CMS. Staining was performed on direct following slides to capture the same tumor areas for analyses. Original pictures with higher quality, including selected images with higher magnification, can be found in Figure S2 (Supplementary Materials). LOVO (CMS1): CAIX expression is rare; positive cells are associated with hypoxic perinecrotic areas (arrow). CAXII is more frequently found and can occur outside of hypoxic tissue, even adjacent to vessels (arrows). SW48 (CMS1): CAIX is rare or absent despite hypoxia. CAXII is not found. SW1463 (CMS2): CAIX is nearly ubiquitously expressed and is found bound to hypoxic areas (arrow below) as well as independent of hypoxia (upper arrow). CAXII-positive cells

can be found, but the signal is confined to the cytoplasmic compartment of colonic epithelial-like cells building ducts (inserted picture). LS1034 (CMS2): CAIX expression is frequent but rarer compared to SW1463 and more associated to hypoxia in comparison to SW1463 tumors. CAXII expression shows the same phenomenon as described for SW1463 (not shown). HT29 (CMS3): CAIX expression is nearly ubiquitous and is found in hypoxic and perinecrotic areas as well as in non-hypoxic areas, even adjacent to vessels (inserted picture). CAXII expression is rarer than CAIX expression but is still found widespread around the tumor mass and is particularly associated with morphological features of goblet cells. Large areas with co-expression of both CAs occur frequently. LS174T (CMS3): CAIX expression is widespread but rarer compared to HT29 tumors and is more associated to hypoxic areas. CAXII is frequently expressed and especially found in tumor areas showing morphological features of goblet cells. Areas with combined CAIX/CAXII expression are frequently found. HCT116 (CMS4): CAIX expression is strong, clearly associated with hypoxia and almost exclusively found in perinecrotic tissue. CAXII expression is missing completely. SW480 (CMS4): CAIX expression is exclusively found in hypoxic tissue. There is no CAXII expression. (Scale bar: 100 μm).

CMS3 tumors represented by xenografts from cell lines HT29 and LS174T showed pronounced differentiation propensity towards goblet-cell-like structures and were overall more hypoxic compared to CMS2 tumors. In HT29 tumors, CAIX expression was nearly ubiquitous. Most areas with expression of CAIX were also hypoxic, but expression of CAIX was also found in non-hypoxic areas. CAXII expression was rarer than CAIX expression but was still found widespread around the tumor mass. Expression of CAXII was almost exclusively found in hypoxic tissue, especially in areas showing morphological features of goblet- and secretory cells. Notably, in hypoxic areas, co-expression of both CAs could be observed. In LS174T tumors, CAIX expression was widespread but rarer compared to HT29 tumors. Most of the CAIX-positive cells were found in hypoxic areas, although expression was also found in non-hypoxic regions. In contrast to HT29, some areas, especially around blood vessels, showed no CAIX expression at all. CAXII expression was rarer than CAIX and almost exclusively found in tumor areas showing morphological features of goblet cells. Some of the cells in those regions showed unclear CAXII signals, similar to the phenomenon described for CMS2 tumors. Hypoxic areas were widespread but rarer than in HT29 tumors. CAIX/CAXII co-expression occurred in hypoxic areas in accordance with HT29 tumors. Thus, as a whole, CMS3 tumors showed the strongest expression of CAIX and CAXII compared to all other types and were particular characterized by occurrence of co-expression of both CAs in same areas and cells, a feature that could not be observed in any tumor outside of the CMS group. Interestingly, the strong combined protein expression of both CAs completely reflected the transcriptomic data (Figure 1c).

CMS4 tumors represented by xenograft tumors from cell lines HCT116 and SW480 showed an undifferentiated phenotype, and large portions of the tumors were hypoxic. In HCT116 tumors, CAIX expression was strong but almost exclusively found in perinecrotic tissue and, therefore, clearly associated with hypoxia. CAXII expression was not found in any of the examined tumors. The tumors were also non-hypoxic apart from the areas adjacent to necrosis. In SW480 tumors, CAIX expression was exclusively found in hypoxic tissue. In contrast to HCT116, some of the perinecrotic tissue showed no expression of CAIX. There was no CAXII expression in any of the examined tumors. Overall, CMS4 tumors showed strong and hypoxia-associated expression of CAIX on a protein level, although lack of CAIX in some hypoxic areas could also be observed. The partly strong expression of CAIX on the protein level in CMS4 tumors was not to be expected based on the transcriptome data, especially when compared to CMS1 tumors (Figure 1c).

To summarize the data obtained in the preclinical CRC tumor models, it can be stated that the models overall do reflect the specific characteristics of CRC regarding expression of CAs and hypoxia provided by the transcriptomic data (Figure 1b); i.e., CAIX and CAXII expression is largely positively correlated to hypoxia, even if not stringent, and both CAs can be occur in a co-expressed manner (Figure 1b). However, considering molecular-subtype-associated expression, there were clear differences between expression on a protein level in preclinical tumors and expression on a transcriptome level in the patient tumor samples. This could be explained by the known fact that transcriptomic expression levels must not necessarily be directly linked to proteomic expression levels and, as such, the high CAIX gene expression level in CMS1 tumors does not necessarily need to translate into high CAIX expression on a protein level. On the other hand, CA expression on transcriptome level does clearly vary among each group and show overlapping areas. Therefore, some of the used preclinical CMS-classified CRC tumor models probably do not reflect the most typical CA expression pattern of the respective group, assuming that each CMS group can be characterized by a specific CA protein expression pattern after all. Strikingly, CMS3 xenograft tumors completely reflected a CAIX/XII expression pattern, which was expected based on the transcriptomic data; hence, they are useful models representing CMS3 group tumors.

2.3. CAIX/CAXII Expression Is Associated with Response to SLC-0111 in Tumor Spheroids

Next, we questioned whether the CAIX/XII protein expression patterns observed in xenograft tumors are reproduced in their respective CMS-classified tumor spheroids in vitro. Tumor spheroids are useful in vitro models representing several important aspects of real tumor tissues, i.e., three-dimensional growth with structural organization and physiologically relevant cell–cell and cell–matrix interactions, establishment of tumor microenvironmental characteristics, such as nutrient gradients, hypoxia and acidosis, as well as drug resistance mechanisms [61,62]. Immunohistochemical analyses of tumor spheroid tissues largely reflected characteristics of xenograft tumors, except few variations. For example, in LOVO spheroids (CMS1), CAXII expression was hardly found, while CAIX was rare or absent in accordance with xenografts. In SW48 spheroids (CMS1), neither CA was observed, nearly reflecting xenografts. Thus, these spheroids represent models with rare or no CAIX/XII expression. Spheroids of SW1463 and LS1034 (CMS2) showed CAIX expression, and the phenomenon of specific differentiated cells with cytoplasmic CAXII signal could also be observed. Strong and combined expression of both CAs was the typical feature of HT29 and LS174T spheroids (CMS3), clearly reflecting characteristics of xenograft tumors. HCT116 spheroids (CMS4) showed typical strong CAIX expression bound to hypoxia and adjacent to the core necrotic area. In SW480 spheroids (CMS4), CAIX was ubiquitously expressed throughout the tissue; thus, the relation of CAIX expression and tumor mass was much higher compared to xenograft tumors.

Using these representative models, we tested the impact of specific CAIX/XII inhibitor SLC-0111 on tumor spheroid growth. To this end, we performed our established spheroid cytotoxicity assay, which is based on stable luciferase expression of tumor cells, enabling monitoring of spheroid growth and response to therapy. Calculated IC_{50} values derived from dose–response curves are summarized in Table 1. There was a differential response among models, reaching from completely insensitive (CMS1 spheroids) to clearly responding (CMS3 spheroids) when exposed to the reported peak plasma levels of SLC-0111 of around 20 μM [48]. Differential response was clearly associated with differential CA expression. Interestingly, the SW480 model (CSM4) was almost as sensitive as our CMS3 models. Together, the data obtained in spheroid models clearly suggest a CMS-related propensity to respond to CA inhibition, with no (CMS1), moderate (CMS2), clear (CMS3), and moderate to clear (CMS4) response. In addition, considering the ability of our preclinical CMS3 models to reflect transcriptomic pattern of CMS3 tumor samples, it can be assumed that patients with CMS3 tumors would most benefit from treatment with CAIX/XII-specific inhibitors such as SLC-0111. Conclusions with regard to the other CMS groups are not possible at this point. It remains to be investigated whether they can also be characterized by specific CA protein expression patterns. Nevertheless, based on the data obtained in these preclinical models, response to CA inhibitors can be assumed in tumors with substantial CAIX expression independent of the CMS.

Table 1. Response of tumor spheroids to SLC-0111. IC_{50} values (μM) ± SD (n ≥ 3).

LOVO	SW48	SW1463	LS1034	HT29	LS174T	HCT116	SW480
>100	98 ± 12	48 ± 12	56 ± 18	21 ± 4	19 ± 4	57 ± 7	23 ± 4
CMS1	CMS1	CMS2	CMS2	CMS3	CMS3	CMS4	CMS4

2.4. Role of CAIX/CAXII Expression for Clonogenic Survival in CMS3 Modeled Cells

The monolayer cultures of our models, analyzed by immunocytochemistry, largely reflected the specific CAXII expression pattern observed in xenograft tumors and spheroids as some LOVO cells and virtually all cells of HT29 and LS174T were positively stained. As expected, the CAIX expression pattern was different. In CMS1 and CMS4 cell lines, CAIX-positive cells were hardly found. HT29 (CMS3) and SW1463 (CMS2) showed strong and ubiquitous expression of CAIX, whereas it was far less in LS174T (CMS3) and LS1034 (CMS2). Notably, HT29 cells were characterized by virtually ubiquitous co-expression of both CAs (Figure 3a), reflecting the most typical feature of CMS3 tumors already on the monolayer level of cell culture. Therefore, it represents a sufficient model to study the importance of expression and activity of both CAs for different tumor functions.

To this end, we analyzed the impact of knockdown or inhibition of CAs on ability of cells to build colonies from single cells under adherent and non-adherent conditions, i.e., colony-forming in six-well plates and single-spheroid-forming in 96-well plates. After establishing stable knockdowns of each CA and of both combined, successful depletion was controlled by means of Western blot analysis of protein lysates prepared from monolayer- and spheroids culture (Figure 3b). Aside from the impact of knockdowns, treatment of HT29 mock cells with CAIX/XII inhibitor SLC-0111 slightly reduced CAIX level but showed no influence on CAXII levels. In the colony-forming assay, only small effects but no significant changes in colony numbers could be observed in any of the knockdown variants (Figure 3d). However, treatment with SLC-0111 resulted in significant decrease in colony number in mock cells (Figure 3d). Mock cells treated with SLC-0111 also showed a significant decrease in colony size, while none of the other knockdown variants showed significant reduction in colony size (Figure 3e). The amount of spheroids formed from single cells was significantly decreased in cells with combined CAIX/CAXII knockdown as well as in mock cells treated with SLC-0111 (Figure 3c). Thus, combined reduction in CAIX/CAXII expression affected clonogenic survival in part, whereas combined inhibition of activity of CAs led to effective reduction in cell survival in each condition.

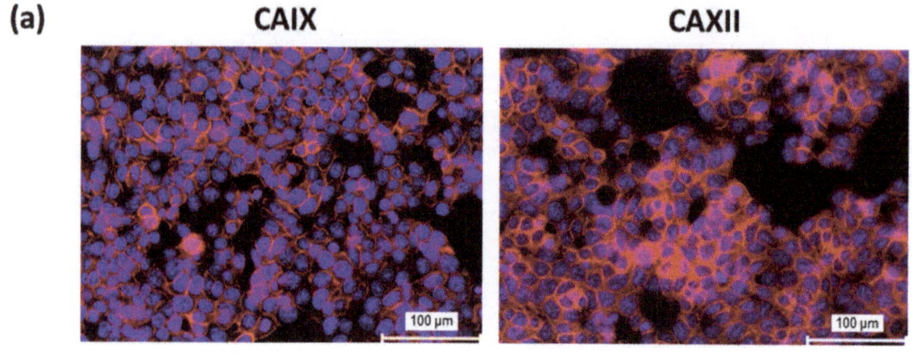

Figure 3. *Cont.*

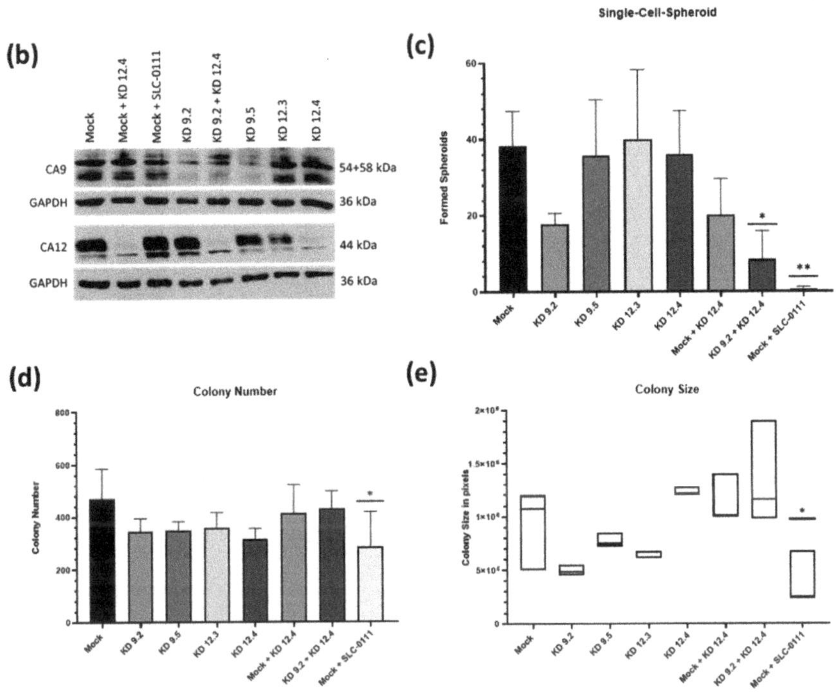

Figure 3. Impact of CAIX/CAXII expression on tumor biological functions of HT29 cells (**a**): immunocytochemistry of monolayer culture showing ubiquitous expression of both CAIX (left) and CAXII (right) (scale bar: 100 μm). (**b**): Western blot analyses of CAIX/CAXII expression in mock and knockdown cells. Two different shRNAs were used for each CA. (**c**): Spheroid forming from single-cell level was significantly reduced in cells with the combined CAIX/CAXII knockdown and mock cells treated with SLC-0111 (mean with SD, n $\geq$ 3). (**d**): Significant reduction in colony number in mock cells treated with SLC-0111 (mean with SD, n $\geq$ 3). (**e**): Significant reduction in colony size in mock cells treated with SLC-0111 (median with range, n $\geq$ 3). The clonogenicity assay and single-cell spheroid assay were repeated three or more times per condition in separate trials. (*: $p \leq 0.05$, **: $p \leq 0.01$).

2.5. Impact of CAIX/CAXII Expression on Chemotherapy in CMS3 Spheroids

Using the HT29 spheroid model, we investigated the impact of different knockdowns and inhibition of CAIX/CAXII by means of co-treatment with SLC-0111 on efficacy of chemotherapeutic drugs 5-fluorouracil (5-FU), irinotecan and oxaliplatin. There were no significant differences in IC_{50} between mock cells and knockdown variants, although a tendency of sensitization towards 5-FU and oxaliplatin could be concluded from slight shifting of knockdown curves, in particular of the combined knockdown. The rather small effects of knockdowns, already observed in the single-cell assays, could be explained by probably sufficient residual expression of CAIX and CAXII despite knockdown resulting in no substantial alteration of the microenvironment.

Co-treatment with SLC-0111 led to clear sensitizing effects towards chemotherapy (Figure 4). For both oxaliplatin and irinotecan, a significant reduction in IC_{50} was achieved by adding SLC-0111. Furthermore, analyses of selected and clinically relevant concentrations around the IC_{50} values revealed a significant reduction in spheroid viability at 0.1 μM of oxaliplatin and at 0.1 μM as well as 1 μM of irinotecan. Clear, but not significant, reduction in spheroid viability could also be observed at 1 μM of oxaliplatin. Together, this showed that additional SLC-0111 treatment is capable of increasing cytotoxicity, especially at low concentrations of the chemotherapeutic agents (Figure 4a,b). Sensitizing effects were also observed with regard to 5-FU, but the characteristic flattened course of curves precluded clear calculation of IC_{50} values and their comparison.

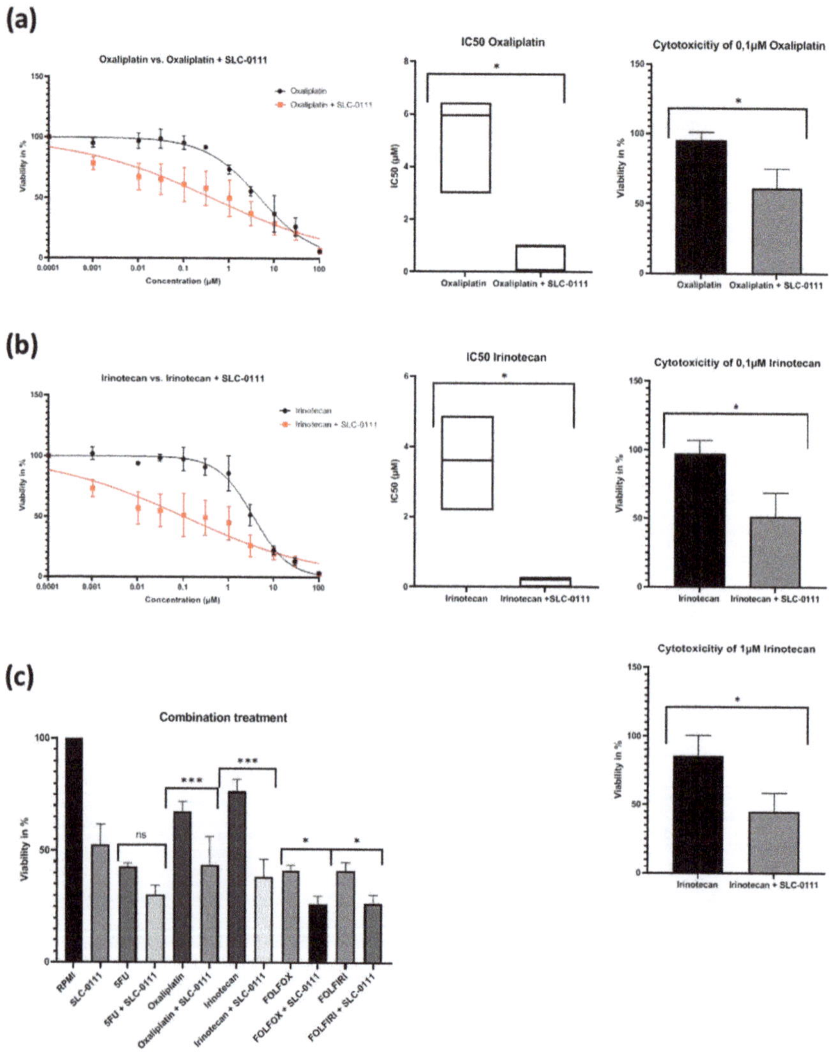

Figure 4. Impact of SLC-0111 co-treatment on chemotherapy in HT29 spheroids. (**a**): Co-treatment of oxaliplatin and SLC-0111 results in significant decrease in IC_{50} value (µM, median with range, n $\geq$ 3) and cell viability at oxaliplatin concentration of 0.1 µM (mean with SD, n $\geq$ 3). (**b**): Co-treatment of irinotecan and SLC-0111 results in significant decrease in IC_{50} value (µM, median with range, n $\geq$ 3) as well as cell viability at irinotecan concentrations of 0.1 µM and 1 µM, respectively (both mean with SD, n $\geq$ 3). (**c**): Additional treatment with 20 µM SLC-0111 results in significant decrease in cell viability in spheroids treated with 0.1 µM irinotecan and 0.1 µM oxaliplatin as well as spheroids treated with combinations of chemotherapeutic agents reflecting the FOLFOX (1 µM 5-FU; 0.1 µM Oxaliplatin) or FOLFIRI (1 µM 5-FU; 0.1 µM irinotecan) regimens (all mean with SD, n $\geq$ 3). (*: $p \leq 0.05$, ***: $p \leq 0.001$).

As standard protocols for chemotherapy of CRC consist of combination schemes (FOLFOX: 5FU + oxaliplatin; FOLFIRI: 5FU + irinotecan), we furthermore investigated impact of co-treatment with SLC-0111 on combination chemotherapy. For this purpose, we used defined drug concentrations and analyzed dual and triple combinations of drugs (Figure 4c). Co-treatment with SLC-0111 led to significant improvement in chemotherapeutic efficacy in the dual therapy scheme, except for 5-FU, which was in accordance with results obtained from calculations of dose–response curves. As expected, the triple therapy regimens were also clearly more effective. Notably, a significant sensitizing effect resulted from addition of SLC-0111. Together, this suggests that inhibition of CAIX/CAXII by co-treatment with specific inhibitors can improve efficacy of chemotherapy independent of the used therapy protocols.

3. Discussion

With the recent introduction of consensus molecular subgroups (CMS), the focus on developing treatment strategies for a more individual and specific therapy for each patient in addition to the already existing therapies is as urgent as ever before. Therefore, new therapeutic targets have to be found and already existing ones have to be reexamined for their usefulness for individual therapy. Two proteins belonging to the second of the aforementioned categories are carbonic anhydrase IX and carbonic anhydrase XII. It is well documented in the literature that cancer cells modify their tumor microenvironment as a result of their otherwise unsustainable growth by changing extracellular pH values, leading to several effects benefiting the tumor and hindering the anti-cancer therapy [2,5,9,10,14], e.g., inhibition of immune function [6], selection of more malignant tumor cells [7], degradation of the extracellular matrix [8], angiogenesis [9] and reduction in efficacy of chemotherapy [10]. Tumor-associated carbonic anhydrases CAIX and CAXII were discovered over two decades ago [19,21]. Their unique niche in the enzyme family of the carbonic anhydrases has long made them interesting targets for anti-cancer therapy as their expression in healthy tissue is by far not as widespread as the expression of other members of their enzyme family, e.g., carbonic anhydrase II or IV. Their importance in creation and retention of this unique microenvironment has already been reported on [14–16]. As such, our goal in this study was to examine the potential of CAIX/CAXII inhibition for treatment of CRC. In order to examine the expression patterns of CAIX/CAXII in the different CMS groups, we analyzed the expression of the CAs in transcriptome data of human tumor samples and standardized cell-line-derived preclinical models of the different CMS groups.

Analyzing the transcriptome data of both CA9 and CA12 in both healthy and cancerous human tissue, we observed that CA12 was downregulated in CRC compared to healthy tissue. This finding was consistent with previous literature [24,63]. The opposite was observed for CA9 as CA9-expression increased in cancer tissue when compared to healthy tissue [23,24]. It is important to note that expression of both CAs was highest in CMS3 compared to the other CMS groups. It is known that CMS3 tumors show an increase in gene expression of goblet cell marker genes and show the highest amount of goblet cells out of all CMS groups [64]. The goblet cells, along with other cells involved in secretion and water absorption, also make up a large part of the cells showing CA expression in healthy tissue of the large intestine [65]. Several authors have previously stated that expression of tumor-associated CAs was likely to be linked to the origin of the cancer cell itself, among other factors, which might also explain why expression of CAIX in both the models of CMS2 and CMS3 was not exclusively found in hypoxic tissue [65,66].

While the transcriptome data were consistent with positive correlation of hypoxia to CA expression, there were also significant differences in our findings between protein and transcriptome level, as mentioned above: While patient tumor samples classified into the CMS1 group showed the highest expression of CA9 and CA12 (together with CMS3) on a transcriptome level, neither our immunohistochemical analysis of spheroid and xenograft tumor tissue nor our analyses in the 2D models showed any abundance of CA expression close to that of the CMS3 group. Aside from CMS1, the transcriptome data for CMS2 tumors showed less expression of CA9, while the xenograft tumors showed high abundance of expression of CAIX. This indicates that there is a complex multifactorial mechanism behind regulation of CAIX and CAXII expression. Although HIF-1 plays the major role in regulation of expression of CAIX and CAXII, it is far from the only protein involved as other mechanisms, including MORC2 and non-coding RNA, were reported on recently [66–70]. Additionally, di Fiore et al. recently reported on the currently known posttranslational modification mechanisms of CAIX and CAXII and stated that these modifications partly have a yet unknown impact on the function of CAIX/CAXII, indicating that there may also be fluctuation in CAIX/CAXII efficacy present on a protein level [71].

To our knowledge, no work has been published yet in which expression of CAIX and CAXII in the different CMS groups of CRC was systematically analyzed by means of immunohistochemistry. As described above, the individual strength of expression of CAIX/CAXII varied from cell line to cell line, yet the pattern in which CAIX and CAXII were expressed was close to the same when comparing the two cell lines of each CMS group with each other. However, expression patterns greatly varied between the different CMS groups, thus suggesting that the distinct molecular features that lead to classification of CRC into different CMS groups also play a huge part in expression of carbonic anhydrase IX and XII. We further validated our findings of differentiated expression with the help of the 3D spheroid model, in which we were able to show that: (1) close to no differences in expression patterns exist between the spheroid and xenograft model and (2) the spheroid model itself as a basic 3D model already shows distinction from the results of our 2D models. As such, the spheroid model proved to be a sufficient 3D model to assess the importance of CAIX and CAXII for tumor function. It is important to note that this distinction is also due to the different conditions the cancer cells are exposed to as, e.g., there is an abundance of oxygen and nutrients in the 2D-models, which are not present in our 3D-models. Therefore, the 2D- and 3D-models are to be compared with caution.

As described above, the effect of sole SLC-0111 treatment on spheroid viability differed between the different cell models of CRC depending on the CMS group. While SLC-0111 treatment of SW48 and LOVO of CMS1 had no effect on spheroid viability, HT29 and LS174T spheroids of CMS3 showed clear response to treatment with SLC-0111. The cytotoxic effects on spheroid models of CMS2 and CMS4 were stronger than in CMS1 but weaker than in CMS3. These findings led us to the conclusion that efficacy of SLC-0111 is very much dependent on the expression status of CAIX and CAXII in the tumor mass as LOVO and SW48 cells showed little to no expression of neither CAIX nor CAXII in both spheroids and xenograft tumors. The findings in CMS1 also proved that SLC-0111 itself probably has no effect on cell viability of cancer cells outside of inhibition of CAIX and CAXII. The IC_{50} value for these two models was outside the used concentrations in our assay, which already far exceeded the dosage previously tested as safe for humans in a phase 1-study conducted on SLC-0111 [48]. In sharp contrast, the HT29 and LS174T models, which we identified as the tumor models with the strongest expression of CAIX and CAXII, showed the strongest reduction in spheroid viability and, therefore, clear response to treatment with SLC-0111. This further proved that SLC-0111 efficacy is closely linked to the CAIX/XII expression status of the tumor.

These findings suggest that SLC-0111 could be a potent weapon for treatment of CRC, but its usage is limited to tumors with high expression of CAIX and CAXII. Therefore, a histopathological examination of the patient's tumor or other procedures, such as PET imaging [72,73], are necessary in order to deduce if therapy with a CAIX/XII inhibitor such as SLC-0111 will be beneficial for the patient due to high expression of CAIX and/or CAXII. These findings may explain the results of the CRC patients in the phase 1 study for SLC-0111, in which only one of the enrolled CRC patients showed a period of stable disease upon treatment with SLC-0111 [48], which could be due to a tumor with high expression of CAIX and CAXII, most probably a CMS3 tumor. As such, our preclinical data demonstrate the benefit of SLC-0111 treatment but also suggest that far from all patients would benefit from additional therapy with such a CA inhibitor. The findings also revealed very differentiated expression of CAIX and CAXII in the different CMS groups, further confirming that assessing the CMS groups in CRC patients is very important to provide the patient with the most beneficial additional anti-tumor treatment, e.g., in the form of SLC-0111 if the CAIX/CAXII expression in the tumor mass is high, as in a CMS3 tumor. Interestingly, the described "metabolic dysregulation" signature of CMS3 tumors [49] could explain the importance of CAIX/CAXII for these tumor types, resulting in pronounced sensitivity towards inhibition of both CAs.

Aside from the aforementioned 3D models, we also used immunocytochemistry in order to examine the expression status of CAIX and CAXII in the different CMS groups on a 2D monolayer level. In these models, HT29 was the only cell line to show strong staining for both CAIX and CAXII. Further, we observed several differences between our 2D- and 3D models. The most striking difference was observed for the LS174T cell line, which showed only little expression of CAIX compared to HT29 in our monolayer models but showed widespread expression in xenograft tumors. CAIX expression in the LS174T cell line was mostly found in hypoxic tissue, and the monolayer model does not provide a severely hypoxic environment; however, this alone cannot be the reason for the lack of CAIX expression as HT29 cells showed a high abundance of CAIX in the 2D models although also exhibiting a strong association between CAIX and hypoxic tissue. In addition, both cell lines of CMS 4 showed close to no CAIX expression in our 2D models, although expression was present in a moderate and sometimes even high abundance in both the xenograft tumors and spheroids. We observed the opposite trend in the LOVO cell line as we observed less expression of CAXII in 3D models but more expression of CAXII in the monolayer culture. Therefore, we assume that expression of CAIX/CAXII is not only influenced by oxygenation of the tissue but also relies on other factors, such as tumor architecture, cell differentiation and cell–cell-interaction. These examples show that a 2D approach of analyzing expression of CAIX and CAXII does not necessarily reflect the real expression status in a tumor and that 3D models, therefore, appear to be more realistic models of CA expression in the tumor mass.

As described above, HT29 cells were found to have the highest expression of CAIX and CAXII in both our 2D- and 3D models; thus, they represent the most typical feature of CMS3 tumors in vitro already. Therefore, we decided to use HT29 cells as our model to examine the importance of CAIX and CAXII for tumor function. The effects of the knockdowns of CAIX and CAXII were not significant in any of our trials. It is important to note, however, that these knockdowns did have a negative effect on the tumor cells in our single-cell spheroid assay and our clonogenicity assay, respectively. Therefore, we concluded that reducing the expression of these genes by a certain amount had a negative impact on tumor function, but not a significant one. Treating HT29 cells with CA inhibitor SLC-0111, however, showed a significant impact on a variety of tumor functions. The ability of HT29 cells regarding clonogenic survival on a monolayer level was impaired severely as both number and size of the colonies were reduced by over 90%. Similar findings were also reported by Parks et al., who showed that disruption of CAIX severely reduced clonogenic proliferation in LS174T cells [13]. SLC-0111 treatment also resulted in a significant reduction in spheroid forming in our single-cell spheroid assay. The assay was designed in order to simulate metastasis at a single-cell stage, shortly after a cell implanted itself into the new tissue. It is important to note that HT29 cells, in which both CAs were knocked down, also showed significant reduction in ability of spheroid forming. This indicates that CIX and CAXII may be especially important for cancer cells in the very early stages of cancer development, in which cell–cell-interaction and surrounding soft tissue are limited or not yet present and decreased protein expression or protein inhibition of CAIX/CAXII hinders the cancer cell's survival in that early stage. CRC is a special type of cancer regarding metastasis as the latency between the diagnosis of the primary tumor and the appearance of metastasis is relatively small [74]. Therefore, it is very important to further understand the mechanisms behind the metastatic process and to continue to find marker and target proteins for metastasis in order to prevent the cancer from spreading out of its organ of origin [75].

One of the largest challenges in the medicamentous approach of cancer treatment is the resistance of tumors towards chemotherapeutic agents. Cancer cells are capable of several mechanisms that can reduce chemotherapeutic efficacy. Alteration in extracellular pH to more acidic values is commonly observed in chemotherapy-resistant tumors and leads to reduced intracellular accumulation of the chemotherapeutic agent due to a shift in the concentration gradient between the acid and base form of the drug [5,76]. The literature has already reported on pH-dependent uptake of irinotecan, a drug commonly used in medicamentous therapy of CRC [77]. Tumor-associated carbonic anhydrase IX and XII play a major role in establishing this pH disbalance by forming a functional complex, a so called "Metabolon", which facilitates extrusion of acidic metabolites created in the cellular metabolism to the extracellular space [5,15,16,78,79]. The literature has already shown that disruption of the CAIX pathway leads to a reversal of the described pH-altering effect [80]. P-glycoprotein-mediated extrusion of chemotherapeutic agents has long been established to be a mechanism by which tumor cells are able to avoid otherwise lethal concentrations of chemotherapeutic therapy. Interestingly, CAXII has been reported to be associated with p-glycoprotein (Pgp) in several tumors, and it was shown that chemotherapeutic efficacy can be increased by means of CAXII inhibition [30,81–83]. Although HT29 cells show high expression of CAXII, expression of Pgp in this cell line is low [81]. As such, this manner of chemotherapeutic resistance plays a minor role at best in our HT29 model but may play a major role in other tumors expressing CAXII, on which the literature has already reported before [84,85]. Aside from the aforementioned mechanisms, there are many more ways described in the literature by which CRC cells may be able to become resistant against basic chemotherapeutic drugs 5-FU, irinotecan and oxaliplatin [86]. Our findings in the co-treatment trials suggest that a combined treatment of SLC-0111 plus the conventional chemotherapeutic agents could be beneficial in treating CRC if the tumor presents itself with a high expression status of the tumor-associated carbonic anhydrases. The increase in cytotoxicity was especially strong in the HT29 cells treated with irinotecan and oxaliplatin as, without SLC-0111, both irinotecan and oxaliplatin only showed little effect on cell viability in concentrations that might be administered to humans. These findings proved that the combination of the normally used chemotherapeutic agents and SLC-0111 is superior to chemotherapy alone in our model. Additive effects were observed for 5-FU. Therefore, treatment with the commonly used FOLFOX or FOLFIRI regimens in combination with SLC-0111 is entirely viable and showed a significant decrease in spheroid viability in our combination treatment trials and might, therefore, be an option for chemotherapeutic treatment of CRC if expression of CAIX and CAXII is high in the patient's tumor.

4. Materials and Methods

4.1. Transcriptomic Analyses of CRC Samples

In order to assess transcriptomic data of CA9 and CA12, as well as HIF1α in human tumor samples we used the platform Cancertool [60]. Cancertool is an online bioinformatic platform performing expression-, correlation- and gene-enrichment analyses based on seven different colorectal cancer transcriptomic datasets: Colonomics, GSE44076; Jorissen et al., GSE14333; Kemper et al., GSE33113; Laibe et al., GSE37892; Marisa et al., GSE39582; Roepman et al., GSE42284; TCGA, cBioPortal. Detailed information about the data sources and methods used by Cancertool can be found on the website: http://genomics.cicbiogune.es/CANCERTOOL/index.html (accessed on 13 March 2023). In addition, using these datasets, the patient's tumor data were classified and separated into the CMS groups using the CMScaller [59], in order to gain differentiated analysis on the expression patterns of both CA9 and CA12 in the different CMS groups. The classification was performed accordingly to the instructions provided on https://github.com/peterawe/CMScaller (accessed on

13 March 2023). Boxplots were generated with BoxPlotR (http://shiny.chemgrid.org/boxplotr/, (accessed on 13 March 2023)).

4.2. Preclinical Tumor Models

The preclinical CRC model used in this study was established in the context of a separate study and comprises luciferase-expressing variants of the cell lines SW48, LOVO, SW1463, LS1034, HT29, LS174T, HCT116 and SW480, derived nude mouse xenograft tumors and spheroids representing the four molecular subtypes according to the CMS classification system. Detailed characterization of the model including generation of the luciferase-expressing variants of cell lines, transcriptomic data and CMS classification using the CMScaller [59] will be published in a separate paper. CMS classification was in accordance to the wild-type CRC cell lines, which were previously characterized by Berg et al. [56]. The new generated luciferase-expressing cell lines were re-authenticated at the DSMZ-German Collection of Microorganisms and Cell Cultures GmbH (Braunschweig, Germany) in 2020/2021.

The monolayer cells were cultivated in RPMI1640 medium (Sigma-Aldrich, St. Louis, MO, USA), to which 10% fetal bovine serum (Biowest, Nuaillé, France) and 1% penicillin-streptomycin (Sigma-Aldrich) was added. For generation of single tumor spheroids, tumor cells resuspended in culture medium were seeded into 96-well plates, which were coated before with 0.7% agarose (SeaKem® GTG™ Agarose, Lonza, Basel, Switzerland). Cell lines HT29-Luc, DLD-1-Luc, LS174T-Luc, LS1034-Luc and SW1463-Luc were able to form compact spheroids within 2 days. For the cell lines LOVO -Luc, SW48-Luc, COLO205-Luc, HCT116-Luc and SW480-Luc, culture medium was supplemented with 10 µg/mL collagen I (Ibidi GmbH, Gräfelfing, Germany) to support spheroid formation. Completely compacted spheroids were formed within 7 days.

Stable knockdowns of CA9 and CA12 in HT29 cells were generated by lentiviral transduction using ready-to-use prepared virus particles from Sigma-Aldrich. The single knockdowns KD 9.2", "KD 9.5", "KD 12.3", "KD 12.4"were obtained with puromycin selection. As a control for the puromycin-based vectors, HT29 cells were transduced with an empty puromycin vector ("Mock"). The vectors of the more potent knockdown of CA9 (KD 9.2) and CA12 (KD 12.4) were chosen to be combined to double to generate a combined CA9/CA12 knockdown. While the aforementioned KD 9.2 vector was used, for the KD 12.4 knockdown a GFP labelled vector was used. FACS sorting was carried out at the local facility (Core facility, Center for Basic Medical Research (ZMG) of the Medical Faculty, Martin Luther University). As a control for the GFP-generated double-knockdown, HT29 mock cells were also transduced with the KD 12.4/GFP-vector ("Mock + KD 12.4"). Therefore, in both the "KD9.2 + KD12.4", as well as the "Mock + KD12.4" samples, CA12 was knocked down using the same GFP-vector. For the Mock + SLC-0111 group, the aforementioned, puromycin-selected "Mock" cells were used.

4.3. Immunohistochemistry of Xenograft Tumors and Spheroids

The xenograft tumors samples used in this study for immunohistochemical analyses were prepared in the context of a separate study during the generation of our preclinical CRC model (see above). In order to be able to analyze intratumoral hypoxia, the mice were treated with Hypoxyprobe™ (Hypoxyprobe, Inc., Burlington, MA, USA) 90 min prior to tumor extraction. After extraction, the tumors/organs were fixated in formalin, before they were processed using the Microm STP120 and then embedded in paraffin using the Microm EC350-1 (both Thermo Fischer Scientific, Waltham, MA, USA). For immunohistochemical analyses, the tumors were cut into 4 µm wide samples. Following deparaffinization and rehydration, the samples were first subjected to peroxidase-blocking and then to protein-blocking. The used reagents were, if not stated otherwise, obtained from Dako (Agilent Technologies, Santa Clara, CA, USA). For the latter a 3% BSA (Carl Roth, Karlsruhe, Germany) in PBS solution was used. The samples were treated with anti-CAIX, anti-CAXII (AB184006 1:1000 and AB195233 1:100; both Abcam, Cambridge,

UK) or anti-PAb2627AP (1:200, Hypoxyprobe, Inc.) primary antibodies, diluted in a 1% BSA in PBS solution, for three hours. They were then treated with biotinylated secondary antibodies (SC-2491, diluted 1:800 in PBS; Santa Cruz Biotechnology, Dallas, TX, USA) for one hour. After treatment with a Streptavidin-HRP-conjugate for 30 min, the samples were exposed to the chromogen 3,3′-diaminobenzidine (DAB) for five minutes, before all samples were stained with hematoxylin for one minute and then dehydrated. Images of the samples were taken on an Axiolab microscope equipped with Axiocam 503 color (Zeiss Group, Oberkochen, Germany).

4.4. Immunocytochemistry on Monolayer Culture

The used cells were seeded onto one well of a chamber slide (Nunc™ Lab-Tek™ II Chamber Slide™ System, Thermo Fischer Scientific). After 24 h of incubation the medium was removed and the plates were rinsed twice with PBS for two minutes each. The cells were then fixated using methanol (Sigma-Aldrich) for 15 min. The methanol was then washed away using PBS three times for 10 min, before the cells were treated with the same 3% BSA in PBS to achieve protein blocking. After that the cells were incubated with primary antibody (CAIX, AB184006 1:000 and CAXII, AB195233 1:175; both Abcam) diluted in a 1% BSA in PBS solution for two hours at room temperature. Following the incubation with the primary antibody, the plates were washed with a 1% BSA in PBS trice for 10 min each to remove excess antibodies. In the following step the plates were incubated with the secondary antibody (A-11036, 1:800, Thermo Fischer Scientific) diluted in PBS at room temperature in the dark for one hour. Excess antibody was then removed by washing the plates with PBS trice for 10 min each. During the washing steps, the plates were kept in the dark. In the next step the cells were stained with 4′,6-diamidino-2-phenylindole (DAPI, Sigma-Aldrich, stock solution 1mg/mL), which was diluted 1:1000 in PBS before, for three minutes. Following the staining process, the cells were washed twice with PBS for 5 min each in the dark. The plates were coated with mounting medium (Fluoromount-G™ Mounting Medium, Thermo Fischer Scientific) and then covered with a coverslip. Images of the samples were taken using the BioRevo BZ-9000 (Keyence Deutschland GmbH, Neu-Isenburg, Germany).

4.5. Protein isolation/Western Blot Analysis

Protein lysates of monolayer cells and spheroids were prepared using RIPA buffer. The RIPA buffer further contained 100mM phenylmethylsulfonyl fluoride, 10mM dithiothreitol (Bio-Rad Laboratories, Hercules, CA, USA), Protease-Inhibitor-Cocktail and Benzonase (both Sigma-Aldrich). After incubation on ice for 45 min, the lysates were then centrifuged at 15,000× g for 15 min at 4 °C. The protein concentration of the probes were then measured using the Bio Photometer D30 (Eppendorf SE, Hamburg, Germany) via Bradford method. The used Bradford reagent was obtained from Bio-Rad. The probes were adjusted to get equal concentrations.

The used Western blot equipment was, if not stated otherwise, obtained from Bio-Rad. Protein samples were separated using SDS-PAGE and subsequently transferred on a nitrocellulose blotting membrane for overnight blotting. Blocking of the membranes was accomplished using a 5% powdered milk (Carl Roth) in phosphate buffered saline (PBS) with 0.1% Tween (Sigma-Aldrich) solution, which was also used for dilution of the primary antibodies. The membranes were then treated with either anti-CAIX (M75 1:2000 (Bioscience Slovakia, Bratislava, Slovak Republic), anti-CAXII (AB195233 1:5000, Abcam) or anti-GAPDH (14C10, 1:5000, Cell Signaling Technology, Danvers, MA, USA) primary antibodies for two hours, before treatment with a horseradish-peroxidase (HRP) conjugated secondary antibody (SC-2357 or SC-516102; Santa Cruz Biotechnology) took place for one hour. The secondary antibodies were diluted 1:5000 in a 0.1% Tween in PBS solution, the same solution, which was also used to wash the membranes between and after antibody exposure. After treatment with a chemiluminescence detection reagent (ECL™ Western

Blotting Detection Reagent, Sigma-Aldrich), images of the membranes were taken using the ImageQuant LAS 4000 (Cytiva Europe GmbH, Freiburg, Germany).

4.6. Clonogenicity Assay

Cell suspensions were first diluted to 500 cells in 5 mL of RPMI medium. The cells were then seeded onto six-well-plates. Ten days after seeding, the medium was removed and the plates were rinsed with PBS. The colonies were then fixated for 10 min using ethanol and afterwards stained with a 1% crystal violet (Sigma-Aldrich) in PBS solution for 10 min as well. The number and size of the colonies were analyzed using ImageJ (8-bit image; subtract background: 100 pixel, threshold: 212, analyze 3 particles: 40-3000, display results, summarize, add to manager, exclude on edges).

4.7. Single-Cell Spheroid Assay

Cell suspension were diluted to 100 cells in 20 mL of RPMI-medium and then seeded onto 96-well-plates (200 µL per well), which were previously coated with an 0.7% agarose (SeaKem GTG Agarose, Lonza Group, Basel, Switzerland) in PBS solution, in order to inhibit adhesion and therefore guarantee spheroid forming starting at a single-cell level. The formed spheroids were counted after 15 days.

4.8. Spheroid Cytotoxicity Assay

The cells were seeded onto a 0.7% agarose-coated, 96-well-plates and were incubated for two or seven days, depending on the model, to allow the undisturbed forming of compact spheroids, before treatment with chemotherapeutic agents/SLC-0111 was carried out. For the SLC-0111 response trial, the different spheroid models were exposed to the carbonic anhydrase inhibitor SLC-0111 (SLC-0111, Hycultec GmbH, Beutelsbach, Germany) at increasing concentrations (0.01, 0.03, 0.1, 0.3, 1.0, 3.0, 10.0, 30.0, 100.0 µM). For the co-treatment trials, spheroids were treated with SLC-0111 and one chemotherapeutic agent. Both drugs were diluted in 50 µL RPMI, and, while the chemotherapeutic agent was used in increasingly high concentrations (0.01, 0.03, 0.1, 0.3, 1.0, 3.0, 10.0, 30.0, 100.0 µM for oxaliplatin and irinotecan; 0.1, 0.3, 1.0, 3.0, 10.0, 30.0, 100.0, 300.0, 1000.0 µM for 5-fluorouracil (5-FU)), SLC-0111 was used in a concentration of 20 µM for all wells. Spheroids only treated with SLC-0111 served as a control to examine the sole efficacy of the CA inhibitor. The control group was treated with RPMI, in which DMSO was diluted, so that it could serve as a negative control for the effect of DMSO on cell viability. For the combination treatment with single doses, chemotherapeutic agents were used in a concentration of 1 µM for oxaliplatin and irinotecan or 10 µM for 5-FU. Spheroids were either treated with a monotherapy of one chemotherapeutic agent or a dual-therapy consisting of both agents (FOLFOX/FOLFIRI regiment). Each of the treatments served as an individual control group for the spheroids treated with the monotherapy/dual-therapy plus 20 µM of SLC-0111.

To measure and quantify the viability of the treated spheroids, 20 µL of D-luciferin (PerkinElmer, Waltham, MA, USA), diluted 1:1000 in RPMI-medium, was added to all wells 7 days after start of treatments. After 15 min of incubation in the dark, measurements were carried out on a Microplate Reader Tecan Spark (Tecan Group, Männedorf, Schwitzerland) using luminescence measurement with an exposure time of 1000 ms/well. The mean value of the eight values of each column was calculated and all values are given as % of untreated control. At least 3 independent experiments were performed and summarized as mean values with standard deviation.

4.9. Statistical Analysis

Statistical analysis was carried out using GraphPad Prism 8. The used *p*-value format for all analyses was NEJM. For the clonogenicity assay, the absolute values of colony numbers and colony size were examined for statistical significance by means of ordinary one-way ANOVA (assumed Gaussian distribution; assumed equal SDs) using the mock group as control for follow-up testing for statistical significance. Ordinary one-way ANOVA (assumed Gaussian distribution; assumed equal SDs) was also used for the single-cell spheroid assay, comparing the absolute values of each treatment to the control group mock. For the SLC-0111 response trial, an extrapolated curve was calculated via equation: [inhibitor] vs. response - variable slope to calculate the different IC_{50} values of each trial. The mean values and the SD of all measured IC_{50} values were then calculated for each spheroid model. For the co-treatment trials, IC_{50} values in both the control and the inhibitor group were calculated via equation: [inhibitor] vs. normalized response - variable slope. The IC_{50} values of both groups (Gaussian distribution assumed) were then compared using a Welch test or an unpaired *t*-test, depending if the variances were significantly different or not. Additionally, the relative viability of the spheroids in the control and inhibitor groups were compared at two different chemotherapeutic concentrations. The concentration was chosen closest to the peak plasma concentration of the respective agent in vivo (10 μM 5-FU; 1 μM irinotecan/oxaliplatin) and the second concentration was chosen at 10% of the first concentration (1 μM 5-FU; 0.1 μM irinotecan/oxaliplatin). The relative spheroid viability was compared using a Welch test or an unpaired *t*-test, as described above. For the combination treatment, the control groups and their respective inhibitor groups were compared to each other using traditional one-way ANOVA (assumed Gaussian distribution; assumed equal SDs).

5. Conclusions

This section is not mandatory but can be added to the manuscript if the discussion is unusually long or complex.

Supplementary Materials: The following supporting information can be downloaded at: https://www.mdpi.com/article/10.3390/ijms24065797/s1.

Author Contributions: Conceptualization, T.M.; investigation, A.R., S.B., M.S.S., J.L. and T.M.; formal analysis, A.R. and M.S.S.; writing—original draft preparation, A.R. and T.M.; writing—review and editing, L.P.M. and T.M. All authors have read and agreed to the published version of the manuscript.

Funding: This research received no external funding.

Institutional Review Board Statement: The xenograft tumor materials investigated in this study were generated in the context of an animal study, which was approved by the Laboratory Animal Care Committee of Sachsen-Anhalt, Germany (approval code: 203.h-42502-2-1250 MLU).

Informed Consent Statement: Not applicable.

Data Availability Statement: The datasets used and/or analyzed in the current study are available from the corresponding author upon reasonable request.

Acknowledgments: We thank Franziska Reipsch for her excellent technical assistance.

Conflicts of Interest: The authors declare no conflict of interest.

References

1. Bosco, M.C.; D'Orazi, G.; Del Bufalo, D. Targeting hypoxia in tumor: A new promising therapeutic strategy. *J. Exp. Clin. Cancer Res.* **2020**, *39*, 8. [CrossRef] [PubMed]
2. Chiche, J.; Brahimi-Horn, M.C.; Pouyssegur, J. Tumour hypoxia induces a metabolic shift causing acidosis: A common feature in cancer. *J. Cell. Mol. Med.* **2010**, *14*, 771–794. [CrossRef]
3. Nakazawa, M.S.; Keith, B.; Simon, M.C. Oxygen availability and metabolic adaptations. *Nat. Rev. Cancer* **2016**, *16*, 663–673. [CrossRef] [PubMed]
4. Swietach, P. What is pH regulation, and why do cancer cells need it? *Cancer Metastasis Rev.* **2019**, *38*, 5–15. [CrossRef] [PubMed]
5. Corbet, C.; Feron, O. Tumour acidosis: From the passenger to the driver's seat. *Nat. Rev. Cancer* **2017**, *17*, 577–593. [CrossRef] [PubMed]
6. Lardner, A. The effects of extracellular pH on immune function. *J. Leukoc. Biol.* **2001**, *69*, 522–530. [CrossRef]
7. Fang, J.S.; Gillies, R.D.; Gatenby, R.A. Adaptation to hypoxia and acidosis in carcinogenesis and tumor progression. *Semin. Cancer Biol.* **2008**, *18*, 330–337. [CrossRef]
8. Glunde, K.; Guggino, S.E.; Solaiyappan, M.; Pathak, A.P.; Ichikawa, Y.; Bhujwalla, Z.M. Extracellular acidification alters lysosomal trafficking in human breast cancer cells. *Neoplasia* **2003**, *5*, 533–545. [CrossRef]

9. Peppicelli, S.; Bianchini, F.; Calorini, L. Extracellular acidity, a "reappreciated" trait of tumor environment driving malignancy: Perspectives in diagnosis and therapy. *Cancer Metastasis Rev.* **2014**, *33*, 823–832. [CrossRef]
10. Barar, J.; Omidi, Y. Dysregulated pH in Tumor Microenvironment Checkmates Cancer Therapy. *Bioimpacts* **2013**, *3*, 149–162. [CrossRef]
11. Taylor, S.; Spugnini, E.P.; Assaraf, Y.G.; Azzarito, T.; Rauch, C.; Fais, S. Microenvironment acidity as a major determinant of tumor chemoresistance: Proton pump inhibitors (PPIs) as a novel therapeutic approach. *Drug Resist. Updat.* **2015**, *23*, 69–78. [CrossRef] [PubMed]
12. Wojtkowiak, J.W.; Verduzco, D.; Schramm, K.J.; Gillies, R.J. Drug resistance and cellular adaptation to tumor acidic pH microenvironment. *Mol. Pharm.* **2011**, *8*, 2032–2038. [CrossRef]
13. Parks, S.K.; Cormerais, Y.; Durivault, J.; Pouyssegur, J. Genetic disruption of the pHi-regulating proteins Na^+/H^+ exchanger 1 (SLC9A1) and carbonic anhydrase 9 severely reduces growth of colon cancer cells. *Oncotarget* **2017**, *8*, 10225–10237. [CrossRef] [PubMed]
14. Asgharzadeh, M.R.; Barar, J.; Pourseif, M.M.; Eskandani, M.; Jafari Niya, M.; Mashayekhi, M.R.; Omidi, Y. Molecular machineries of pH dysregulation in tumor microenvironment: Potential targets for cancer therapy. *Bioimpacts* **2017**, *7*, 115–133. [CrossRef] [PubMed]
15. Becker, H.M.; Deitmer, J.W. Transport Metabolons and Acid/Base Balance in Tumor Cells. *Cancers* **2020**, *12*, 899. [CrossRef]
16. Morgan, P.E.; Pastorekova, S.; Stuart-Tilley, A.K.; Alper, S.L.; Casey, J.R. Interactions of transmembrane carbonic anhydrase, CAIX, with bicarbonate transporters. *Am. J. Physiol. Cell Physiol.* **2007**, *293*, C738–C748. [CrossRef]
17. Kroemer, G.; Pouyssegur, J. Tumor cell metabolism: Cancer's Achilles' heel. *Cancer Cell* **2008**, *13*, 472–482. [CrossRef]
18. Mekhail, K.; Gunaratnam, L.; Bonicalzi, M.E.; Lee, S. HIF activation by pH-dependent nucleolar sequestration of VHL. *Nat. Cell Biol.* **2004**, *6*, 642–647. [CrossRef]
19. Pastorek, J.; Pastorekova, S.; Callebaut, I.; Mornon, J.P.; Zelnik, V.; Opavsky, R.; Zat'ovicova, M.; Liao, S.; Portetelle, D.; Stanbridge, E.J.; et al. Cloning and characterization of MN, a human tumor-associated protein with a domain homologous to carbonic anhydrase and a putative helix-loop-helix DNA binding segment. *Oncogene* **1994**, *9*, 2877–2888.
20. Pastorekova, S.; Gillies, R.J. The role of carbonic anhydrase IX in cancer development: Links to hypoxia, acidosis, and beyond. *Cancer Metastasis Rev.* **2019**, *38*, 65–77. [CrossRef] [PubMed]
21. Tureci, O.; Sahin, U.; Vollmar, E.; Siemer, S.; Gottert, E.; Seitz, G.; Parkkila, A.K.; Shah, G.N.; Grubb, J.H.; Pfreundschuh, M.; et al. Human carbonic anhydrase XII: cDNA cloning, expression, and chromosomal localization of a carbonic anhydrase gene that is overexpressed in some renal cell cancers. *Proc. Natl. Acad. Sci. USA* **1998**, *95*, 7608–7613. [CrossRef] [PubMed]
22. Waheed, A.; Sly, W.S. Carbonic anhydrase XII functions in health and disease. *Gene* **2017**, *623*, 33–40. [CrossRef] [PubMed]
23. Saarnio, J.; Parkkila, S.; Parkkila, A.K.; Haukipuro, K.; Pastorekova, S.; Pastorek, J.; Kairaluoma, M.I.; Karttunen, T.J. Immunohistochemical study of colorectal tumors for expression of a novel transmembrane carbonic anhydrase, MN/CA IX, with potential value as a marker of cell proliferation. *Am. J. Pathol.* **1998**, *153*, 279–285. [CrossRef] [PubMed]
24. Kivela, A.J.; Parkkila, S.; Saarnio, J.; Karttunen, T.J.; Kivela, J.; Parkkila, A.K.; Bartosova, M.; Mucha, V.; Novak, M.; Waheed, A.; et al. Expression of von Hippel-Lindau tumor suppressor and tumor-associated carbonic anhydrases IX and XII in normal and neoplastic colorectal mucosa. *World J. Gastroenterol.* **2005**, *11*, 2616–2625. [CrossRef]
25. Wykoff, C.C.; Beasley, N.; Watson, P.H.; Campo, L.; Chia, S.K.; English, R.; Pastorek, J.; Sly, W.S.; Ratcliffe, P.; Harris, A.L. Expression of the hypoxia-inducible and tumor-associated carbonic anhydrases in ductal carcinoma in situ of the breast. *Am. J. Pathol.* **2001**, *158*, 1011–1019. [CrossRef]
26. Beasley, N.J.; Wykoff, C.C.; Watson, P.H.; Leek, R.; Turley, H.; Gatter, K.; Pastorek, J.; Cox, G.J.; Ratcliffe, P.; Harris, A.L. Carbonic anhydrase IX, an endogenous hypoxia marker, expression in head and neck squamous cell carcinoma and its relationship to hypoxia, necrosis, and microvessel density. *Cancer Res.* **2001**, *61*, 5262–5267. [PubMed]
27. Liao, S.Y.; Aurelio, O.N.; Jan, K.; Zavada, J.; Stanbridge, E.J. Identification of the MN/CA9 protein as a reliable diagnostic biomarker of clear cell carcinoma of the kidney. *Cancer Res.* **1997**, *57*, 2827–2831. [CrossRef] [PubMed]
28. Wang, J.; Fang, R.; Wang, L.; Chen, G.; Wang, H.; Wang, Z.; Zhao, D.; Pavlov, V.N.; Kabirov, I.; Wang, Z.; et al. Identification of Carbonic Anhydrase IX as a Novel Target for Endoscopic Molecular Imaging of Human Bladder Cancer. *Cell. Physiol. Biochem.* **2018**, *47*, 1565–1577. [CrossRef]
29. Mboge, M.Y.; Mahon, B.P.; McKenna, R.; Frost, S.C. Carbonic Anhydrases: Role in pH Control and Cancer. *Metabolites* **2018**, *8*, 19. [CrossRef] [PubMed]
30. Kopecka, J.; Campia, I.; Jacobs, A.; Frei, A.P.; Ghigo, D.; Wollscheid, B.; Riganti, C. Carbonic anhydrase XII is a new therapeutic target to overcome chemoresistance in cancer cells. *Oncotarget* **2015**, *6*, 6776–6793. [CrossRef]
31. Koukourakis, M.I.; Giatromanolaki, A.; Sivridis, E.; Simopoulos, K.; Pastorek, J.; Wykoff, C.C.; Gatter, K.C.; Harris, A.L. Hypoxia-regulated carbonic anhydrase-9 (CA9) relates to poor vascularization and resistance of squamous cell head and neck cancer to chemoradiotherapy. *Clin. Cancer Res.* **2001**, *7*, 3399–3403. [PubMed]
32. Yang, J.S.; Lin, C.W.; Chuang, C.Y.; Su, S.C.; Lin, S.H.; Yang, S.F. Carbonic anhydrase IX overexpression regulates the migration and progression in oral squamous cell carcinoma. *Tumor Biol.* **2015**, *36*, 9517–9524. [CrossRef] [PubMed]
33. Nasu, K.; Yamaguchi, K.; Takanashi, T.; Tamai, K.; Sato, I.; Ine, S.; Sasaki, O.; Satoh, K.; Tanaka, N.; Tanaka, Y.; et al. Crucial role of carbonic anhydrase IX in tumorigenicity of xenotransplanted adult T-cell leukemia-derived cells. *Cancer Sci.* **2017**, *108*, 435–443. [CrossRef]

34. Zavada, J.; Zavadova, Z.; Pastorek, J.; Biesova, Z.; Jezek, J.; Velek, J. Human tumour-associated cell adhesion protein MN/CA IX: Identification of M75 epitope and of the region mediating cell adhesion. *Br. J. Cancer* **2000**, *82*, 1808–1813. [CrossRef] [PubMed]
35. Svastova, E.; Zilka, N.; Zat'ovicova, M.; Gibadulinova, A.; Ciampor, F.; Pastorek, J.; Pastorekova, S. Carbonic anhydrase IX reduces E-cadherin-mediated adhesion of MDCK cells via interaction with beta-catenin. *Exp. Cell Res.* **2003**, *290*, 332–345. [CrossRef]
36. Robertson, N.; Potter, C.; Harris, A.L. Role of carbonic anhydrase IX in human tumor cell growth, survival, and invasion. *Cancer Res.* **2004**, *64*, 6160–6165. [CrossRef] [PubMed]
37. Chiche, J.; Ilc, K.; Laferriere, J.; Trottier, E.; Dayan, F.; Mazure, N.M.; Brahimi-Horn, M.C.; Pouyssegur, J. Hypoxia-inducible carbonic anhydrase IX and XII promote tumor cell growth by counteracting acidosis through the regulation of the intracellular pH. *Cancer Res.* **2009**, *69*, 358–368. [CrossRef] [PubMed]
38. Korkeila, E.; Talvinen, K.; Jaakkola, P.M.; Minn, H.; Syrjanen, K.; Sundstrom, J.; Pyrhonen, S. Expression of carbonic anhydrase IX suggests poor outcome in rectal cancer. *Br. J. Cancer* **2009**, *100*, 874–880. [CrossRef]
39. Kim, J.I.; Choi, K.U.; Lee, I.S.; Choi, Y.J.; Kim, W.T.; Shin, D.H.; Kim, K.; Lee, J.H.; Kim, J.Y.; Sol, M.Y. Expression of hypoxic markers and their prognostic significance in soft tissue sarcoma. *Oncol. Lett.* **2015**, *9*, 1699–1706. [CrossRef]
40. Furjelova, M.; Kovalska, M.; Jurkova, K.; Horacek, J.; Carbolova, T.; Adamkov, M. Carbonic anhydrase IX: A promising diagnostic and prognostic biomarker in breast carcinoma. *Acta Histochem.* **2014**, *116*, 89–93. [CrossRef]
41. Chen, Y.; Li, X.; Wu, S.; Xu, G.; Zhou, Y.; Gong, L.; Li, Z.; Yang, D. Expression of HIF-1alpha and CAIX in nasopharyngeal carcinoma and their correlation with patients' prognosis. *Med. Oncol.* **2014**, *31*, 304. [CrossRef] [PubMed]
42. Yoo, C.W.; Nam, B.H.; Kim, J.Y.; Shin, H.J.; Lim, H.; Lee, S.; Lee, S.K.; Lim, M.C.; Song, Y.J. Carbonic anhydrase XII expression is associated with histologic grade of cervical cancer and superior radiotherapy outcome. *Radiat. Oncol.* **2010**, *5*, 101. [CrossRef] [PubMed]
43. Du, Y.; Xin, Z.; Liu, T.; Xu, P.; Mao, F.; Yao, J. Overexpressed CA12 has prognostic value in pancreatic cancer and promotes tumor cell apoptosis via NF-kappaB signaling. *J. Cancer Res. Clin. Oncol.* **2021**, *147*, 1557–1564. [CrossRef] [PubMed]
44. Chia, S.K.; Wykoff, C.C.; Watson, P.H.; Han, C.; Leek, R.D.; Pastorek, J.; Gatter, K.C.; Ratcliffe, P.; Harris, A.L. Prognostic significance of a novel hypoxia-regulated marker, carbonic anhydrase IX, in invasive breast carcinoma. *J. Clin. Oncol.* **2001**, *19*, 3660–3668. [CrossRef]
45. Giatromanolaki, A.; Koukourakis, M.I.; Sivridis, E.; Pastorek, J.; Wykoff, C.C.; Gatter, K.C.; Harris, A.L. Expression of hypoxia-inducible carbonic anhydrase-9 relates to angiogenic pathways and independently to poor outcome in non-small cell lung cancer. *Cancer Res.* **2001**, *61*, 7992–7998.
46. Divgi, C.R.; Bander, N.H.; Scott, A.M.; O'Donoghue, J.A.; Sgouros, G.; Welt, S.; Finn, R.D.; Morrissey, F.; Capitelli, P.; Williams, J.M.; et al. Phase I/II radioimmunotherapy trial with iodine-131-labeled monoclonal antibody G250 in metastatic renal cell carcinoma. *Clin. Cancer Res.* **1998**, *4*, 2729–2739.
47. Angeli, A.; Carta, F.; Nocentini, A.; Winum, J.Y.; Zalubovskis, R.; Akdemir, A.; Onnis, V.; Eldehna, W.M.; Capasso, C.; Simone, G.; et al. Carbonic Anhydrase Inhibitors Targeting Metabolism and Tumor Microenvironment. *Metabolites* **2020**, *10*, 412. [CrossRef]
48. McDonald, P.C.; Chia, S.; Bedard, P.L.; Chu, Q.; Lyle, M.; Tang, L.; Singh, M.; Zhang, Z.; Supuran, C.T.; Renouf, D.J.; et al. A Phase 1 Study of SLC-0111, a Novel Inhibitor of Carbonic Anhydrase IX, in Patients With Advanced Solid Tumors. *Am. J. Clin. Oncol.* **2020**, *43*, 484–490. [CrossRef]
49. Guinney, J.; Dienstmann, R.; Wang, X.; de Reynies, A.; Schlicker, A.; Soneson, C.; Marisa, L.; Roepman, P.; Nyamundanda, G.; Angelino, P.; et al. The consensus molecular subtypes of colorectal cancer. *Nat. Med.* **2015**, *21*, 1350–1356. [CrossRef]
50. Okita, A.; Takahashi, S.; Ouchi, K.; Inoue, M.; Watanabe, M.; Endo, M.; Honda, H.; Yamada, Y.; Ishioka, C. Consensus molecular subtypes classification of colorectal cancer as a predictive factor for chemotherapeutic efficacy against metastatic colorectal cancer. *Oncotarget* **2018**, *9*, 18698–18711. [CrossRef]
51. Lee, M.S.; Menter, D.G.; Kopetz, S. Right Versus Left Colon Cancer Biology: Integrating the Consensus Molecular Subtypes. *J. Natl. Compr. Canc. Netw.* **2017**, *15*, 411–419. [CrossRef] [PubMed]
52. Li, X.; Ling, A.; Kellgren, T.G.; Lundholm, M.; Lofgren-Burstrom, A.; Zingmark, C.; Rutegard, M.; Ljuslinder, I.; Palmqvist, R.; Edin, S. A Detailed Flow Cytometric Analysis of Immune Activity Profiles in Molecular Subtypes of Colorectal Cancer. *Cancers* **2020**, *12*, 3440. [CrossRef] [PubMed]
53. Trinh, A.; Ladrach, C.; Dawson, H.E.; Ten Hoorn, S.; Kuppen, P.J.K.; Reimers, M.S.; Koopman, M.; Punt, C.J.A.; Lugli, A.; Vermeulen, L.; et al. Tumour budding is associated with the mesenchymal colon cancer subtype and RAS/RAF mutations: A study of 1320 colorectal cancers with Consensus Molecular Subgroup (CMS) data. *Br. J. Cancer* **2018**, *119*, 1244–1251. [CrossRef]
54. Lenz, H.J.; Argiles, G.; Yoshino, T.; Tejpar, S.; Ciardiello, F.; Braunger, J.; Salnikov, A.V.; Gabrielyan, O.; Schmid, R.; Hofler, J.; et al. Association of Consensus Molecular Subtypes and Molecular Markers with Clinical Outcomes in Patients with Metastatic Colorectal Cancer: Biomarker Analyses from LUME-Colon 1. *Clin. Color. Cancer* **2021**, *20*, 84–95.e8. [CrossRef] [PubMed]
55. Stintzing, S.; Wirapati, P.; Lenz, H.J.; Neureiter, D.; Fischer von Weikersthal, L.; Decker, T.; Kiani, A.; Kaiser, F.; Al-Batran, S.; Heintges, T.; et al. Consensus molecular subgroups (CMS) of colorectal cancer (CRC) and first-line efficacy of FOLFIRI plus cetuximab or bevacizumab in the FIRE3 (AIO KRK-0306) trial. *Ann. Oncol.* **2019**, *30*, 1796–1803. [CrossRef]
56. Berg, K.C.G.; Eide, P.W.; Eilertsen, I.A.; Johannessen, B.; Bruun, J.; Danielsen, S.A.; Bjornslett, M.; Meza-Zepeda, L.A.; Eknaes, M.; Lind, G.E.; et al. Multi-omics of 34 colorectal cancer cell lines—A resource for biomedical studies. *Mol. Cancer* **2017**, *16*, 116. [CrossRef] [PubMed]

57. Linnekamp, J.F.; Hooff, S.R.V.; Prasetyanti, P.R.; Kandimalla, R.; Buikhuisen, J.Y.; Fessler, E.; Ramesh, P.; Lee, K.; Bochove, G.G.W.; de Jong, J.H.; et al. Consensus molecular subtypes of colorectal cancer are recapitulated in in vitro and in vivo models. *Cell Death Differ.* **2018**, *25*, 616–633. [CrossRef]
58. Sveen, A.; Bruun, J.; Eide, P.W.; Eilertsen, I.A.; Ramirez, L.; Murumagi, A.; Arjama, M.; Danielsen, S.A.; Kryeziu, K.; Elez, E.; et al. Colorectal Cancer Consensus Molecular Subtypes Translated to Preclinical Models Uncover Potentially Targetable Cancer Cell Dependencies. *Clin. Cancer Res.* **2018**, *24*, 794–806. [CrossRef]
59. Eide, P.W.; Bruun, J.; Lothe, R.A.; Sveen, A. CMScaller: An R package for consensus molecular subtyping of colorectal cancer pre-clinical models. *Sci. Rep.* **2017**, *7*, 16618. [CrossRef]
60. Cortazar, A.R.; Torrano, V.; Martin-Martin, N.; Caro-Maldonado, A.; Camacho, L.; Hermanova, I.; Guruceaga, E.; Lorenzo-Martin, L.F.; Caloto, R.; Gomis, R.R.; et al. CANCERTOOL: A Visualization and Representation Interface to Exploit Cancer Datasets. *Cancer Res.* **2018**, *78*, 6320–6328. [CrossRef]
61. Costa, E.C.; Moreira, A.F.; de Melo-Diogo, D.; Gaspar, V.M.; Carvalho, M.P.; Correia, I.J. 3D tumor spheroids: An overview on the tools and techniques used for their analysis. *Biotechnol. Adv.* **2016**, *34*, 1427–1441. [CrossRef] [PubMed]
62. Nunes, A.S.; Barros, A.S.; Costa, E.C.; Moreira, A.F.; Correia, I.J. 3D tumor spheroids as in vitro models to mimic in vivo human solid tumors resistance to therapeutic drugs. *Biotechnol. Bioeng.* **2019**, *116*, 206–226. [CrossRef] [PubMed]
63. Kivela, A.; Parkkila, S.; Saarnio, J.; Karttunen, T.J.; Kivela, J.; Parkkila, A.K.; Waheed, A.; Sly, W.S.; Grubb, J.H.; Shah, G.; et al. Expression of a novel transmembrane carbonic anhydrase isozyme XII in normal human gut and colorectal tumors. *Am. J. Pathol.* **2000**, *156*, 577–584. [CrossRef] [PubMed]
64. Miller, S.A.; Ghobashi, A.H.; O'Hagan, H.M. Consensus molecular subtyping of colorectal cancers is influenced by goblet cell content. *Cancer Genet.* **2021**, *254–255*, 34–39. [CrossRef]
65. Liao, S.Y.; Lerman, M.I.; Stanbridge, E.J. Expression of transmembrane carbonic anhydrases, CAIX and CAXII, in human development. *BMC Dev. Biol.* **2009**, *9*, 22. [CrossRef]
66. Kaluz, S.; Kaluzova, M.; Liao, S.Y.; Lerman, M.; Stanbridge, E.J. Transcriptional control of the tumor- and hypoxia-marker carbonic anhydrase 9: A one transcription factor (HIF-1) show? *Biochim. Biophys. Acta* **2009**, *1795*, 162–172. [CrossRef]
67. Franke, C.M.; Gu, V.W.; Grimm, B.G.; Cassady, V.C.; White, J.R.; Weigel, R.J.; Kulak, M.V. TFAP2C regulates carbonic anhydrase XII in human breast cancer. *Oncogene* **2020**, *39*, 1290–1301. [CrossRef]
68. Shao, Y.; Li, Y.; Zhang, J.; Liu, D.; Liu, F.; Zhao, Y.; Shen, T.; Li, F. Involvement of histone deacetylation in MORC2-mediated down-regulation of carbonic anhydrase IX. *Nucleic Acids Res.* **2010**, *38*, 2813–2824. [CrossRef]
69. Shen, T.; Xia, W.; Min, S.; Yang, Z.; Cheng, L.; Wang, W.; Zhan, Q.; Shao, F.; Zhang, X.; Wang, Z.; et al. A pair of long intergenic non-coding RNA LINC00887 variants act antagonistically to control Carbonic Anhydrase IX transcription upon hypoxia in tongue squamous carcinoma progression. *BMC Biol.* **2021**, *19*, 192. [CrossRef]
70. Vergara, D.; Ravaioli, S.; Fonzi, E.; Adamo, L.; Damato, M.; Bravaccini, S.; Pirini, F.; Gaballo, A.; Barbano, R.; Pasculli, B.; et al. Carbonic Anhydrase XII Expression Is Modulated during Epithelial Mesenchymal Transition and Regulated through Protein Kinase C Signaling. *Int. J. Mol. Sci.* **2020**, *21*, 715. [CrossRef]
71. Di Fiore, A.; Supuran, C.T.; Scaloni, A.; De Simone, G. Post-translational modifications in tumor-associated carbonic anhydrases. *Amino Acids* **2022**, *54*, 543–558. [CrossRef]
72. Lau, J.; Zhang, Z.; Jenni, S.; Kuo, H.T.; Liu, Z.; Vullo, D.; Supuran, C.T.; Lin, K.S.; Benard, F. PET Imaging of Carbonic Anhydrase IX Expression of HT-29 Tumor Xenograft Mice with (68)Ga-Labeled Benzenesulfonamides. *Mol. Pharm.* **2016**, *13*, 1137–1146. [CrossRef] [PubMed]
73. Zhang, Z.; Lau, J.; Zhang, C.; Colpo, N.; Nocentini, A.; Supuran, C.T.; Benard, F.; Lin, K.S. Design, synthesis and evaluation of (18)F-labeled cationic carbonic anhydrase IX inhibitors for PET imaging. *J. Enzyme Inhib. Med. Chem.* **2017**, *32*, 722–730. [CrossRef] [PubMed]
74. Nguyen, D.X.; Bos, P.D.; Massague, J. Metastasis: From dissemination to organ-specific colonization. *Nat. Rev. Cancer* **2009**, *9*, 274–284. [CrossRef]
75. Stein, U.; Walther, W.; Arlt, F.; Schwabe, H.; Smith, J.; Fichtner, I.; Birchmeier, W.; Schlag, P.M. MACC1, a newly identified key regulator of HGF-MET signaling, predicts colon cancer metastasis. *Nat. Med.* **2009**, *15*, 59–67. [CrossRef] [PubMed]
76. Webb, B.A.; Chimenti, M.; Jacobson, M.P.; Barber, D.L. Dysregulated pH: A perfect storm for cancer progression. *Nat. Rev. Cancer* **2011**, *11*, 671–677. [CrossRef] [PubMed]
77. Kobayashi, K.; Bouscarel, B.; Matsuzaki, Y.; Ceryak, S.; Kudoh, S.; Fromm, H. pH-dependent uptake of irinotecan and its active metabolite, SN-38, by intestinal cells. *Int. J. Cancer* **1999**, *83*, 491–496. [CrossRef]
78. Deitmer, J.W.; Becker, H.M. Transport metabolons with carbonic anhydrases. *Front. Physiol.* **2013**, *4*, 291. [CrossRef]
79. Svastova, E.; Hulikova, A.; Rafajova, M.; Zat'ovicova, M.; Gibadulinova, A.; Casini, A.; Cecchi, A.; Scozzafava, A.; Supuran, C.T.; Pastorek, J.; et al. Hypoxia activates the capacity of tumor-associated carbonic anhydrase IX to acidify extracellular pH. *FEBS Lett.* **2004**, *577*, 439–445. [CrossRef]
80. Chafe, S.C.; Vizeacoumar, F.S.; Venkateswaran, G.; Nemirovsky, O.; Awrey, S.; Brown, W.S.; McDonald, P.C.; Carta, F.; Metcalfe, A.; Karasinska, J.M.; et al. Genome-wide synthetic lethal screen unveils novel CAIX-NFS1/xCT axis as a targetable vulnerability in hypoxic solid tumors. *Sci. Adv.* **2021**, *7*, eabj0364. [CrossRef]
81. Kopecka, J.; Rankin, G.M.; Salaroglio, I.C.; Poulsen, S.A.; Riganti, C. P-glycoprotein-mediated chemoresistance is reversed by carbonic anhydrase XII inhibitors. *Oncotarget* **2016**, *7*, 85861–85875. [CrossRef] [PubMed]

82. Parkkila, S.; Parkkila, A.K.; Saarnio, J.; Kivela, J.; Karttunen, T.J.; Kaunisto, K.; Waheed, A.; Sly, W.S.; Tureci, O.; Virtanen, I.; et al. Expression of the membrane-associated carbonic anhydrase isozyme XII in the human kidney and renal tumors. *J. Histochem. Cytochem.* **2000**, *48*, 1601–1608. [CrossRef] [PubMed]
83. Podolski-Renic, A.; Dinic, J.; Stankovic, T.; Jovanovic, M.; Ramovic, A.; Pustenko, A.; Zalubovskis, R.; Pesic, M. Sulfocoumarins, specific carbonic anhydrase IX and XII inhibitors, interact with cancer multidrug resistant phenotype through pH regulation and reverse P-glycoprotein mediated resistance. *Eur. J. Pharm. Sci.* **2019**, *138*, 105012. [CrossRef] [PubMed]
84. Salaroglio, I.C.; Mujumdar, P.; Annovazzi, L.; Kopecka, J.; Mellai, M.; Schiffer, D.; Poulsen, S.A.; Riganti, C. Carbonic Anhydrase XII Inhibitors Overcome P-Glycoprotein-Mediated Resistance to Temozolomide in Glioblastoma. *Mol. Cancer Ther.* **2018**, *17*, 2598–2609. [CrossRef]
85. Von Neubeck, B.; Gondi, G.; Riganti, C.; Pan, C.; Parra Damas, A.; Scherb, H.; Erturk, A.; Zeidler, R. An inhibitory antibody targeting carbonic anhydrase XII abrogates chemoresistance and significantly reduces lung metastases in an orthotopic breast cancer model in vivo. *Int. J. Cancer* **2018**, *143*, 2065–2075. [CrossRef]
86. Hammond, W.A.; Swaika, A.; Mody, K. Pharmacologic resistance in colorectal cancer: A review. *Ther. Adv. Med. Oncol.* **2016**, *8*, 57–84. [CrossRef] [PubMed]

International Journal of
Molecular Sciences

MDPI

Article

New 4,5-Diarylimidazol-2-ylidene–iodidogold(I) Complexes with High Activity against Esophageal Adenocarcinoma Cells

Sebastian W. Schleser [1,†], Hindole Ghosh [2,†], Gerald Hörner [3], Jonathan Seib [1], Sangita Bhattacharyya [2], Birgit Weber [3], Rainer Schobert [1], Prasad Dandawate [2] and Bernhard Biersack [1,*]

1 Organic Chemistry 1, University of Bayreuth, Universitätsstrasse 30, 95440 Bayreuth, Germany
2 Cancer Biology, University of Kansas Medical Center, 3901 Rainbow Boulevard, Kansas City, KS 66160, USA
3 Inorganic Chemistry IV, University of Bayreuth, Universitätsstrasse 30, 95440 Bayreuth, Germany
* Correspondence: bernhard.biersack@uni-bayreuth.de or bernhard.biersack@yahoo.com; Tel.: +49-921-552673
† These authors contributed equally to this work.

Abstract: Inspired by the vascular-disrupting agent combretastatin A-4 and recently published anticancer active *N*-heterocyclic carbene (NHC) complexes of Au(I), a series of new iodidogold(I)–NHC complexes was synthesized and characterized. The iodidogold(I) complexes were synthesized by a route involving van Leusen imidazole formation and *N*-alkylation, followed by complexation with Ag_2O, transmetalation with chloro(dimethylsulfide)gold(I) [Au(DMS)Cl], and anion exchange with KI. The target complexes were characterized by IR spectroscopy, ^{1}H and ^{13}C NMR spectroscopy, and mass spectrometry. The structure of **6c** was validated via single-crystal X-ray diffraction. A preliminary anticancer screening of the complexes using two esophageal adenocarcinoma cell lines showed promising nanomolar activities for certain iodidogold(I) complexes accompanied with apoptosis induction, as well as *c*-Myc and cyclin D1 suppression in esophageal adenocarcinoma cells treated with the most promising derivative **6b**.

Keywords: gold; carbene ligands; metal-based drugs; anticancer agents; esophageal cancer

Citation: Schleser, S.W.; Ghosh, H.; Hörner, G.; Seib, J.; Bhattacharyya, S.; Weber, B.; Schobert, R.; Dandawate, P.; Biersack, B. New 4,5-Diarylimidazol-2-ylidene–iododogold(I) Complexes with High Activity against Esophageal Adenocarcinoma Cells. *Int. J. Mol. Sci.* **2023**, *24*, 5738. https://doi.org/10.3390/ijms24065738

Academic Editor: Anindita Das

Received: 10 February 2023
Revised: 11 March 2023
Accepted: 15 March 2023
Published: 17 March 2023

1. Introduction

Esophageal cancer (EC) ranks seventh worldwide in terms of incidence (604,100 new cases) and sixth in overall mortality (544,076 deaths) [1]. Especially high morbidity and mortality rates for esophageal cancer are observed in East, Central, and South Asia, South and East Africa, and Northwest Europe. Moreover, these alarming numbers are expected to rise by 35–37% until 2030 [2]. Considering the poor prognosis of EC patients, frequently resulting from late diagnoses, this tumor is a considerable health issue for developed and developing countries [2,3]. EC is a rapidly growing cancer with a poor five-year survival rate of <20% [4]. The biology of EC is hardly understood compared to other cancers and typically shows extremely aggressive clinical features upon diagnosis [5]. Hence, it is one of the most challenging cancers to treat [6]. EC is divided into two histological subtypes [squamous cell carcinoma (ESCC) and adenocarcinoma (EAC)], which have diverse etiologies. ESCC is the most common EC worldwide, while EAC accounts for roughly two-thirds of EC cases in western countries [7]. The incidence of EAC is rapidly increasing across high-income countries due to obesity, an increase in gastroesophageal reflux disease (GERD), and Barrett's esophagus (BE) [8]. Current treatment options for esophageal cancer include endoscopic and surgical treatments and chemoradiotherapy based on platinum complexes such as cisplatin [9,10]. Recently, the combination of platinum-based chemo(radio)therapy with immune checkpoint inhibitors revealed promising results in patients suffering from EC [11]. Thus, platinum complexes continue to be highly relevant for the management of EC, and metal-based drugs with other metals than platinum may have great potential for the treatment of this disease in the future.

While platinum complexes are clinically applied for the treatment of various solid tumors, gold complexes such as auranofin (**1a**) are being used for the treatment of rheumatoid arthritis ("chrysotherapy") (Figure 1) [12]. However, as early as 1986, auranofin was investigated as an active compound in a study with esophageal carcinoma [13]. Auranofin revealed anticancer properties in preclinical studies based on its considerable inhibition of thioredoxin reductase (TrxR) associated with oxidative stress by the formation of reactive oxygen species (ROS), leading to increased apoptosis induction [12]. In addition, auranofin was described as an inhibitor of proteasomal deubiquitinase in cancer cells [13,14]. The mitochondrial Mia40/CHCHD4 pathway playing a crucial role in the oxidation of freshly imported cysteine-rich proteins in mitochondria was identified as a target of auranofin in fungi, which might be relevant in vigorously proliferating cancer cells too [15,16]. Furthermore, derivatives of auranofin with improved anticancer properties were described [17,18]. Recent developments in the field of auranofin-derived gold compounds led to highly active thiolatopurine complexes showing strong DNA damaging effects aside from TrxR inhibition, apoptosis induction, ROS formation, and antiangiogenic effects [19]. Advances in the research of *N*-heterocyclic carbene (NHC) gold complexes established a prospering class of metal-based drugs, which showed eminent anticancer activities based on TrxR inhibition and DNA interaction, adding well to the currently available arsenal of anticancer platinum complexes and the leading gold(I) complex auranofin [20–24]. Another emphasis was laid on the investigation of the effects on cytoskeletal structures, especially actin filaments and microtubules, by gold(I)–NHC complexes with 4,5-diarylimidazole-2-ylidene ligands derived from the tubulin-binding natural product combretastatin A-4 [25–27]. Meanwhile, the disruption of the actin cytoskeleton was described for cancer cells treated with auranofin too [28]. Recently, the relevance of the iodido ligand of neutral 4,5-dianisylimidazole-2-ylidene–iodidogold(I) complexes was proved, which revealed promising in vitro and in vivo antihepatoma activities of the iodido complex **1b** when compared with analogous gold(I)–NHC complexes bearing halo ligands (bromido), pseudohalo ligands (isocyanato), or acetato ligands (Figure 1) [29].

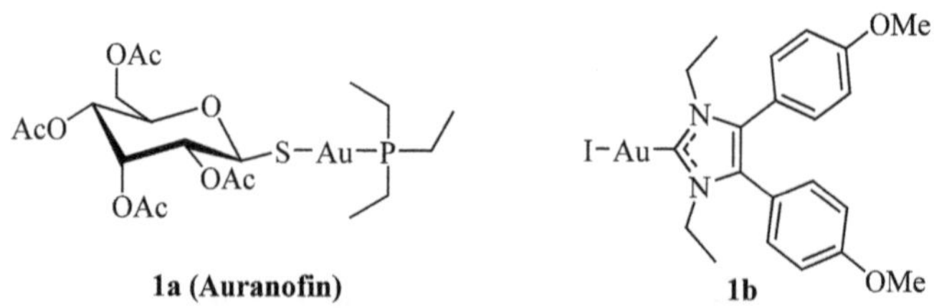

1a (Auranofin) **1b**

Figure 1. Structures of auranofin (**1a**) and the NHC–gold(I) complex **1b**.

With iodido being the optimal ligand for neutral 4,5-dianisylimidazol-2-ylidene–gold(I) complexes in hepatoma, subtle changes in the 4,5-dianisylimidazole-based NHC ligand might lead to iodidogold(I) complexes with improved activities against various tumor entities. This work studied the impact of various alterations of the NHC ligand of **1b** on the activity against invasive EAC cells.

2. Results

2.1. Chemistry

The synthesis of the new iodidogold(I) complexes **6a–n** was carried out following procedures from the literature (Scheme 1) [25–27,29]. The TosMIC reagent **2** was prepared as described before [30]. Reagent **2** was treated with $EtNH_2$ and the corresponding aryl aldehyde to form the 1-ethyl-4,5-diarylimidazoles **3**. *N*-Alkylation with ethyl iodide led to the imidazolium iodides **4**. The reaction of **4** with Ag_2O followed by transmetalation of the NHC–silver(I) intermediates with chloro(dimethylsulfide)gold(I) [Au(DMS)Cl] resulted in the chloridogold(I) complexes **5**. Finally, the chlorido ligands of **5** were replaced by iodido ligands upon reaction with KI, thus, generating the target complexes **6a–n**.

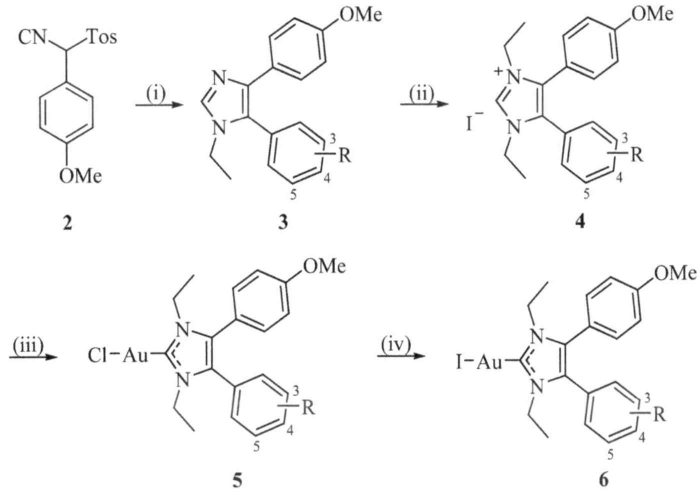

Residues:

a: R = 4-F	**h:** R = 3,4-$(MeO)_2$
b: R = 4-Cl	**i:** R = 3-F, 4-MeO
c: R = 4-Br	**j:** R = 3-Br, 4-MeO
d: R = 4-I	**k:** R = 3,4,5-$(MeO)_3$
e: R = 3,5-F_2	**l:** R = 3-Cl, 4,5-$(MeO)_2$
f: R = 3,5-Cl_2	**m:** R = 3-Br, 4,5-$(MeO)_2$
g: R = 3,5-Br_2	**n:** R = 3-I, 4,5-$(MeO)_2$

Scheme 1. Reagents and conditions: (i) aryl aldehyde, 2M $EtNH_2$/THF, AcOH, K_2CO_3, EtOH/DME, reflux, 7 h; (ii) EtI, MeCN, reflux, 24 h; (iii) (a) Ag_2O, CH_2Cl_2, r.t., 24 h, (b) [Au(DMS)Cl], CH_2Cl_2, r.t., 24 h; (iv) KI, acetone, r.t., 24 h.

In addition to the diethylimidazol-2-ylidene complexes **1b** and **6a–o**, the new **1b**-analogous dimethylimidazol-2-ylidene complex **6o** was prepared from the known chloridogold(I) precursor **5o** for comparison purposes (Scheme 2) [24].

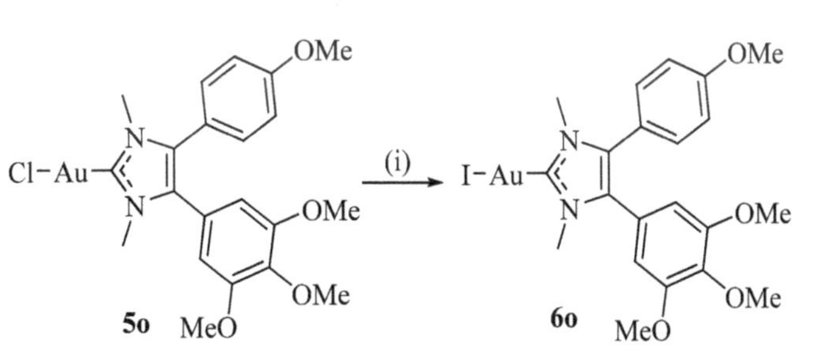

Scheme 2. Reagents and conditions: (i) KI, acetone, r.t., 24 h.

The stability of complexes **5e** and **6e** was studied by ^{1}H NMR spectroscopy in DMSO-d_6 upon addition of 5% D_2O (Figure 2). After 24 h, new signals of the *N*-ethyl groups appeared, which grew with prolonged incubation. These new signals can be assigned to

DMSO coordination and hydrolysis products of the iodidogold(I) complexes. However, the gold complexes possess considerable stability in aqueous solvents even after longer incubation since only a tiny fraction of complexes underwent hydrolysis by that time.

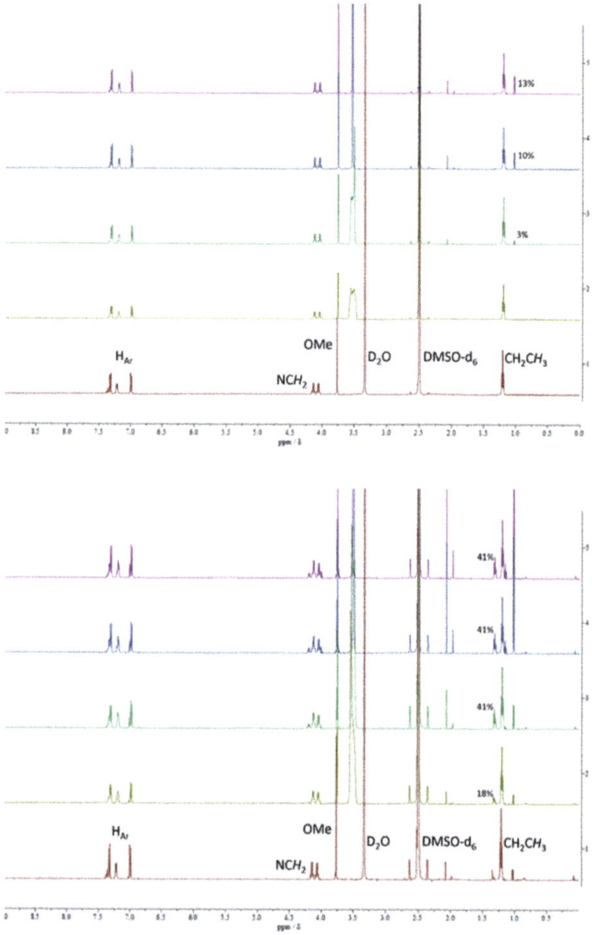

Figure 2. ^{1}H-NMR spectra of **5e** (top) and **6e** (bottom) in DMSO-d$_6$ after 0 h (red), +5% D_2O 0 h (yellow), 24 h (green), 48 h (blue), and 72 h (purple) with assignment. The hydrolysis product is given as a percentage averaged from the integrals of the additional peaks. These were set in a 1:1 ratio to the product peaks, assuming the same number of protons and approximately the same molar mass.

Single crystals suitable for X-ray diffraction structure elucidation were obtained via slow diffusion of *n*-hexane into a saturated solution of **6c** in $CHCl_3$ at 4 °C. The **6c** crystallizes in the monoclinic space group $P2_1/n$ with Z = 4. A plot of the molecular structure of **6c** is given in Figure 3A. The linearity of the I–Au–C vector and the bond distance between Au and the carbene carbon are well in agreement with values from the literature [31]. Selected bond lengths and angles are given in the caption of Figure 3; crystallographic details are assembled in Table 1. A view along crystallographic axis *a* reveals the formation of pseudo-dimers in a head-to-tail arrangement with a short Au–Au non-bonded contact of d(Au–Au) = 3.806 Å (Figure 3B).

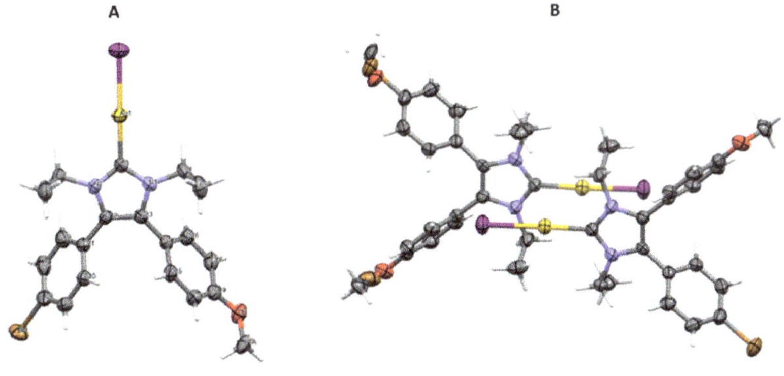

Figure 3. (**A**) Crystal structure of the iodidogold(I) complex **6c**. Selected bond lengths (Å) and angles (°): Au1–I1, 2.5567(6); Au1–C3, 2.002(7); N1–C3, 1.351(8); N2–C3, 1.342(8); C3–Au1–I1, 174.30(18); N2–C3–Au1, 128.4(5); N1–C3–Au1, 126.4(4); N2–C3–N1, 104.9(6); (**B**) pseudo-dimer of **6c** viewed along crystallographic axis *a* (ellipsoids plotted at 50% probability); positional disorder in one of the complex modules causes an apparent overlap of methoxy and bromo substituents; crystallographic data of the structure have been deposited at the Cambridge Crystallographic Data Centre: CCDC-2235874.

Table 1. Crystallographic data of the iodidogold complex **6c**.

	6c
CCDC number	2235874
Sum formula	C20.16 H21.47 Au Br0.84 I N2 O1.16
M/gmol^{-1}	701.56
Crystal system	monoclinic
Space group	$P2_1/n$ (Nr. 14)
Crystal description	Pale yellow block
Crystal size/mm	0.24 × 0.16 × 0.13
a/Å	9.8361(3)
b/Å	13.1351(6)
c/Å	16.7031(5)
$\alpha/°$	90
$\beta/°$	101.576(2)
$\gamma/°$	90
V/Å^3	2114.11(13)
Z	4
$\rho_{\text{calculated}}/\text{gcm}^{-3}$	2.204
μ/mm^{-1}	10.020
$\theta_{\text{range}}/°$	2.7–28.7
Radiation	Mo-K$_\alpha$ (0.71073 Å)
T/K	170(2)
Measured reflections	16653
Independent reflections	5028
Reflections with $I > 2\ s(I)$	3475

Table 1. *Cont.*

	6c
R_{int}	0.078
Restraints and parameters	18 and 268
R_1 $[F^2 > 2\,s\,(F^2)]$	0.0440
$wR_2(F^2)$	0.0980
GooF (S)	0.980
$\rho_{residual}$ (largest peak and hole)/eÅ^{-3}	0.920, −1.884

2.2. Activity against Esophageal Cancer

Initially, the antiproliferative activities of complexes **6a–o** against cell lines (FLO-1 and SK-GT-4) were evaluated using the hexosaminidase assay (Table 2 and Figure 4A). The known iodido complex **1b** and the chlorido complexes **5a–g** were analyzed for comparison. Except for complex **6l**, all iodidogold(I) complexes **6** exhibited moderate to high activities against FLO-1 cells, while the SK-GT-4 cells were less sensitive in most cases. Complexes **6b**, **6i**, and **6m** were the most active compounds that inhibited the growth of EAC cells in a dose- and time-dependent manner, with IC_{50} values in the nanomolar concentration range of 0.26–0.4 µM and 0.12–0.45 µM in the cases of SK-GT-4 and FLO-1, respectively. These complexes were also distinctly more active than the known complex **1b**. In analogy to the high activity of the 4-chlorophenyl derivative **6b**, which was the most active compound of this series, the 3,5-dichlorophenyl derivative **6f** was distinctly more active against the SK-GT-4 cells than its close congeners **6e** and **6g**. In contrast, the 3-chloro-4,5-dimethoxyphenyl analog **6l** was inactive. The chloridogold(I) complexes **5** were generally less active than their iodidogold(I) analogs, except for complexes **5a** and **5g**, which were more active against SK-GT-4 cells than the iodido complexes **6a** and **6g**. In addition, complex **6o**, the *N,N*-dimethyl analog of **6k**, was more active than **6k**.

Table 2. Inhibitory activities (IC_{50} values) of compounds **5a–g** and **6a–o**. Complex **1b** served as positive control. IC_{50} values are presented in µM concentrations at the 72 h time-point.

Compound	SK-GT-4	FLO-1
1b	1.31 ± 0.39	0.95 ± 0.15
5a	3.9 ± 0.71	1.5 ± 0.17
5b	>40	29.8 ± 0.52
5c	>40	27.33 ± 0.83
5d	>40	29.2 ± 0.52
5e	>40	28.13 ± 0.83
5f	>40	4.5 ± 0.28
5g	23.2 ± 0.92	4.78 ± 0.35
6a	15.33 ± 8.96	0.51 ± 0.01
6b	0.26 ± 0.09	0.12 ± 0.01
6c	>10	0.5 ± 0.02
6d	1.1 ± 0.06	0.58 ± 0.07
6e	>10	1.0 ± 0.06
6f	3.78 ± 1.25	0.9 ± 0.04
6g	23.53 ± 7.71	1.4 ± 0.10
6h	7.4 ± 0.50	0.55 ± 0.03
6i	0.325 ± 0.06	0.3 ± 0.02
6j	1.2 ± 0.08	0.625 ± 0.06
6k	13.4 ± 4.10	4.07 ± 2.21
6l	>10	>10
6m	0.4 ± 0.04	0.45 ± 0.09
6n	9.2 ± 1.71	0.95 ± 0.03
6o	2.9 ± 0.26	1.43 ± 0.25

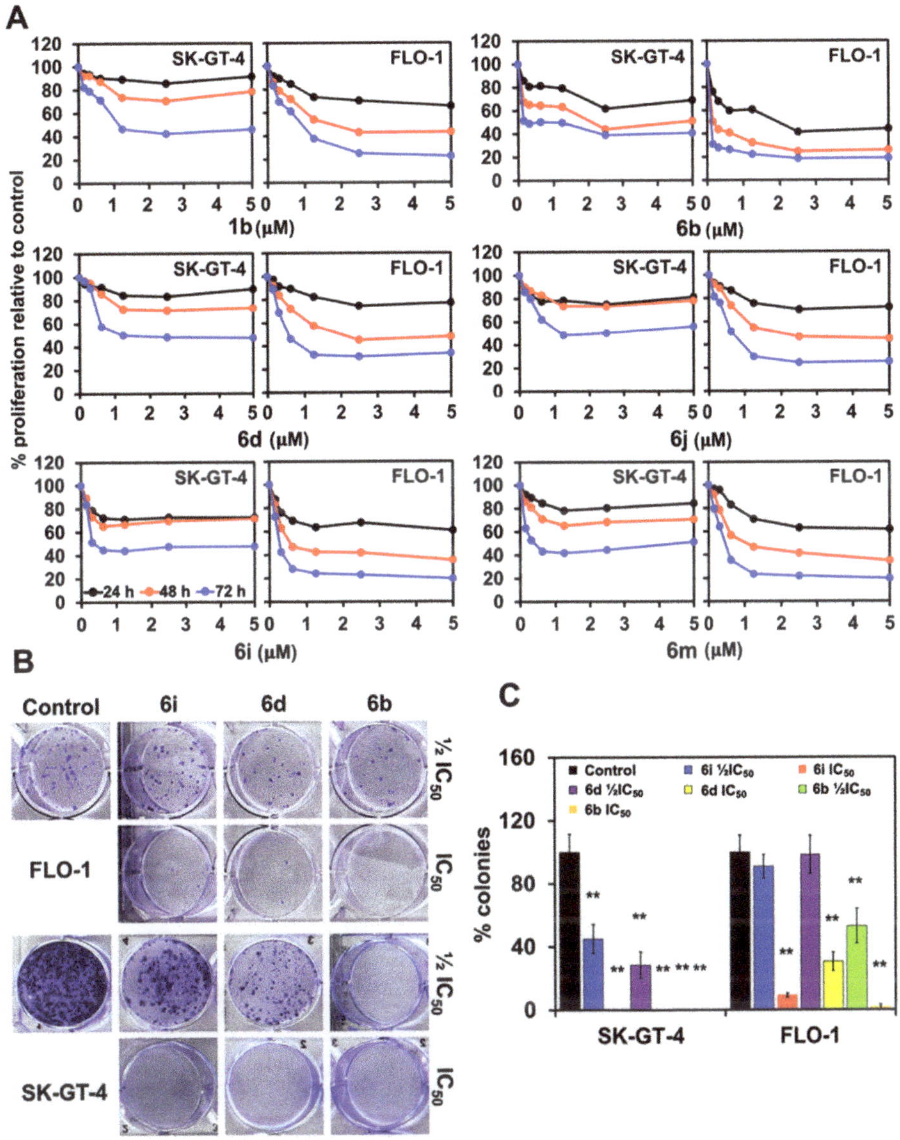

Figure 4. Gold complexes **6b**, **6d**, and **6i** inhibited the proliferation and colony formation in EAC cell lines. (**A**) Gold complexes inhibited the proliferation of SK-GT-4 and FLO-1 cells in a time- and dose-dependent manner. Gold complexes (semi-IC_{50} and IC_{50} concentrations) inhibited colony formation both in size (**B**) and number (**C**) in EAC cell lines. ** $p < 0.01$.

Compounds **6b**, **6d**, and **6i** were selected for further evaluation of their inhibitory activity against EAC cell lines (SK-GT-4 and FLO-1). The colony formation was performed to understand the long-term effect of these gold complexes on EAC cells. Complex **6b** showed the strongest colony formation suppression (i.e., reduced colony size and number) at IC_{50} and semi-IC_{50} concentrations. It completely inhibited colony formation in terms of size and number ($p < 0.01$) at its IC_{50} dose after treatment for 72 h (Figure 4B,C). In addition, at IC_{50} doses, the colony formation was almost completely inhibited by complex **6i**, while

6d was less inhibitory. These data suggested that the anticancer effects of the tested gold complexes are irreversible.

Several studies showed cancer stem cells (CSCs) are involved in tumor initiation, aggressiveness, and drug resistance in EAC. Hence, targeting CSCs is an attractive strategy to treat [32]. It was observed that CSCs form spheroids in ultra-low attachment plates. Hence, we used a spheroid formation assay to assess the effects of gold complexes on EAC CSCs. The complexes **6b**, **6d**, and **6i** inhibited spheroid formation (both size and numbers) by EAC cells (Figure 5A,B). Complexes **6b** and **6d** showed the strongest inhibitory effects on spheroids. Next, the highly antiproliferative complexes **6b**, **6d**, and **6i** (IC_{50} concentration, 72 h time-point) were selected and tested for their effects on the cell cycle of EAC cells using flow cytometry (Figure 6A–D). All three compounds induced cell cycle arrest in SK-GT-4 and FLO-1 cells. Specifically, treatment with compounds **6b** and **6d** induced G0–G1 cell cycle arrest in SK-GT-4 cells ($p < 0.01$), while compound **6i** led to accumulation of FLO-1 cells in the sub-G0 phase of the cell cycle. The accumulation of cells in the sub-G0 phase after gold complex treatment can be the consequence of fragmented DNA, indicating the cytotoxic effects on EAC cell lines. To understand the mechanistic changes in proteins due to cell cycle arrest, we performed a Western blot to study the levels of cyclin D1 and cMyc. Cyclin D1 is known to drive cell cycle progression, while *c*-Myc regulates cyclin D1 to induce proliferation and tumor growth [33]. Moreover, these complexes suppressed the expression of *c*-Myc and cyclin D1, which is in line with their strong cell death/apoptosis induction (Figure 6E). Hence, we further studied the ability of **6b**, **6d**, and **6i** (IC_{50} concentration) to induce apoptosis in EAC cell lines using the Annexin V/PI assay and flow cytometry. Cell populations treated with the gold complexes (IC_{50} concentration) for 72 h showed increased percentages of apoptotic (especially late apoptotic) and necrotic cells when compared with untreated control populations (Figure 7A–D, $p < 0.01$). In line with this finding, the number of viable cells was reduced in the treated populations. The Western blot analysis showed that all three gold(I) complexes (IC_{50} concentration) increased cleaved PARP as a sign of apoptosis after 72 h. The expression of the anti-apoptotic factors Bcl-XL, Bcl-2, and Mcl-1 was suppressed in cells treated with the gold complexes. No significant differences in pro-apoptotic Bax expression were observed compared to untreated cells (Figure 7E). These data suggest that gold complexes induced apoptosis by inhibiting the apoptotic proteins involved in cancer cell survival and, hence, can be used in the combination with current chemotherapeutic agents.

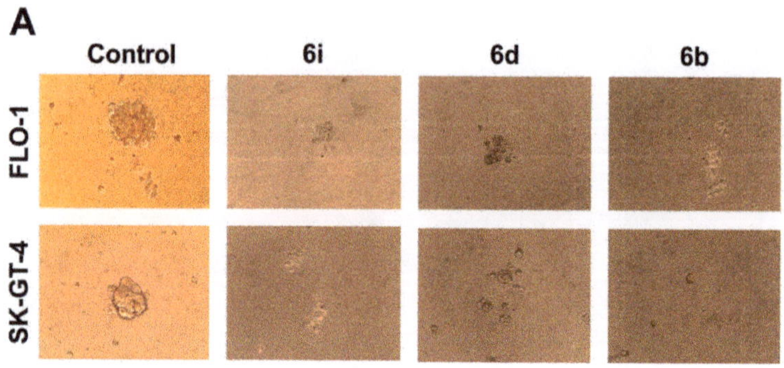

Figure 5. *Cont.*

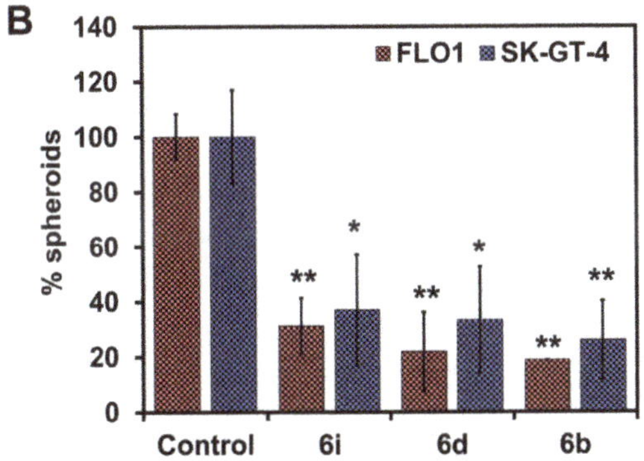

Figure 5. Inhibition of EAC cell spheroid formation by test compounds **6b**, **6d**, and **6i** (IC_{50} concentrations). The gold complexes reduced (**A**) size and (**B**) number of spheroids. The spheroids were photographed at 10× magnification. * $p < 0.05$, ** $p < 0.01$.

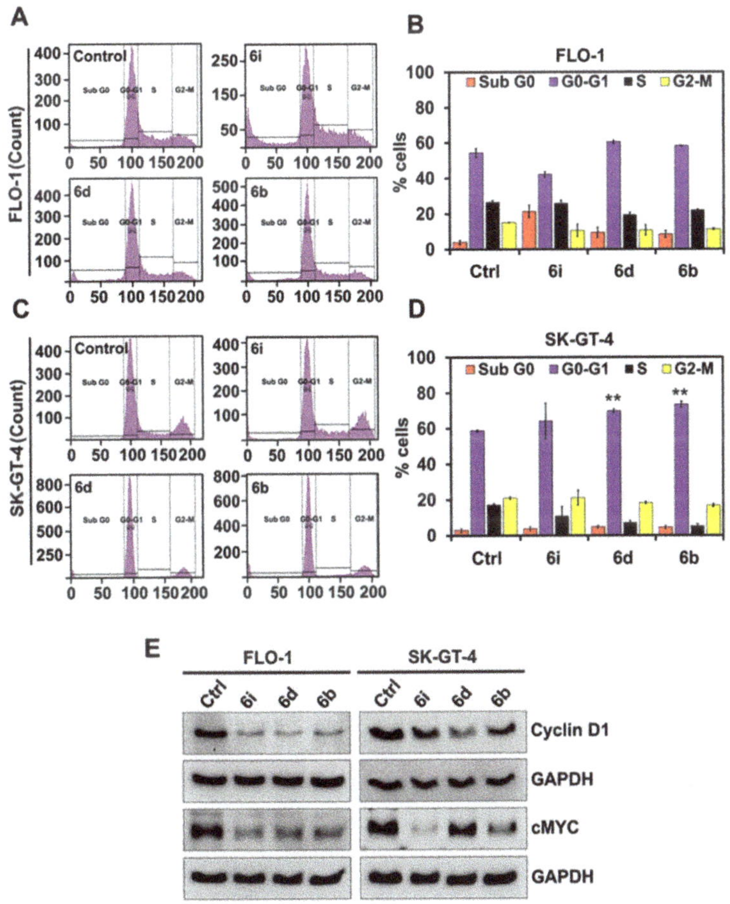

Figure 6. Gold complexes induced cell cycle arrest in EAC cells. (**A–D**) Gold complexes **6b**, **6d**, and **6i** (IC_{50} concentrations) induced G0–G1 cell cycle arrest in SK-GT-4 cells. (**E**) Gold complexes **6b**, **6d**, and **6i** (IC_{50} concentrations) inhibited cyclin D1 and c-Myc expression in SK-GT-4 and FLO-1 cells. ** $p < 0.01$.

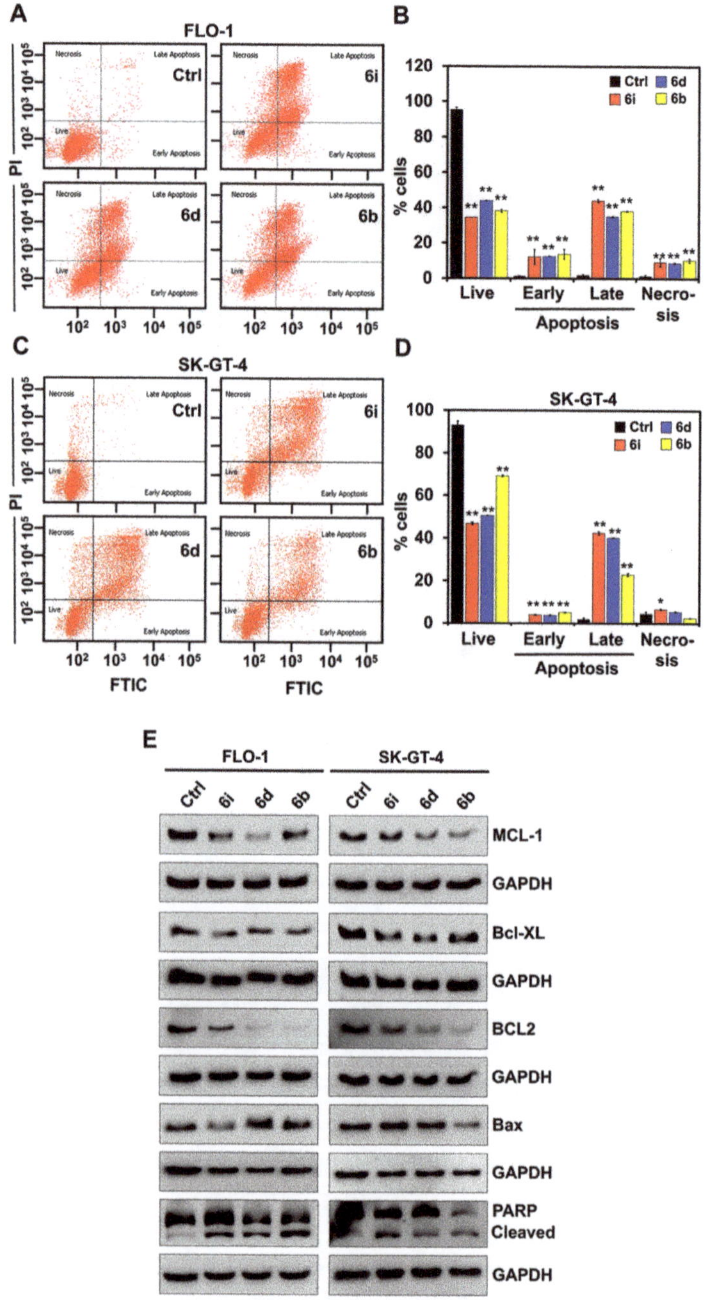

Figure 7. Gold complexes induced apoptosis in EAC cells. (**A**–**D**) Gold complexes **6b**, **6d**, and **6i** (IC_{50} concentrations) induced late-phase apoptosis in SK-GT-4 and FLO-1 cells, as assessed by an Annexin-PI assay using flow cytometry. (**E**) Cell lysates from EAC cells, when treated with complexes **6b**, **6d**, and **6i** (IC_{50} concentrations), showed significant cleavage of PARP compared to untreated controls. The treatment also reduced anti-apoptotic markers Bcl-XL, Mcl-1, and Bcl-2. * $p < 0.05$, ** $p < 0.01$.

3. Discussion

New derivatives of the published 4,5-dianisylimidazol-2-ylidene–iodidogold(I) complex **1b** were prepared by changing the modification of one of the ligand's anisyl residues. The synthesis of the new iodidogold(I)–NHC complexes **6a–o** was straightforward and based on previously published works by our groups and by Bian and coworkers [25–27,29]. In this way, many highly antitumoral iodidogold(I)–NHC complexes were identified. The antiproliferative activities of several new iodidogold(I) complexes against two invasive EAC cell lines were superior to the activity of the known complex **1b**. They were also much more active than their chloridogold(I) precursors. Neutral chloridogold(I)–NHC complexes were often reported to be less anticancer active than analogous cationic NHC–gold(I) complexes [34,35]. Chloridogold–NHC complexes were casually developed as selective antiparasitic agents due to their relatively low toxicity to human cells [36]. However, with the exchange of the chlorido ligand for an iodido ligand, the cytotoxicity of neutral iodidogold(I) complexes reached excellent IC_{50} values in cancer cells, being in the active concentration range of known cationic triphenylphosphinogold(I)–NHC complexes and biscarbene–gold(I) complexes. Future studies will reveal how far neutral iodidogold(I)–NHC complexes and cationic NHC–gold(I) complexes differ in their performance in cancers in terms of activity, cellular localization/accumulation, and mechanisms of action. Nevertheless, it is noteworthy that, already, slight modifications of one of the phenyl rings of the 4-anisyl-5-arylimidazole-based NHC ligand system applied in this study can lead to strong changes in the anticancer activity of the tested iodidogold(I) complexes. For instance, while the 4-bromophenyl derivative **6c** was virtually inactive against the SK-GT-4 cells, its close 4-chlorophenyl analog **6b** is the most active complex identified in this study. In addition, while complexes **1b**, **6b**, **6d**, **6i**, and **6m** showed high antiproliferative activities against FLO-1 cells and SK-GT-4 cells, the FLO-1 cells were much more sensitive to certain complexes such as **6a**, **6c**, **6e** and **6g** than the SK-GT-4 cells. FLO-1 cells are p53-mutant cells, and the gold complexes may take advantage of the absence of p53, the "guardian of the genome", to kill the FLO-1 cells. Analogously, higher sensitivities of p53-knockout HCT-116 colon carcinoma cells, when compared with p53-wildtype HCT-116 cells, were observed only recently for various NHC–gold(I) complexes and phosphinogold(I) complexes [19,35].

Apoptosis is strongly induced by the most promising iodidogold(I) complex **6b**. This is in line with previous reports about the induction of apoptosis by auranofin (**1a**) and complex **1b** in tumor cells [12,29]. The downregulation of Mcl-1 was described before in association with apoptosis induction in FLO-1 and SK-GT-4 cells [37]. In contrast to the described G2/M arrest of HepG2 hepatoma cells caused by **1b**, the new complexes **6b**, **6d**, and **6i** exhibited no cell cycle arrest in EAC cells. Instead, high sub-G1 levels generated by these complexes indicate a strong preference to induce cell death. The observed suppression of cyclin D1 and *c*-Myc by the complexes **6b**, **6d**, and **6i** might prohibit the ability of treated cells to enter the proper cell cycle, including mitosis, and directly paves the way to the induction of cell death instead. Both factors are relevant for EAC progression in FLO-1 cells [38,39]. In addition, the suppression of *c*-Myc and/or cyclin D1 is important for the treatment of other cancer diseases, which broadens the therapeutic scope of the newly discovered gold complexes [40,41].

The suppression of the formation of colonies and spheroids by EAC cells by complexes **6b**, **6d**, and **6i** is another positive attribute. The downregulation of the stem cell marker CD44 is a hint at an efficient targeting of esophageal cancer stem-like cells (CSCs) and the inhibition of mesenchymal features of EAC cells associated with metastasis formation [42]. EMT reversal in FLO-1 and SK-GT-4 EAC cells can also suppress paracrine effects and the production of exosomes [43].

Since platinum complexes are crucial components of currently applied clinical therapies of EAC, future studies with gold(I)–NHC complexes will probably shed more light on common and distinctive modes of action of platinum and gold complexes. The binding to cysteine and selenocysteine proteins (e.g., thioredoxin reductase) may play a role in the mode of action of the new iodidogold(I) complexes. The relevance of selenium was also highlighted by the suppression of selenium-binding protein 1 as a sign of EAC formation [44]. The combination of gold complexes with HDAC inhibitors might be promising when considering the known effects of HDAC inhibition on thioredoxin and thioredoxin-interacting protein in EAC [45,46]. In addition, aurora kinase inhibitors revealed promising anticancer effects in combination with docetaxel or cisplatin in p53-mutant FLO-1 cells. Thus, they might also be suitable combination partners for the treatment of EAC together with active iodidogold(I)–NHC complexes [47,48]. *c*-Myc inhibitors might also be suitable combination partners [40]. Moreover, our groups have recently identified highly active *c*-Myb inhibitors [49–51]. This transcription factor was found to be upregulated in EAC and crucial for the immune escape of EAC cells via the miR-145-5p/SPOP/PD-L1 axis [52–54]. Hence, a combination of **6b** with a potent *c*-Myb inhibitor seems worth being investigated in EAC cells.

4. Materials and Methods

4.1. General Procedures

Column chromatography: silica gel 60 (230–400 mesh, Merck, Darmstadt, Germany). Melting points (uncorrected), Electrothermal 9100 (Thermo Fisher Scientific, Geel, Belgium); IR spectra, Perkin-Elmer Spectrum One FT-IR spectrophotometer with ATR sampling unit (Perkin-Elmer, Rodgau, Germany); NMR spectra, Bruker Avance 300/500 spectrometer (Bruker, Billerica, MA, USA); chemical shifts (δ) are given in parts per million (ppm) downfield from tetramethylsilane as internal standard; mass spectra, Thermo Finnigan MAT 8500 (EI, Thermo Finnigan, San Jose, CA, USA).

4.2. Materials

Starting compounds and reagents were obtained from abcr (Karlsruhe, Germany), Sigma-Aldrich (Darmstadt, Germany) and TCI (Zwijndrecht, Belgium). Compound **1b** was prepared according to a literature procedure, and analytical data of the newly prepared compound were in line with published data [24]. The known intermediates **3k–m** and **4k–m**, **5k**, and **5o** were also prepared following procedures from the literature [25–27].

4.3. Synthesis

4.3.1. Synthesis of Imidazoles 3—Typical Procedure

Benzaldehyde derivatives (1.00 equiv.) were dissolved in EtOH (15.0 mL/mmol). Then, 2 M EtNH2/THF (5.00 equiv.) and AcOH (10.0 equiv.) were added, and the reaction mixture was stirred under reflux for 2 h. After cooling to room temperature, the anisyl-TosMIC reagent **2** (1.50 equiv.) was dissolved in DME (5.00 mL/mmol) and added to the reaction mixture together with K2CO3 (4.00 equiv.). The reaction mixture was then stirred again under reflux for 5 h. The solvent was evaporated, and the residue was taken up in ethyl acetate and water. The organic phase was washed with brine, dried over MgSO4, and filtered, and the filtrate was removed in vacuum. The residue was purified by column chromatography (silica gel 60). The products were obtained as yellow to off-white solids or oils with yields of 55–100%.

3a: yield: 100% (quant.); R_f = 0.22 (ethyl acetate); ^{1}H NMR (500 MHz, $CDCl_3$) δ 7.62 (s, 1H, H^{ar}), 7.43 (t, J = 1.9 Hz, 1H, H^{ar}), 7.36 (dt, J = 8.9 Hz, 2.6 Hz, 2H, H^{ar}), 7.23 (d, J = 1.9 Hz, 2H, H^{ar}), 6.80 (dt, J = 8.9 Hz, 2.6 Hz, 2H, H^{ar}), 3.84 (q, J = 7.3 Hz, 2H, NCH_2), 3.78 (s, 3H, OCH_3), 1.30 (t, J = 7.3 Hz, 3H, CH_2CH_3); ^{13}C NMR (126 MHz, $CDCl_3$) δ 162.84 (d, $J_{C\text{-}F}$ = 248.8 Hz, C^{ar}), 158.57 (NCN), 139.25 (C^{ar}), 136.55 (C^{ar}), 135.59 (C^{ar}), 134.20 (C^{ar}), 132.70 (d, $J_{C\text{-}F}$ = 8.6 Hz, C^{ar}), 129.12 (C^{ar}), 128.81 (C^{ar}), 127.99 (C^{ar}), 126.62 (C^{ar}), 124.39 (C^{ar}), 116.23 (d, $J_{C\text{-}F}$ = 21.7 Hz, C^{ar}), 113.81 (C=C), 55.22 (OCH_3), 40.31 (NCH_2), 16.42 (CH_2CH_3).

3b: yield: 100% (quant.); R_f = 0.21 (ethyl acetate); ^{1}H NMR (500 MHz, $CDCl_3$) δ 7.61 (s, 1H, H^{ar}), 7.43 (dq, J = 8.8, 2.3 Hz, 2H, H^{ar}), 7.38–7.34 (m, 2H, H^{ar}), 7.28 (d, J = 2.0 Hz, 1H, H^{ar}), 6.78–6.75 (m, 2H, H^{ar}), 3.82 (q, J = 7.3 Hz, 2H, NCH_2), 3.76 (s, 3H, OCH_3), 1.28–1.25 (m, 3H CH_2CH_3); ^{13}C NMR (126 MHz, $CDCl_3$) δ 158.31 (NCN), 138.56 (C^{ar}), 136.18 (C^{ar}),

134.64 (C^{ar}), 132.15 (C^{ar}), 132.11 (C^{ar}), 129.58 (C^{ar}), 129.41 (C^{ar}), 127.87 (C^{ar}), 127.60 (C^{ar}), 127.16 (C^{ar}), 125.97 (C^{ar}), 113.52 (C=C) 55.19 (OCH_3), 40.14 (NCH_2), 16.46 (CH_2CH_3).

3c: yield: 79%; R_f = 0.20 (ethyl acetate); ^{1}H NMR (500 MHz, $CDCl_3$) δ 7.60–7.58 (m, 2H, H^{ar}), 7.38–7.34 (m, 2H, H^{ar}), 7.23–7.19 (m, 2H, H^{ar}), 6.78–6.76 (m, 2H, H^{ar}), 3.85–3.81 (m, 2H, NCH_2), 3.77 (s, 3H, OCH_3), 1.29–1.25 (m, 3H CH_2CH_3); ^{13}C NMR (126 MHz, $CDCl_3$) δ 158.33 (NCN), 136.22 (C^{ar}), 132.42 (C^{ar}), 132.36 (C^{ar}), 132.12 (C^{ar}), 127.88 (C^{ar}), 127.59 (C^{ar}), 122.86 (C^{ar}), 114.50, 113.67 (C=C), 55.19 (OCH_3), 40.15 (NCH_2), 16.51 (CH_2CH_3), 16.47 (CH_2CH_3).

3d: yield: 79%; R_f = 0.18 (ethyl acetate); ^{1}H NMR (500 MHz, $CDCl_3$) δ 7.79–7.75 (m, 2H, H^{ar}), 7.38–7.34 (m, 2H, H^{ar}), 7.09–7.04 (m, 2H, H^{ar}), 6.79–6.75 (m, 2H, H^{ar}), 3.82 (q, J = 7.3 Hz, 2H, NCH_2), 3.76 (s, 3H, OCH_3), 1.25 (t, J = 7.3 Hz, 4H CH_2CH_3); ^{13}C NMR (126 MHz, $CDCl_3$) δ 158.33 (NCN), 138.29 (C^{ar}), 136.25 (C^{ar}), 132.56 (C^{ar}), 130.63 (C^{ar}), 127.91 (C^{ar}), 127.13 (C^{ar}), 126.08 (C^{ar}), 113.68 (C=C), 55.20 (OCH_3), 40.15 (NCH_2), 16.48 (CH_2CH_3).

3e: yield: 64%; R_f = 0.24 (ethyl acetate); ^{1}H NMR (500 MHz, $CDCl_3$) δ 7.61 (s, 1H, H^{ar}), 7.38–7.34 (m, 2H, H^{ar}), 6.87 (dq, J = 6.0, 1.6 Hz, 2H, H^{ar}), 6.80–6.77 (m, 2H, H^{ar}), 3.86 (q, J = 7.3 Hz, 2H, NCH_2), 3.78 (s, 3H, OCH_3), 1.29 (t, J = 7.3 Hz, 3H CH_2CH_3); ^{13}C NMR (126 MHz, $CDCl_3$) δ 164.23 (dd, $J_{C\text{-}F}$ = 250,2, 13.2 Hz), 162.24 (d, $J_{C\text{-}F}$ = 13.2 Hz), 158.54 (NCN), 139.13 (C^{ar}), 136.54 (C^{ar}), 134.30 (t, $J_{C\text{-}F}$ = 10.2 Hz, C^{ar}), 128.04 (C^{ar}), 126.71 (C^{ar}), 113.84–113.66 (m, C^{ar}), 104.25 (t, $J_{C\text{-}F}$ = 25.2 Hz), 55.21 (OCH_3), 40.29 (NCH_2), 16.42 (CH_2CH_3).

3f: yield: 55%; R_f = 0.29 (ethyl acetate); ^{1}H NMR (500 MHz, $CDCl_3$) δ 7.62 (s, 1H, H^{ar}), 7.43 (t, J = 1.9 Hz, 1H, H^{ar}), 7.38–7.33 (m, 2H, H^{ar}), 7.23 (d, J = 1.9 Hz, 2H, H^{ar}), 6.83–6.77 (m, 2H, H^{ar}), 3.84 (q, J = 7.3 Hz, 2H, NCH_2), 3.78 (s, 3H, OCH_3), 1.30 (t, J = 7.3 Hz, 3H CH_2CH_3); ^{13}C NMR (126 MHz, $CDCl_3$) δ 158.57 (NCN), 139.25 (C^{ar}), 136.55 (C^{ar}), 135.59 (C^{ar}), 134.20 (C^{ar}), 129.12 (C^{ar}), 128.81 (C^{ar}), 127.99 (C^{ar}), 126.62 (C^{ar}), 124.39 (C^{ar}), 113.81 (C=C), 55.22 (OCH_3), 40.31 (NCH_2), 16.42 (CH_2CH_3).

3g: yield: 64%; R_f = 0.25 (ethyl acetate); ^{1}H NMR (500 MHz, $CDCl_3$) δ 7.74 (t, J = 1.8 Hz, 1H, H^{ar}), 7.61 (s, 1H, H^{ar}), 7.43 (d, J = 1.8 Hz, 2H, H^{ar}), 7.39–7.34 (m, 2H, H^{ar}), 6.84–6.77 (m, 2H, H^{ar}), 3.84 (q, J = 7.3 Hz, 2H, NCH_2), 3.78 (s, 3H, OCH_3), 1.30 (t, J = 7.3 Hz, 3H CH_2CH_3); ^{13}C NMR (126 MHz, $CDCl_3$) δ 158.57 (NCN), 139.28 (C^{ar}), 136.54 (C^{ar}), 134.77 (C^{ar}), 134.23 (C^{ar}), 132.38 (C^{ar}), 127.97 (C^{ar}), 126.61 (C^{ar}), 124.17 (C^{ar}), 123.44 (C^{ar}), 113.82 (C=C), 55.23 (OCH_3), 40.31 (NCH_2), 16.42 (CH_2CH_3).

3h: yield: 76%; R_f = 0.27 (ethyl acetate); υ_{max}(ATR)/cm^{-1} 2935, 2836, 1612, 1582, 1559, 1518, 1500, 1462, 1414, 1338, 1317, 1294, 1242, 1167, 1137, 1105, 1024, 953, 865, 835, 815, 798, 765, 743, 663; ^{1}H NMR (300 MHz, $CDCl_3$) δ 7.53 (s, 1H, H^{ar}), 7.39 (d, J = 8.9 Hz, 2H, H^{ar}), 6.9–6.8 (m, 2H, H^{ar}), 6.77 (s, 1H, H^{ar}), 6.70 (d, J = 8.9 Hz, 2H, H^{ar}), 3.89 (s, 3H, OCH_3), 3.8–3.7 (m, 5H, NCH_2, OCH_3), 3.69 (s, 3H, OCH_3), 1.22 (t, J = 7.3 Hz, 3H, CH_2CH_3); ^{13}C NMR (75.5 MHz, $CDCl_3$) δ 157.9 (C^{ar}), 149.1 (C^{ar}), 137.6 (C^{ar}), 135.4 (C^{ar}), 131.9 (C^{ar}), 128.4 (C^{ar}), 127.4 (C^{ar}), 127.0 (C^{ar}), 123.2 (C^{ar}), 114.3 (C^{ar}), 113.6 (C^{ar}), 113.4 (C^{ar}), 111.4 (C^{ar}), 55.8 (OCH_3), 55.7 (OCH_3), 55.0 (OCH_3), 39.8 (NCH_2), 16.3 (CH_2CH_3); m/z (%) 338 (100) [M^+], 323 (27), 308 (23).

3i: yield: 70%; R_f = 0.28 (ethyl acetate); υ_{max}(ATR)/cm^{-1} 2976, 2938, 2838, 1615, 1580, 1520, 1499, 1462, 1442, 1422, 1339, 1299, 1267, 1245, 1213, 1174, 1132, 1105, 1024, 953, 880, 835, 818, 799, 761, 744, 706, 662; ^{1}H NMR (300 MHz, $CDCl_3$) δ 7.54 (s, 1H, H^{ar}), 7.35 (d, J = 9.0 Hz, 2H, H^{ar}), 7.1–7.0 (m, 3H, H^{ar}), 6.72 (d, J = 9.0 Hz, 2H, H^{ar}), 3.89 (s, 3H, OCH_3), 3.76 (q, J = 7.3 Hz, 2H, NCH_2), 3.71 (s, 3H, OCH_3), 1.22 (t, J = 7.3 Hz, 3H, CH_2CH_3); ^{13}C NMR (75.5 MHz, $CDCl_3$) δ 158.1 (C^{ar}), 153.8–150.5 (m, C^{ar}), 147.8 (C^{ar}), 138.2 (C^{ar}), 135.7 (C^{ar}), 132.0 (C^{ar}), 127.9–127.3 (m, C^{ar}), 127.0 (C^{ar}), 125.7 (C^{ar}), 123.4 (C^{ar}), 118.4–118.2 (m, C^{ar}), 114.4 (C^{ar}), 113.6–113.4 (m, C^{ar}), 56.0 (OCH_3), 55.0 (OCH_3), 39.9 (NCH_2), 16.3 (CH_2CH_3); m/z (%) 326 (100) [M^+], 311 (43), 119 (17).

3j: yield: 50%; R_f = 0.31 (ethyl acetate); υ_{max}(ATR)/cm^{-1} 3117, 3066, 3019, 2981, 2942, 2902, 2839, 1612, 1577, 1560, 1512, 1485, 1462, 1386, 1372, 1343, 1308, 1285, 1242, 1194, 1173, 1147, 1118, 1104, 1056, 1043, 1025, 1018, 951, 900, 834, 819, 808, 796, 744, 703, 676, 653; ^{1}H NMR (300 MHz, $CDCl_3$) δ 7.55 (s, 1H, H^{ar}), 7.51 (s, 1H, H^{ar}), 7.4–7.3 (m, 2H, H^{ar}), 7.3–7.2

(m, 1H, H^{ar}), 7.0–6.9 (m, 1H, H^{ar}), 6.8–6.7 (m, 2H, H^{ar}), 3.92 (s, 3H, OCH_3), 3.9–3.7 (m, 5H, NCH_2, OCH_3), 1.3–1.2 (m, 3H, CH_2CH_3); ^{13}C NMR (75.5 MHz, $CDCl_3$) δ 159.7 (C^{ar}), 158.2 (C^{ar}), 156.0 (C^{ar}), 138.3 (C^{ar}), 135.8 (C^{ar}), 135.5 (C^{ar}), 135.3 (C^{ar}), 132.0 (C^{ar}), 131.3 (C^{ar}), 127.6 (C^{ar}), 127.2 (C^{ar}), 125.5 (C^{ar}), 124.5 (C^{ar}), 114.4 (C^{ar}), 113.6 (C^{ar}), 113.5 (C^{ar}), 112.1 (C^{ar}), 112.0 (C^{ar}), 56.2 (OCH_3), 55.1 (OCH_3), 40.0 (NCH_2), 16.4 (CH_2CH_3); *m/z* (%) 388 (100) [M^+], 386 (100) [M^+], 375 (25), 373 (27), 308 (92), 293 (34), 119 (42).

3n: yield: 64%; R_f = 0.30 (ethyl acetate); υ_{max}(ATR)/cm^{-1} 2959, 2930, 2844, 1611, 1595, 1557, 1541, 1513, 1463, 1445, 1425, 1408, 1393, 1355, 1334, 1315, 1295, 1273, 1247, 1236, 1197, 1182, 1155, 1108, 1073, 1024, 1000, 955, 873, 856, 844, 826, 797, 754, 742, 712, 687, 660; ^1H NMR (300 MHz, $CDCl_3$) δ 7.57 (s, 1H, H^{ar}), 7.40 (d, *J* = 9.0 Hz, 2H, H^{ar}), 7.34 (s, 1H, H^{ar}), 6.8–6.7 (m, 3H, H^{ar}), 3.90 (s, 3H, OCH_3), 3.82 (q, *J* = 7.3 Hz, 2H, NCH_2), 3.76 (s, 3H, OCH_3), 3.73 (s, 3H, OCH_3), 1.29 (t, *J* = 7.3 Hz, 3H, CH_2CH_3); ^{13}C NMR (75.5 MHz, $CDCl_3$) δ 158.3 (C^{ar}), 152.7 (C^{ar}), 149.1 (C^{ar}), 138.3 (C^{ar}), 135.9 (C^{ar}), 132.3 (C^{ar}), 132.0 (C^{ar}), 128.8 (C^{ar}), 127.7 (C^{ar}), 127.1 (C^{ar}), 125.4 (C^{ar}), 115.4 (C^{ar}), 114.4 (C^{ar}), 113.6 (C^{ar}), 92.7 (C^{ar}), 60.5 (OCH_3), 56.0 (OCH_3), 55.1 (OCH_3), 40.1 (NCH_2), 16.5 (CH_2CH_3); *m/z* (%) 464 (1) [M^+], 330 (2), 239 (12), 210 (12), 135 (83), 57 (98), 43 (100).

4.3.2. Synthesis of Diethylimidazolium Iodides 4—Typical Procedure

Imidazoles **3** (1.00 equiv.) were dissolved in acetonitrile (60.0 mL/mmol) and treated with ethyl iodide (55.0 equiv.). The reaction mixture was stirred at 85 °C for 24–70 h. The reaction was concentrated in vacuum, and the residue was dissolved in a small amount of CH_2Cl_2 and dropped into Et_2O, leading to the precipitation of the product. The solvent was decanted, and the residue was dried in vacuum. The products were obtained as brown or off-white solids or oils in yields of 74–91%.

4a: yield: 100% (quant.); ^1H NMR (500 MHz, $CDCl_3$) δ 10.28 (s, 1H, NC*H*N), 7.33–7.29 (m, 2H, H^{ar}), 7.22–7.18 (m, 2H, H^{ar}), 7.11–7.07 (m, 2H, H^{ar}), 6.91–6.87 (m, 2H, H^{ar}), 4.27 (dq, *J* = 7.3, 4.7 Hz, 4H, NCH_2), 3.79 (s, 3H, OCH_3), 1.50 (dt, *J* = 7.4, 1.6 Hz, 6H, CH_2CH_3); ^{13}C NMR (126 MHz, $CDCl_3$) δ 163.61 (d, $J_{C\text{-}F}$ = 252.0 Hz, C^{ar}), 161.02 (NCN), 132.78 (d, $J_{C\text{-}F}$ = 8.6 Hz, C^{ar}), 131.98 (C^{ar}), 116.62 (d, $J_{C\text{-}F}$ = 21.8 Hz, C^{ar}), 114.78 (C=C), 55.41 (OCH_3), 43.48 (vd, *J* = 14.5 Hz, NCH_2), 15.75 (CH_2CH_3).

4b: yield: 85%; ^1H NMR (500 MHz, $CDCl_3$) δ 10.29 (s, 1H, NC*H*N), 7.38–7.35 (m, 2H, H^{ar}), 7.28–7.25 (m, 2H, H^{ar}), 7.22–7.19 (m, 2H, H^{ar}), 6.91–6.87 (m, 2H, H^{ar}), 4.30–4.22 (m, 4H, NCH_2), 3.79 (s, 3H, OCH_3), 1.50 (dt, *J* = 7.3, 3.9 Hz, 6H, CH_2CH_3); ^{13}C NMR (126 MHz, $CDCl_3$) δ 161.08 (NCN), 136.71 (C^{ar}), 132.00 (d, *J* = 2.3 Hz, C^{ar}), 129.64 (C^{ar}), 123.60 (C^{ar}), 116.35 (C^{ar}), 114.83 (C=C), 55.42 (OCH_3), 43.57 (vd, *J* = 22.7 Hz, NCH_2), 15.78 (CH_2CH_3).

4c: yield: 91%; ^1H NMR (500 MHz, $CDCl_3$) δ 10.30 (s, 1H, NC*H*N), 7.54–7.51 (m, 2H, H^{ar}), 7.22–7.18 (m, 4H, H^{ar}), 6.91–6.89 (m, 2H, H^{ar}), 4.29–4.25 (m, 4H, NCH_2), 3.79 (s, 3H, OCH_3), 1.52–1.48 (m, 6H, CH_2CH_3); ^{13}C NMR (126 MHz, $CDCl_3$) δ 161.09 (NCN), 132.60 (C^{ar}), 132.17 (C^{ar}), 132.00 (C^{ar}), 131.92 (C^{ar}), 125.03 (C^{ar}), 124.09 (C^{ar}), 114.84 (C=C), 114.74 (C=C), 55.42 (OCH_3), 43.66 (NCH_2), 43.47 (NCH_2), 15.78 (CH_2CH_3).

4d: yield: 80%; ^1H NMR (500 MHz, $CDCl_3$) δ 10.27 (s, 1H, NC*H*N), 7.74–7.71 (m, 2H, H^{ar}), 7.23–7.17 (m, 2H, H^{ar}), 7.07–7.02 (m, 2H, H^{ar}), 6.89 (dd, *J* = 8.8, 3.6 Hz, 2H, H^{ar}), 4.30–4.23 (m, 4H, NCH_2), 3.79 (s, 3H, OCH_3), 1.66–1.30 (m, 6H, CH_2CH_3); ^{13}C NMR (126 MHz, $CDCl_3$) δ 161.08 (NCN), 138.50 (C^{ar}), 132.16 (C^{ar}), 132.01 (C^{ar}), 131.92 (C^{ar}), 124.65 (C^{ar}), 114.84 (C=C), 114.74 (C=C), 55.43 (OCH_3), 43.56 (vd, *J* = 23.6 Hz, NCH_2), 15.79 (CH_2CH_3).

4e: yield: 74%; ^1H NMR (500 MHz, $CDCl_3$) δ 10.27 (s, 1H, NC*H*N), 7.23–7.18 (m, 2H, H^{ar}), 6.90–6.81 (m, 5H, H^{ar}), 4.24 (dq, *J* = 23.8, 7.3 Hz, 4H, NCH_2), 3.76 (s, 3H, OCH_3), 1.48 (dt, *J* = 18.7, 7.3 Hz, 6H, CH_2CH_3); ^{13}C NMR (126 MHz, $CDCl_3$) δ 169.95 (s, C^{ar}), 162,9 (dd, $J_{C\text{-}F}$ = 251.1, 13.2 Hz, C^{ar}) 160.58 (NCN), 131.71 (C^{ar}), 130.92 (s, C^{ar}), 128.48 (s, C^{ar}), 118.73 (s, C^{ar}), 114.61 (C=C), 114.26–113.91 (m, C^{ar}), 105.81 (t, $J_{C\text{-}F}$ = 25.0 Hz, C^{ar}), 55.44 (OCH_3), 43.71 (vd, *J* = 32.2 Hz, NCH_2), 15.72 (CH_2CH_3).

4f: yield: 70%; ^1H NMR (500 MHz, $CDCl_3$) δ 10.38 (s, 1H, NC*H*N), 7.44 (t, *J* = 1.9 Hz, 1H, H^{ar}), 7.27–7.23 (m, 4H, H^{ar}), 6.97–6.93 (m, 2H, H^{ar}), 4.29 (dq, *J* = 9.3, 7.3 Hz, 4H, NCH_2),

3.83 (s, 3H, OCH_3), 1.57 (t, *J* = 7.3 Hz, 3H, CH_2CH_3), 1.52 (t, *J* = 7.3 Hz, 3H, CH_2CH_3); ^{13}C NMR (126 MHz, $CDCl_3$) δ 161.30 (NCN), 136.00 (C^{ar}), 132.87 (C^{ar}), 131.99 (C^{ar}), 130.76 (C^{ar}), 129.00 (C^{ar}), 128.18 (C^{ar}), 115.83 (C^{ar}), 114.97 (C=C), 55.46 (OCH_3), 43.71 (vd, *J* = 27.7 Hz, NCH_2), 15.77 (CH_2CH_3), 15.75 (CH_2CH_3).

4g: yield: 89%; ^{1}H NMR (500 MHz, $CDCl_3$) δ 10.36 (s, 1H, NCHN), 7.74 (t, *J* = 1.8 Hz, 1H, H^{ar}), 7.42 (s, 2H, H^{ar}), 7.25–7.22 (m, 2H, H^{ar}), 6.96–6.92 (m, 2H, H^{ar}), 4.27 (*p*, *J* = 7.2 Hz, 4H, NCH_2), 3.82 (s, 3H, OCH_3), 1.55 (t, *J* = 7.3 Hz, 3H, CH_2CH_3), 1.50 (t, *J* = 7.3 Hz, 3H, CH_2CH_3); ^{13}C NMR (126 MHz, $CDCl_3$) δ 161.30 (NCN), 136.21 (C^{ar}), 132.88 (C^{ar}), 132.21 (C^{ar}), 131.98 (C^{ar}), 128.68 (C^{ar}), 123.71 (C^{ar}), 115.82 (C=C), 114.98 (C=C), 55.46 (OCH_3), 43.70 (vd, *J* = 25.9 Hz, NCH_2), 15.77 (CH_2CH_3), 15.74 (CH_2CH_3).

4h: yield: 100% (quant.); υ_{max}(ATR)/cm^{-1} 3438, 3128, 2978, 2937, 2837, 1609, 1596, 1559, 1521, 1507, 1462, 1428, 1386, 1350, 1293, 1250, 1230, 1177, 1139, 1111, 1091, 1019, 883, 840, 805, 765, 740; ^{1}H NMR (300 MHz, $CDCl_3$) δ 10.03 (s, 1H, H^{ar}), 7.16 (d, *J* = 8.8 Hz, 2H, H^{ar}), 6.9–6.8 (m, 4H, H^{ar}), 6.71 (s, 1H, H^{ar}), 4.3–4.1 (m, 4H, NCH_2), 3.79 (s, 3H, OCH_3), 3.72 (s, 3H, OCH_3), 3.68 (s, 3H, OCH_3), 1.5–1.3 (m, 6H, CH_2CH_3); ^{13}C NMR (75.5 MHz, $CDCl_3$) δ 160.6 (C^{ar}), 150.2 (C^{ar}), 149.0 (C^{ar}), 134.8 (C^{ar}), 131.8 (C^{ar}), 131.5 (C^{ar}), 131.4 (C^{ar}), 123.3 (C^{ar}), 116.9 (C^{ar}), 116.7 (C^{ar}), 114.5 (C^{ar}), 113.2 (C^{ar}), 111.2 (C^{ar}), 56.0 (OCH_3), 55.7 (OCH_3), 55.2 (OCH_3), 43.3 (NCH_2), 43.1 (NCH_2), 15.6 (CH_2CH_3); *m/z* (%) 352 (13), 338 (100), 323 (32), 308 (16), 156 (16), 142 (42), 127 (23).

4i: yield: 100% (quant.); υ_{max}(ATR)/cm^{-1} 3452, 3128, 2979, 2935, 2839, 1616, 1601, 1559, 1524, 1508, 1439, 1387, 1351, 1294, 1272, 1251, 1216, 1177, 1133, 1111, 1091, 1041, 1019, 896, 840, 806, 761, 742; ^{1}H NMR (300 MHz, $CDCl_3$) δ 10.04 (s, 1H, H^{ar}), 7.16 (d, *J* = 8.8 Hz, 2H, H^{ar}), 7.1–6.9 (m, 3H, H^{ar}), 6.83 (d, *J* = 8.8 Hz, 2H, H^{ar}), 4.3–4.1 (m, 4H, NCH_2), 3.81 (s, 3H, OCH_3), 3.72 (s, 3H, OCH_3), 1.5–1.4 (m, 6H, CH_2CH_3); ^{13}C NMR (75.5 MHz, $CDCl_3$) δ 160.8 (C^{ar}), 153.4–150.1 (m, C^{ar}), 149.1 (C^{ar}), 135.0 (C^{ar}), 131.9–131.7 (m, C^{ar}), 130.3 (C^{ar}), 127.9–127.3 (m, C^{ar}), 127.3 (C^{ar}), 118.0–117.8 (m, C^{ar}), 116.9 (C^{ar}), 116.3 (C^{ar}), 114.5 (C^{ar}), 113.7 (C^{ar}), 56.1 (OCH_3), 55.2 (OCH_3), 43.3 (NCH_2), 43.2 (NCH_2), 15.6 (CH_2CH_3), 15.5 (CH_2CH_3); *m/z* (%) 355 (3) [M^+], 340 (13), 326 (100), 311 (53), 142 (24), 127 (17).

4j: yield: 100% (quant.); υ_{max}(ATR)/cm^{-1} 3437, 3128, 2977, 2935, 2838, 1608, 1593, 1559, 1518, 1497, 1456, 1386, 1350, 1290, 1250, 1176, 1111, 1091, 1050, 1040, 1017, 964, 915, 846, 829, 805, 771, 743, 721, 682, 655; ^{1}H NMR (300 MHz, $CDCl_3$) δ 10.08 (s, 1H, H^{ar}), 7.40 (s, 1H, H^{ar}), 7.3–7.1 (m, 3H, H^{ar}), 6.9–6.8 (m, 3H, H^{ar}), 4.2–4.1 (m, 4H, NCH_2), 3.82 (s, 3H, OCH_3), 3.73 (s, 3H, H^{ar}), 1.5–1.4 (m, 6H, CH_2CH_3); ^{13}C NMR (75.5 MHz, $CDCl_3$) δ 160.8 (C^{ar}), 160.7 (C^{ar}), 157.1 (C^{ar}), 135.1 (C^{ar}), 134.9 (C^{ar}), 134.7 (C^{ar}), 132.0 (C^{ar}), 131.7 (C^{ar}), 131.5 (C^{ar}), 131.4 (C^{ar}), 130.2 (C^{ar}), 118.1 (C^{ar}), 116.7 (C^{ar}), 116.3 (C^{ar}), 114.6 (C^{ar}), 114.5 (C^{ar}), 112.2 (C^{ar}), 112.0 (C^{ar}), 56.3 (OCH_3), 55.2 (OCH_3), 43.3 (NCH_2), 43.1 (NCH_2), 15.6 (CH_2CH_3), 15.5 (CH_2CH_3); *m/z* (%) 418 (3) [M^+], 416 (3) [M^+], 402 (17), 400 (16), 388 (63), 386 (67), 373 (37), 371 (34), 308 (43), 142 (100), 127 (53).

4n: yield: 100% (quant.); υ_{max}(ATR)/cm^{-1} 2973, 2937, 2837, 1606, 1549, 1515, 1482, 1462, 1413, 1397, 1351, 1293, 1250, 1177, 1149, 1110, 1091, 1022, 998, 892, 840, 799, 775, 724, 687; ^{1}H NMR (300 MHz, $CDCl_3$) δ 10.05 (s, 1H, H^{ar}), 7.3–7.2 (m, 3H, H^{ar}), 6.9–6.8 (m, 3H, H^{ar}), 4.3–4.1 (m, 4H, NCH_2), 3.77 (s, 3H, OCH_3), 3.75 (s, 3H, OCH_3), 3.69 (s, 3H, OCH_3), 1.50 (t, *J* = 7.3 Hz, 3H, CH_2CH_3), 1.43 (t, *J* = 7.3 Hz, 3H, CH_2CH_3); ^{13}C NMR (75.5 MHz, $CDCl_3$) δ 160.9 (C^{ar}), 152.5 (C^{ar}), 150.1 (C^{ar}), 135.3 (C^{ar}), 132.2 (C^{ar}), 131.8 (C^{ar}), 131.5 (C^{ar}), 129.9 (C^{ar}), 122.6 (C^{ar}), 116.4 (C^{ar}), 115.5 (C^{ar}), 114.4 (C^{ar}), 114.1 (C^{ar}), 92.5 (C^{ar}), 60.5 (OCH_3), 56.4 (OCH_3), 55.2 (OCH_3), 43.5 (NCH_2), 43.2 (NCH_2), 15.6 (CH_2CH_3), 15.4 (CH_2CH_3); *m/z* (%) 464 (1) [M^+], 330 (2), 239 (12), 210 (12), 135 (83), 57 (98), 43 (100).

4.3.3. Synthesis of Chloridogold(I) Complexes 5—Typical Procedure

Imidazolium salts **4** (1.00 equiv.) were dissolved in CH_2Cl_2 (30 mL/mmol) and Ag_2O (0.60 equiv.) was added. The reaction mixture was stirred at room temperature for 6 h at r.t. in the dark. Thereupon, AuCl(SMe2) (1.10 equiv.) was added, and the reaction mixture was stirred at room temperature for 24 h in the dark. The mixture was filtered through Celite®, the filtrate was concentrated in vacuum, and the product was recrystallized from CH_2Cl_2/

n-pentane or CH_2Cl_2/*n*-hexane. The products were obtained as colorless or off-white solids or gums in yields of 72–96%.

5a: yield: 96%; mp 140 °C; υ_{max}(ATR)/cm^{-1} 2971, 2839, 1635, 1599, 1574, 1519, 1503, 1459, 1414, 1371, 1344, 1315, 1291, 1247, 1221, 1175, 1157, 1109, 1095, 1055, 1026, 1014, 960, 834, 821, 810, 788, 738, 724, 691, 658; ^{1}H NMR (500 MHz, $CDCl_3$) δ 7.30–7.16 (m, 2H, H^{ar}), 7.12–7.08 (m, 2H, H^{ar}), 7.07–7.03 (m, 2H, H^{ar}), 6.89–6.85 (m, 2H, H^{ar}), 4.16 (dq, *J* = 7.2 Hz, 3.8 Hz, 4H, NCH_2), 3.80 (s, 3H, OCH_3), 1.29 (t, *J* = 7.2 Hz, 6H, CH_2CH_3); ^{13}C NMR (125 MHz, $CDCl_3$) δ 169.1 (NCN), 163.2 (d, $J_{C\text{-}F}$ = 245.7 Hz, C^{ar}), 160.4 (C^{ar}), 132.6 (d, $J_{C\text{-}F}$ = 8.5 Hz, C^{ar}), 131.9 (C^{ar}), 131.4 (C^{ar}), 130.0 (C^{ar}), 124.0 (C^{ar}), 119.5 (C^{ar}), 116.4 (C^{ar}), 116.2 (C^{ar}), 114.5 (C=C), 55.42 (OCH_3), 44.49 (NCH_2), 44.43 (NCH_2), 17.05 (CH_2CH_3); *m*/*z* (%) 568 (34) [M^+], 566 (100) [M^+], 521 (83) [M^+–Cl], 519 (48), 492 (17), 323 (52) [M^+–AuCl].

5b: yield: 84%; mp 186 °C; υ_{max}(ATR)/cm^{-1} 2994, 2971, 2928, 2838, 1635, 1608, 1591, 1574, 1514, 1489, 1459, 1442, 1406, 1365, 1344, 1315, 1290, 1247, 1174, 1108, 1091, 1056, 1028, 1013, 963, 841, 831, 810, 791, 733, 722, 710, 690; ^{1}H NMR (500 MHz, $CDCl_3$) δ 7.35–7.31 (m, 2H, H^{ar}), 7.15–7.12 (m, 2H, H^{ar}), 7.12–7.08 (m, 2H, H^{ar}), 6.90–6.83 (m, 2H, H^{ar}), 4.17 (vdq, *J* = 11 Hz, 7.3 Hz, 4H, NCH_2), 3.80 (s, 3H, OCH_3), 1.29 (t, *J* = 7.1 Hz, 6H, CH_2CH_3); ^{13}C NMR (125 MHz, $CDCl_3$) δ 169.4 (NCN), 160.5 (C^{ar}), 135.7 (C^{ar}), 131.9 (C^{ar}), 131.8 (C^{ar}), 131.5 (C^{ar}), 129.8 (C^{ar}), 129.4 (C^{ar}), 126.5 (C^{ar}), 119.4 (C^{ar}), 114.6 (C=C), 55.43 (OCH_3), 44.55 (NCH_2), 17.06 (CH_2CH_3); *m*/*z* (%) 574 (71) [M^+], 572 (100) [M^+], 537 (72) [M^+–Cl], 535 (74), 339 (62) [M^+–AuCl].

5c: yield: 87%; mp. 183 °C; υ_{max}(ATR)/cm^{-1} 2972, 2828, 2837, 1634, 1606, 1587, 1573, 1513, 1487, 1459, 1405, 1390, 1365, 1314, 1289, 1246, 1174, 1108, 1085, 1071, 1055, 1027, 1009, 963, 840, 829, 809, 789, 736, 723, 704, 690; ^{1}H NMR (500 MHz, $CDCl_3$) δ 7.51–7.47 (m, 2H, H^{ar}), 7.12–7.04 (m, 4H, H^{ar}), 6.87 (dd, *J* = 9.3 Hz, 7.5 Hz, 2H, H^{ar}), 4.17 (dq, *J* = 12.8 Hz, 7.2 Hz, 4H, NCH_2), 3.80 (s, 3H, OCH_3), 1.29 (dt, *J* = 7.2 Hz, 1.7 Hz, 6H, CH_2CH_3); ^{13}C NMR (125 MHz, $CDCl_3$) δ 169.5 (NCN), 160.5 (C^{ar}), 132.3 (C^{ar}), 132.1 (C^{ar}), 131.9 (C^{ar}), 131.4 (C^{ar}), 129.8 (C^{ar}), 127.0 (C^{ar}), 119.3 (C^{ar}), 114.6 (C=C), 114.4 (C=C), 55.43 (OCH_3), 44.56 (NCH_2C), 44.44 (NCH_2), 17.07 (CH_2CH_3); *m*/*z* (%) 618 (100) [M^+], 616 (72) [M^+], 583 (64) [M^+–Cl], 581 (86) [M^+–Cl], 383 (42) [M^+–AuCl], 134 (41).

5d: yield: 86%; mp 188 °C; υ_{max}(ATR)/cm^{-1} 2998, 2970, 2931, 2833, 1635, 1608, 1573, 1512, 1482, 1461, 1414, 1387, 1346, 1315, 1290, 1245, 1175, 1109, 1082, 1061, 1028, 1005, 841, 801, 736, 722, 692; ^{1}H NMR (500 MHz, $CDCl_3$) δ 7.71–7.66 (m, 2H, H^{ar}), 7.11–7.08 (m, 2H, H^{ar}), 6.94–6.91 (m, 2H, H^{ar}), 6.90–6.86 (m, 2H, H^{ar}), 4.17 (dq, *J* = 14 Hz, 7.2 Hz, 4H, NCH_2), 3.80 (s, 3H, OCH_3), 1.29 (dt, *J* = 7.2 Hz, 3.2 Hz, 6H, CH_2CH_3); ^{13}C NMR (125 MHz, $CDCl_3$) δ 169.5 (NCN), 160.5 (C^{ar}), 138.3 (C^{ar}), 132.1 (C^{ar}), 131.9 (C^{ar}), 131.4 (C^{ar}), 130.0, 127.5 (C^{ar}), 119.3 (C^{ar}), 114.5 (C=C), 114.4 (C=C), 55.43 (OCH_3), 44.56 (NCH_2), 44.44 (NCH_2), 17.04 (CH_2CH_3); *m*/*z* (%) 666 (37) [M^+], 664 (100) [M^+], 629 (62) [M^+–Cl], 431 (27) [M^+–AuCl]

5e: yield: 72%; mp 205 °C; υ_{max}(ATR)/cm^{-1} 3974, 2928, 2840, 1618, 1590, 1516, 1463, 1433, 1416, 1366, 1345, 1326, 1292, 1274, 1252, 1176, 1117, 1092, 1053, 1029, 986, 868, 836, 812, 787, 766, 687; ^{1}H NMR (500 MHz, $CDCl_3$) δ 7.13–7.08 (m, 2H, H^{ar}), 6.93–6.88 (m, 2H, H^{ar}), 6.83 (tt, *J* = 8.8 Hz, 2.3 Hz, 1H, H^{ar}), 6.76–6.72 (m, 2H, H^{ar}), 4.22 (q, *J* = 7.2 Hz, 2H, NCH_2), 4.15 (q, ^3J = 7.2 Hz, 2H, NCH_2), 3.82 (s, 3H, OCH_3), 1.31 (dt, *J* = 14 Hz, 7.2 Hz, 6H, CH_2CH_3); ^{13}C NMR (125 MHz, $CDCl_3$) δ 170.1 (NCN), 162.9 (dd, $J_{C\text{-}F}$ = 245.7 Hz, 12.6 Hz, C^{ar}), 160.7 (C^{ar}), 131.9 (C^{ar}), 131.8 (C^{ar}), 131.0 (C^{ar}), 128.6 (C^{ar}), 118.9 (C^{ar}), 114.7 (C=C), 113.9 (C=C), 55.47 (OCH_3), 44.73 (NCH_2), 44.52 (NCH_2), 17.07 (CH_2CH_3); *m*/*z* (%) 576 (33) [M^+], 574 (100) [M^+], 539 (67) [M^+–Cl], 341 (42) [M^+–AuCl].

5f: yield: 87%; mp 260 °C; υ_{max}(ATR)/cm^{-1} 3064, 2974, 2928, 2837, 1640, 1609, 1583, 1558, 1515, 1461, 1440, 1416, 1378, 1345, 1309, 1293, 1254, 1176, 1132, 1119, 1103, 1032, 974, 867, 851, 831, 803, 746, 685; ^{1}H NMR (500 MHz, $CDCl_3$) δ 7.37 (t, *J* = 1.9 Hz, 1H, H^{ar}), 7.13–7.09 (m, 4H, H^{ar}), 6.94–6.88 (m, 2H, H^{ar}), 4.17 (vdq, *J* = 20 Hz, 7.2 Hz, 4H, NCH_2), 3.82 (s, 3H, OCH_3), 1.31 (vdt, *J* = 17 Hz, 7.1 Hz, 6H, CH_2CH_3); ^{13}C NMR (125 MHz, $CDCl_3$) δ 170.1 (NCN), 160.7 (C^{ar}), 135.7 (C^{ar}), 132.1 (C^{ar}), 131.8 (C^{ar}), 131.0 (C^{ar}), 129.8 (C^{ar}), 128.9 (C^{ar}), 128.2 (C^{ar}), 118.7 (C^{ar}), 114.7 (C=C), 55.48 (OCH_3), 44.73 (NCH_2), 44.53 (NCH_2), 17.14

(CH_2CH_3), 17.03 (CH_2CH_3); m/z (%) 610 (32) [M^+], 608 (100) [M^+], 606 (97) [M^+], 573 (33) [M^+–Cl], 571 (74) [M^+–Cl], 569 (46), 375 (26) [M^+–AuCl], 373 (41) [M^+–AuCl].

5g: yield: 92%; mp 266 °C; υ_{max}(ATR)/cm^{-1} 3068, 2972, 2928, 2840, 1604, 1578, 1542, 1514, 1461, 1412, 1376, 1345, 1293, 1254, 1176, 1032, 972, 865, 846, 808, 751, 736, 684; ^{1}H NMR (500 MHz, $CDCl_3$) δ 7.67 (t, J = 1.7 Hz, 1H, H^{ar}), 7.30 (d, J = 1.7 Hz, 2H, H^{ar}), 7.14–7.07 (m, 2H, H^{ar}), 6.95–6.89 (m, 2H, H^{ar}), 4.16 (vdq, J = 18.6 Hz, 7.0 Hz, 4H, NCH_2), 3.82 (s, 3H, OCH_3), 1.31 (vdt, J = 20 Hz, 7.2 Hz, 6H, CH_2CH_3); ^{13}C NMR (125 MHz, $CDCl_3$) δ 170.0 (NCN), 160.7 (C^{ar}), 135.2 (C^{ar}), 132.2 (C^{ar}), 131.9 (C^{ar}), 131.4 (C^{ar}), 128.1 (C^{ar}), 123.5 (C^{ar}), 119.7 (C^{ar}), 115.7 (C=C), 55.49 (OCH_3), 44.72 (NCH_2), 44.53 (NCH_2), 17.15 (CH_2CH_3), 17.03 (CH_2CH_3); m/z (%) 698 (75) [M^+], 696 (100) [M^+], 694 (44) [M^+], 663 (35) [M^+–Cl], 661 (87) [M^+–Cl], 650 (68), 463 (30) [M^+–AuCl], 134 (24).

5h: yield: 94%; mp > 100 °C (dec.); υ_{max}(ATR)/cm^{-1} 2963, 2934, 2836, 1610, 1597, 1520, 1506, 1461, 1417, 1377, 1346, 1324, 1291, 1251, 1171, 1139, 1110, 1089, 1022, 974, 886, 839, 809, 764, 737, 694; ^{1}H NMR (300 MHz, $CDCl_3$) δ 7.2–7.1 (m, 2H, H^{ar}), 6.9–6.7 (m, 4H, H^{ar}), 6.61 (s, 1H, H^{ar}), 4.2–4.1 (m, 4H, NCH_2), 3.85 (3 H, s, OCH_3), 3.78 (3 H, s, OCH_3), 3.73 (3 H, s, OCH_3), 1.4–1.2 (m, 6H, CH_2CH_3); ^{13}C NMR (75.5 MHz, $CDCl_3$) δ 168.7 (NCN), 160.2 (C^{ar}), 149.7 (C^{ar}), 148.9 (C^{ar}), 131.9 (C^{ar}), 131.7 (C^{ar}), 130.9 (C^{ar}), 123.3 (C^{ar}), 120.1(C^{ar}), 119.9 (C^{ar}), 114.4 (C^{ar}), 114.3 (C^{ar}), 113.3 (C^{ar}), 111.2 (C^{ar}), 55.9 (OCH_3), 55.8 (OCH_3), 55.3 (OCH_3), 44.3 (NCH_2), 44.2 (NCH_2), 17.0 (CH_2CH_3), 16.9 (CH_2CH_3); m/z (%) 600 (35) [M^+], 598 (100) [M^+], 563 (23) [M^+–Cl], 561 (17) [M^+–Cl], 365 (26), 50 (24).

5i: yield: 94%; mp >120 °C (dec.); υ_{max}(ATR)/cm^{-1} 2977, 2935, 2840, 1617, 1599, 1575, 1522, 1506, 1459, 1434, 1405, 1380, 1342, 1291, 1272, 1249, 1180, 1135, 1108, 1090, 1048, 1022, 976, 896, 872, 840, 827, 808, 781, 761, 736, 696; ^{1}H NMR (300 MHz, $CDCl_3$) δ 7.08 (d, J = 8.8 Hz, 2H, H^{ar}), 6.9–6.8 (m, 5H, H^{ar}), 4.2–4.1 (m, 4H, NCH_2), 3.86 (s, 3H, OCH_3), 3.79 (s, 3H, OCH_3), 1.4–1.3 (m, 6H, CH_2CH_3); ^{13}C NMR (75.5 MHz, $CDCl_3$) δ 169.0 (NCN), 160.3 (C^{ar}), 150.6–153.3 (m, C^{ar}), 148.3 (C^{ar}), 131.8–131.7 (m, C^{ar}), 131.2 (C^{ar}), 126.9–126.8 (m, C^{ar}), 120.3 (C^{ar}), 119.4 (C^{ar}), 118.3–118.0 (m, C^{ar}), 114.5–114.3 (m, C^{ar}), 113.4 (C^{ar}), 56.1 (OCH_3), 55.3 (OCH_3), 44.3 (NCH_2), 16.9 (CH_2CH_3); m/z (%) 588 (34) [M^+], 586 (100) [M^+], 551 (34) [M^+–Cl], 549 (31) [M^+–Cl], 353 (44) [M^+–AuCl], 50 (26).

5j: yield: 73%; υ_{max}(ATR)/cm^{-1} 2969, 2837, 1608, 1593, 1516, 1495, 1460, 1411, 1378, 1345, 1288, 1249, 1176, 1109, 1089, 1050, 1021, 888, 848, 832, 809, 721, 682; ^{1}H NMR (300 MHz, $CDCl_3$) δ 7.40 (1 H, s, H^{ar}), 7.1–7.0 (m, 3H, H^{ar}), 6.9–6.8 (m, 3H, H^{ar}), 4.2–4.1 (m, 4H, NCH2), 3.87 (s, 3H, OCH_3), 3.79 (s, 3H, OCH_3), 1.3–1.2 (m, 6H, CH_2CH_3); ^{13}C NMR (75.5 MHz, $CDCl_3$) δ 169.0 (NCN), 160.3 (C^{ar}), 156.7 (C^{ar}), 135.1 (C^{ar}), 131.9 (C^{ar}), 131.8 (C^{ar}), 131.3 (C^{ar}), 131.0 (C^{ar}), 129.3 (C^{ar}), 121.2 (C^{ar}), 119.4 (C^{ar}), 114.4 (C^{ar}), 114.3 (C^{ar}), 112.0 (C^{ar}), 111.8 (C^{ar}), 56.3 (OCH_3), 55.3 (OCH_3), 44.3 (NCH_2), 17.0 (CH_2CH_3), 16.9 (CH_2CH_3); m/z (%) 648 (66) [M^+], 646 (48) [M^+], 611 (20) [M^+–Cl], 570 (36), 568 (100), 533 (28), 531 (26), 335 (41), 135 (24), 50 (36).

5l: yield: 75%; mp >110 °C (dec.); υ_{max}(ATR)/cm^{-1} 2973, 2938, 2833, 1608, 1563, 1515, 1492, 1461, 1411, 1319, 1293, 1252, 1175, 1109, 1091, 1043, 1003, 901, 835, 812, 755, 709; ^{1}H NMR (300 MHz, $CDCl_3$) δ 7.11 (d, J = 8.8 Hz, 2H, H^{ar}), 6.88 (d, J = 8.8 Hz, 2H, H^{ar}), 6.85 (s, 1H, H^{ar}), 6.54 (s, 1H, H^{ar}), 4.2–4.1 (m, 4H, NCH_2), 3.85 (s, 3H, OCH_3), 3.80 (s, 3H, OCH_3), 3.70 (s, 3H, OCH_3), 1.34 (t, J = 7.2 Hz, 3H, CH_2CH_3), 1.28 (t, J = 7.2 Hz, 3H, CH_2CH_3); ^{13}C NMR (75.5 MHz, $CDCl_3$) δ 169.3 (NCN), 160.4 (C^{ar}), 153.8 (C^{ar}), 146.2 (C^{ar}), 131.8 (C^{ar}), 131.2 (C^{ar}), 129.6 (C^{ar}), 128.7 (C^{ar}), 123.9 (C^{ar}), 119.3 (C^{ar}), 114.5 (C^{ar}), 114.3 (C^{ar}), 113.2 (C^{ar}), 60.8 (OCH_3), 56.2 (OCH_3), 55.3 (OCH_3), 44.5 (NCH_2), 44.3 (NCH_2), 17.1 (CH_2CH_3), 16.9 (CH_2CH_3); m/z (%) 634 (36) [M^+], 632 (59) [M^+], 586 (100), 567 (33), 549 (34), 353 (52), 134 (33), 50 (100).

5m: yield: 84%; mp >100 °C (dec.); υ_{max}(ATR)/cm^{-1} 2970, 2934, 2833, 1607, 1589, 1555, 1514, 1488, 1461, 1410, 1346, 1319, 1292, 1250, 1176, 1156, 1110, 1090, 1027, 998, 896, 840, 810, 754, 699, 662; ^{1}H NMR (300 MHz, $CDCl_3$) δ 7.11 (d, J = 8.8 Hz, 2H, H^{ar}), 7.02 (s, 1H, H^{ar}), 6.89 (d, J = 8.8 Hz, 2H, H^{ar}), 6.57 (s, 1H, H^{ar}), 4.2–4.1 (m, 4H, NCH_2), 3.84 (s, 3H, OCH_3), 3.80 (s, 3H, OCH_3), 3.69 (s, 3H, OCH_3), 1.4–1.2 (m, 6H, CH_2CH_3); ^{13}C NMR (75.5 MHz, $CDCl_3$) δ 169.3 (NCN), 160.4 (C^{ar}), 153.6 (C^{ar}), 147.2 (C^{ar}), 131.7 (C^{ar}), 131.3 (C^{ar}), 129.4

(C^{ar}), 126.6 (C^{ar}), 124.6 (C^{ar}), 119.3 (C^{ar}), 118.0 (C^{ar}), 114.5 (C^{ar}), 113.9 (C^{ar}), 60.7 (OCH_3), 56.1 (OCH_3), 55.3 (OCH_3), 44.5 (NCH_2), 44.3 (NCH_2), 17.1 (CH_2CH_3), 16.9 (CH_2CH_3); *m/z* (%) 678 (100) [M^+], 676 (73) [M^+], 643 (67) [M^+–Cl], 641 (76) [M^+–Cl], 445 (14) [M^+–AuCl], 443 (14) [M^+–AuCl], 50 (50).

5n: yield: 50%; mp >100 °C (dec.); υ_{max}(ATR)/cm^{-1} 2970, 2933, 2835, 1606, 1585, 1513, 1482, 1460, 1408, 1345, 1317, 1291, 1249, 1176, 1153, 1109, 1089, 1025, 997, 894, 839, 800, 752, 692, 659; ^{1}H NMR (300 MHz, $CDCl_3$) δ 7.23 (s, 1H, H^{ar}), 7.11 (d, *J* = 8.8 Hz, 2H, H^{ar}), 6.88 (d, *J* = 8.8 Hz, 2H, H^{ar}), 6.59 (s, 1H, H^{ar}), 4.2–4.1 (m, 4H, NCH_2), 3.82 (s, 3H, OCH_3), 3.79 (s, 3H, OCH_3), 3.67 (s, 3H, OCH_3), 1.4–1.2 (m, 6H, CH_2CH_3); ^{13}C NMR (75.5 MHz, $CDCl_3$) δ 169.2 (NCN), 160.4 (C^{ar}), 152.4 (C^{ar}), 149.7 (C^{ar}), 132.2 (C^{ar}), 131.9 (C^{ar}), 131.2 (C^{ar}), 129.2 (C^{ar}), 125.4 (C^{ar}), 119.3 (C^{ar}), 115.0 (C^{ar}), 114.4 (C^{ar}), 114.3 (C^{ar}), 92.5 (C^{ar}), 60.5 (OCH_3), 56.3 (OCH_3), 55.3 (OCH_3), 44.4 (NCH_2), 44.3 (NCH_2), 17.1 (CH_2CH_3), 16.9 (CH_2CH_3); *m/z* (%) 726 (7) [M^+], 724 (24) [M^+], 689 (84) [M^+–Cl], 660 (36), 563 (42), 533 (83), 142 (34), 50 (100).

4.3.4. Synthesis of Iodidogold(I) Complexes 6—Typical Procedure

Chlorido-Au(I) complexes **5** (1.00 equiv.) were dissolved in acetone (45.0 mL/mmol) and treated with KI (4.00 equiv.). The reaction mixture was stirred at room temperature for 24 h. The solvent was removed in vacuum, and the residue was resuspended in CH_2Cl_2. The suspension was filtered through Celite® and a plug of silicate, the filtrate was concentrated in vacuum, and the residue was recrystallized from CH_2Cl_2/*n*-pentane or CH_2Cl_2/*n*-hexane. The products were obtained as colorless solids in yields of 63–100%.

6a: yield: 96%; mp 168 °C; υ_{max}(ATR)/cm^{-1} 2972, 2931, 2838, 1631, 1599, 1573, 1519, 1503, 1459, 1412, 1378, 1345, 1314, 1291, 1248, 1223, 1175, 1157, 1109, 1093, 1054, 1041, 1026, 958, 836, 821, 811, 739, 723, 688, 658; ^{1}H NMR (500 MHz, $CDCl_3$) δ 7.22–7.16 (m, 2H, H^{ar}), 7.12–7.08 (m, 2H, H^{ar}), 7.08–7.03 (m, 2H, H^{ar}), 6.89–6.85 (m, 2H, H^{ar}), 4.16 (vdq, *J* = 7.3 Hz, 3.9 Hz, 4H, NCH_2), 3.80 (s, 3H, OCH_3), 1.31 (vdt, *J* = 7.0 Hz, 1.7 Hz, 6H, CH_2CH_3); ^{13}C NMR (125 MHz, $CDCl_3$) δ 179.8 (NCN), 163.2 (d, $J_{C\text{-}F}$ = 245.7 Hz, C^{ar}), 160.4 (C^{ar}), 132.6 (C^{ar}), 132.5 (C^{ar}), 131.9 (C^{ar}), 131.2 (C^{ar}), 129.8 (C^{ar}), 119.5 (C^{ar}), 116.4 (C^{ar}), 116.2 (C^{ar}), 114.5 (C=C), 55.43 (OCH_3), 44.14 (NCH_2), 17.09 (CH_2CH_3); *m/z* (%) 648 (63) [M^+], 521 (100) [M^+–I]; anal. calcd. for $C_{20}H_{21}AuFIN_2O$: C, 37.06; H, 3.27; N, 4.32. Found: C, 37.74; H, 3.31; N, 4.29.

6b: yield: 100% (quant.); mp 175 °C; υ_{max}(ATR)/cm^{-1} 2970, 2828, 2833, 1635, 1607, 1572, 1513, 1489, 1460, 1408, 1377, 1344, 1291, 1248, 1175, 1090, 1025, 1013, 835, 795, 734; ^{1}H-NMR (500 MHz, $CDCl_3$) δ 7.36–7.32 (m, 2H, H^{ar}), 7.17–7.12 (m, 2H, H^{ar}), 7.12–7.08 (m, 2H, H^{ar}), 6.91–6.85 (m, 2H, H^{ar}), 4.17 (vdq, *J* = 11 Hz, 7.3 Hz, 4H, NCH_2), 3.80 (s, 3H, OCH_3), 1.31 (t, *J* = 7.4 Hz, 6H, CH_2CH_3); ^{13}C NMR (125 MHz, $CDCl_3$) δ 180.1 (NCN), 160.5 (C^{ar}), 135.7 (C^{ar}), 131.9 (C^{ar}), 131.8 (C^{ar}), 131.3 (C^{ar}), 129.7 (C^{ar}), 129.4 (C^{ar}), 126.5 (C^{ar}), 119.4 (C^{ar}), 114.6 (C=C), 55.44 (OCH_3), 44.20 (NCH_2), 44.10 (NCH_2), 17.11 (CH_2CH_3); *m/z* (%) 666 (18) [M^+], 664 (53) [M^+], 539 (36) [M^+–I], 537 (100) [M^+–I]; anal. calcd. for $C_{20}H_{21}AuClIN_2O$: C, 36.14; H, 3.18; N, 4.21. Found: C, 36.81; H, 3.15; N, 4.17.

6c: yield: 96%; mp 170 °C; υ_{max}(ATR)/cm^{-1} 2969, 2930, 2836, 1635, 1607, 1586, 1573, 1513, 1485, 1459, 1440, 1416, 1384, 1369, 1345, 1314, 1291, 1245, 1174, 1109, 1099, 1082, 1069, 1028, 1009, 955, 833, 812, 795, 735, 724, 703, 688, 662; ^{1}H NMR (500 MHz, $CDCl_3$) δ 7.51–7.47 (m, 2H, H^{ar}), 7.12–7.06 (m, 4H, H^{ar}), 6.90–6.84 (m, 2H, H^{ar}), 4.18 (vdq, *J* = 12.8 Hz, 7.2 Hz, 4H, NCH_2), 3.81 (s, 3H, OCH_3), 1.31 (vdt, *J* = 7.2 Hz, 2.2 Hz, 6H, CH_2CH_3); ^{13}C NMR (125 MHz, $CDCl_3$) δ 180.2 (NCN), 160.5 (C^{ar}), 132.3 (C^{ar}), 132.1 (C^{ar}), 131.9 (C^{ar}), 131.3 (C^{ar}), 131.0 (C^{ar}), 127.0 (C^{ar}), 119.4 (C^{ar}), 114.6 (C=C), 114.4 (C=C), 55.44 (OCH_3), 44.21 (NCH_2), 44.10 (NCH_2), 17.11 (CH_2CH_3); *m/z* (%) 710 (53) [M^+], 708 (54) [M^+], 660 (23), 583 (92) [M^+–I], 581 (100) [M^+–I], 533 (47); anal. calcd. for $C_{20}H_{21}AuBrIN_2O$: C, 33.87; H, 2.98; N, 3.95. Found: C, 34.71; H, 2.94; N, 4.07.

6d: yield: 91%; mp 143 °C; υ_{max}(ATR)/cm^{-1} 2994, 2968, 2928, 2834, 1634, 1607, 1583, 1512, 1482, 1460, 1415, 1385, 1369, 1345, 1314, 1291, 1245, 1174, 1109, 1082, 1061, 1027, 1005, 956, 831, 812, 794, 735, 722, 697; ^{1}H NMR (500 MHz, $CDCl_3$) δ 7.71–7.68 (m, 2H, H^{ar}), 7.13–7.07 (m, 2H, H^{ar}), 6.97–6.91 (m, 2H, H^{ar}), 6.90–6.83 (m, 2H, H^{ar}), 4.17 (vdq, *J* = 14.2 Hz,

7.1 Hz, 4H, NCH_2), 3.81 (s, 3H, OCH_3), 1.31 (vdt, J = 7.2 Hz, 3.3 Hz, 6H, CH_2CH_3); ^{13}C NMR (125 MHz, $CDCl_3$) δ 180.2 (NCN), 160.5 (C^{ar}), 138.3 (C^{ar}), 132.1 (C^{ar}), 131.9 (C^{ar}), 131.2 (C^{ar}), 129.8 (C^{ar}), 127.5 (C^{ar}), 119.4 (C^{ar}), 114.6 (C=C), 114.4 (C=C), 55.44 (OCH_3), 44.21 (NCH_2), 44.09 (NCH_2), 17.13 (CH_2CH_3), 17.09 (CH_2CH_3); m/z (%) 756 (67) [M^+], 660 (23), 629 (100), 533 (43) [M^+–I], 502 (20); anal. calcd. for $C_{20}H_{21}AuI_2N_2O$: C, 31.77; H, 2.80; N, 3.70. Found: C, 32.26; H, 2.83; N, 3.67.

6e: yield: 93%; mp 164 °C; υ_{max}(ATR)/cm^{-1} 2965, 2928, 2833, 1616, 1589, 1513, 1459, 1446, 1432, 115, 1371, 1293, 1273, 1249, 1176, 1122, 1090, 1052, 1030, 1011, 991, 961, 880, 863, 836, 810, 786, 763, 736, 720, 686; ^{1}H NMR (500 MHz, $CDCl_3$) δ 7.14–7.09 (m, 2H, H^{ar}), 6.92–6.88 (m, 2H, H^{ar}), 6.83 (tt, J = 8.8 Hz, 2.3 Hz, 1H, H^{ar}), 6.77–6.71 (m, 2H, H^{ar}), 4.23 (q, J = 7.4 Hz, 2H, NCH_2), 4.16 (q, J = 7.3 Hz, 2H, NCH_2), 3.82 (s, 3H, OCH_3), 1.33 (dt, J = 14 Hz, 7.2, 6H, CH_2CH_3); ^{13}C NMR (125 MHz, $CDCl_3$) δ 181.7 (NCN), 162.9 (dd, $J_{C\text{-}F}$ = 252.2 Hz, 13.5 Hz, C^{ar}), 160.7 (C^{ar}), 131.9 (C^{ar}), 131.8 (C^{ar}), 131.1 (C^{ar}), 128.4 (C^{ar}), 119.9 (C^{ar}), 114.7 (C=C), 113.9 (C=C), 55.47 (OCH_3), 44.37 (NCH_2), 44.17 (NCH_2), 17.07 (CH_2CH_3); m/z (%) 666 (53) [M^+], 539 (100) [M^+–I]; anal. calcd. for $C_{20}H_{20}AuF_2IN_2O$: C, 36.05; H, 3.03; N, 4.20. Found: C, 36.49; H, 3.00; N, 4.26.

6f: yield: 93%; mp 163 °C; υ_{max}(ATR)/cm^{-1} 3078, 2976, 2952, 2931, 2835, 1631, 1605, 1583, 1558, 1512, 1440, 1415, 1379, 1346, 1305, 1292, 1250, 1175, 1126, 1110, 1090, 1051, 1027, 1007, 972, 888, 866, 850, 826, 802, 744, 682; ^{1}H NMR (500 MHz, $CDCl_3$) δ 7.37 (t, J = 1.9 Hz, 1H, 1H, H^{ar}), 7.15–7.10 (m, 4H, H^{ar}), 6.91 (d, J = 8.4 Hz, 2H, H^{ar}), 4.17 (vdq, J = 20 Hz, 7.4 Hz, 4H, NCH_2), 3.82 (s, 3H, OCH_3), 1.32 (vdt, J = 17.8 Hz, 7.1 Hz, 6H, CH_2CH_3); ^{13}C NMR (125 MHz, $CDCl_3$) δ 180.5 (NCN), 160.7 (C^{ar}), 135.6 (C^{ar}), 132.0 (C^{ar}), 131.9 (C^{ar}), 131.0 (C^{ar}), 129.7 (C^{ar}), 128.9 (C^{ar}), 128.1 (C^{ar}), 118.8 (C^{ar}), 114.7 (C=C), 55.48 (OCH_3), 44.36 (NCH_2), 44.18 (NCH_2), 17.18 (CH_2CH_3), 17.08 (CH_2CH_3); m/z (%) 700 (33) [M^+], 698 (52) [M^+], 573 (64) [M^+–I], 571 (100) [M^+–I]; anal. calcd. for $C_{20}H_{20}AuCl_2IN_2O$: C, 34.36; H, 2.88; N, 4.01. Found: C, 34.68; H, 2.85; N, 4.04.

6g: yield: 83%; mp 177 °C; υ_{max}(ATR)/cm^{-1} 3064, 2972, 2931, 2833, 1628, 1605, 1579, 1544, 1512, 1460, 1438, 1409, 1377, 1344, 1306, 1292, 1250, 1175, 1102, 1050, 1026, 971, 888, 845, 806, 752, 734, 682; ^{1}H NMR (500 MHz, $CDCl_3$) δ 7.68 (t, J = 1.8 Hz, 1H, H^{ar}), 7.31 (d, J = 1.7 Hz, 2H, H^{ar}), 7.12 (d, J = 8.7 Hz, 2H, H^{ar}), 6.91 (d, J = 8.7 Hz, 2H, H^{ar}), 4.17 (vdq, J = 17 Hz, 7.1 Hz, 4H, NCH_2), 3.83 (s, 3H, OCH_3), 1.32 (vdt, J = 21 Hz, 7.2 Hz, 6H, CH_2CH_3); ^{13}C NMR (125 MHz, $CDCl_3$) δ 180.7 (NCN), 160.7 (C^{ar}), 135.2 (C^{ar}), 132.2 (C^{ar}), 132.0 (C^{ar}), 131.9 (C^{ar}), 131.5 (C^{ar}), 127.9 (C^{ar}), 123.4 (C^{ar}), 118.7 (C^{ar}), 114.7 (C=C), 55.50 (OCH_3), 44.36 (NCH_2), 44.18 (NCH_2), 17.20 (CH_2CH_3), 17.07 (CH_2CH_3); m/z (%) 790 (23) [M^+], 788 (52) [M^+], 786 (25) [M^+], 663 (66) [M^+–I], 661 (100) [M^+–I], 659 (54) [M^+–I]; anal. calcd. for $C_{20}H_{20}AuBr_2IN_2O$: C, 30.48; H, 2.56; N, 3.55. Found: C, 30.81; H, 2.52; N, 3.58.

6h: yield: 63%; mp 181–182 °C; υ_{max}(ATR)/cm^{-1} 2989, 2950, 2931, 2900, 1610, 1598, 1582, 1520, 1506, 1460, 1417, 1379, 1362, 1346, 1309, 1294, 1252, 1237, 1174, 1136, 1090, 1022, 975, 889, 877, 833, 819, 808, 797, 778, 762, 736, 718, 687, 666; ^{1}H NMR (300 MHz, $CDCl_3$) δ 7.11 (d, J = 8.8 Hz, 2H, H^{ar}), 6.9–6.7 (m, 4H, H^{ar}), 6.62 (s, 1H, H^{ar}), 4.2–4.1 (m, 4H, NCH_2), 3.85 (s, 3H, OCH_3), 3.77 (s, 3H, OCH_3), 3.72 (s, 3H, OCH_3), 1.4–1.2 (m, 6H, CH_2CH_3); ^{13}C NMR (75.5 MHz, $CDCl_3$) δ 179.3 (NCN), 160.1 (C^{ar}), 149.6 (C^{ar}), 148.9 (C^{ar}), 131.7 (C^{ar}), 130.7 (C^{ar}), 130.5 (C^{ar}), 123.2 (C^{ar}), 120.1 (C^{ar}), 119.8 (C^{ar}), 114.3 (C^{ar}), 113.3 (C^{ar}), 111.1 (C^{ar}), 55.9 (OCH_3), 55.8 (OCH_3), 55.3 (OCH_3), 44.0 (NCH_2), 43.9 (NCH_2), 17.1 (CH_2CH_3), 16.9 (CH_2CH_3); m/z (%) 690 (27) [M^+], 660 (7), 564 (100) [M^+–I], 547 (13), 533 (17), 282 (8); anal. calcd. for $C_{22}H_{26}AuIN_2O_3$: C, 38.28; H, 3.80; N, 4.06. Found: C, 38.72; H, 3.83; N, 4.03.

6i: yield: 79%; mp 168–169 °C; υ_{max}(ATR)/cm^{-1} 2971, 2938, 2838, 1602, 1574, 1520, 1505, 1459, 1442, 1416, 1371, 1345, 1316, 1302, 1290, 1269, 1245, 1227, 1175, 1133, 1121, 1109, 1089, 1047, 1025, 1009, 977, 961, 898, 887, 837, 807, 777, 762, 736, 720, 687, 666; ^{1}H NMR (300 MHz, $CDCl_3$) δ 7.09 (d, J = 8.5 Hz, 2H, H^{ar}), 6.9–6.8 (m, 5H, H^{ar}), 4.2–4.1 (m, 4H, NCH_2), 3.86 (s, 3H, OCH_3), 3.78 (s, 3H, OCH_3), 1.4–1.2 (m, 6H, CH_2CH_3); ^{13}C NMR (75.5 MHz, $CDCl_3$) δ 179.6 (NCN), 160.3 (C^{ar}), 153.5–150.2 (m, C^{ar}), 148.5 (C^{ar}), 131.8–131.7 (m, C^{ar}), 131.0 (C^{ar}), 129.4 (C^{ar}), 126.9 (C^{ar}), 120.3–120.2 (m, C^{ar}), 119.4 (C^{ar}), 118.3–118.0 (m, (C^{ar})), 114.4–114.3 (m, C^{ar}), 113.4 (C^{ar}), 56.1 (OCH_3), 55.3 (OCH_3), 44.0 (NCH_2), 43.9

(NCH_2), 16.9 (CH_2CH_3); m/z (%) 678 (48) [M^+], 660 (7), 552 (100) [M^+–I], 276 (12); anal. calcd. for $C_{21}H_{23}AuFIN_2O_2$: C, 37.19; H, 3.42; N, 4.13. Found: C, 37.74; H, 3.34; N, 4.28.

6j: yield: 74%; υ_{max}(ATR)/cm^{-1} 2969, 2931, 2835, 1608, 1593, 1573, 1516, 1494, 1410, 1378, 1345, 1288, 1249, 1175, 1147, 1109, 1089, 1050, 1020, 968, 888, 847, 831, 808, 722, 681; ^{1}H NMR (300 MHz, $CDCl_3$) δ 7.40 (s, 1H, H^{ar}), 7.1–7.0 (m, 3H, H^{ar}), 6.9–6.8 (m, 3H, H^{ar}), 4.2–4.1 (m, 4H, NCH_2), 3.87 (s, 3H, OCH_3), 3.78 (s, 3H, OCH_3), 1.4–1.2 (m, 6H, CH_2CH_3); ^{13}C NMR (75.5 MHz, $CDCl_3$) δ 179.6 (NCN), 160.3 (C^{ar}), 156.5 (C^{ar}), 135.0 (C^{ar}), 131.7 (C^{ar}), 131.1 (C^{ar}), 131.0 (C^{ar}), 129.1 (C^{ar}), 121.2 (C^{ar}), 119.8 (C^{ar}), 119.4 (C^{ar}), 114.4 (C^{ar}), 114.3 (C^{ar}), 111.9 (C^{ar}), 56.3 (OCH_3), 55.3 (OCH_3), 44.0 (NCH_2), 43.9 (NCH_2), 17.0 (CH_2CH_3), 16.9 (CH_2CH_3); m/z (%) 740 (59) [M^+], 738 (61) [M^+], 660 (76), 613 (73) [M^+–I], 611 (71) [M^+–I], 533 (100); anal. calcd. for $C_{21}H_{23}AuBrIN_2O_2$: C, 34.12; H, 3.14; N, 3.79. Found: C, 34.47; H, 3.20; N, 3.85.

6k: mp 207–208 °C; υ_{max}(ATR)/cm^{-1} 3000, 2959, 2934, 2868, 2835, 1607, 1579, 1515, 1502, 1459, 1412, 1379, 1347, 1306, 1295, 1240, 1177, 1125, 1050, 1030, 1004, 976, 885, 859, 840, 827, 811, 785, 763, 736, 692, 670; ^{1}H NMR (300 MHz, $CDCl_3$) δ 7.17 (d, J = 8.8 Hz, 2H, H^{ar}), 6.92 (d, J = 8.8 Hz, 2H, H^{ar}), 6.40 (s, 2H, H^{ar}), 4.3–4.1 (m, 4H, NCH_2), 3.87 (s, 3H, OCH_3), 3.84 (s, 3H, OCH_3), 3.80 (s, 6H, OCH_3), 1.5–1.3 (m, 6H, CH_2CH_3); ^{13}C NMR (75.5 MHz, $CDCl_3$) δ 179.7 (NCN), 160.3 (C^{ar}), 153.3 (C^{ar}), 138.8 (C^{ar}), 131.8 (C^{ar}), 130.8 (C^{ar}), 130.6 (C^{ar}), 123.0 (C^{ar}), 119.8 (C^{ar}), 114.3 (C^{ar}), 107.8 (C^{ar}), 60.9 (OCH_3), 56.4 (OCH_3), 56.2 (OCH_3), 55.3 (OCH_3), 44.1 (NCH_2), 43.9 (NCH_2), 17.2 (CH_2CH_3), 17.0 (CH_2CH_3); m/z (%) 720 (46) [M^+], 594 (100) [M^+–I]; anal. calcd. for $C_{23}H_{28}AuIN_2O_4$: C, 38.35; H, 3.92; N, 3.89. Found: C, 38.63; H, 3.88; N, 3.94.

6l: yield: 58%; mp 192–193 °C; υ_{max}(ATR)/cm^{-1} 2972, 2935, 2833, 1607, 1592, 1563, 1515, 1492, 1460, 1411, 1317, 1294, 1270, 1251, 1175, 1157, 1123, 1108, 1091, 1075, 1041, 1001, 976, 903, 870, 834, 811, 787, 755, 708, 687, 665; ^{1}H NMR (300 MHz, $CDCl_3$) δ 7.12 (d, J = 8.7 Hz, 2H, H^{ar}), 6.88 (d, J = 8.7 Hz, 2H, H^{ar}), 6.85 (s, 1H, H^{ar}), 6.56 (s, 1H, H^{ar}), 4.2–4.1 (m, 4H, NCH_2), 3.85 (s, 3H, OCH_3), 3.80 (s, 3H, OCH_3), 3.70 (s, 3H, OCH_3), 1.35 (t, J = 7.2 Hz, 3H, CH_2CH_3), 1.29 (t, J = 7.2 Hz, 3H, CH_2CH_3); ^{13}C NMR (75.5 MHz, $CDCl_3$) δ 179.9 (NCN), 160.4 (C^{ar}), 153.8 (C^{ar}), 146.2 (C^{ar}), 131.8 (C^{ar}), 131.1 (C^{ar}), 129.4 (C^{ar}), 128.7 (C^{ar}), 124.0 (C^{ar}), 123.8 (C^{ar}), 119.3 (C^{ar}), 114.4 (C^{ar}), 114.3 (C^{ar}), 113.2 (C^{ar}), 60.8 (OCH_3), 56.2 (OCH_3), 55.3 (OCH_3), 44.0 (NCH_2), 43.9 (NCH_2), 17.2 (CH_2CH_3), 16.9 (CH_2CH_3); m/z (%) 726 (20) [M^+], 724 (48) [M^+], 660 (13), 598 (100) [M^+–I], 534 (23), 142 (16), 127 (8); anal. calcd. for $C_{22}H_{25}AuClIN_2O_3$: C, 36.46; H, 3.48; N, 3.87. Found: C, 36.70; H, 3.43; N, 3.84.

6m: yield: 73%; mp 194–195 °C; υ_{max}(ATR)/cm^{-1} 2967, 2941, 2835, 1607, 1588, 1558, 1514, 1489, 1460, 1410, 1344, 1318, 1291, 1266, 1250, 1234, 1179, 1156, 1110, 1091, 1075, 1037, 1024, 991, 899, 862, 843, 813, 789, 753, 738, 698, 665; ^{1}H NMR (300 MHz, $CDCl_3$) δ 7.12 (d, J = 8.8 Hz, 2H, H^{ar}), 7.02 (s, 1H, H^{ar}), 6.88 (d, J = 8.8 Hz, 2H, H^{ar}), 6.60 (s, 1H, H^{ar}), 4.2–4.1 (m, 4H, NCH_2), 3.84 (s, 3H, OCH_3), 3.80 (s, 3H, OCH_3), 3.69 (s, 3H, OCH_3), 1.35 (t, J = 7.2 Hz, 3H, CH_2CH_3), 1.23 (t, J = 7.3 Hz, 3H, CH_2CH_3); ^{13}C NMR (75.5 MHz, $CDCl_3$) δ 179.8 (NCN), 160.4 (C^{ar}), 153.6 (C^{ar}), 147.2 (C^{ar}), 131.8 (C^{ar}), 131.1 (C^{ar}), 129.2 (C^{ar}), 126.6 (C^{ar}), 124.6 (C^{ar}), 119.3 (C^{ar}), 117.9 (C^{ar}), 114.4 (C^{ar}), 113.9 (C^{ar}), 60.7 (OCH_3), 56.1 (OCH_3), 55.3 (OCH_3), 44.1 (NCH_2), 44.0 (NCH_2), 17.2 (CH_2CH_3), 16.9 (CH_2CH_3); m/z (%) 770 (48) [M^+], 768 (46) [M^+], 725 (21), 690 (22), 643 (100) [M^+–I], 641 (99) [M^+–I], 563 (42), 142 (93), 127 (40), 94 (27), 43 (65); anal. calcd. for $C_{22}H_{25}AuBrIN_2O_3$: C, 34.35; H, 3.28; N, 3.64. Found: C, 34.70; H, 3.33; N, 3.60.

6n: yield: 71%; mp 118–120 °C; υ_{max}(ATR)/cm^{-1} 2966, 2932, 2834, 1606, 1584, 1550, 1513, 1482, 1460, 1409, 1344, 1316, 1292, 1250, 1176, 1151, 1110, 1091, 1075, 1029, 995, 895, 863, 838, 800, 751, 689, 658; ^{1}H NMR (300 MHz, $CDCl_3$) δ 7.24 (s, 1H, H^{ar}), 7.12 (d, J = 8.8 Hz, 2H, H^{ar}), 6.88 (d, J = 8.8 Hz, 2H, H^{ar}), 6.62 (s, 1H, H^{ar}), 4.2–4.1 (m, 4H, NCH_2), 3.81 (s, 3H, OCH_3), 3.80 (s, 3H, OCH_3), 3.67 (s, 3H, OCH_3), 1.35 (t, J = 7.2 Hz, 3H, CH_2CH_3), 1.28 (t, J = 7.2 Hz, 3H, CH_2CH_3); ^{13}C NMR (75.5 MHz, $CDCl_3$) δ 179.8 (NCN), 160.3 (C^{ar}), 152.3 (C^{ar}), 149.6 (C^{ar}), 132.2 (C^{ar}), 131.8 (C^{ar}), 131.0 (C^{ar}), 129.0 (C^{ar}), 125.5 (C^{ar}), 119.3 (C^{ar}), 115.0 (C^{ar}), 114.4 (C^{ar}), 114.3 (C^{ar}), 92.5 (C^{ar}), 60.5 (OCH_3), 56.0 (OCH_3), 55.3 (OCH_3), 44.1 (NCH_2), 43.9 (NCH_2), 17.2 (CH_2CH_3), 16.9 (CH_2CH_3); m/z (%) 816 (52) [M^+], 689 (87) [M^+–I], 660

(23), 564 (12), 533 (100), 345 (13), 142 (75), 127 (27); anal. calcd. for $C_{22}H_{25}AuI_2N_2O_3$: C, 32.37; H, 3.09; N, 3.43. Found: C, 32.58; H, 3.16; N, 3.38.

6o: yield: 44%; mp 178–179 °C; υ_{max}(ATR)/cm^{-1} 3005, 2963, 2938, 2831, 1607, 1579, 1517, 1502, 1455, 1404, 1318, 1293, 1236, 1173, 1123, 1018, 998, 913, 843, 814, 790, 766, 736, 683, 665; ^{1}H NMR (300 MHz, $CDCl_3$) δ 7.11 (d, *J* = 8.8 Hz, 2H, H^{ar}), 6.87 (d, *J* = 8.8 Hz, 2H, H^{ar}), 6.36 (s, 2H, H^{ar}), 3.83 (s, 3H, OCH_3), 3.79 (s, 3H, OCH_3), 3.76 (s, 3H, OCH_3), 3.73 (s, 3H, OCH_3), 3.71 (s, 3H, NCH_3), 3.70 (s, 3H, NCH_3); ^{13}C NMR (75.5 MHz, $CDCl_3$) δ 180.8 (NCN), 160.3 (C^{ar}), 153.3 (C^{ar}), 138.8 (C^{ar}), 131.7 (C^{ar}), 131.3 (C^{ar}), 122.8 (C^{ar}), 119.6 (C^{ar}), 114.4 (C^{ar}), 110.0 (C^{ar}), 107.7 (C^{ar}), 60.9 (OCH_3), 56.2 (OCH_3), 55.3 (OCH_3), 36.4 (NCH_3), 36.3 (NCH_3); *m*/*z* (%) 692 (47) [M^+], 566 (100) [M^+–I], 355 (28), 339 (23), 283 (19), 142 (63), 127 (27); anal. calcd. for $C_{21}H_{24}AuIN_2O_4$: C, 36.43; H, 3.49; N, 4.05. Found: C, 36.65; H, 3.43; N, 4.10.

4.4. Crystal Structure Analysis of 6c

X-ray structure analysis of single crystals of the complex **6c** was performed on a Stoe StadiVari diffractometer equipped with a graphite-monochromated Mo-K_α (λ = 0.71073 Å) radiation source and an Oxford Cryostream low-temperature unit. A suitable single crystal of **6c** was embedded in inert perfluorinated oil (Fomblin® YR-1800) and mounted on a nylon loop before collecting data at 170(2) K.

Data were corrected for Lorentz and polarization effects; a spherical absorption correction was applied. The structures were solved by direct method SHELXT 2014/5 and refined by full-matrix least-squares procedures on $F_0^2 - F_c^2$ with SHELXL 2018/3, interfaced by WinGX [55–57]. All non-hydrogen atoms were refined with anisotropic displacement parameters. Hydrogen atoms were calculated in idealized positions with fixed displacement parameters during refinement. Occupational disorder of bromine and the phenyl-appended methoxy group was resolved in one of the two symmetry-independent units of **6c**. Mercury was used for structure illustrations/graphical output [58].

4.5. Anticancer Activity

4.5.1. Cell Line and Culture Conditions

EAC cells FLO-1 and SK-GT-4 (a gift from Dr. Shrikant Anant's lab, University of Kansas Medical Center, Kansas City), were cultured in complete DMEM (4.5 g/L glucose, sodium pyruvate and L-glutamine, Corning, MA). The complete DMEM was prepared by adding fetal bovine serum (10% FBS, heat-inactivated, Sigma-Aldrich, MO) and 1% antibiotic-antimycotic solution (Corning, MA). EAC cells were cultured in 5% CO_2 at 37 °C. All procedures were performed according to the standard guidelines and regulations and as per the manufacturers' instructions.

4.5.2. Proliferation Assay

A total of 5000 EAC cells/well (SK-GT-4 and FLO-1) were plated in a 96-well plate using complete DMEM. After 24 h of plating, EAC cells were treated with gold complexes at different concentrations. After 72 h, the medium was removed, and cell viability was measured using the hexosaminidase enzymatic assay [59]. The percentage of inhibition was calculated by comparing cell viability after compound treatment with controls.

4.5.3. Colony Formation Assay

A total of 500 cells/well of EAC were plated in 6-well plates. After 24 h, the EAC cells were treated with IC_{50} and semi-IC_{50} concentrations of **6b**, **6d**, and **6i**. Compounds containing media were replaced after 72 h with complete DMEM to remove the test compounds. The cells were grown for 10–12 days to form colonies. The resulting colonies were washed and fixed using a 10% formalin solution. After 20 min, the formalin was removed, and the fixed cells were washed and stained with 1% crystal violet solution in 10% ethanol. After staining, colonies were washed to remove crystal violet, dried, counted, and compared to controls [60]. We scanned the stained and dried 6-well plates using a Canon Image RUNNER Advance scanner to make figures.

4.5.4. Cell Cycle Analysis

A total of 200,000 EAC cells (SK-GT-4 and FLO-1) per well were plated in 6-well plates. After 24 h, EAC cells were treated with IC_{50} and semi-IC_{50} concentrations of compounds **6b**, **6d**, and **6i**. After 72 h, EAC cells were washed, resuspended in PBS, and fixed using an ice-cold fixing solution (70% ethanol in PBS), followed by storage overnight at 4 °C. The next day, EAC cells were centrifuged, washed with PBS, resuspended, permeabilized, and stained with FxCycleTM PI/RNase staining solution (Invitrogen). The cell cycle was studied by flow cytometry using an FACS Calibur analyzer (Becton Dickinson, Mountain View, CA, USA). The experimental datasets were plotted using ModFit LT™ software (Verity Software House, Topsham, ME, USA).

4.5.5. Apoptosis Assay

A total of 200,000 EAC cells/well were plated in a 6-well plate in complete DMEM, and, after 24 h, the EAC cells were treated with IC_{50} concentrations of compounds **6b**, **6d**, and **6i**. After 72 h, cells were trypsinized, washed and stained using the Annexin V-FITC Early Apoptosis Detection Kit (Cell Signaling Technology#6592) following the manufacturer's instructions, and studied by flow cytometry.

4.5.6. Spheroid Formation Assay

A total of 500 EAC cells were plated in an ultra-low attachment 96-well plate (96-well, Corning, Lowell, MA, USA) in spheroid medium prepared from serum-free DMEM supplemented with heparin salt (4 µg/mL), EGF (20 ng/mL), FGF (20 ng/mL), 1% antibiotic-antimycotic, and B27 supplement. After 2 days, spheroids were treated with IC_{50} and semi-IC_{50} concentrations of **6b**, **6d**, and **6i**. After 7 days, spheroids were counted and imaged [60].

4.5.7. Western Blot Analysis

A total of 500,000 cells of EAC cell lines were plated in a 10 cm cell culture Petri dish, and, after 24 h, cells were incubated with IC_{50} concentrations of **6b**, **6d**, and **6i** for 72 h. Cells were washed and lysed in lysis buffer with phosphatase and protease inhibitor (Roche), followed by sonification. The protein lysate was centrifuged at 6000 rpm for 10 min at 4 °C. Protein determination was performed using the Pierce BCA protein assay kit to estimate protein contents. A total of 50 µg of protein from each group was subjected to gel electrophoresis and further transferred onto polyvinylidene difluoride membranes (Millipore, Bedford, MA, USA) at 90 V for 2 h under cold conditions. These PVDF membranes were then blocked for 1 h by using 5% milk in TBST, washed with TBST, and incubated with primary antibodies at 4 °C overnight. The next day, membranes were washed using TBST to remove primary antibodies and incubated with respective secondary anti-mouse and anti-rabbit antibodies (Cell Signaling Technology, anti-mouse#7076, anti-rabbit#7074) for 1 h. The proteins were identified by using the GE Health Care chemiluminescence system (Piscataway, NJ, USA), imaged using the Bio-Rad ChemiDoc-XRS+ instrument, and processed by image lab. Antibodies for detecting cyclin D1 (CST#2922), Bcl-XL (CST#2762), Bcl-2 (CST#4223), c-Myc (CST#9402), Bax (CST#2772), Mcl-1 (CST#4572), and PARP (CST#9542) were bought from Cell Signaling Technology (Beverly, MA, USA), and GAPDH (G-9) was purchased from Santa Cruz Biotech Inc. (Santa Cruz, CA, USA).

4.5.8. Statistical Analysis

All values are shown as the mean ± SD. Experimental data were examined using an unpaired two-tailed t-test by comparing to the corresponding control group. A probability value of less than 0.05 was considered as statistically significant (* $p < 0.05$, ** $p < 0.01$).

5. Conclusions

The NHC ligand system of a previously published anticancer active iodidogold(I)–NHC complex was successfully optimized in terms of anticancer properties, and high activities against EAC cells were achieved for several new gold complexes. These compounds induced programmed cell death and suppressed colony and spheroid formation by EAC cells at low doses. Together with their promising suppressive effects on cyclin D1 and *c*-Myc expression, there exists a considerable potential of iodidogold(I)–NHC complexes as new candidates for the treatment of problematic tumor diseases such as EAC. Deeper investigations of the mechanisms of action of these gold compounds will provide more information about their prospects as new anticancer drugs. An extension of compound testing to other tumor entities than EAC will also be of great interest given the described anticancer properties of the newly discovered gold complexes.

Author Contributions: Conceptualization, S.W.S., P.D. and B.B.; methodology, B.W., R.S., P.D. and B.B.; software, G.H.; validation and formal analysis, S.W.S., H.G., G.H., J.S. and B.B.; investigation, S.W.S., H.G., G.H., J.S., S.B. and B.B.; resources, B.W., R.S. and P.D.; data curation, S.W.S., H.G., G.H., P.D. and B.B.; writing—original draft preparation, B.B.; writing—review and editing, S.W.S., B.W., R.S. and P.D.; supervision, B.W., R.S., P.D. and B.B.; project administration, B.B.; funding acquisition, B.W., R.S. and P.D. All authors have read and agreed to the published version of the manuscript.

Funding: R.S. thanks the Deutsche Forschungsgemeinschaft for financial support (grant Scho 402/12-2). P.D. is supported by grants from the Department of Defense (CA201135) and the American Gastroenterological Association Research Scholar Award (2021RSA-12522), the Lied Pre-Clinical Research Pilot Grant Program, and funds from the University of Kansas Cancer Center. The Flow Cytometry Core Laboratory is sponsored, in part, by the NIH COBRE program of the NCRR P20 RR016443 and The University of Kansas Cancer Center P30CA168524-01 grants. The research reported in this publication was partly supported by the KUMC Research Institute. The content is solely responsibility of the authors and does not necessary represent the official views of the KUMC Research Institute.

Institutional Review Board Statement: Not applicable.

Informed Consent Statement: Not applicable.

Data Availability Statement: Original data can be requested from the corresponding author.

Acknowledgments: G.H. and B.W. thank Frank Heinemann (University Erlangen) for helpful advice concerning the treatment of the crystallographic disorder.

Conflicts of Interest: The authors declare no conflict of interest.

References

1. Sung, H.; Ferlay, J.; Siegel, R.L.; Laversanne, M.; Soerjomataram, I.; Jemal, A.; Bray, F. Global Cancer Statistics 2020: GLOBOCAN Estimates of Incidence and Mortality Worldwide for 36 Cancers in 185 Countries. *CA Cancer J. Clin.* **2021**, *71*, 209–249. [CrossRef] [PubMed]
2. Uhlenhopp, D.J.; Then, E.O.; Sunkara, T.; Gaduputi, V. Epidemiology of esophageal cancer: Update in global trends, etiology and risk factors. *Clin. J. Gastroenterol.* **2020**, *13*, 1010–1021. [CrossRef] [PubMed]
3. Njeij, B.; McCarty, T.R.; Birk, J.W. Trends in esophageal cancer survival in United States adults from 1973 to 2009: A SEER database analysis. *J. Gastroenterol. Hepatol.* **2016**, *31*, 1141–1146. [CrossRef] [PubMed]
4. Siegel, R.L.; Miller, K.D.; Fuchs, H.E.; Jemal, A. Cancer Statistics, 2021. *CA Cancer J. Clin.* **2021**, *71*, 7–33. [CrossRef] [PubMed]
5. Ratajczak, T.; Kumar, G.; Albandar, H.; Redlien, P.; Yacyshyn, S.; Markert, R. Metabolic characteristics of esophageal carcinoma in a veteran population. *J. Clin. Oncol.* **2016**, *34*, 4023. [CrossRef]
6. Xu, Q.L.; Li, H.; Zhu, Y.J.; Xu, G. The treatments and postoperative complications of esophageal cancer: A review. *J. Cardiothorac. Surg.* **2020**, *15*, 163. [CrossRef]
7. Patel, N.; Benipal, B. Incidence of Esophageal Cancer in the United States from 2001–2015: A United States Cancer Statistics Analysis of 50 States. *Cureus* **2018**, *10*, e3709. [CrossRef] [PubMed]
8. Blot, W.J.; Tarone, R.E. Esophageal Cancer. In *Cancer Epidemiology and Prevention*, 4th ed.; Thun, M., Linet, M.S., Cerhan, J.R., Haiman, C.A., Schottenfeld, D., Eds.; Oxford University Press: New York, USA, 2017; pp. 579–592.

9. Wang, H.; Li, S.; Liu, T.; Chen, J.; Dang, J. Neoadjuvant immune checkpoint inhibitor in combination with chemotherapy or chemoradiotherapy in resectable esophageal cancer: A systematic review and meta-analysis. *Front. Immunol.* **2022**, *13*, 998620. [CrossRef] [PubMed]
10. Shaw, C.F., III. Gold-based therapeutic agents. *Chem. Rev.* **1999**, *99*, 2589–2600. [CrossRef]
11. Harris, G.J.; Turner, J.N.; Von Hoff, D.D. Growth of carcinoma of the esophagus and gastroesophageal junction in a human tumor cloning assay. *Cancer Drug Deliv.* **1986**, *3*, 273–278. [CrossRef]
12. Onodera, T.; Momose, I.; Kawada, M. Potential anticancer activity of auranofin. *Chem. Pharm. Bull.* **2019**, *67*, 186–191. [CrossRef]
13. Liu, N.; Li, X.; Huang, H.; Zhao, C.; Liao, S.; Yang, C.; Liu, S.; Song, W.; Lu, X.; Lan, X.; et al. Clinically used antirheumatic agent auranofin is a proteasomal deubiquitinase inhibitor and inhibits tumor growth. *Oncotarget* **2014**, *5*, 5453–5471. [CrossRef]
14. Huang, H.; Liao, Y.; Liu, N.; Hua, X.; Cai, J.; Yang, C.; Long, H.; Zhao, C.; Chen, X.; Lan, X.; et al. Two clinical drugs deubiquitinase inhibitor auranofin and aldehyde dehydrogenase inhibitor disulfiram trigger synergistic anti-tumor effects in vitro and in vivo. *Oncotarget* **2016**, *7*, 2796–2808. [CrossRef]
15. Thangamani, S.; Maland, M.; Mohammad, H.; Pascuzzi, P.E.; Avramova, L.; Koehler, C.M.; Hazbun, T.R.; Seleem, M.N. Repurposing approach identifies auranofin with broad spectrum antifungal activity that targets Mia40-Erv1 pathway. *Front. Cell. Infect. Microbiol.* **2017**, *7*, 4. [CrossRef]
16. Dickson-Murray, E.; Nedara, K.; Modjtahedi, N.; Tokatlidis, K. The Mia40/CHCHD4 oxidative folding system: Redox regulation and signaling in the mitochondrial intermembrane space. *Antioxidants* **2021**, *10*, 592. [CrossRef] [PubMed]
17. Landini, I.; Massai, L.; Cirri, D.; Gamberi, T.; Paoli, P.; Messori, L.; Mini, E.; Nobili, S. Structure-activity relationships in a series of auranofin analogues showing remarkable antiproliferative properties. *J. Inorg. Biochem.* **2020**, *208*, 111079. [CrossRef] [PubMed]
18. Massai, L.; Cirri, D.; Marzo, T.; Messori, L. Auranofin and its analogs as prospective agents for the treatment of colorectal cancer. *Cancer Drug Resist.* **2022**, *5*, 1–14. [CrossRef]
19. Kober, L.; Schleser, S.W.; Bär, S.I.; Schobert, R. Revisiting the anticancer properties of phosphane(9-ribosylpurine-6-thiolato)gold(I) complexes and their 9H-purine precursors. *J. Biol. Inorg. Chem.* **2022**, *27*, 731–745. [CrossRef]
20. Schmidt, C.; Karge, B.; Misgeld, R.; Prokop, A.; Brönstrup, M.; Ott, I. Biscarbene gold(I) complexes: Structure-activity-relationships regarding antibacterial effects, cytotoxicity, TrxR inhibition and cellular bioavailability. *Med. Chem. Commun.* **2017**, *8*, 1681–1689. [CrossRef]
21. Schmidt, C.; Albrecht, L.; Balasupramaniam, S.; Misgeld, R.; Karge, B.; Brönstrup, M.; Prokop, A.; Baumann, K.; Reichl, S.; Ott, I. A gold(I) biscarbene complex with improved activity as a TrxR inhibitor and cytotoxic drug: Comparative studies with different gold metallodrugs. *Metallomics* **2019**, *11*, 533–545. [CrossRef] [PubMed]
22. Schuh, E.; Pflüger, C.; Citta, A.; Folda, A.; Rigobello, M.P.; Bindoli, A.; Casini, A.; Mohr, F. Gold(I) carbene complexes causing thioredoxin 1 and thioredoxin 2 oxidation as potential anticancer agents. *J. Med. Chem.* **2012**, *55*, 5518–5528. [CrossRef] [PubMed]
23. Wragg, D.; de Almeida, A.; Bonsignore, R.; Kühn, F.E.; Leoni, S.; Casini, A. On the mechanism of gold/BHC compounds binding to DNA G-quadruplexes: Combined metadynamics and biophysical methods. *Angew. Chem. Int. Ed.* **2018**, *57*, 14524–14528. [CrossRef] [PubMed]
24. Tialiou, A.; Chin, J.; Keppler, B.K.; Reithofer, M.R. Current developments of *N*-heterocyclic carbene (Au(I)/Au(III)) complexes toward cancer treatment. *Biomedicines* **2022**, *10*, 1417. [CrossRef]
25. Kaps, L.; Biersack, B.; Müller-Bunz, H.; Mahal, K.; Münzner, J.; Tacke, M.; Mueller, T.; Schobert, R. Gold(I)-NHC complexes of antitumoral diarylimidazoles: Structures, cellular uptake routes and anticancer activities. *J. Inorg. Biochem.* **2012**, *106*, 52–58. [CrossRef]
26. Münzner, J.; Biersack, B.; Kaps, L.; Schobert, R.; Sasse, F. Synergistic "gold effects" of anti-vascular 4,5-diarylimidazol-2-ylidene gold(I) carbene complexes. *Int. J. Clin. Pharmacol. Ther.* **2013**, *51*, 44–46. [CrossRef]
27. Muenzner, J.K.; Biersack, B.; Kalie, H.; Andronache, I.C.; Kaps, L.; Schuppan, D.; Sasse, F.; Schobert, R. Gold(I) biscarbene complexes derived from vascular-disrupting combretastatin A-4 address different targets and show antimetastatic potential. *ChemMedChem* **2014**, *9*, 1195–1204. [CrossRef]
28. Hatem, E.; El Banna, N.; Heneman-Masurel, A.; Baille, D.; Vernis, L.; Riquier, S.; Golinelli-Cohen, M.-P.; Guittet, O.; Vallières, C.; Camadro, J.-M.; et al. Novel insights into redox-based mechanisms for auranofin-induced rapid cancer cell death. *Cancers* **2022**, *14*, 4864. [CrossRef]
29. Bian, M.; Fan, R.; Jiang, G.; Wang, Y.; Lu, Y.; Liu, W. Halo and pseudohalo gold(I)-NHC complexes derived from 4,5-diarylimidazoles with excellent in vitro and in vivo anticancer activities against HCC. *J. Med. Chem.* **2020**, *63*, 9197–9211. [CrossRef]
30. Plante, J.P.; Glass, T.E. Shape-selective fluorescent sensing ensemble using a tweezer-type metalloreceptor. *Org. Lett.* **2006**, *8*, 2163–2166. [CrossRef] [PubMed]
31. Benitez, D.; Shapiro, N.D.; Tkatchouk, E.; Wang, Y.; Goddard III, W.A.; Toste, F.D. A bonding model for gold(I) carbene complexes. *Nat. Chem.* **2009**, *1*, 482–486. [CrossRef]
32. Das, P.K.; Islam, F.; Smith, R.A.; Lam, A.K. Therapeutic strategies against cancer stem cells in esophageal carcinomas. *Front. Oncol.* **2021**, *10*, 598957. [CrossRef] [PubMed]
33. Liao, D.J.; Thakur, A.; Wu, J.; Biliran, H.; Sarkar, F.H. Perspectives on c-Myc, Cyclin D1, and their interaction in cancer formation, progression, and response to chemotherapy. *Crit. Rev. Oncog.* **2007**, *13*, 93–158. [CrossRef] [PubMed]

34. Muenzner, J.K.; Biersack, B.; Albrecht, A.; Rehm, T.; Lacher, U.; Milius, W.; Casini, A.; Zhang, J.-J.; Ott, I.; Brabec, V.; et al. Ferrocenyl-coupled *N*-heterocyclic carbene complexes of gold(I): A successful approach to multinuclear anticancer drugs. *Chem. Eur. J.* **2016**, *22*, 18953–18962. [CrossRef]
35. Bär, S.I.; Gold, M.; Schleser, S.W.; Rehm, T.; Bär, A.; Köhler, L.; Carnell, L.R.; Biersack, B.; Schobert, R. Guided antitumoural drugs: (imidazol-2-ylidene)(L)gold(I) complexes seeking cellular targets controlled by the nature of ligand L. *Chem. Eur. J.* **2021**, *27*, 5003–5010. [CrossRef] [PubMed]
36. Paloque, L.; Hemmert, C.; Valentin, A.; Gornitzka, H. Synthesis, characterization, and antileishmanial activities of gold(I) complexes involving quinoline functionalized *N*-heterocyclic carbenes. *Eur. J. Med. Chem.* **2015**, *94*, 22–29. [CrossRef]
37. Tong, Z.; Chatterjee, D.; Deng, D.; Veeranki, O.; Mejia, A.; Ajani, J.A.; Hofstetter, W.; Lin, S.; Guha, S.; Kopetz, S.; et al. Antitumor effects of cyclin dependent kinase 9 inhibition in esophageal adenocarcinoma. *Oncotarget* **2017**, *8*, 28696–28710. [CrossRef]
38. Su, W.; Guo, C.; Wang, L.; Wang, Z.; Yang, X.; Niu, F.; Tzou, D.; Yang, X.; Huang, X.; Wu, J.; et al. LncRNA *MIR22HG* abrogation inhibits proliferation and induces apoptosis in esophageal adenocarcinoma cells via activation of the STAT3/*c*-Myc/FAK signaling. *Aging* **2019**, *11*, 4587–4596. [CrossRef]
39. Vangamudi, B.; Zhu, S.; Soutto, M.; Belkhiri, A.; El-Rifai, W. Regulation of β-catenin by t-DARPP in upper gastrointestinal cancer cells. *Mol. Cancer* **2011**, *10*, 32. [CrossRef] [PubMed]
40. Madden, S.K.; De Araujo, A.D.; Gerhardt, M.; Fairlie, D.P.; Mason, J.M. Taking the Myc out of cancer: Toward therapeutic strategies to directly inhibit *c*-Myc. *Mol. Cancer* **2021**, *20*, 3. [CrossRef]
41. Montalto, F.I.; De Amicis, F. Cyclin D1 in cancer: A molecular connection for cell cycle control, adhesion and invasion in tumor and stroma. *Cells* **2020**, *9*, 2648. [CrossRef]
42. Taylor, C.; Loomans, H.A.; Le Bras, G.F.; Koumangoye, R.B.; Romero-Morales, A.I.; Quast, L.L.; Zaika, A.I.; El-Rifai, W.; Andl, T.; Andl, C.D. Activin a signaling regulates cell invasion and proliferation in esophageal adenocarcinoma. *Oncotarget* **2015**, *6*, 34228–34244. [CrossRef] [PubMed]
43. Chen, C.; Ma, Z.; Jiang, H. EMT participates in the regulation of exosomes secretion and function in esophageal cancer cells. *Technol. Cancer Res. Treat.* **2021**, *20*, 1–11. [CrossRef] [PubMed]
44. Silvers, A.L.; Lin, L.; Bass, A.J.; Chen, G.; Wang, Z.; Thomas, D.G.; Lin, J.; Giordano, T.J.; Orringer, M.B.; Beer, D.G.; et al. Decreased selenium-binding protein 1 in esophageal adenocarcinoma results from posttranscriptional and epigenetic regulation and affects chemosensitivity. *Clin. Cancer Res.* **2010**, *16*, 2009–2021. [CrossRef] [PubMed]
45. Ahrens, T.D.; Timme, S.; Ostendorp, J.; Bogatyreva, L.; Hoeppner, J.; Hopt, U.T.; Hauschke, D.; Werner, M.; Lassmann, S. Response of esophageal cancer cells to epigenetic inhibitors is mediated via altered thioredoxin activity. *Lab. Investig.* **2016**, *96*, 307–316. [CrossRef]
46. Feingold, P.L.; Surman, D.R.; Brown, K.; Xu, Y.; McDuffie, L.A.; Shukla, V.; Reardon, E.S.; Crooks, D.R.; Trepel, J.B.; Lee, S.; et al. Induction of thioredoxin-interacting protein by a histone deacetylase inhibitor, entinostat, is associated with DNA damage and apoptosis in esophageal adenocarcinoma. *Mol. Cancer Ther.* **2018**, *17*, 2013–2023. [CrossRef] [PubMed]
47. Sehdev, V.; Katsha, A.; Ecsedy, J.; Zaika, A.; Belkhiri, A.; El-Rifai, W. The combination of alisertib, an investigational aurora kinase A inhibitor, and docetaxel promotes cell death and reduces tumor growth in preclinical cell models of upper gastrointestinal adenocarcinomas. *Cancer* **2013**, *15*, 904–914. [CrossRef]
48. Sehdev, V.; Peng, D.; Soutto, M.; Washington, M.K.; Revetta, F.; Ecsedy, J.; Zaika, A.; Rau, T.T.; Schneider-Stock, R.; Belkhiri, A.; et al. The aurora kinase A inhibitor MLN8237 enhances cisplatin-induced cell death in esophageal adenocarcinoma cells. *Mol. Cancer Ther.* **2012**, *11*, 763–774. [CrossRef]
49. Yusenko, M.V.; Biyanee, A.; Frank, D.; Köhler, L.H.F.; Andersson, M.K.; Khandanpour, C.; Schobert, R.; Stenman, G.; Biersack, B.; Klempnauer, K.-H. Bcr-TMP, a novel nanomolar-active compound that exhibits both MYB- and microtubule-inhibitory activity. *Cancers* **2022**, *14*, 43. [CrossRef]
50. Köhler, L.H.F.; Reich, S.; Yusenko, M.; Klempnauer, K.-H.; Shaikh, A.H.; Ahmed, K.; Begemann, G.; Schobert, R.; Biersack, B. A new naphthopyran derivative combines c-Myb inhibition, microtubule-targeting effects, and antiangiogenic properties. *ACS Med. Chem. Lett.* **2022**, *13*, 1783–1790. [CrossRef]
51. Köhler, L.H.F.; Reich, S.; Yusenko, M.; Klempnauer, K.-H.; Begemann, G.; Schobert, R.; Biersack, B. Multimodal 4-arylchromene derivatives with microtubule-destabilizing, anti-angiogenic, and MYB-inhibitory activities. *Cancer Drug Resist.* **2023**, *6*, 59–77. [CrossRef]
52. Brabender, J.; Lord, R.V.; Danenberg, K.D.; Metzger, R.; Schneider, P.M.; Park, J.M.; Salonga, D.; Groshen, S.; Tsao-Wei, D.D.; DeMeester, T.R.; et al. Increased *c*-Myb mRNA expression in Barrett's esophagus and Barrett's-associated adenocarcinoma. *J. Surg. Res.* **2001**, *99*, 301–306. [CrossRef] [PubMed]
53. Peng, D.; Guo, Y.; Chen, H.; Zhao, S.; Washington, K.; Hu, T.; Shyr, Y.; El-Rifai, W. Integrated molecular analysis reveals complex interactions between genomic and epigenomic alterations in esophageal adenocarcinomas. *Sci. Rep.* **2017**, *7*, 40729. [CrossRef]
54. Zhang, L.; Wang, X.; Li, Y.; Han, J.; Gao, X.; Li, S.; Wang, F. *c*-Myb facilitates immune escape of esophageal adenocarcinoma cells through the miR-145-5p/SPOP/PD-L1 axis. *Clin. Transl. Med.* **2021**, *11*, e464. [CrossRef] [PubMed]
55. Sheldrick, G.M. SHELXT—Integrated space-group and crystal-structure determination. *Acta Crystallogr. A* **2015**, *71*, 3–8. [CrossRef]
56. Sheldrick, G.M. Crystal structure refinement with SHELX. *Acta Crystallogr. C Struct. Chem.* **2015**, *71*, 3–8. [CrossRef] [PubMed]
57. Farrugia, L.J. WinGX and ORTEP for Windows: An update. *J. Appl. Cryst.* **2012**, *45*, 849–854. [CrossRef]

58. Macrae, C.F.; Sovago, I.; Cottrell, S.J.; Galek, P.T.A.; McCabe, P.; Pidcock, E.; Platings, M.; Shields, G.P.; Stevens, J.S.; Towler, M.; et al. Mercury 4.0: From visualization to analysis, design and prediction. *J. Appl. Cryst.* **2020**, *53*, 226–235. [CrossRef]
59. Landegren, U. Measurement of cell numbers by means of the endogenous enzyme hexosaminidase. Applications to detection of lymphokines and cell surface antigens. *J. Immunol. Methods* **1984**, *67*, 379–388. [CrossRef]
60. Dandawate, P.; Subramaniam, D.; Panovich, P.; Standing, D.; Krishnamachary, B.; Kaushik, G.; Thomas, S.M.; Dhar, A.; Weir, S.J.; Jensen, R.A.; et al. Cucurbitacin B and I inhibits colon cancer growth by targeting the Notch signaling pathway. *Sci. Rep.* **2020**, *10*, 1290. [CrossRef]

International Journal of *Molecular Sciences*

MDPI

Review

Anticancer Small-Molecule Agents Targeting Eukaryotic Elongation Factor 1A: State of the Art

Han Zhang, Jiayou Cai, Siqi Yu, Bin Sun and Weicheng Zhang *

The State Key Laboratory of Medicinal Chemical Biology, College of Pharmacy, and Tianjin Key Laboratory of Molecular Drug Research, Nankai University, Tianjin 300353, China
* Correspondence: zhangweicheng@nankai.edu.cn

Abstract: Eukaryotic elongation factor 1A (eEF1A) canonically delivers amino acyl tRNA to the ribosomal A site during the elongation stage of protein biosynthesis. Yet paradoxically, the oncogenic nature of this instrumental protein has long been recognized. Consistently, eEF1A has proven to be targeted by a wide assortment of small molecules with excellent anticancer activity, among which plitidepsin has been granted approval for the treatment of multiple myeloma. Meanwhile, metarrestin is currently under clinical development for metastatic cancers. Bearing these exciting advances in mind, it would be desirable to present a systematic up-to-date account of the title topic, which, to the best of our knowledge, has thus far been unavailable in the literature. The present review summarizes recent advances in eEF1A-targeting anticancer agents, both naturally occurring and synthetically crafted, with regard to their discovery or design, target identification, structure–activity relationship, and mode of action. Their structural diversity and differential eEF1A-targeting mechanisms warrant continuing research in pursuit of curing eEF1A-driven malignancy.

Keywords: anticancer; eEF1A; mechanism; macrocycle; heterocycle

Citation: Zhang, H.; Cai, J.; Yu, S.; Sun, B.; Zhang, W. Anticancer Small-Molecule Agents Targeting Eukaryotic Elongation Factor 1A: State of the Art. *Int. J. Mol. Sci.* **2023**, *24*, 5184. https://doi.org/10.3390/ijms24065184

Academic Editor: Bernhard Biersack

Received: 16 February 2023
Revised: 5 March 2023
Accepted: 7 March 2023
Published: 8 March 2023

1. Introduction

The majority of cancers arise from accumulated somatic mutations, which over time, transform the cell into a state of malignancy. This is characterized by uncontrolled proliferation, aggressive invasion into surrounding normal tissues, and ultimately lethal metastasis at distant vital organs. Genomic instability and phenotypic heterogeneity inherent in each malignant tissue [1] render the treatment of cancer extremely challenging [2]. Today, apart from surgical operation, a variety of promising therapeutic strategies have been developed, including but not limited to radiotherapy [3], chemotherapy [4], biological therapy [5], immunotherapy [6], and microbial-based therapy [7,8]. From this ever-growing curative armamentarium, targeted chemotherapy with small-molecule or biomacromolecular agents is an indispensable measure [9–11].

Eukaryotic elongation factor 1A (eEF1A, formerly termed eEF-1α) is an essential GTPase evolutionarily conserved across diverse eukaryotes [12]. As the second most abundant intracellular protein after actin, it is localized extensively in the cytoplasm and nucleus [13–15]. The canonical function of eEF1A is to deliver amino acyl tRNAs to the ribosomal A site during the elongation stage of protein synthesis [16]. Strikingly, beyond this housekeeping role for the translational machinery and many other moonlighting functions [17–20], mounting evidence has pointed to a causal link between eEF1A and malignancy [21–23], suggesting that eEF1A may serve as not only an out-of-control translational cofactor [24–26] but also a signal transducer woven into a network of protumorigenic pathways [27–32]. Consistent with this revelation, structurally distinct small-molecule anticancer agents continue to emerge with a proven eEF1A-targeting mechanism [33]. Early examples include didemnin B [34], plitidepsin (dehydrodidemnin B) [35], tamandarin A [36], cytotrienin A [37], ansatrienin B [38], narciclasine [39], and synthetic flavonoids [40]. Among them, plitidepsin was approved in Australia for combined treatment of relapsed/refractory multiple myeloma with dexamethasone [41], thus providing initial proof of principle that eEF1A inhibitors can achieve the desired therapeutic efficacy with safety. In recent years, more intriguing compounds of this kind have been discovered and actively investigated. As a notable case, metarrestin is currently in a phase I clinical trial for the treatment of metastatic solid tumors [42]. Though a number of reviews have been published on protein-synthesis inhibitors [43–46], to the best of our knowledge, no systematic survey has ever been conducted on eEF1A-targeting agents. Hence, the present review delineates the state of the art on eEF1A-targeting small-molecule anticancer agents with a special focus on those actively studied over the recent years, covering their discovery or design, target identification, structure–activity relationship (SAR), and mode of action.

2. Recent Advances in Anticancer eEF1A-Targeting Agents

2.1. Didemnins and Tamandarins

Didemnins are a family of marine cyclic depsipeptides with strong anticancer, antiviral, and immunosuppressive activities [47]. Since their initial discovery in the early 1980s [48], these macrocycles have become the subject of intense research over the last four decades. Among them, didemnin B (**1**, Figure 1) and plitidepsin (**2**) have entered multiple clinical trials. Gratifyingly, plitidepsin was approved in Australia for treating multiple myeloma [41]. With nearly identical architectures but subtly different macrocyclic backbones (highlighted red), tamandarins such as tamandarin A (**3**) were discovered from a different colony of marine ascidian [49]. As already reviewed in multiple comprehensive monographs [36,50–56], these compounds will not be reiterated here. Instead, a brief update is presented below on their anticancer mode of action.

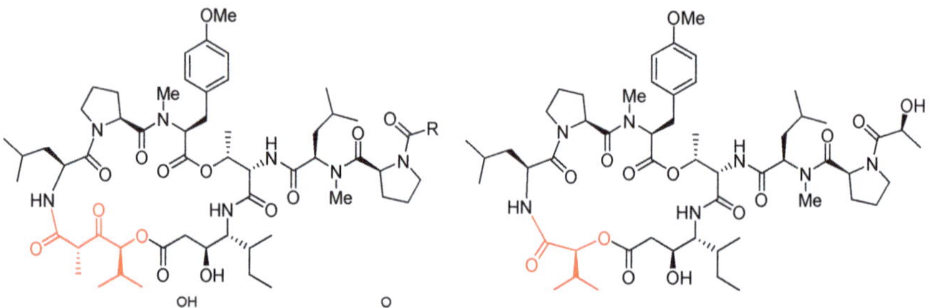

didemnin B (**1**, R =), plitidepsin (**2**, R =) tamandarin A (**3**)

Figure 1. Structures of didemnin B (**1**), plitidepsin (**2**), and tamandarin A (**3**).

Having developed functional signature ontology (FUSION) maps for drug discovery and mechanistic elucidation [57], White et al. revisited the mechanism of **1** and found that it induces rapid and extensive apoptosis in sensitive cancer cell lines through concomitant inhibition of palmitoyl-protein thioesterase 1 (PPT1) and eEF1A1 [58]. Independently, Galmarini et al. showed that eEF1A2 is the specific binding target of **2** with a measured dissociation constant (K_D) of 80 nM [35]. Since translation inhibition cannot account per se for the observed antiproliferative effect of **2**, it was suspected that this drug impacts non-canonical functions of eEF1A2. Indeed, double-stranded RNA (dsRNA)-dependent protein kinase (PKR) was later identified as a novel binding partner of eEF1A2 [30]. In this case, eEF1A2 interacts directly with PKR to block its pro-apoptotic activity and boost tumor survival. In the presence of **2**, however, PKR was disengaged from eEF1A2, thereby regaining its kinase activity to initiate extrinsic apoptosis through activation of MAPK and NF-κB signaling cascades [30]. More recently, Martinez-Leal et al. reported that **2** induces endoplasmic reticulum (ER) stress in HeLa cells by activating the multipronged unfolded protein response (UPR) in a characteristic pattern [59]. Working simultaneously as an ER stress inducer and an autophagy inhibitor, **2** was combined with bortezomib to synergistically block proteasomal degradation and autophagy, thereby exacerbating accumulation of misfolded proteins that originate from plitidepsin-induced oxidative stress. Such elevated proteotoxic stress led to apoptosis both in vitro (in MM1S multiple myeloma cells) and in vivo (in SCID mice xenografted with RPMI-8226 multiple myeloma cells). This study shows the promise of combined anticancer therapy using plitidepsin and proteasomal inhibitors such as bortezomib in a clinical setting.

2.2. *Cytotrienin A and Ansatrienin B*

Cytotrienin A (**4**, Figure 2) was initially isolated from the culture broth of soil-dwelling *Streptomyces* sp. RK95-74 [60]; this compound has strong cytotoxicity (IC_{50} = 7.7 nM) against human leukemia cell line HL-60 [61]. Despite early mechanistic studies [62–64], its target remained elusive until Pelletier et al. identified **4** as a translation inhibitor through high-throughput screening [37]. Their finding is that akin to didemnin B, cytotrienin A modulates eEF1A-dependent loading of aa-tRNA to the ribosome, most likely by stabilizing the eEF1A/GTP/aa-tRNA assembly positioned at the ribosomal A site. Thus, without release of eEF1A from the ribosome, translation elongation stops. Further insight into this compound's mode of action came from a 2015 study led by Taunton, who aimed to seek out the target of a potent antiproliferative ternatin derivative **7** (ternatin-4, cross-refer to the following section) [38]. With the help of photoaffinity labeling, they were able to capture the binding partner with a ternatin-based probe. Interestingly, the photolabeled protein is a ternary complex comprising eEF1A, GTP, and aa-tRNA rather than eEF1A alone. Subsequent competitive-binding experiments noted that the photolabeling efficiency is diminished dose-dependently with the addition of didemnin B or ansatrienin B (**5**, a close side-chain analogue of cytotrienin A shown in Figure 2). Therefore, it was concluded that ternatin, didemnin B, and cytotrienin A/ansatrienin B may share a binding hot spot on the eEF1A surface, probably located near A399, as indicated by resistance-conferring mutation experiments.

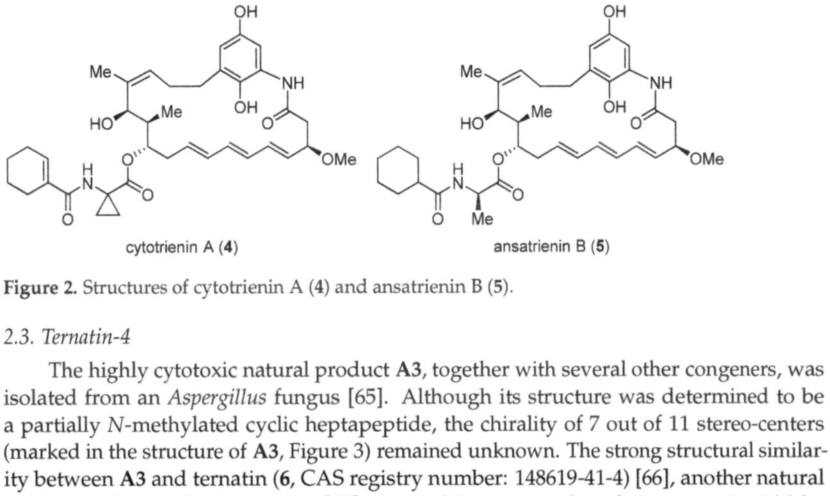

Figure 2. Structures of cytotrienin A (4) and ansatrienin B (5).

2.3. *Ternatin-4*

The highly cytotoxic natural product **A3**, together with several other congeners, was isolated from an *Aspergillus* fungus [65]. Although its structure was determined to be a partially *N*-methylated cyclic heptapeptide, the chirality of 7 out of 11 stereo-centers (marked in the structure of **A3**, Figure 3) remained unknown. The strong structural similarity between **A3** and ternatin (**6**, CAS registry number: 148619-41-4) [66], another natural product with anti-obesity activity [67], inspired Taunton et al. to design ternatin-4 (**7**) by incorporating the dehydromethyl leucine and pipecolic acid residues of **A3** (highlighted red in the structure of **7**, Figure 3) into **6** [38]. The resulting hybrid molecule **7** attained more than 10-fold enhancement of potency over the parent compound **6** (IC_{50} 4.6 nM vs. 71 nM against HCT-116 cancer cell line), thus solving all but one stereo-configuration of **A3**. Wondering the molecular target of ternatins, they developed a bifunctional photoaffinity probe **10**. Under UV irradiation, its photolabile diazirine subunit at residue 4 (highlighted red) decomposes into a highly reactive carbene that instantaneously crosslinks to the nearby binding protein. The alkyne at residue 6 (highlighted blue) will then connect to a fluorescent reporter via click cycloaddition, thus tagging the photolabeled target for characterization. In this way, eEF1A was captured and confirmed as the target of ternatins.

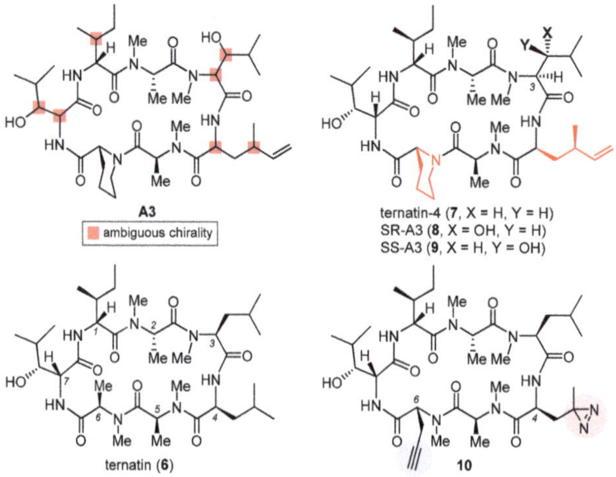

Compound	IC_{50} (nM)*	target residence time (s)
7	4.6	56
8	0.9	82
9	2.7	51

* against HCT-116 cancer cell line

Figure 3. Structures of natural products **A3** and ternatin (**6**), and synthetic analogues ternatin-4 (**7**), SR-A3 (**8**), SS-A3 (**9**), and photoaffinity probe **10**. The inset table compares anticancer activity and target residence time of **7–9**.

Based on the verified structure of ternatin-4 (**7**), a β-hydroxyl group was introduced at its residue 3 to obtain two epimers of **A3**, namely SR-A3 (**8**) and SS-A3 (**9**). An improved second-generation total synthesis allowed quick access to both compounds [68]. Hence, the identity of natural product **A3** was established as SR-A3 (**8**). Of special note, despite a minor difference in the side chain of residue 3, **8** was found to display a more prolonged duration of growth inhibition action than **7** and **9**. Single-molecule fluorescence resonance energy transfer (smFRET) imaging corroborated this observation, while further quantification through in vitro chase experiments confirmed enhanced drug–target residence time (table inset in Figure 3) and rebinding kinetics of **8**. Finally, preclinical evaluation of **8** vis-a-vis **7** was carried out in an aggressive Myc-driven mouse lymphoma model. Compared with its des-hydroxyl variant ternatin-4 (**7**), SR-A3 (**8**) significantly reduced tumor burden while extending the survival of the treated Eμ-Myc mice [68]. This work highlights the importance of side-chain modification in macrocyclic drug discovery and also makes a good case that the drug–target interaction can be more precisely characterized using the drug–target residence time model [69]. More recently, with the help of smFRET imaging and cryogenic electron microscopy (cryo-EM), Taunton and collaborators demonstrated that in spite of sharing a common eEF1A-binding site, ternatin-4 (**7**) and didemnin B exhibit differential inhibition dynamics in that the former traps the eEF1A/GDP/aa-tRNA ternary complex on the ribosome in a more reversible fashion than does the latter [70]. Their in-depth mechanistic investigation also revealed that by trapping eEF1A at the ribosomal A site, ternatin-4 induces ubiquitination and degradation of eEF1A on stalled ribosomes through a previously unknown surveillance pathway for translation quality control [71].

2.4. *Nannocystin A*

Nannocystin A (**11**, Figure 4) is a 21-membered cyclic depsipeptide isolated independently by Brönstrup et al. [72] and Hoepfner et al. [73] from the myxobacteria of the *Nannocystis* genus. Brönstrup et al. found that **11** is a strong inducer of apoptosis, as such inhibiting the growth of multiple cancer cell lines with low nanomolar IC_{50} values [72]. Meanwhile, another team led by Hoepfner pinned down its target through a combination of genetic and chemoproteomic approaches [73]. In brief, initial haploinsufficiency profiling and mutagenesis experiments implied eEF1A as the most likely target. To verify direct binding, they set up affinity chromatography with the semisynthetic probe **12** (Figure 4). During elution, this immobilized nannocystin sequestered eEF1A1 and eEF1A2 out of the 3644 proteins comprising the HCT-116 cell lysates. Moreover, it competed with unbound nannocystin A (**11**) and didemnin B for binding to eEF1A. Hence, eEF1A was determined to be the target of nannocystin A [73].

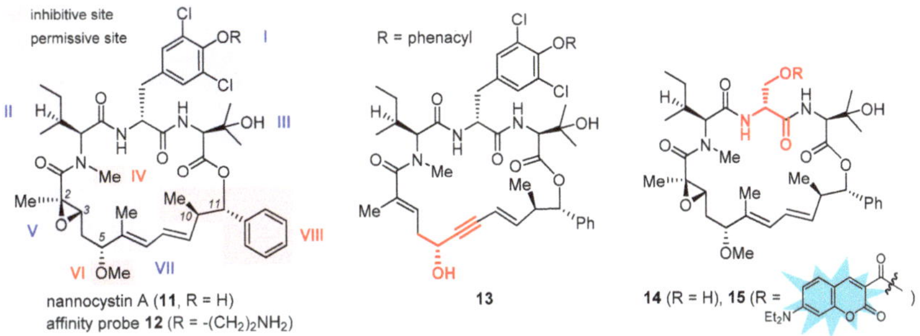

Figure 4. Structures of naturally occurring nannocystin A (**11**), semisynthetic affinity probe **12**, as well as synthetic macrocyclic propargylic alcohol **13**, serine-containing intermediate **14**, and coumarin-tagged fluorescent probe **15**. The SAR illustrated in the structure of **11** consists of inhibitive (red) and permissive (blue) sites: the former must be reserved for high activity, whereas the latter can tolerate moderate change.

The firsthand structure–activity relationship of nannocystins was derived from isolated and semisynthetic nannocystins [72,73], which indicates that modification at the tyrosine phenol moiety (subdomain **I** in the structure of **11**, Figure 4) or the side chain of *N*-Me-L-isoleucine (subdomain **II**) is tolerated. To obtain comprehensive SAR, nevertheless, total synthesis is a must. Thus far, seven routes have been reported for the total syntheses of nannocystin A (**11**) or its 2*E*-alkene surrogate nannocystin Ax (**16**, structure shown in Table 1) [74–81], each involving a distinct macrocyclization reaction as the key strategic step [82]. With the dual purpose of (1) total synthesis and (2) SAR validation concerning the binding role of the polyketide C5-C7 region, Fürstner et al. devised a motif-oriented strategy so that the macrocyclic propargylic alcohol **13** (shown in Figure 4) underwent post-macrocyclization elaboration [83] into an array of novel analogues besides nannocystin Ax (**16**) [80]. It was found that the 5*R*-methoxy ether (subdomain **VI**), instead of the neighboring C6-C7 (*E*)-alkene (subdomain **VII**), must be reserved for high activity. As an illustration, Table 1 compares the anticancer activity of four pairs of nannocystin derivatives with or without methylation at the C5-OH group (R = Me or H). Clearly, removal of this moiety causes a drastic reduction in potency (**16** vs. **17**, **18** vs. **19**, **20** vs. **21**, **22** vs. **23**); on the other hand, changing the C6-methyl (**16**) to a fluorine (**18**) or hydrogen (**20**) atom, or even curtailing the (*Z*) alkene to an alkyne (**22**), has insignificant impact on activity.

Following a total synthesis of nannocystin A [76] via Heck macrocyclization [84,85], we prepared a diversity of non-natural nannocystins modified at different sites. Our findings demonstrated that (1) the (2*R*, 3*S*)-epoxide (subdomain **V**) may be substituted for a 2*E*-alkene without compromising activity [86], (2) the side chain of β-OH-L-valine (subdomain **III**) is tolerant of minor change [86], and (3) the polyketide C9-C10 segment including its entire (10*R*, 11*S*) stereo-chemistry (subdomain **VIII**) is a key determinant of potency [87,88]. In parallel, He et al. synthesized more variants via Heck macrocyclization too and observed that removal of the *N*-methyl moiety (subdomain **IV**) incurred a dramatic loss of activity [89]. Taken together, the SAR of nannocystins is illustrated in the structure of **11** (Figure 4).

Table 1. Antiproliferative activity of nannocystin derivatives **16–23** against human colorectal carcinoma HCT-116 cells [80].

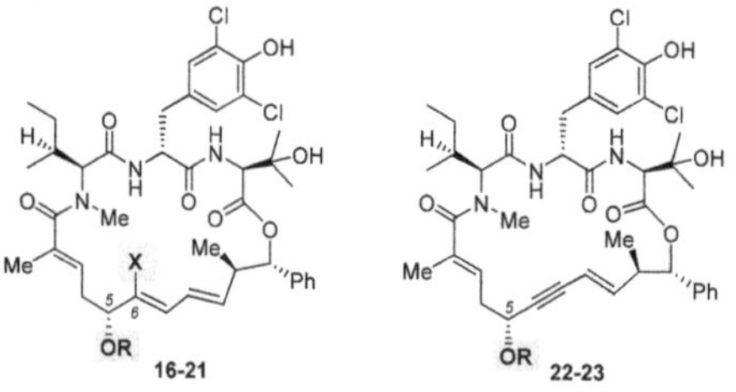

Compound	X	R	IC_{50} (nM)
Nannocystin Ax (**16**)	Me	Me	0.8
17		H	198
18	F	Me	1.5
19		H	1345
20	H	Me	4.3
21		H	1549
22	—	Me	22.2
23		H	1761

Aside from target elucidation and SAR profiling, the exact mechanism of nannocystins is a subject of enduring interest [90–92] given the poorly understood role of eEF1A in tumorigenesis. Chen et al. showed that the antimetastatic effect of nannocystin Ax in lung cancer cells is attributable to its interference with the TGF-β/Smad signaling pathway [90]. Their result is in step with another study uncovering the promigratory ability of eEF1A2 to promote lung adenocarcinoma metastasis [32]. The additional finding that the regulation of TGFβI (TGF β receptor I) by nannocystin Ax occurs at the transcriptional rather than the (post-)translational level implied the presence of an alternative mechanism independent of eEF1A inhibition [90]. Therefore, similar to the case of plitidepsin [30], the possibility that nannocystins impact certain protumorigenic pathway(s) mediated by eEF1A cannot be ruled out. Recently, we designed a serine-incorporating nannocystin **14** (Figure 4) to leverage a post-macrocyclization diversification strategy for efficient side-chain variation [92]. Thus obtained SAR concurred with the general trend depicted in Figure 4 and further informed the development of a coumarin-conjugated fluorescent probe **15**. With good permeability into the cancer cells, this probe was localized to the ER, as visualized by confocal fluorescence microscopy, which implies that nannocystins act on eEF1A predominantly at the ER-bound ribosome. Our result is in good agreement with the latest work by Förster et al. capturing eEF1A associated with the ribosome at the ER membrane by the use of cryo-electron tomography [93], thereby shedding light on the intracellular mode of action of nannocystins.

2.5. Metarrestin

Perinucleolar compartment (PNC) is a heritable multicomponent dynamic subnuclear organelle located at the periphery of the nucleolus of eukaryotic cells and uniquely associated with metastatic cancer cells [94]. Huang et al. found that PNC prevalence, defined as the percentage of non-apoptotic and non-mitotic cells harboring at least one PNC, is a pan-cancer prognostic marker positively correlated with metastatic capacity [95,96]. Subsequent screening of clinically approved anticancer drugs led to the observation that some of these drugs are capable of reducing PNC prevalence via specific on-target inhibition in lieu of promiscuous toxicity [97]. Having confirmed the existence of mechanism-specific PNC disassemblers with clinical efficacy, this proof-of-concept study supported taking PNC prevalence reduction as a phenotypic screening marker to discover antimetastatic drugs. To this end, a metastatic prostate cancer cell line PC3M with a PNC prevalence of 75% to 85% was engineered to stably express green fluorescent protein (GFP)-fused polypyrimidine-tract-binding protein (PTB) [98]. PTB is an essential PNC marker routinely measured by immunohistochemistry, which is unfortunately inconvenient for automated screening. But now with the self-reporting fluorescent PC3M-GFP-PTB cell line at hand, they were able to establish an image-based high-throughput, high-content assay (HCA) primed for spotting compounds able to reduce PNC prevalence by 50% [99]. Aiming at antimetastasis, the initial hits underwent secondary assays to select for invasion inhibition while excluding those acting via apoptosis induction, DNA intercalation, general cytotoxicity, or cell-cycle arrest [100]. By means of this multistage screening protocol, two leads were eventually identified out of 140,800 structurally diverse compounds from the NIH Molecular Libraries Small Molecule Repository (MLSMR) due to their outstanding PNC-disassembling efficiency and low cytotoxicity, thus setting the stage for the ensuing medicinal chemistry campaign [101].

After a preliminary exploration, pyrrolopyrimidine **24** (Figure 5) was favored over the other lead (structure not shown) for systematic optimization. Robust synthetic methods were next developed to access a broad variety of analogues evaluated for PNC disassembly and drug-like properties as well. The SAR trends are summarized in the structure of **24** (Figure 5). Specifically, (1) the *N*-3 substitution at the subdomain **I** prefers a linear alkyl chain bearing a hydroxy, ether, or amine, and conformational constraint with a cyclohexyl ring gives rise to the highest potency; (2) the *N*-7 position at the subdomain **II** tolerates a benzyl, phenethyl, or 4-methoxylphenyl group, but the presence of an alkyl substituent diminishes the potency significantly; (3) the unsubstituted C5 and C6 phenyl rings at the subdomain **III** are indispensable for high potency. While deducing the above trends, multi-round optimization finally yielded metarrestin (**25**), which possesses a superior selectivity window between PNC reduction and cell viability compared with the classic anticancer drugs doxorubicin and camptothecin (table inset in Figure 5) [101]. The in vitro performance of **25** was smoothly translated into in vivo efficacy in three mouse models of human cancer, where it suppressed metastatic invasion with concomitant reduction in PNC prevalence in the cancer cells of primary and metastasized tumors, offering a remarkable survival advantage to the treated animals [100]. After an in-depth evaluation of its pharmacokinetics [102,103] and safety [104], this drug has been advanced into a phase I clinical trial for the treatment of metastatic solid tumors [42].

The excellent antimetastatic capability of **25** prompted Huang et al. to investigate its mechanism, with the primary conclusion that the drug disrupts PNC assembly by blocking RNA polymerase I transcription [100]. Further seeking the binding target of **25**, they designed a biotin-conjugated probe **26** (Figure 5) that is likewise efficacious in disassembling PNC. Affinity purification with **26** combined with competition experiments using untagged metarrestin identified eEF1A2 as the binding target. The metarrestin–eEF1A2 interaction was confirmed by cellular thermal shift assay. Subsequent experiments along this line of research showed that (1) eEF1A2 enhances PNC assembly and metastatic progression; and (2) eEF1A2, at least in part, mediates the PNC-elimination function of metarrestin [100]. Whereas further details await elucidation, it was believed that metarrestin interferes with certain non-translational functions of eEF1A2. Recently, Jin et al. developed a proteolysis-targeting chimera (PROTAC) [105–107] strategy by tethering metarrestin with various ligands for the von Hippel–Lindau (VHL) E3 ligase [108]. Thus obtained heterobifunctional molecules were designed to recruit eEF1A2, the binding target of metarrestin, to the ubiquitin/proteasome system (UPS) for selective degradation. As one of these first-in-class eEF1A2 degraders, **27** (Figure 5) was shown to degrade eEF1A2 in three cancer cells in a dose-dependent manner, thus holding promise for the treatment of eEF1A2-mediated carcinogenesis.

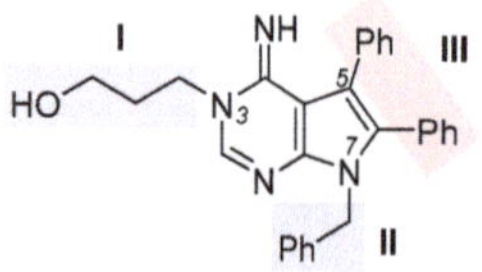

24

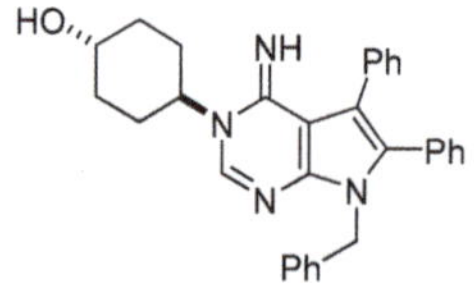

metarrestin (25)

Compound	PNC reduction IC_{50} (μM)	cytotoxicity IC_{50} (μM)	selectivity window*
24	1.88	10.86	5.8
25	0.20	7.65	38.3
doxorubicin	0.86	0.66	0.76
camptothecin	0.19	0.48	2.5

* cytotoxicity IC_{50} / PNC reduction IC_{50}

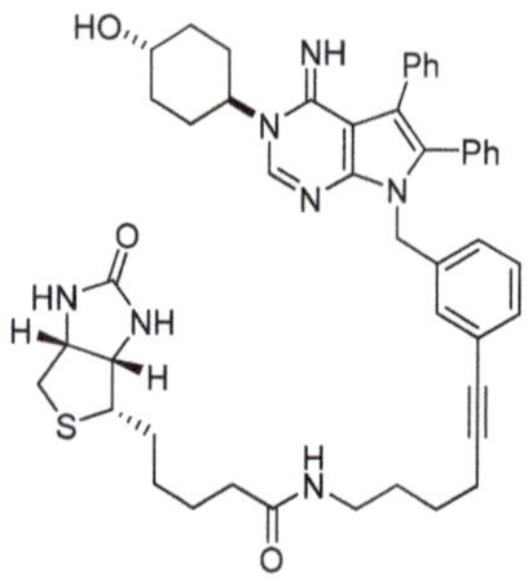

26

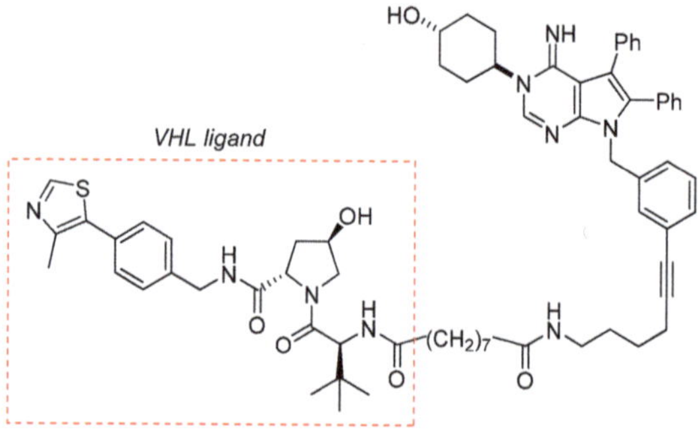

27

Figure 5. Structures of the initial hit **24**, final drug candidate metarrestin (**25**), biotin-labeled affinity probe **26**, and metarrestin-based PROTAC **27**. The inset table compares the selectivity window between PNC reduction and cytotoxicity of **24**, **25**, doxorubicin, and camptothecin.

2.6. 2-Phenyl-3-Hydroxy-4(1H)-Quinolinones

2-Phenyl-3-hydroxy-4(1*H*)-quinolinones (abbreviated as 3-HQs) such as **28** (Figure 6) are aza-analogues of previously reported eEF1A-targeting anticancer flavonoids [40]. Based on a homology model of human eEF1A1, Hlavac et al. carried out docking calculations to identify the binding site for gamendazole, a known eEF1A1 inhibitor for male contraception [109]. Encouragingly, they found that these 3-HQs fit into the same gamendazole-binding site on the surface of eEF1A1. Such direct interaction between eEF1A1 and 3-HQs was verified through pull-down assay using biotinylated 3-HQ derivatives [110]. Having validated the constructed eEF1A1 model, the authors performed virtual screening of in silico designed 3-HQs with varying substituents R^1, R^2, and R^3 (illustrated in the structure of **28**). The six highest-scored and synthetically accessible compounds were chosen for wet-lab preparation. Their binding to eEF1A1 was quantitatively characterized with isothermal titration calorimetry (ITC), which provided thermodynamic information consistent with docking calculation results. Biological evaluation discovered **29**, one of these rationally designed eEF1A1 inhibitors, with optimal inhibitory activity against several cancer cell lines but low toxicity toward a normal cell line [110].

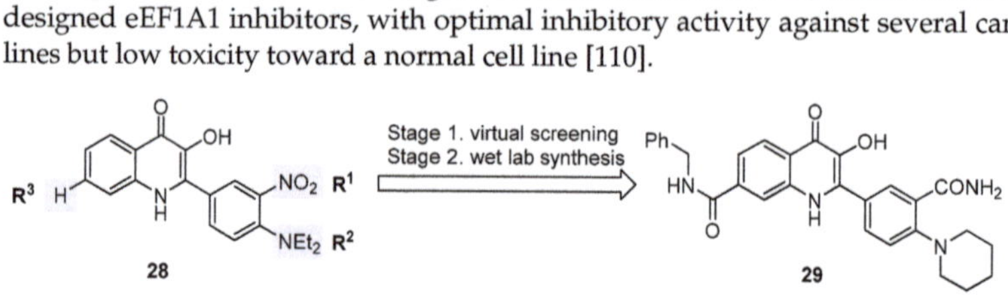

Figure 6. Structures of eEF1A1-targeting anticancer 3-HQ derivative **29** and its prototype **28**.

2.7. Cordyheptapeptide A

Cordyheptapeptide A (**30**), a partially *N*-methylated cyclic heptapeptide with anticancer activity, was originally isolated from the insect pathogenic fungus *Cordyceps* [111,112]. Although a solution-phase total synthesis of **30** was reported before [113], its SAR and mechanism of action were unclear. Lokey et al. developed a high-throughput solid-phase peptide synthesis (SPPS) to access a library of side-chain- and backbone-modified analogues [114]. They observed the following SAR trends: (1) all side chains are critical to its antiproliferative activity [115]; (2) halogenation at the aromatic side chain of residue 2 or 5 deteriorates activity at varying degrees; (3) whereas removal of the *N*-methyl moiety at residue 2 or 6 impairs activity, this is not the case for residue 4, for which changing sarcosine to glycine tends to improve activity, and when coupled with *ortho*-fluorination at residue 5, such *N*-demethylation brings about equipotent variant **31** with a 39-fold improvement in aqueous solubility (table inset in Figure 7). According to molecular dynamics simulations, the enhancement in activity stems from more conformational flexibility of its glycine-carrying scaffold, which is accordingly more accessible to target-binding conformations than the parent natural product **30**.

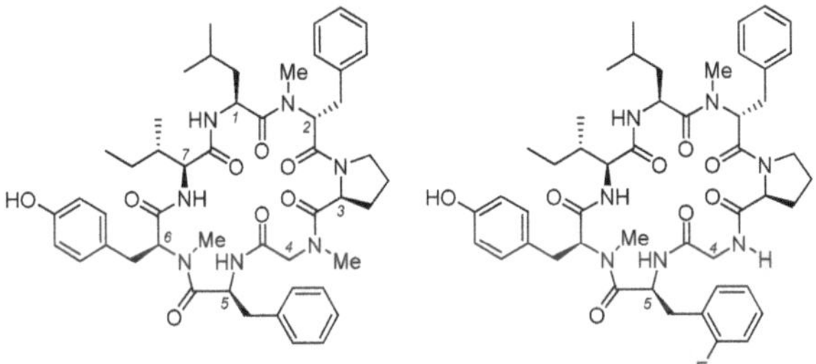

cordyheptapeptide A (**30**)   **31**

Compound	IC_{50} (μM)*	Solubility (μM)
30	0.2	4
31	0.2	156

* against HCT-116 cancer cell line

Figure 7. Structures of natural cordyheptapeptide A (**30**) and synthetic analogue **31**, with the inset table comparing their IC_{50} values against HCT-116 cancer cell line and aqueous solubility.

To find out the mechanism of action of **30**, the authors determined its cytotoxicity profile via the NCI60 human tumor cell line assay [114]. Analyzed by the COMPARE algorithm, this profile was best correlated with that of phyllanthoside, a known eukaryotic protein-synthesis inhibitor [116]. Consistently, cytological profiling (CP) [117] indicated that **30** clustered most closely with protein-synthesis inhibitors such as didemnin B and ternatin but deviated significantly from microtubule inhibitors and poly (ADP-ribose) polymerase (PARP) inhibitors. Combining the results from both phenotypic experiments, **30** is quite likely a protein-synthesis inhibitor. This inference was confirmed by bioorthogonal noncanonical amino acid tagging (BONCAT) [118], which proved that the agent primarily blocks protein synthesis and has a secondary influence on DNA synthesis. Suspecting its target to be eEF1A, **30** was evaluated in the HCT-116 cancer cells with a point mutation of eEF1A (A399V). Previously, the same mutation was reported to confer resistance to eEF1A-targeting didemnin B [119], ternatin [38], and nannocystin A [73]. Indeed, the activity dropped remarkably in the mutant cells, thus providing genetic evidence that supports eEF1A as the target of **30**.

2.8. BE-43547A_2

Isolated from *Streptomyces* sp. in 1998, BE-43547A_1 (**32**, Figure 8), BE-43547A_2 (**33**), and other congeners are a series of macrocyclic depsipeptides differing in the C21 side chain [120]. These compounds belong to the amidopentadienoate-containing cyclolipodepsipeptide (APD-CLD) natural products that feature an electrophilic 4-amido-2,4-pentadienoate (APD, highlighted red) functionality as well as a lipophilic side chain [121]. Poulsen et al. developed the first total synthesis of *ent*-**32** and re-isolated the authentic **32** from the fermentation broth of a BE-43547-producing microorganism, thereby establishing the absolute stereo-configuration of the BE-43547 family [122]. Importantly, **32** and **33** exhibited superior hypoxia-selective cytotoxicity in PANC-1 pancreatic cancer cells than rakicidin A, another APD-CLD natural product they investigated earlier [123–126]. Shortly after this work, Chen et al. developed a total synthesis of **33** and reported that this agent selectively targets pancreatic cancer stem cells (PCSCs) [127,128]. Preparation of more analogues [129,130] led to the following SAR results (illustrated in Figure 8): (1) the exocyclic alkene at C8 within the APD unit must be reserved; (2) the macrolide cannot be changed to the corresponding macrolactam, or in other words, the O35 cannot be replaced with a nitrogen atom; (3) the (*S*)-hydroxyl group at C15 is critical to activity; (4) a lipophilic side chain at C21 is necessary but variable.

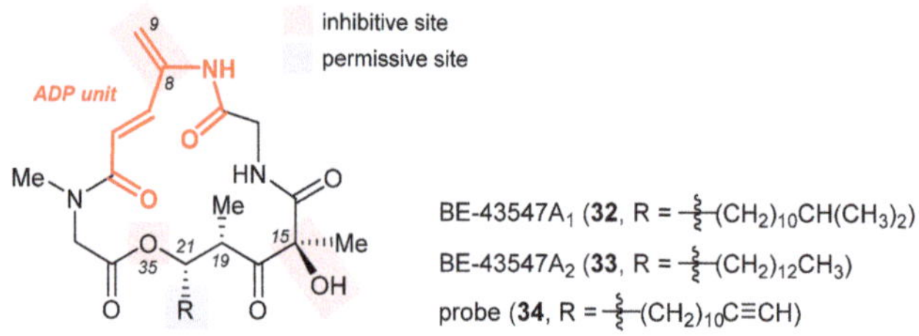

Figure 8. Structures of BE-43547A$_1$ (**32**), BE-43547A$_2$ (**33**), and the clickable probe **34**. The SAR illustrated in their structures consists of inhibitive (red) and permissive (blue) sites: the former must be reserved for high activity, whereas the latter can tolerate moderate change.

The excellent hypoxia-selective toxicity of APD-CLD natural products such as rakicidin A and the BE-43547A members intrigued Poulsen et al. to elucidate their mechanism of action, which is distinguished from conventional hypoxia-activated compounds. They showed that these APD-CLDs induced rapid and hypoxia-selective impairment of mitochondrial structure and function, thereby driving a peculiar form of non-apoptotic cancer cell death in a hypoxic milieu [131]. To discover the molecular target of **33**, Chen et al. synthesized a clickable probe **34** [130]. Biotinylation of **34** via in situ click cycloaddition followed by pull-down assay indicated eEF1A as a binding target, and the isoform was determined to be eEF1A1 via immunoblotting. Having located the cysteine234 residue of eEF1A1 as the most probable binding site of **33** according to LC-MS/MS analysis, they engineered three types of pancreatic cancer cells with (1) eEF1A1 knockdown (KD), (2) eEF1A1 recovered from KD (RE-KD), and (3) C234-mutant eEF1A1 constructed from KD (RE-C234S). As shown in Table 2, the in vitro cytotoxicity of **33** against these three and the wild-type (WT) pancreatic cancer cells, as well as its in vivo anticancer efficacy in the four corresponding mouse xenograft models, provided concrete evidence favoring the Cys234 residue of eEF1A1 as the binding site for **33** [130]. Furthermore, it was shown that eEF1A1 plays a significant role in regulating pancreatic cancer cell stemness, its levels positively correlated with pancreatic cancer progression and negatively affecting patient survival.

Table 2. In vitro cytotoxicity of **33** against different types of pancreatic cancer cells (WT, KD, RE-KD, RE-C234S) and its in vivo anticancer efficacy in the corresponding mouse xenograft models [130].

Pancreatic Cancer Cells	In Vitro Cytotoxicity IC_{50} (μM)	In Vivo Tumor Inhibition Rate (%)
WT	1.33	98.8
KD	11.62	20.7
RE-KD	0.80	93.2
RE-C234S	11.61	18.3

3. Conclusions and Future Perspectives

Translational control with small-molecule agents represents an emerging direction for anticancer drug discovery [132–134]. As proofs of principle, hitherto two such drugs have gained approval for clinical use, namely homoharringtonine and plitidepsin. The former is the first protein translation inhibitor for the treatment of chronic myeloid leukemia, which works by fitting into the ribosomal A site so as to block access by the charged tRNA [135]. Pertaining to the present subject, the latter is a specific inhibitor of eEF1A that, at least in part, disrupts its canonical function of assisting translation elongation to combat multiple myeloma [35]. The whole ensemble of known eEF1A-targeting small-molecule agents is compiled in Table 3, alongside a brief summary of their anticancer effectiveness and selectivity. Intriguingly, though sharing the same molecular target, these compounds manifested differential antiproliferative profiles. For example, narciclasine [39], synthetic flavonoids [40], and ternatin-4 [68] within this category displayed preferential anticancer potency against human melanoma, breast carcinoma, and colorectal carcinoma cells, respectively. Presumably, such contrasting selectivity may partially stem from their distinct chemotypes that influence the drug–target interaction as well as the consequent therapeutic outcome in a subtle but profound way, as revealed very recently by Taunton et al. on the inhibitory mode of action of didemnin B and ternatin-4 at the single-molecule level [70]. Overall, the current review outlines the remarkable progress achieved over the recent years in eEF1A-targeting anticancer agents, which are structurally distinct macrocycles and heterocycles, either naturally occurring, developed based on the hit from high-throughput screening, or rationally designed. Their development status is summarized in Figure 9. For the time being, metarrestin is under clinical development as an unprecedented modality for controlling cancer metastasis [42], whereas metarrestin-based PROTACs have also been disclosed for selective degradation of eEF1A2 [108]. Albeit beyond the scope of this review, it is also worth noting that plitidepsin has now entered a clinical trial as a potential anti-SARS-CoV-2 drug [136] since its target eEF1A turned out to be a crucial host protein co-opted by virus to infect human cells [137,138].

In spite of the aforementioned advances, however, there remains much to learn about the exact mechanisms of action of these targeted agents. A more fundamental question lies in the oncogenic mechanism of eEF1A, a multitalented protein capable of both translation elongation and a myriad of moonlighting duties. As evidenced by an illuminating study on plitidepsin [30], it seems indeed viable for malignant cells to exploit certain non-canonical functions of eEF1A for survival. Moreover, the involvement of eEF1A1 in aggressive castration-resistant prostate cancer (CRPC) [139] and non-small cell lung cancer (NSCLC) metastasis [140] has been demonstrated through its complexation with actin and the eEF1A1/MDM2/MTBP signaling axis, respectively. These latest results showcase the potential opportunities of designing next-generation magic bullets [141] that act upon the manifold oncogenic functions of eEF1A, selectively extinguishing the malignant while sparing the normal. It is foreseeable that ongoing investigation of these mechanistic aspects, along with the expanding repertoire of eEF1A-targeting agents, will position us at a better forefront against cancer and other eEF1A-driven diseases as well.

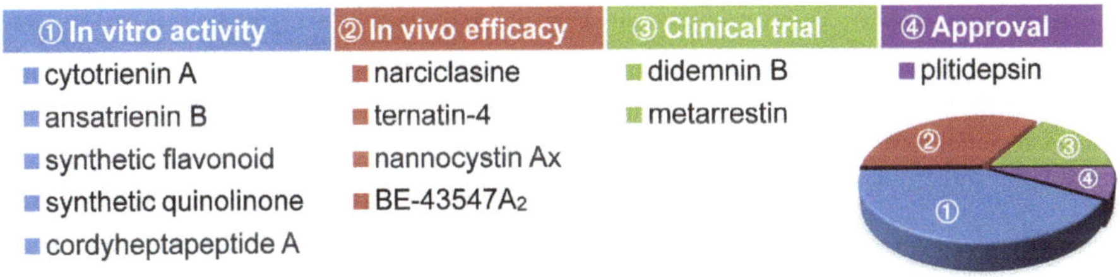

Figure 9. Development status of eEF1A-targeting small-molecule anticancer agents, among which narciclasine, ternatin-4, nannocystin Ax, and BE-43547A$_2$ have been validated in at least one in vivo preclinical model, didemnin B and metarrestin have been or currently are being studied in clinical trials, and plitidepsin has gained approval for clinical use. These three groups of compounds are at more advanced development stages than the leftmost group, making up over half of the whole collection of anticancer eEF1A inhibitors discernible from the accompanying pie chart. Their validated efficacy in preclinical models or clinical cohorts lends concrete support to the principle of targeting eEF1A with viable anticancer selectivity.

Table 3. Summary of eEF1A-targeting small-molecule agents on their anticancer efficacy and selectivity.

Compound	Anticancer Efficacy and Selectivity	Reference
Didemnins	• potent against human cancer cell lines from different tissues • in vivo validated in several preclinical models • didemnin B (**1**) and plitidepsin (**2**) have been studied in multiple clinical trials • plitidepsin (**2**, aplidin®) has been approved in Australia for combined treatment of relapsed/refractory multiple myeloma with dexamethasone	[36,41,50–56]
Tamandarins	• potent against human cancer cell lines from different tissues	[36,49]
Cytotrienin A (**4**)	• potent against human leukemia HL-60 cells (IC_{50} = 7.7 nM) and lung carcinoma A549 cells (IC_{50} = 0.1 μM)	[61,64]
Ansatrienin B (**5**)	• potent against three human pancreatic cancer cell lines (IC_{50} range, 0.17–1.69 μM)	[142]
Narciclasine	• potent against five human melanoma cell lines (IC_{50}~40 nM) • in vivo validated in a mice xenograft model of brain metastatic melanoma	[39]
Synthetic flavonoids	• potent against human breast cancer cell lines (IC_{50} range, 1–50 μM for MDA-MB231)	[40]
Ternatin-4 (**7**)	• potent against human colorectal carcinoma HCT-116 cells (IC_{50} = 4.6 nM) • in vivo validated in an aggressive Myc-driven mouse lymphoma model	[68]
Nannocystin A (**11**)	• potent against 472 cancer cell lines (IC_{50} range, 5–500 nM) • nannocystin Ax in vivo validated in an HCT-116-derived xenograft zebrafish model	[73,91]
Metarrestin (**25**)	• excellent antimetastatic selectivity over cytotoxicity (cytotoxicity IC_{50}/PNC reduction IC_{50} = 38.3) • in vivo validated in several preclinical models • currently in a phase I clinical trial for the treatment of metastatic solid tumors	[42,100–104]
Synthetic quinolinones	• potent against human cancer cell lines from different tissues (IC_{50} range, 0.56–50 μM)	[110]
Cordyheptapeptide A (**30**)	• potent against human colorectal carcinoma HCT-116 cells (IC_{50} = 0.2 μM)	[114]
BE-43547A_2 (**33**)	• potent against human pancreatic carcinoma PANC-1 cells (IC_{50} = 0.87 μM) • remarkable hypoxia-selective toxicity against human leukemia K562 cells and breast carcinoma MCF-7 cells (selective index = 28 and 79, respectively) • selectively targets pancreatic cancer stem cells (PCSCs) • in vivo validated in a pancreatic cancer xenograft mouse model	[127,130]

Author Contributions: Conceptualization and funding acquisition, W.Z.; data curation, H.Z. and J.C.; resources, S.Y.; visualization, B.S.; writing—original draft preparation, W.Z.; writing—review and editing, W.Z., H.Z., J.C., S.Y. and B.S. All authors have read and agreed to the published version of the manuscript.

Funding: This work is generously supported by the National Natural Science Foundation of China (No. 21772101) and the Natural Science Foundation of Tianjin City (Grant No. 17JCYBJC28400).

Conflicts of Interest: The authors declare no conflict of interest.

References

1. Burrell, R.A.; McGranahan, N.; Bartek, J.; Swanton, C. The causes and consequences of genetic heterogeneity in cancer evolution. *Nature* **2013**, *501*, 338–345. [CrossRef] [PubMed]
2. Kitano, H. Cancer as a robust system: Implications for anticancer therapy. *Nat. Rev. Cancer* **2004**, *4*, 227–235. [CrossRef] [PubMed]
3. Barker, H.E.; Paget, J.T.; Khan, A.A.; Harrington, K.J. The tumour microenvironment after radiotherapy: Mechanisms of resistance and recurrence. *Nat. Rev. Cancer* **2015**, *15*, 409–425. [CrossRef] [PubMed]
4. Avendaño, C.; Menendez, J.C. *Medicinal Chemistry of Anticancer Drugs*, 2nd ed.; Elsevier: Amsterdam, The Netherlands, 2015.
5. Papież, M.A.; Krzyściak, W. Biological therapies in the treatment of cancer—Update and new directions. *Int. J. Mol. Sci.* **2021**, *22*, 11694. [CrossRef]
6. Kaiser, M.; Semeraro, M.D.; Herrmann, M.; Absenger, G.; Gerger, A.; Renner, W. Immune aging and immunotherapy in cancer. *Int. J. Mol. Sci.* **2021**, *22*, 7016. [CrossRef]
7. Diwan, D.; Cheng, L.; Usmani, Z.; Sharma, M.; Holden, N.; Willoughby, N.; Sangwan, N.; Baadhe, R.R.; Liu, C.; Gupta, V.K. Microbial cancer therapeutics: A promising approach. *Semin. Cancer Biol.* **2022**, *86*, 931–950. [CrossRef]
8. Sawant, S.S.; Patil, S.M.; Gupta, V.; Kunda, N.K. Microbes as medicines: Harnessing the power of bacteria in advancing cancer treatment. *Int. J. Mol. Sci.* **2020**, *21*, 7575. [CrossRef]
9. Zhong, L.; Li, Y.; Xiong, L.; Wang, W.; Wu, M.; Yuan, T.; Yang, W.; Tian, C.; Miao, Z.; Wang, T. Small molecules in targeted cancer therapy: Advances, challenges, and future perspectives. *Signal Transduct. Target Ther.* **2021**, *6*, 201. [CrossRef]
10. Ohishi, T.; Kaneko, M.K.; Yoshida, Y.; Takashima, A.; Kato, Y.; Kawada, M. Current Targeted Therapy for Metastatic Colorectal Cancer. *Int. J. Mol. Sci.* **2023**, *24*, 1702. [CrossRef]
11. Lau, K.H.; Tan, A.M.; Shi, Y. New and emerging targeted therapies for advanced breast cancer. *Int. J. Mol. Sci.* **2022**, *23*, 2288. [CrossRef]

12. Negrutskii, B.S.; El'skaya, A.V. Eukaryotic translation elongation factor 1α: Structure, expression, functions, and possible role in aminoacyl-tRNA channeling. *Prog. Nucleic Acid Res. Mol. Biol.* **1998**, *60*, 47–78.
13. Sanders, J.; Brandsma, M.; Janssen, G.M.C.; Dijk, J.; Moeller, W. Immunofluorescence studies of human fibroblasts demonstrate the presence of the complex of elongation factor-1βγδ in the endoplasmic reticulum. *J. Cell Sci.* **1996**, *109*, 1113–1117. [CrossRef] [PubMed]
14. Kjær, S.; Wind, T.; Ravn, P.; Østergaard, M.; Clark, B.F.C.; Nissim, A. Generation and epitope mapping of high-affinity scFv to eukaryotic elongation factor 1A by dual application of phage display. *Eur. J. Biochem.* **2001**, *268*, 3407–3415. [CrossRef]
15. Migliaccio, N.; Ruggiero, I.; Martucci, N.M.; Sanges, C.; Arbucci, S.; Tatè, R.; Rippa, E.; Arcari, P.; Lamberti, A. New insights on the interaction between the isoforms 1 and 2 of human translation elongation factor 1A. *Biochimie* **2015**, *118*, 1–7. [CrossRef] [PubMed]
16. Dever, T.E.; Dinman, J.D.; Green, R. Translation elongation and recoding in eukaryotes. *Cold Spring Harb. Perspect. Biol.* **2018**, *10*, a032649. [CrossRef] [PubMed]
17. Mateyak, M.K.; Kinzy, T.G. eEF1A: Thinking outside the ribosome. *J. Biol. Chem.* **2010**, *285*, 21209–21213. [CrossRef]
18. Sasikumar, A.N.; Perez, W.B.; Kinzy, T.G. The many roles of the eukaryotic elongation factor 1 complex. *Wiley Interdiscip. Rev. RNA* **2012**, *3*, 543–555. [CrossRef]
19. Li, D.; Wei, T.; Abbott, C.M.; Harrich, D. The unexpected roles of eukaryotic translation elongation factors in RNA virus replication and pathogenesis. *Microbiol. Mol. Biol. Rev.* **2013**, *77*, 253–266. [CrossRef]
20. Carriles, A.A.; Mills, A.; Muñoz-Alonso, M.-J.; Gutiérrez, D.; Domínguez, J.M.; Hermoso, J.A.; Gago, F. Structural Cues for Understanding eEF1A2 Moonlighting. *ChemBioChem* **2021**, *22*, 374–391. [CrossRef]
21. Thornton, S.; Anand, N.; Purcell, D.; Lee, J. Not just for housekeeping: Protein initiation and elongation factors in cell growth and tumorigenesis. *J. Mol. Med.* **2003**, *81*, 536–548. [CrossRef]
22. Lamberti, A.; Caraglia, M.; Longo, O.; Marra, M.; Abbruzzese, A.; Arcari, P. The translation elongation factor 1A in tumorigenesis, signal transduction and apoptosis: Review article. *Amino Acids* **2004**, *26*, 443–448. [CrossRef]
23. Abbas, W.; Kumar, A.; Herbein, G. The eEF1A Proteins: At the Crossroads of Oncogenesis, Apoptosis, and Viral Infections. *Front. Oncol.* **2015**, *5*, 75. [CrossRef] [PubMed]
24. Hussey, G.S.; Chaudhury, A.; Dawson, A.E.; Lindner, D.J.; Knudsen, C.R.; Wilce, M.C.; Merrick, W.C.; Howe, P.H. Identification of an mRNP Complex Regulating Tumorigenesis at the Translational Elongation Step. *Mol. Cell* **2011**, *41*, 419–431. [CrossRef] [PubMed]
25. Liu, S.; Hausmann, S.; Carlson, S.M.; Fuentes, M.E.; Francis, J.W.; Pillai, R.; Lofgren, S.M.; Hulea, L.; Tandoc, K.; Lu, J.; et al. METTL13 Methylation of eEF1A Increases Translational Output to Promote Tumorigenesis. *Cell* **2019**, *176*, 491–504. [CrossRef] [PubMed]
26. Kobayashi, D.; Tokuda, T.; Sato, K.; Okanishi, H.; Nagayama, M.; Hirayama-Kurogi, M.; Ohtsuki, S.; Araki, N. Identification of a specific translational machinery via TCTP-EF1A2 interaction regulating NF1-associated tumor growth by affinity purification and data-independent mass spectrometry acquisition (AP-DIA). *Mol. Cell. Proteom.* **2019**, *18*, 245–262. [CrossRef] [PubMed]
27. Amiri, A.; Noei, F.; Jeganathan, S.; Kulkarni, G.; Pinke, D.E.; Lee, J.M. eEF1A2 activates Akt and stimulates Akt-dependent actin remodeling, invasion and migration. *Oncogene* **2007**, *26*, 3027–3040. [CrossRef] [PubMed]
28. Li, Z.; Qi, C.F.; Shin, D.M.; Zingone, A.; Newbery, H.J.; Kovalchuk, A.L.; Abbott, C.M.; Morse, H.C.I. *Eef1a2* promotes cell growth, inhibits apoptosis and activates JAK/STAT and AKT signaling in mouse plasmacytomas. *PLoS ONE* **2010**, *5*, e10755. [CrossRef]
29. Pellegrino, R.; Calvisi, D.F.; Neumann, O.; Kolluru, V.; Wesely, J.; Chen, X.; Wang, C.; Wuestefeld, T.; Ladu, S.; Elgohary, N.; et al. EEF1A2 inactivates p53 by way of PI3K/AKT/mTOR-dependent stabilization of MDM4 in hepatocellular carcinoma. *Hepatology* **2014**, *59*, 1886–1899. [CrossRef]
30. Losada, A.; Munoz-Alonso, M.J.; Martinez-Diez, M.; Gago, F.; Dominguez, J.M.; Martinez-Leal, J.F.; Galmarini, C.M. Binding of eEF1A2 to the RNA-dependent protein kinase PKR modulates its activity and promotes tumor cell survival. *Br. J. Cancer* **2018**, *119*, 1410–1420. [CrossRef]
31. Itagaki, K.; Sasada, M.; Iyoda, T.; Imaizumi, T.; Haga, M.; Kuga, A.; Inomata, H.; Fukai, F.; Miyazaki, S.; Kondo, Y.; et al. Exposure of the cryptic de-adhesive site FNIII14 in fibronectin molecule and its binding to membrane-type eEF1A induce migration and invasion of cancer cells via β1-integrin inactivation. *Am. J. Cancer Res.* **2020**, *10*, 3990–4004.
32. Jia, L.; Ge, X.; Du, C.; Chen, L.; Zhou, Y.; Xiong, W.; Xiang, J.; Li, G.; Xiao, G.; Fang, L.; et al. EEF1A2 interacts with HSP90AB1 to promote lung adenocarcinoma metastasis via enhancing TGF-β/SMAD signalling. *Brit. J. Cancer* **2021**, *124*, 1301–1311. [CrossRef]
33. Mills, A.; Gago, F. On the Need to Tell Apart Fraternal Twins eEF1A1 and eEF1A2, and Their Respective Outfits. *Int. J. Mol. Sci.* **2021**, *22*, 6973; Human eFE1A has two predominant isoforms eEF1A1 and eEF1A2. According to the literature, some agents were reported to target eEF1A without knowing whether it is eEF1A1, eEF1A2, or both, whereas others target only one isoform. For a recent discussion about these two isoforms. [CrossRef]
34. Crews, C.M.; Collins, J.L.; Lane, W.S.; Snapper, M.L.; Schreiber, S.L. GTP-dependent binding of the antiproliferative agent didemnin to elongation factor 1α. *J. Biol. Chem.* **1994**, *269*, 15411–15414. [CrossRef]
35. Losada, A.; Munoz-Alonso, M.J.; Garcia, C.; Sanchez-Murcia, P.A.; Martinez-Leal, J.F.; Dominguez, J.M.; Lillo, M.P.; Gago, F.; Galmarini, C.M. Translation Elongation Factor eEF1A2 is a Novel Anticancer Target for the Marine Natural Product Plitidepsin. *Sci. Rep.* **2016**, *6*, 35100. [CrossRef] [PubMed]

36. Lee, J.; Currano, J.N.; Carroll, P.J.; Joullie, M.M. Didemnins, tamandarins and related natural products. *Nat. Prod. Rep.* **2012**, *29*, 404–424. [CrossRef] [PubMed]
37. Lindqvist, L.; Robert, F.; Merrick, W.; Kakeya, H.; Fraser, C.; Osada, H.; Pelletier, J. Inhibition of translation by cytotrienin A–a member of the ansamycin family. *RNA* **2010**, *16*, 2404–2413. [CrossRef] [PubMed]
38. Carelli, J.D.; Sethofer, S.G.; Smith, G.A.; Miller, H.R.; Simard, J.L.; Merrick, W.C.; Jain, R.K.; Ross, N.T.; Taunton, J. Ternatin and improved synthetic variants kill cancer cells by targeting the elongation factor-1A ternary complex. *eLife* **2015**, *4*, e10222. [CrossRef] [PubMed]
39. Van Goietsenoven, G.; Hutton, J.; Becker, J.P.; Lallemand, B.; Robert, F.; Lefranc, F.; Pirker, C.; Vandenbussche, G.; Van Antwerpen, P.; Evidente, A.; et al. Targeting of eEF1A with Amaryllidaceae isocarbostyrils as a strategy to combat melanomas. *FASEB J.* **2010**, *24*, 4575–4584. [CrossRef]
40. Yao, N.; Chen, C.Y.; Wu, C.Y.; Motonishi, K.; Kung, H.J.; Lam, K.S. Novel flavonoids with antiproliferative activities against breast cancer cells. *J. Med. Chem.* **2011**, *54*, 4339–4349. [CrossRef]
41. Jimenez, P.C.; Wilke, D.V.; Branco, P.C.; Bauermeister, A.; Rezende-Teixeira, P.; Gaudencio, S.P.; Costa-Lotufo, L.V. Enriching cancer pharmacology with drugs of marine origin. *Brit. J. Pharmacol.* **2020**, *177*, 3–27. [CrossRef]
42. Metarrestin (ML-246) in Subjects with Metastatic Solid Tumors. 2020. Available online: https://ClinicalTrials.gov/show/NCT04222413 (accessed on 2 March 2023).
43. Dmitriev, S.E.; Vladimirov, D.O.; Lashkevich, K.A. A Quick Guide to Small-Molecule Inhibitors of Eukaryotic Protein Synthesis. *Biochemistry (Moscow)* **2020**, *85*, 1389–1421. [CrossRef] [PubMed]
44. Brönstrup, M.; Sasse, F. Natural products targeting the elongation phase of eukaryotic protein biosynthesis. *Nat. Prod. Rep.* **2020**, *37*, 752–762. [CrossRef]
45. Burgers, L.D.; Fuerst, R. Natural products as drugs and tools for influencing core processes of eukaryotic mRNA translation. *Pharmacol. Res.* **2021**, *170*, 105535. [CrossRef]
46. Fan, A.; Sharp, P.P. Inhibitors of Eukaryotic Translational Machinery as Therapeutic Agents. *J. Med. Chem.* **2021**, *64*, 2436–2465. [CrossRef]
47. Zhang, J.-N.; Xia, Y.-X.; Zhang, H.-J. Natural cyclopeptides as anticancer agents in the last 20 years. *Int. J. Mol. Sci.* **2021**, *22*, 3973. [CrossRef] [PubMed]
48. Rinehart, K.L., Jr.; Gloer, J.B.; Hughes, R.G., Jr.; Renis, H.E.; McGovren, J.P.; Swynenberg, E.B.; Stringfellow, D.A.; Kuentzel, S.L.; Li, L.H. Didemnins: Antiviral and antitumor depsipeptides from a caribbean tunicate. *Science* **1981**, *212*, 933–935. [CrossRef] [PubMed]
49. Vervoort, H.; Fenical, W.; de Epifanio, R. Tamandarins A and B: New cytotoxic depsipeptides from a Brazilian ascidian of the family Didemnidae. *J. Org. Chem.* **2000**, *65*, 782–792. [CrossRef]
50. Leisch, M.; Egle, A.; Greil, R. Plitidepsin: A potential new treatment for relapsed/refractory multiple myeloma. *Future Oncol.* **2019**, *15*, 109–120. [CrossRef]
51. Chun, H.G.; Davies, B.; Hoth, D.; Suffness, M.; Plowman, J.; Flora, K.; Grieshaber, C.; Leyland-Jones, B. Didemnin B. The first marine compound entering clinical trials as an antineoplastic agent. *Investig. New Drugs* **1986**, *4*, 279–284.
52. Vera, M.D.; Joullie, M.M. Natural products as probes of cell biology: 20 years of didemnin research. *Med. Res. Rev.* **2002**, *22*, 102–145. [CrossRef]
53. Le Tourneau, C.; Raymond, E.; Faivre, S. Aplidine: A paradigm of how to handle the activity and toxicity of a novel marine anticancer poison. *Curr. Pharm. Des.* **2007**, *13*, 3427–3439. [CrossRef]
54. Muñoz-Alonso, M.J.; González-Santiago, L.; Martínez, T.; Losada, A.; Galmarini, C.M.; Muñoz, A. The mechanism of action of plitidepsin. *Curr. Opin. Investig. Drugs* **2009**, *10*, 536–542. [PubMed]
55. Danu, A.; Willekens, C.; Ribrag, V. Plitidepsin: An orphan drug. *Expert Opin. Orphan Drugs* **2013**, *1*, 569–580. [CrossRef]
56. Alonso-Alvarez, S.; Pardal, E.; Sanchez-Nieto, D.; Navarro, M.; Caballero, M.D.; Mateos, M.V.; Martin, A. Plitidepsin: Design, development, and potential place in therapy. *Drug Des. Dev. Ther.* **2017**, *11*, 253–264. [CrossRef]
57. Potts, M.B.; Kim, H.S.; Fisher, K.W.; Hu, Y.; Carrasco, Y.P.; Bulut, G.B.; Ou, Y.-H.; Herrera-Herrera, M.L.; Cubillos, F.; Mendiratta, S.; et al. Using functional signature ontology (FUSION) to identify mechanisms of action for natural products. *Sci. Signal.* **2013**, *6*, ra90. [CrossRef] [PubMed]
58. Potts, M.B.; McMillan, E.A.; Rosales, T.I.; Kim, H.S.; Ou, Y.-H.; Toombs, J.E.; Brekken, R.A.; Minden, M.D.; MacMillan, J.B.; White, M.A. Mode of action and pharmacogenomic biomarkers for exceptional responders to didemnin B. *Nat. Chem. Biol.* **2015**, *11*, 401–408. [CrossRef]
59. Losada, A.; Berlanga, J.J.; Molina-Guijarro, J.M.; Jimenez-Ruiz, A.; Gago, F.; Aviles, P.; de Haro, C.; Martinez-Leal, J.F. Generation of endoplasmic reticulum stress and inhibition of autophagy by plitidepsin induces proteotoxic apoptosis in cancer cells. *Biochem. Pharmacol.* **2020**, *172*, 113744. [CrossRef]
60. Kakeya, H.; Zhang, H.P.; Kobinata, K.; Onose, R.; Onozawa, C.; Kudo, T.; Osada, H. Cytotrienin A, a novel apoptosis inducer in human leukemia HL-60 cells. *J. Antibiot.* **1997**, *50*, 370–372. [CrossRef]
61. Zhang, H.P.; Kakeya, H.; Osada, H. Novel triene-ansamycins, cytotrienins A and B, inducing apoptosis on human leukemia HL-60 cells. *Tetrahedron Lett.* **1997**, *38*, 1789–1792. [CrossRef]
62. Kakeya, H.; Onose, R.; Osada, H. Caspase-mediated activation of a 36-kDa myelin basic protein kinase during anticancer drug-induced apoptosis. *Cancer Res.* **1998**, *58*, 4888–4894. [PubMed]

63. Watabe, M.; Kakeya, H.; Onose, R.; Osada, H. Activation of MST/Krs and c-Jun N-terminal kinases by different signaling pathways during cytotrienin A-induced apoptosis. *J. Biol. Chem.* **2000**, *275*, 8766–8771. [CrossRef]
64. Yamada, Y.; Taketani, S.; Osada, H.; Kataoka, T. Cytotrienin A, a translation inhibitor that induces ectodomain shedding of TNF receptor 1 via activation of ERK and p38 MAP kinase. *Eur. J. Pharmacol.* **2011**, *667*, 113–119. [CrossRef]
65. Blunt, J.; Cole, T.; Munro, M.; Sun, L.; Weber, J.-F.R.; Ramasamy, K.; Abu Bakar, H.; Abdul Majeed, A.B.B. Bioactive Compounds Derived from Endophytic Aspergillus Fungus Strain Isolated from Garcinia Scortechinii. WO2010062159A1, 3 June 2010.
66. Briggs, L.H.; Locker, R.H. 459. Flavonols from the bark of Melicope ternata. Part I. The isolation of four new flavonols, meliternatin, meliternin, ternatin, and wharangin. *J. Chem. Soc.* **1949**, 2157–2162, The term ternatin also refers to a flavone-type natural product, 5-dihydroxy-3,3′,7,8-tetramethoxyflavone (CAS registry number: 571-71-1) first reported in 1949. [CrossRef]
67. Shimokawa, K.; Mashima, I.; Asai, A.; Yamada, K.; Kita, M.; Uemura, D. (−)-Ternatin, a highly N-methylated cyclic heptapeptide that inhibits fat accumulation: Structure and synthesis. *Tetrahedron Lett.* **2006**, *47*, 4445–4448. [CrossRef]
68. Wang, H.-Y.; Yang, H.; Holm, M.; Tom, H.; Oltion, K.; Al-Khdhairawi, A.A.Q.; Weber, J.-F.F.; Blanchard, S.C.; Ruggero, D.; Taunton, J. Synthesis and single-molecule imaging reveal stereospecific enhancement of binding kinetics by the antitumour eEF1A antagonist SR-A3. *Nat. Chem.* **2022**, *14*, 1443–1450. [CrossRef]
69. Copeland, R.A. The drug–target residence time model: A 10-year retrospective. *Nat. Rev. Drug Discov.* **2016**, *15*, 87–95. [CrossRef]
70. Juette, M.F.; Carelli, J.D.; Rundlet, E.J.; Brown, A.; Shao, S.; Ferguson, A.; Wasserman, M.R.; Holm, M.; Taunton, J.; Blanchard, S.C. Didemnin B and ternatin-4 differentially inhibit conformational changes in eEF1A required for aminoacyl-tRNA accommodation into mammalian ribosomes. *eLife* **2022**, *11*, e81608. [CrossRef] [PubMed]
71. Oltion, K.; Carelli, J.D.; Yang, T.; See, S.K.; Wang, H.-Y.; Kampmann, M.; Taunton, J. An E3 ligase network engages GCN1 to promote the degradation of translation factors on stalled ribosomes. *Cell* **2023**, *186*, 346–362. [CrossRef] [PubMed]
72. Hoffmann, H.; Kogler, H.; Heyse, W.; Matter, H.; Caspers, M.; Schummer, D.; Klemke-Jahn, C.; Bauer, A.; Penarier, G.; Debussche, L.; et al. Discovery, Structure Elucidation, and Biological Characterization of Nannocystin A, a Macrocyclic Myxobacterial Metabolite with Potent Antiproliferative Properties. *Angew. Chem. Int. Ed.* **2015**, *54*, 10145–10148. [CrossRef]
73. Krastel, P.; Roggo, S.; Schirle, M.; Ross, N.T.; Perruccio, F.; Aspesi, P., Jr.; Aust, T.; Buntin, K.; Estoppey, D.; Liechty, B.; et al. Nannocystin A: An Elongation Factor 1 Inhibitor from Myxobacteria with Differential Anti-Cancer Properties. *Angew. Chem. Int. Ed.* **2015**, *54*, 10149–10154. [CrossRef]
74. Liao, L.; Zhou, J.; Xu, Z.; Ye, T. Concise Total Synthesis of Nannocystin A. *Angew. Chem. Int. Ed.* **2016**, *55*, 13263–13266. [CrossRef] [PubMed]
75. Huang, J.; Wang, Z. Total Syntheses of Nannocystins A and A0, Two Elongation Factor 1 Inhibitors. *Org. Lett.* **2016**, *18*, 4702–4705. [CrossRef]
76. Yang, Z.; Xu, X.; Yang, C.H.; Tian, Y.; Chen, X.; Lian, L.; Pan, W.; Su, X.; Zhang, W.; Chen, Y. Total Synthesis of Nannocystin A. *Org. Lett.* **2016**, *18*, 5768–5770. [CrossRef] [PubMed]
77. Zhang, Y.H.; Liu, R.; Liu, B. Total synthesis of nannocystin Ax. *Chem. Commun.* **2017**, *53*, 5549–5552. [CrossRef] [PubMed]
78. Poock, C.; Kalesse, M. Total Synthesis of Nannocystin Ax. *Org. Lett.* **2017**, *19*, 4536–4539. [CrossRef]
79. Liu, Q.; Hu, P.; He, Y. Asymmetric Total Synthesis of Nannocystin A. *J. Org. Chem.* **2017**, *82*, 9217–9222. [CrossRef]
80. Meng, Z.; Souillart, L.; Monks, B.; Huwyler, N.; Herrmann, J.; Mueller, R.; Furstner, A. A "Motif-Oriented" Total Synthesis of Nannocystin Ax. Preparation and Biological Assessment of Analogues. *J. Org. Chem.* **2018**, *83*, 6977–6994. [CrossRef]
81. Wang, Z. The Chemical Syntheses of Nannocystins. *Synthesis* **2019**, *51*, 2252–2260. [CrossRef]
82. Zhang, W. From Target-Oriented to Motif-Oriented: A Case Study on Nannocystin Total Synthesis. *Molecules* **2020**, *25*, 5327. [CrossRef]
83. Fürstner, A. Lessons from Natural Product Total Synthesis: Macrocyclization and Postcyclization Strategies. *Acc. Chem. Res.* **2021**, *54*, 861–874. [CrossRef]
84. Paul, D.; Das, S.; Saha, S.; Sharma, H.; Goswami, R.K. Intramolecular Heck Reaction in Total Synthesis of Natural Products: An Update. *Eur. J. Org. Chem.* **2021**, *2021*, 2057–2076. [CrossRef]
85. Zhang, W. Heck macrocyclization in natural product total synthesis. *Nat. Prod. Rep.* **2021**, *38*, 1109–1135. [CrossRef]
86. Tian, Y.; Xu, X.; Ding, Y.; Hao, X.; Bai, Y.; Tang, Y.; Zhang, X.; Li, Q.; Yang, Z.; Zhang, W.; et al. Synthesis and biological evaluation of nannocystin analogues toward understanding the binding role of the (2*R*,3*S*)-Epoxide in nannocystin A. *Eur. J. Med. Chem.* **2018**, *150*, 626–632. [CrossRef] [PubMed]
87. Tian, Y.; Ding, Y.; Xu, X.; Bai, Y.; Tang, Y.; Hao, X.; Zhang, W.; Chen, Y. Total synthesis and biological evaluation of nannocystin analogues modified at the polyketide phenyl moiety. *Tetrahedron Lett.* **2018**, *59*, 3206–3209. [CrossRef]
88. Tian, Y.; Wang, J.; Liu, W.; Yuan, X.; Tang, Y.; Li, J.; Chen, Y.; Zhang, W. Stereodivergent total synthesis of Br-nannocystins underpinning the polyketide (10*R*,11*S*) configuration as a key determinant of potency. *J. Mol. Struct.* **2019**, *1181*, 568–578. [CrossRef]
89. Liu, Q.; Yang, X.; Ji, J.; Zhang, S.-L.; He, Y. Novel nannocystin A analogues as anticancer therapeutics: Synthesis, biological evaluations and structure-activity relationship studies. *Eur. J. Med. Chem.* **2019**, *170*, 99–111. [CrossRef]
90. Sun, C.; Liu, R.; Xia, M.; Hou, Y.; Wang, X.; Lu, J.-J.; Liu, B.; Chen, X. Nannocystin Ax, a natural elongation factor 1α inhibitor from Nannocystis sp., suppresses epithelial-mesenchymal transition, adhesion and migration in lung cancer cells. *Toxicol. Appl. Pharmacol.* **2021**, *420*, 115535. [CrossRef]

91. Hou, Y.; Liu, R.; Xia, M.; Sun, C.; Zhong, B.; Yu, J.; Ai, N.; Lu, J.-J.; Ge, W.; Liu, B.; et al. Nannocystin ax, an eEF1A inhibitor, induces G1 cell cycle arrest and caspase-independent apoptosis through cyclin D1 downregulation in colon cancer in vivo. *Pharmacol. Res.* **2021**, *173*, 105870. [CrossRef] [PubMed]
92. Zhang, H.; Tian, Y.; Yuan, X.; Xie, F.; Yu, S.; Cai, J.; Sun, B.; Shan, C.; Zhang, W. Site-directed Late-Stage Diversification of Macrocyclic Nannocystins Facilitating Anticancer SAR and Mode of Action Studies. *RSC Med. Chem.* **2023**, *14*, 299–312. [CrossRef]
93. Gemmer, M.; Chaillet, M.L.; van Loenhout, J.; Cuevas Arenas, R.; Vismpas, D.; Gröllers-Mulderij, M.; Koh, F.A.; Albanese, P.; Scheltema, R.A.; Howes, S.C.; et al. Visualization of translation and protein biogenesis at the ER membrane. *Nature* **2023**, *614*, 160–167. [CrossRef]
94. Pollock, C.; Huang, S. The perinucleolar compartment. *Cold Spring Harbor Perspect. Biol.* **2010**, *2*, a000679. [CrossRef] [PubMed]
95. Kamath, R.V.; Thor, A.D.; Wang, C.; Edgerton, S.M.; Slusarczyk, A.; Leary, D.J.; Wang, J.; Wiley, E.L.; Jovanovic, B.; Wu, Q.; et al. Perinucleolar compartment prevalence has an independent prognostic value for breast cancer. *Cancer Res.* **2005**, *65*, 246–253. [CrossRef] [PubMed]
96. Norton, J.T.; Pollock, C.B.; Wang, C.; Schink, J.C.; Kim, J.J.; Huang, S. Perinucleolar compartment prevalence is a phenotypic pancancer marker of malignancy. *Cancer* **2008**, *113*, 861–869. [CrossRef]
97. Norton, J.T.; Wang, C.; Gjidoda, A.; Henry, R.W.; Huang, S. The Perinucleolar Compartment Is Directly Associated with DNA. *J. Biol. Chem.* **2009**, *284*, 4090–4101. [CrossRef]
98. Huang, S.; Deerinck, T.J.; Ellisman, M.H.; Spector, D.L. The Dynamic Organization of the Perinucleolar Compartment in the Cell Nucleus. *J. Cell Biol.* **1997**, *137*, 965–974. [CrossRef] [PubMed]
99. Norton, J.T.; Titus, S.A.; Dexter, D.; Austin, C.P.; Zheng, W.; Huang, S. Automated High-Content Screening for Compounds That Disassemble the Perinucleolar Compartment. *J. Biomol. Screen.* **2009**, *14*, 1045–1053. [CrossRef]
100. Frankowski, K.J.; Wang, C.; Patnaik, S.; Schoenen, F.J.; Southall, N.; Li, D.; Teper, Y.; Sun, W.; Kandela, I.; Hu, D.; et al. Metarrestin, a perinucleolar compartment inhibitor, effectively suppresses metastasis. *Sci. Transl. Med.* **2018**, *10*, eaap8307. [CrossRef]
101. Frankowski, K.J.; Patnaik, S.; Wang, C.; Southall, N.; Dutta, D.; De, S.; Li, D.; Dextras, C.; Lin, Y.-H.; Bryant-Connah, M.; et al. Discovery and Optimization of Pyrrolopyrimidine Derivatives as Selective Disruptors of the Perinucleolar Compartment, a Marker of Tumor Progression toward Metastasis. *J. Med. Chem.* **2022**, *65*, 8303–8331. [CrossRef]
102. Vilimas, T.; Wang, A.Q.; Patnaik, S.; Hughes, E.A.; Singleton, M.D.; Knotts, Z.; Li, D.; Frankowski, K.; Schlomer, J.J.; Guerin, T.M.; et al. Pharmacokinetic evaluation of the PNC disassembler metarrestin in wild-type and Pdx1-Cre; LSL-Kras$^{G12D/+}$; Tp53$^{R172H/+}$ (KPC) mice, a genetically engineered model of pancreatic cancer. *Cancer Chemoth. Pharm.* **2018**, *82*, 1067–1080. [CrossRef]
103. Padilha, E.C.; Shah, P.; Wang, A.Q.; Singleton, M.D.; Hughes, E.A.; Li, D.; Rice, K.A.; Konrath, K.M.; Patnaik, S.; Marugan, J. Metabolism and pharmacokinetics characterization of metarrestin in multiple species. *Cancer Chemother. Pharm.* **2020**, *85*, 805–816. [CrossRef]
104. Bourdi, M.; Rudloff, U.; Patnaik, S.; Marugan, J.; Terse, P.S. Safety assessment of metarrestin in dogs: A clinical candidate targeting a subnuclear structure unique to metastatic cancer cells. *Regul. Toxicol. Pharm.* **2020**, *116*, 104716. [CrossRef]
105. Békés, M.; Langley, D.R.; Crews, C.M. PROTAC targeted protein degraders: The past is prologue. *Nat. Rev. Drug Discov.* **2022**, *21*, 181–200. [CrossRef]
106. Pedrucci, F.; Pappalardo, C.; Marzaro, G.; Ferri, N.; Ferlin, A.; De Toni, L. Proteolysis Targeting Chimeric Molecules: Tuning Molecular Strategies for a Clinically Sound Listening. *Int. J. Mol. Sci.* **2022**, *23*, 6630. [CrossRef] [PubMed]
107. Yao, T.; Xiao, H.; Wang, H.; Xu, X. Recent Advances in PROTACs for Drug Targeted Protein Research. *Int. J. Mol. Sci.* **2022**, *23*, 10328. [CrossRef]
108. Jin, J.; Kabir, M.; Sun, N.; Kaniskan, H.U. Preparation of Heterobifunctional Compounds as Degraders of eEF1A2. WO2022159650A1, 28 July 2022.
109. Tash, J.S.; Chakrasali, R.; Jakkaraj, S.R.; Hughes, J.; Smith, S.K.; Hornbaker, K.; Heckert, L.L.; Ozturk, S.B.; Hadden, M.K.; Kinzy, T.G.; et al. Gamendazole, an orally active indazole carboxylic acid male contraceptive agent, targets HSP90AB1 (HSP90BETA) and EEF1A1 (eEF1A), and stimulates *Il1a* transcription in rat sertoli cells. *Biol. Reprod.* **2008**, *78*, 1139–1152. [CrossRef] [PubMed]
110. Burglová, K.; Rylová, G.; Markos, A.; Prichystalova, H.; Soural, M.; Petracek, M.; Medvedikova, M.; Tejral, G.; Sopko, B.; Hradil, P.; et al. Identification of Eukaryotic Translation Elongation Factor 1-α 1 Gamendazole-Binding Site for Binding of 3-Hydroxy-4(1*H*)-quinolinones as Novel Ligands with Anticancer Activity. *J. Med. Chem.* **2018**, *61*, 3027–3036. [CrossRef] [PubMed]
111. Isaka, M.; Srisanoh, U.; Lartpornmatulee, N.; Boonruangprapa, T. ES-242 derivatives and cycloheptapeptides from *Cordyceps* sp. strains BCC 16173 and BCC 16176. *J. Nat. Prod.* **2007**, *70*, 1601–1604. [CrossRef]
112. Rukachaisirikul, V.; Chantaruk, S.; Tansakul, C.; Saithong, S.; Chaicharernwimonkoon, L.; Pakawatchai, C.; Isaka, M.; Intereya, K. A cyclopeptide from the insect pathogenic fungus *Cordyceps* sp. BCC 1788. *J. Nat. Prod.* **2006**, *69*, 305–307. [CrossRef]
113. Kumar, S.; Dahiya, R.; Khokra, S.L.; Mourya, R.; Chennupati, S.V.; Maharaj, S. Total synthesis and pharmacological investigation of cordyheptapeptide A. *Molecules* **2017**, *22*, 682. [CrossRef]
114. Klein, V.G.; Bray, W.M.; Wang, H.-Y.; Edmondson, Q.; Schwochert, J.; Ono, S.; Naylor, M.R.; Turmon, A.C.; Faris, J.H.; Okada, O.; et al. Identifying the Cellular Target of Cordyheptapeptide A and Synthetic Derivatives. *ACS Chem. Biol.* **2021**, *16*, 1354–1364. [CrossRef] [PubMed]

115. Naylor, M.R.; Ly, A.M.; Handford, M.J.; Ramos, D.P.; Pye, C.R.; Furukawa, A.; Klein, V.G.; Noland, R.P.; Edmondson, Q.; Turmon, A.C.; et al. Lipophilic Permeability Efficiency Reconciles the Opposing Roles of Lipophilicity in Membrane Permeability and Aqueous Solubility. *J. Med. Chem.* **2018**, *61*, 11169–11182. [CrossRef]
116. Chan, J.; Khan, S.N.; Harvey, I.; Merrick, W.; Pelletier, J. Eukaryotic protein synthesis inhibitors identified by comparison of cytotoxicity profiles. *RNA* **2004**, *10*, 528–543. [CrossRef]
117. Schulze, C.J.; Bray, W.M.; Woerhmann, M.H.; Stuart, J.; Lokey, R.S.; Linington, R.G. "Function-first" lead discovery: Mode of action profiling of natural product libraries using image-based screening. *Chem. Biol.* **2013**, *20*, 285–295. [CrossRef] [PubMed]
118. Dieterich, D.C.; Link, A.J.; Graumann, J.; Tirrell, D.A.; Schuman, E.M. Selective identification of newly synthesized proteins in mammalian cells using bioorthogonal noncanonical amino acid tagging (BONCAT). *Proc. Natl. Acad. Sci. USA* **2006**, *103*, 9482–9487. [CrossRef] [PubMed]
119. Sánchez-Murcia, P.A.; Cortés-Cabrera, Á.; Gago, F. Structural rationale for the cross-resistance of tumor cells bearing the A399V variant of elongation factor eEF1A1 to the structurally unrelated didemnin B, ternatin, nannocystin A and ansatrienin B. *J. Comput. Aided Mol. Des.* **2017**, *31*, 915–928. [CrossRef]
120. Nishioka, H.; Nakajima, S.; Nagashima, M.; Kojiri, K.; Suda, H. BE-43547 Series Substances, Their Manufacture with Streptomyces Species, and Their Use as Antitumor Agents. JP10147594A, 2 June 1998.
121. Poulsen, T.B. Total Synthesis of Natural Products Containing Enamine or Enol Ether Derivatives. *Acc. Chem. Res.* **2021**, *54*, 1830–1842. [CrossRef]
122. Villadsen, N.L.; Jacobsen, K.M.; Keiding, U.B.; Weibel, E.T.; Christiansen, B.; Vosegaard, T.; Bjerring, M.; Jensen, F.; Johannsen, M.; Tørring, T.; et al. Synthesis of *ent*-BE-43547A_1 reveals a potent hypoxia-selective anticancer agent and uncovers the biosynthetic origin of the APD-CLD natural products. *Nat. Chem.* **2017**, *9*, 264–272. [CrossRef]
123. Poulsen, T.B. A concise route to the macrocyclic core of the rakicidins. *Chem. Commun.* **2011**, *47*, 12837–12839. [CrossRef] [PubMed]
124. Clement, L.L.; Tsakos, M.; Schaffert, E.S.; Scavenius, C.; Enghild, J.J.; Poulsen, T.B. The amido-pentadienoate-functionality of the rakicidins is a thiol reactive electrophile–development of a general synthetic strategy. *Chem. Commun.* **2015**, *51*, 12427–12430. [CrossRef] [PubMed]
125. Tsakos, M.; Clement, L.L.; Schaffert, E.S.; Olsen, F.N.; Rupiani, S.; Djurhuus, R.; Yu, W.; Jacobsen, K.M.; Villadsen, N.L.; Poulsen, T.B. Total synthesis and biological evaluation of rakicidin A and discovery of a simplified bioactive analogue. *Angew. Chem. Int. Ed.* **2016**, *55*, 1030–1035. [CrossRef]
126. Tsakos, M.; Jacobsen, K.M.; Yu, W.; Villadsen, N.L.; Poulsen, T.B. The rakicidin family of anticancer natural products–synthetic strategies towards a new class of hypoxia-selective cytotoxins. *Synlett* **2016**, *27*, 1898–1906. [CrossRef]
127. Sun, Y.; Ding, Y.; Li, D.; Zhou, R.; Su, X.; Yang, J.; Guo, X.; Chong, C.; Wang, J.; Zhang, W.; et al. Cyclic Depsipeptide BE-43547A_2: Synthesis and Activity against Pancreatic Cancer Stem Cells. *Angew. Chem. Int. Ed.* **2017**, *56*, 14627–14631. [CrossRef]
128. Sun, Y.; Su, X.; Zhou, R.; Wang, D.; Zhao, Y.; Jiang, Y.; Wang, L.; Chen, Y. Total synthesis of BE-43547A_2. *Tetrahedron* **2018**, *74*, 5955–5964. [CrossRef]
129. Sun, Y.; Zhou, R.; Xu, H.; Wang, D.; Su, X.; Wang, C.; Ding, Y.; Wang, L.; Chen, Y. Syntheses and biological evaluation of BE-43547A_2 analogs modified at O35 ester and C15-OH sites. *Tetrahedron* **2019**, *75*, 1808–1818. [CrossRef]
130. Liu, C.; Wang, L.; Sun, Y.; Zhao, X.; Chen, T.; Su, X.; Guo, H.; Wang, Q.; Xi, X.; Ding, Y.; et al. Probe Synthesis Reveals Eukaryotic Translation Elongation Factor 1 Alpha 1 as the Anti-Pancreatic Cancer Target of BE-43547A_2. *Angew. Chem. Int. Ed.* **2022**, *61*, e202206953. [CrossRef] [PubMed]
131. Jacobsen, K.M.; Villadsen, N.L.; Toerring, T.; Nielsen, C.B.; Salomon, T.; Nielsen, M.M.; Tsakos, M.; Sibbersen, C.; Scavenius, C.; Nielsen, R.; et al. APD-Containing Cyclolipodepsipeptides Target Mitochondrial Function in Hypoxic Cancer Cells. *Cell Chem. Biol.* **2018**, *25*, 1337–1349. [CrossRef]
132. Silvera, D.; Formenti, S.C.; Schneider, R.J. Translational control in cancer. *Nat. Rev. Cancer* **2010**, *10*, 254–266. [CrossRef]
133. Bhat, M.; Robichaud, N.; Hulea, L.; Sonenberg, N.; Pelletier, J.; Topisirovic, I. Targeting the translation machinery in cancer. *Nat. Rev. Drug Discov.* **2015**, *14*, 261–278. [CrossRef] [PubMed]
134. Chu, J.; Pelletier, J. Therapeutic opportunities in eukaryotic translation. *Cold Spring Harbor Perspect. Biol.* **2018**, *10*, a032995. [CrossRef] [PubMed]
135. Gandhi, V.; Plunkett, W.; Cortes, J.E. Omacetaxine: A Protein Translation Inhibitor for Treatment of Chronic Myelogenous Leukemia. *Clin. Cancer Res.* **2014**, *20*, 1735–1740. [CrossRef]
136. Varona, J.F.; Landete, P.; Lopez-Martin, J.A.; Estrada, V.; Paredes, R.; Guisado-Vasco, P.; Fernandez de Orueta, L.; Torralba, M.; Fortun, J.; Vates, R.; et al. Preclinical and randomized phase I studies of plitidepsin in adults hospitalized with COVID-19. *Life Sci. Alliance* **2022**, *5*, e202101200. [CrossRef] [PubMed]
137. White, K.M.; Rosales, R.; Yildiz, S.; Kehrer, T.; Miorin, L.; Moreno, E.; Jangra, S.; Uccellini, M.B.; Rathnasinghe, R.; Coughlan, L.; et al. Plitidepsin has potent preclinical efficacy against SARS-CoV-2 by targeting the host protein eEF1A. *Science* **2021**, *371*, 926–931. [CrossRef] [PubMed]
138. Sachse, M.; Tenorio, R.; de Castro, I.F.; Muñoz-Basagoiti, J.; Perez-Zsolt, D.; Raïch-Regué, D.; Rodon, J.; Losada, A.; Avilés, P.; Cuevas, C. Unraveling the antiviral activity of plitidepsin against SARS-CoV-2 by subcellular and morphological analysis. *Antivir. Res.* **2022**, *200*, 105270. [CrossRef]

139. Bosutti, A.; Dapas, B.; Grassi, G.; Bussani, R.; Zanconati, F.; Giudici, F.; Bottin, C.; Pavan, N.; Trombetta, C.; Scaggiante, B. High eEF1A1 Protein Levels Mark Aggressive Prostate Cancers and the In Vitro Targeting of eEF1A1 Reveals the eEF1A1–actin Complex as a New Potential Target for Therapy. *Int. J. Mol. Sci.* **2022**, *23*, 4143. [CrossRef] [PubMed]
140. Wu, A.; Tang, J.; Guo, Z.; Dai, Y.; Nie, J.; Hu, W.; Liu, N.; Ye, C.; Li, S.; Pei, H. Long non-coding RNA CRYBG3 promotes lung cancer metastasis via activating the eEF1A1/MDM2/MTBP axis. *Int. J. Mol. Sci.* **2021**, *22*, 3211. [CrossRef] [PubMed]
141. Dugger, S.A.; Platt, A.; Goldstein, D.B. Drug development in the era of precision medicine. *Nat. Rev. Drug Discov.* **2018**, *17*, 183–196. [CrossRef]
142. Li, H.; Chen, S.; Wang, J.; Zhang, M.; Wu, W.; Liu, W.; Sun, P. Ansafurantrienins, Unprecedented Ansatrienin Derivatives Formed via Photocatalytic Intramolecular [3+ 2] Oxidative Cycloaddition. *Org. Lett.* **2022**, *24*, 592–596. [CrossRef]

International Journal of *Molecular Sciences*

MDPI

Review

Synthetic Small Molecule Modulators of Hsp70 and Hsp40 Chaperones as Promising Anticancer Agents

Bianca Nitzsche [1], Michael Höpfner [1] and Bernhard Biersack [2,*]

[1] Institute for Physiology, Charité—Universitätsmedizin Berlin, Corporate Member of Freie Universität Berlin and Humboldt Universität zu Berlin, Charitéplatz 1, 10117 Berlin, Germany

[2] Organische Chemie 1, Universität Bayreuth, Universitätsstrasse 30, 95440 Bayreuth, Germany

* Correspondence: bernhard.biersack@uni-bayreuth.de or bernhard.biersack@yahoo.com; Tel.: +49-921-552673

Abstract: A class of chaperones dubbed heat shock protein 70 (Hsp70) possesses high relevance in cancer diseases due to its cooperative activity with the well-established anticancer target Hsp90. However, Hsp70 is closely connected with a smaller heat shock protein, Hsp40, forming a formidable Hsp70-Hsp40 axis in various cancers, which serves as a suitable target for anticancer drug design. This review summarizes the current state and the recent developments in the field of (semi-)synthetic small molecule inhibitors directed against Hsp70 and Hsp40. The medicinal chemistry and anticancer potential of pertinent inhibitors are discussed. Since Hsp90 inhibitors have entered clinical trials but have exhibited severe adverse effects and drug resistance formation, potent Hsp70 and Hsp40 inhibitors may play a significant role in overcoming the drawbacks of Hsp90 inhibitors and other approved anticancer drugs.

Keywords: anticancer agents; heat shock proteins; Hsp70; Hsp40

Citation: Nitzsche, B.; Höpfner, M.; Biersack, B. Synthetic Small Molecule Modulators of Hsp70 and Hsp40 Chaperones as Promising Anticancer Agents. *Int. J. Mol. Sci.* **2023**, *24*, 4083. https://doi.org/10.3390/ijms24044083

Academic Editor: Maurizio Battino

Received: 28 January 2023
Revised: 9 February 2023
Accepted: 11 February 2023
Published: 17 February 2023

1. Introduction

Stress factors such as heat lead to defensive and protective cellular responses, enabling the cell to cope with the stress [1]. The emergence of chromosomal puffs as a defined heat shock response in *Drosophila* flies was observed for the first time in 1962 by Ferruccio Ritossa [2]. In 1974, the increased expression of certain heat shock proteins in response to heat and other stress factors was discovered [3]. These heat shock proteins (Hsps) are classified by their molecular weights in kDa (e.g., Hsp90, Hsp70, etc.) and function as chaperones in order to protect important proteins from degradation, to control the quality of protein folding, and to deliver misfolded or damaged proteins to the proteasome for disposal (protein triage), thus vouchsafing cell viability under these conditions [4]. This multi-chaperone system ("epichaperome") plays an important role in various cancer diseases [5]. Canonical functions of chaperones are linked with the ubiquitin-proteasome system (UPS) and chaperone-mediated autophagy (CMA), while non-canonical functions affect the immune system, including inflammatory and autoimmune mechanisms [6]. Thus, targeting heat shock proteins is a promising strategy to combat cancer.

The Hsp family is subdivided into Hsp110 (HSPH), Hsp90 (HSPC), Hsp70 (HSPA), Hsp40 (DNAJ), small Hsps (HSPB), and the chaperonin family proteins Hsp60/Hsp10 (HSPD/E) and TRiC (CCT) [7]. Hsp90 has already become a valuable chaperone drug target. Prominent examples of anticancer active Hsp90 inhibitors are the natural products (geldanamycin and radicicol) as well as their (semi-)synthetic derivatives, 17-AAG/tanespimycin and ganetespib, the latter have reached clinical trials [8]. Hsp90 regulates the activity and stability of crucial transcription factors such as the tumor suppressor p53 and the androgen receptor (AR), while Hsp90 activity itself is regulated by posttranslational modifications (e.g., lysine acetylation under control by HDAC6) and by other heat shock proteins such as Hsp70 [9]. The protein folding by Hsp90 and Hsp70 is ATP-dependent, while Hsp70 has a crucial function in the protection of cells against various stress factors, including enhanced cell survival. In addition,

Hsp70 forms complexes with Hsp90 with the help of HOP (Hsp70–Hsp90 organizing protein) in order to exert its housekeeping activities [10–12]. There is growing evidence that Hsp70 inhibitors have the potential to overcome Hsp90 inhibitor resistance [13]. The organelle-specific members of the Hsp70 protein family, such as mitochondrial mortalin (mtHsp70, Grp75) and endoplasmic reticulum/ER-based Grp78, were also identified as possible anticancer drug targets due to their crucial roles in cell proliferation and survival in various cancers [14,15]. In addition, extracellular Hsps attracted increased attention as cancer targets and biomarkers [16].

Client polypeptides are transferred to Hsp70 by the smaller co-chaperone, Hsp40. Hsp40 prevents the aggregation of unfolded polypeptides, but there are also folded proteins among the clients of Hsp40. There are numerous Hsp40 isoforms, also dubbed J-domain proteins (DNAJs), that are structurally different from Hsp70 and Hsp90 proteins [17]. Nevertheless, DNAJ/Hsp40 and Hsp70 proteins form a tight Hsp70/Hsp40 complex in order to fulfill their protein-folding functions [17,18]. Hsp40 binds to the nucleotide-binding domain (NBD) of Hsp70 and accelerates the ATPase activity of Hsp70 enormously [18]. Thus, small-molecule inhibitors of the Hsp70-Hsp40 axis have considerable potential as anticancer drug candidates.

In this review, the current state and recent developments in the field of synthetic Hsp70 and Hsp40 inhibitors are discussed.

2. Hsp70 Inhibitors

The mechanisms of action of proteins of the Hsp70 family and their roles in various cancer diseases were thoroughly reviewed recently [10–12]. In line with the mounting knowledge of these proteins, Hsp70 inhibitors are a continuously growing class of compounds, which can be subdivided into Hsp70 and Hsc70 inhibitors on the one hand and inhibitors targeting organelle-specific Hsp70 proteins such as Grp78 and mortalin on the other hand.

2.1. *Hsp70i and Hsc70 Inhibitors*

Inducible Hsp70 (also abbreviated as Hsp70i) and consecutive Hsc70 inhibitors can be classified according to their binding mode into N-terminal nucleotide binding domain/NBD-targeting inhibitors, C-terminal substrate binding domain/SBD-targeting inhibitors, and allosteric inhibitors (Table 1).

In 2009, the adenosine-derivative **1** (Figure 1, VER-155008) was identified as a selective NBD-targeting Hsc70/Hsp70 inhibitor (IC_{50} = 0.5 μM), which showed antiproliferative effects on HCT-116 colon carcinoma cells (GI_{50} = 5.0 μM), reduced Raf-1 and Her2 protein levels, and enhanced the apoptosis induction by the Hsp90 inhibitors 17-AAG and VER-82160 in HCT-116 cells [19,20]. Compound **1** also revealed promising effects on non-small cell lung cancer (NSCLC), such as inhibition of NSCLC proliferation and cell cycle arrest (increased G0/G1 cell percentage) [21]. In addition, compound **1** inhibited pleural mesothelioma cell proliferation and colony formation, which are associated with G1 cell cycle arrest, suppressed phospho-Akt, and induction of macroautophagy [22]. In LNCaP95 prostate cancer cells, the Hsp70 inhibitor **1** induced apoptosis and suppressed the expression of the full-length androgen receptor (AR-FL) and of the androgen receptor splice variant 7 (AR-V7), which are associated with castration-resistant prostate cancer (CRPC) [23]. These AR-suppressing effects were correlated with an inhibition of YB-1 phosphorylation by compound **1**, followed by reduced nuclear translocation of YB-1. Another study using the AR-positive LNCaP and the AR-negative PC-3 prostate cancer cell lines showed that compound **1** was antiproliferative and pro-apoptotic in both cell lines, albeit the pro-apoptotic effects were higher in the AR-positive cells [24]. Compound **1** downregulated AR expression and induced G1 cell cycle arrest in the LNCaP cells. Moreover, a distinct suppression was observed for Hsp27 in PC-3 cells and HOP and Hsp90β in both cell lines treated with compound **1**. In MCF-7 breast cancer cells, compound **1** induced apoptosis associated with mitochondrial damage, and the anticancer activity of compound **1** was reduced by heat shock [25]. Anaplastic thyroid carcinoma (APC) is the most lethal thyroid cancer with high drug resistance, but compound **1** was able to induce paraptosis in APC cells dependent on de novo protein synthesis [26]. In a panel of glioma cells (1321N1, GOS-3, and U87-MG), compound **1** showed higher antiproliferative activities (IC_{50} = 12–13 μM) than the approved drug temozolomide (IC_{50} = 135–180 μM) associated with downregulation of Akt kinase activity and the modulation of certain miRNAs, e.g., the upregulation of miR-215 and miR-194-5p [27]. Compound **1** was evaluated in muscle-invasive bladder cancer (MIBC) models and induced apoptosis along with inhibition of MIBC cell proliferation and migration [28]. The activity of compound **1** against MIBC was associated with the suppression of protein members of p53/Rb, PI3K, and SWI/SFW signaling. Especially strong degrading effects of compound **1** were observed on the demethylase KDMA6 and the histone acetyltransferase EP300, both members of the histone modification pathway.

Due to these promising anticancer effects of compound **1**, its combination with various other anticancer drugs was investigated. In combination with the Hsp90 inhibitor radicicol, compound **1** was able to enhance the APC cell killing activity of radicicol, accompanied by suppressed heat shock cognate 70/Hsc70, Akt, and survival [29]. The combination of compound **1** with the Hsp90 inhibitor STA9090 was evaluated in MIBC models and was more efficient than single compound therapy [28]. However, the combination of compound **1** with doxorubicin in canine osteosarcoma (OSA) cells showed no improvements. Already, compound **1** alone displayed strong apoptosis induction, inhibition of colony formation, and antiproliferative activities against OSA cells based on Hsp70 inhibition as well as Akt suppression and BAG1 degradation [30]. In contrast to that, compound **1** showed synergy effects in combination with the Hsp90 inhibitor 17-AAD and sensitized A549 NSCLC cells to radiation therapy [21]. The combination of compound **1** with manumycin A, an anticancer active antibiotic that upregulates Hsp70 in cancer cells, sensitized lung tumor cells to manumycin A treatment [31]. Micelles of compound **1** together with gold nanorods were successfully tested as mild-temperature photothermal therapy in colon cancer, leading to strong colon tumor growth in vivo at a temperature of 45 °C [32]. Compound **1** exhibited considerable activity against multiple myeloma (MM) cells (IC_{50} = 1.7 μM for OPM2, 3.0 μM for RPMI 8226, and 6.5 μM for MM.1S cells), and the combination of compound **1** with the proteasome inhibitor bortezomib displayed synergy effects in terms of apoptosis induction in MM cells, which was associated with suppression of anti-apoptotic Bcl-2, Bcl-xL, and Mcl-1 and upregulation of pro-apoptotic NOXA and Bim [33]. In addition, the ER stress marker CHOP (CCAAT-enhancer binding protein homologous protein) was induced by this combination treatment. The natural product shikonin was described as a proteasome inhibitor and necroptosis inducer in MM cells while it also upregulated Hsp70, and, thus, the combination with compound **1** enhanced shikonin-induced MM cell death [34]. Compound **1** also exhibited promising effects on acute myeloid leukemia (AML) cells alone (induction of apoptosis, inhibition of cell proliferation, and colony formation) and in combination with the Hsp90 inhibitor 17-DMAG (additive antiproliferative and pro-apoptotic activity) [35]. The release of AML cell growth factors and regulators such as TNF-α, VEGF, IL-3, IL-1β, and IL-1 receptor antagonist was strongly suppressed upon treatment with compound **1**. A new structurally related 6,8,9-trisubstituted purine derivative was recently disclosed that induced apoptosis and senescence in luminal A subtype MCF-7 breast carcinoma cells [36].

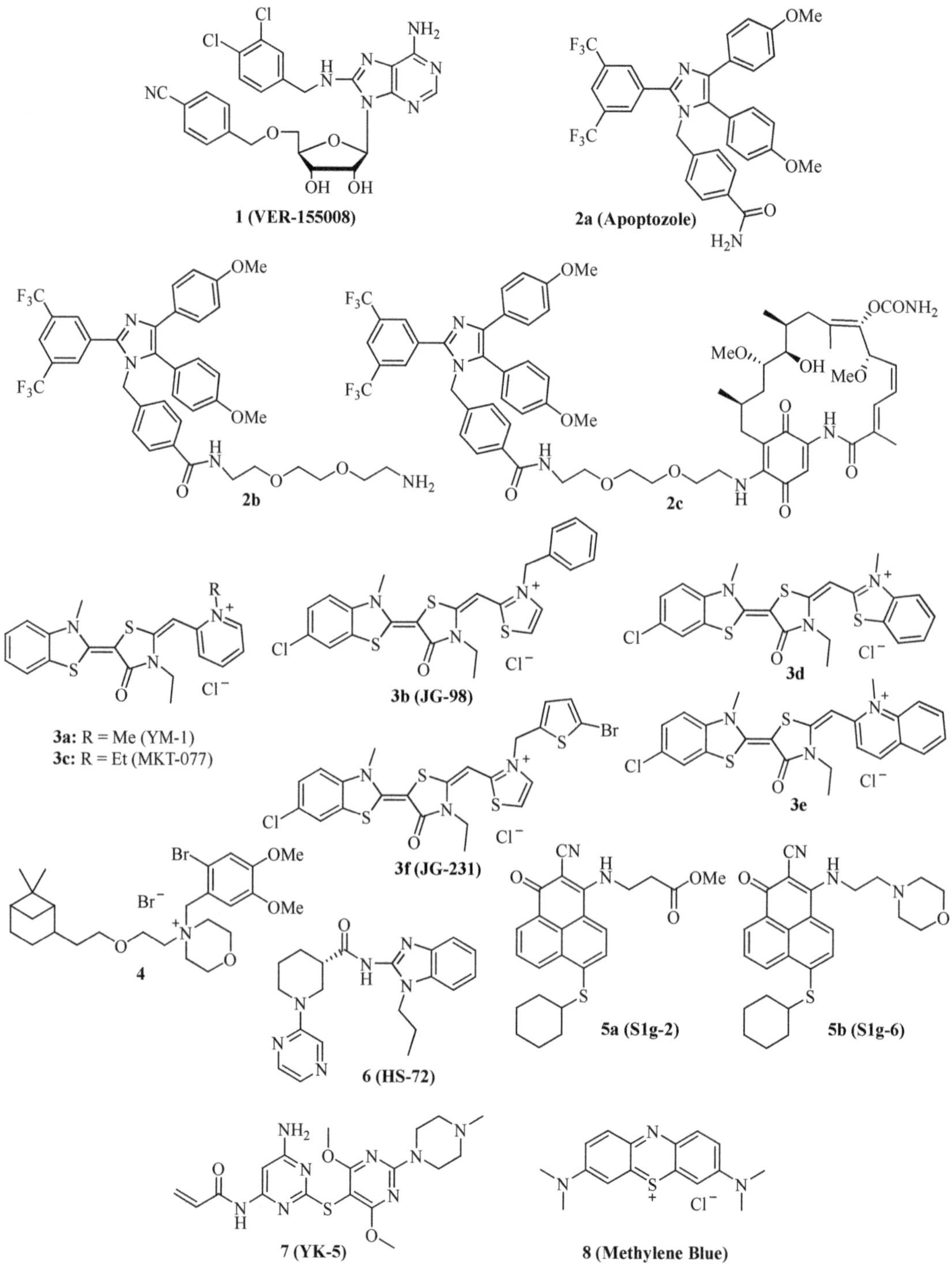

Figure 1. Structures of NBD-binding and interfering compounds **1**–**8**.

The substituted imidazole derivative **2a** (Figure 1, apoptozole) was discovered as a pro-apoptotic inhibitor of Hsc70 (K_D = 210 nM) and Hsp72 (K_D = 140 nM) in 2008, which showed high tumor cell growth inhibitory activities with GI_{50} values in the nanomolar concentration range (GI_{50} = 220 nM for SK-OV-3, 250 nM for HCT-15, and 130 nM for A549 cells) [37]. Compound **2a** was identified as an inhibitor of the Hsp70 ATPase by affinity chromatography upon conjugation of the amino-ethyloxy modified apoptozole derivative **2b** (Figure 1) with a resin, leading to a reduced interaction of Hsp70 with APAF-1, while no affinity to Hsp40, Hsp60, or Hsp90 was observed [38]. Time-dependent antiproliferative activity was checked, and amenable IC_{50} values (0.8 μM for A549, 0.8 μM for HeLa, 0.7 μM for MDA-MB-231, and 0.7 μM for HepG2 cells) were obtained after 72 h of incubation, while the combination with doxorubicin led to a sensitization of A549 and HeLa cells to doxorubicin treatment. At doses of 4 mg/kg/day (i.p.) given for two weeks, compound **2b** reduced the tumor growth of A549, RKO, and HeLa tumor xenografts by 61%, 65%, and 68%, respectively; in the latter case, the combination with doxorubicin led to more pronounced tumor growth inhibition (81% tumor growth reduction). The anti-leukemia properties of compound **2b** and its hybrid molecules, such as compound **2c** (Figure 1), with the Hsp90 inhibitor geldanamycin (**2b** and geldanamycin connected by ethylene glycol-based linker systems) were studied [39]. Compounds **2b**, hybrid **2c**, and other related hybrids induced apoptosis in a caspase-dependent way, however, the hybrids were more active against leukemia cells than their parent compounds **2b** and geldanamycin. Compound **2c** inhibited autophagy, but it also induced apoptosis in HeLa cancer cells based on its selective inhibition of the lysosomal Hsp70 and the degradation of lysosomal membranes associated with cathepsin release followed by caspase activation [40].

Several synthetic allosteric inhibitors of the Hsp70 ATPase domain were described. The rhodacyanine class of the Hsp70-inhibitory dye compounds was established by the discovery of compound **3a** (Figure 1, YM-1), in particular, by the blocking of the Hsp70 interaction with the nucleotide exchange factor (NEF) Bag3 by compound **3a** (IC_{50} = 4.8 μM) [41]. **3a** binds to an allosteric binding site, stabilizing ADP-bound Hsp70 with a weak Bag3 affinity. Consequently, compound **3a** suppressed FoxM1 and HIF1α pathways in MCF-7 and HeLa cells, which is unique for Hsp70-Bag3 inhibition since other Hsp70 inhibitors (e.g., the natural flavonoid myricetin) did not show such mechanistic effects. In MCF-7 breast carcinoma xenografts, compound **3a** (25 mg/kg every second day for 6 days, i.p.) inhibited tumor growth associated with induction of p21 and suppression of FoxM1 and survivin. The activity of compound **3a** was also investigated in glioma models, and it sensitized U251 and U343 glioma cells to treatment with the Bcl-2 inhibitors (-)-gossypol (AT-101) and ABT-737 [42]. Analogously, **3a** sensitized apoptosis-resistant and chemo-resistant breast cancer cells to drug treatment [43]. In doxorubicin-resistant BT-549^{r}DOX cells, compound **3a** suppressed Mcl-1 and showed synergistic antiproliferative effects in combination with doxorubicin. In addition to compound **3a**, further close Hsp70-inhibitory analogs such as compound **3b** (Figure 1, JG-98) with thiazolium moieties were described, which were discovered during the search for optimized Hsp70 inhibitors derived from compound **3a** and the mortalin (mitochondrial Hsp70) inhibitor MKT-077 (Figure 1, compound **3c**, see below). Compound **3b** exhibited prolonged microsomal half-lives and at least 3-fold higher antiproliferative activities against MDA-MB-231 triple-negative breast cancer (TNBC; EC_{50} = 0.4 μM) and hormone-sensitive MCF-7 breast cancer cells (EC_{50} = 0.7 μM) than compounds **3a** (EC_{50} = 2.0 for MDA-MB-231 and 5.2 for MCF-7 cells) and **3c** (EC_{50} = 1.4 for MDA-MB-231 and 2.2 for MCF-7 cells) [44]. Compound **3b** induced apoptosis in MDA-MB-231 cells by caspase activation and enhanced p62 oligomerization as a hint at forming autophagosomes. Compound **3b** also inhibited the Hsp70-Bag3 interaction, which was accompanied by FoxM1 suppression and upregulation of p21 and p27 in treated MCF-7 cells [45]. Doses of 3 mg/kg (every second day for six days) of compound **3b** inhibited the growth of MCF-7 breast carcinoma xenografts in vivo. In addition to breast cancer cell lines MCF-7 and MDA-MB-231, further cancer cell lines sensitive to compound **3b** were identified (with EC_{50} < 1 μM), such as HeLa, A375, HT-29, SKOV-3, Jurkat, MM1.R, INA6,

RPMI-8226, JJN-3, and U266. Mechanistically, compound **3b** exerted its antiproliferative activities against breast cancers both in Bag3-dependent (via ERK activation) and Bag3-independent ways (via suppressed Akt and *c*-Myc) [46]. Combinations with the proteasome inhibitors MG132 and bortezomib enhanced the antitumor activity of compound **3b** in breast cancer models both in vitro and in vivo (5 mg/kg of compound **3b** plus 1 mg/kg of bortezomib in MDA-MB-231 TNBC xenografts). Synergy effects in breast cancer cells were also observed in combination with α-amanitin (RNA-polymerase II inhibitor), LY294002 (Akt inhibitor), and sunitinib (RTK inhibitor). In addition, the infiltration of tumor bodies by tumor-associated macrophages (TAMs) was inhibited by compound **3b** based on the inactivation of tumor stromal cell Hsp70 proteins [47]. Similar to compound **1**, compound **3b** also sensitized lung cancer cells to treatment with manumycin A [31]. In a recent effort, new rhodacyanine analogs of compound **3b** with benzo-fused *N*-heterocycle moieties were described, with benzothiazolium compound **3d** and quinolinium compound **3e** (Figure 1) as the most promising compounds exhibiting high antiproliferative activity against TNBC cells (IC_{50} = 0.24 μM for **3d** and 0.37 μM for **3e** in MDA-MB-231 cells), accompanied by a considerable selectivity since non-malignant MCF-10A cells with low levels of Hsp70 were much less sensitive to treatment with **3d** [48]. Both compounds are very stable, with half-lives of more than 2 h in microsomes. Apoptosis upon caspase-activation was induced by compounds **3d** and **3e** in MDA-MB-468 breast cancer cells, while both compounds led to autophagy both in MDA-MB-231 and MDA-MB-468 cells. In addition to FoxM1, survivin, HuR, and Akt suppression, compounds **3d** and **3e** also degraded KRAS in MDA-MB-231 and MDA-MB-468 cells. Further **3b**/JG-98-derived benzothiazole rhodacyanines were described, culminating in the discovery of the bromothienyl analog **3f** (JG-231), which showed high activity against MCF-7 (IC_{50} = 0.12 μM) and MDA-MB-231 breast cancer cells (IC_{50} = 0.25 μM), disruption of Bag3 interaction, amenable microsomal stability (half-life of more than 60 min), degradation of Akt and HuR in MCF-7 xenografts, amenable in vivo pharmacokinetics parameters and in vivo tumor growth inhibition of MDA-MB-231 xenografts at doses of 4 mg/kg (i.p.) [49].

The cationic spasmolytic drug pinaverium bromide (Figure 1, **4**) was repurposed as an inhibitor of constitutively activated Hsc70. Compound **4** inhibited cell proliferation (IC_{50} of ca. 10 μM) of A2058 melanoma cells and induced apoptosis in these cells [50]. Its binding site was located at the NBD and linker domains of Hsc70.

S1g-2 (Figure 1, **5a**) was identified as an inhibitor of Hsp70-Bim interaction (IC_{50} = 0.4 μM) in CML cells by screening a Bcl-2 inhibitor library [51]. Its allosteric Hsp70-binding site is near the binding site of the rhodacyanines **3**. Compound **5a** selectively induces apoptosis in CML cells by suppressing oncoprotein clients such as Akt, Raf-1, eIF4E, and RPS16. Hsp70-Bim interaction protected BCR-ABL-independent TKI-resistant CML cells from apoptosis, and, thus, treatment with compound **5a** can overcome the resistance of a highly problematic CML type. The close analog compound **5b** (Figure 1, S1g-6), which has a morpholino side chain replacing the unstable ester side chain of **5a**, was recently described as a new sub-micromolar Hsp70-Bim interaction inhibitor [52]. Compound **5b** also induced apoptosis in cancer cells and suppressed Akt and Raf-1.

The benzimidazole derivative **6** (Figure 1, HS-72) selectively inhibits inducible Hsp70 (Hsp70i), which is in stark contrast to its low affinity for the closely related constitutively activated Hsc70 [53]. At doses of 25 and 50 μM, compound **6** inhibited the proliferation of BT474, MCF-7, and SkBr3 breast cancer cells, and the proteins Her2 and Akt were degraded by compound **6** in BT474 and MCF-7 breast cancer cells. In the Her2-overexpressing MMTV-*neu* spontaneous breast tumor mouse model, compound **6** (20 mg/kg biweekly for 21 days, i.p.) was well tolerated and inhibited breast tumor growth, leading to prolonged survival of treated mice.

The 2,5′-thiodipyrimidine **7** (Figure 1, YK-5) was identified as an irreversible inhibitor of a new allosteric site of the Hsp70 NBD domain and binds covalently to a cysteine residue of the binding site via its reactive acrylamide moiety [54]. It is highly active against Kasumi-1 AML cells (IC_{50} = 0.9 μM) and SkBr3 breast cancer cells (IC_{50} = 0.8 μM) and a potent apoptosis inducer by caspase-3/7 activation (IC_{50} = 1.2 μM) in MOLM13 AML cells. **7** degraded Her2 and Raf-1 in SkBr3 breast cancer cells as a consequence of Hsp70 inhibition.

The dye methylene blue (Figure 1, **8**) showed manifold biological activities, and, thus, it was also identified as an inhibitor of the Hsp70 ATPase, leading to a rapid suppression of Tau protein in neurodegenerative cell models [55]. In HeLa cervix carcinoma cells expressing poly-glutaminylated AR (AR112Q), compound **8** inhibited the Hsp70-mediated degradation of AR112Q [56]. In A375 and G361 metastatic melanoma cells, compound **8** suppressed the heat shock response (downregulation of Hsp27, Hsp70, and Hsc70), induced ROS formation, and caused glutathione depletion at a concentration of 10 μM [57]. Geldanamycin is an Hsp90 inhibitor, which increases Hsp70 expression, but compound **8** (10 μM) was able to suppress geldanamycin-induced Hsp70 expression in A375 melanoma cells. Hence, 10 μM of compound **8** also sensitized A375 melanoma cells to geldanamycin treatment, but it also sensitized these cells to treatment with etoposide and doxorubicin. Compound **8** possessed antiproliferative properties against A549 NSCLC cells and induced early apoptosis, yet it enhanced the degradation of N-terminal AR fragments as well as autophagy [58]. In mice, compound **8** inhibited benzo[a]pyrene induced lung carcinogenesis and suppressed Hsp70 as well as the tumor biomarkers ADA and LDH.

The structures of competitive and allosteric NBD binders are shown in Figure 1.

Table 1. Effects of NBF-targeting Hsp70 inhibitors on cancers.

Compound	Cancer Model(s)	Effects
1 (VER-155008)	HCT-116 colon carcinoma, A549 NSCLC, pleural mesothelioma, LNCaP95 prostate carcinoma, anaplastic thyroid carcinoma, glioma (1321N1, GOS-3, U87-MG), muscle-invasive bladder cancer, osteosarcoma, multiple myeloma, acute myeloid leukemia	Selective Hsp70/Hsc70 inhibition, antiproliferative, suppression of Her2 and Raf-1, G1 cell cycle arrest, sensitization to 17-AAG and radiation, suppression of Akt and phospho-Akt, macroautophagy induction, suppression of AR-FL and ARV7, apoptosis and paraptosis induction, upregulation of miR-215 and miR-194-5p, degradation of KDMA6 and EP300, degradation of BAG1, upregulation of CHOP, suppression of VEGF release by leukemia cells, synergy effects with drugs (manumycin A, bortezomib, shikonin, 17-DMAG) and PDT [19–35]
2a (Apoptozole)	SK-OV-3 ovarian carcinoma, HCT-15 colon carcinoma, A549 NSCLC	Suppression of Hsp70-APAF-1, antiproliferative, pro-apoptotic [37]
2b, **2c**	HeLa cervix carcinoma, MDA-MB-231 breast carcinoma, HepG2 hepatoma, A549 NSCLC, RKO colon carcinoma, leukemia	Suppression of Hsp70-APAF-1, antiproliferative, sensitization to doxorubicin, in vivo tumor growth inhibition of A549, RKO, and HeLa xenografts, apoptosis induction, autophagy inhibition, cathepsin release [38–40]
3a (YM-1)	MCF-7 breast carcinoma, HeLa cervix carcinoma, U251 and U343 glioma, doxorubicin-resistant BT-549^{r}DOX breast carcinoma	Inhibition of Hsp70-Bag3, suppression of FoxM1 and HIF1α pathways, in vivo inhibition of MCF-7 tumor growth, induction of p21, suppression of FoxM1 and surviving, sensitization of glioma to (−)-gossypol (AT-101) and ABT-737, suppression of Mcl-1, synergistic antiproliferative effects with doxorubicin [41–43]
3b (JG-98)	Triple-negative MDA-MB-231 and hormone sensitive MCF-7 breast cancer, lung cancer, miscellaneous	Inhibition of Hsp70-Bag3, FoxM1 suppression, upregulation of p21 and p27, sensitization of breast cancer to bortezomib in vivo, inhibition of TAM infiltration, sensitization of lung cancer to manumycin A [31,44–47]
3d, **3e**	Triple-negative breast cancer (e.g., MDA-MB-231, MDA-MB-468)	Stable, antiproliferative, tumor-selective, induction of apoptosis and autophagy, degradation of KRAS, suppression of FoxM1, survivin, HuR, and Akt [48]

Table 1. *Cont.*

Compound	Cancer Model(s)	Effects
3f (JG-231)	MCF-7 and MDA-MB-231 breast cancer	Stable, antiproliferative, inhibition of Hsp70-Bag3, degradation of Akt and HuR, tumor growth inhibition in vivo [49]
4 (Pinaverium bromide)	A2058 melanoma	Apoptosis induction [50]
5a (S1g-2)	CML	Inhibition of Hsp70-Bim, apoptosis induction, suppression of Akt, Raf-1, eIF4E and RPS16 [51]
5b (S1g-6)	Miscellaneous	Inhibition of Hsp70-Bim, apoptosis induction, degradation of Akt and Raf-1 [52]
6 (HS-72)	BT474, MCF-7 and SkBr3 breast carcinoma, Her2-overexpressing MMTV-*neu* spontaneous breast tumor mouse model	Selective Hsp70i inhibition, antiproliferative, Her2 and Akt degradation, tumor growth inhibition and prolonged survival in vivo [53]
7 (YK-5)	Kasumi-1 AML, SkBr3 breast carcinoma, MOLM13 AML	Antiproliferative, apoptosis induction, Her2 and Raf-1 degradation [54]
8 (Methylene Blue)	AR112Q-expressing HeLa cervix carcinoma, A375 and G361 melanoma, A549 NSCLC	Heat shock response suppression, ROS formation, glutathione depletion, suppressed geldanamycin-induced Hsp70, sensitization of cancer cells to geldanamycin, etoposide and doxorubicin, apoptosis induction, inhibition of lung carcinogenesis in vivo [56–58]

Some Hsp70 inhibitors with SBD-targeting properties were described (Figure 2). Compound **9a** (Figure 2, 2-phenylethynesulfonamide, PES, pifithrin-μ) is a prominent example, which was thoroughly studied for its Hsp70-related effects in various cancers. Compound **9a** showed antiproliferative activities against various osteosarcoma, breast, and pancreatic carcinoma cell lines at IC_{50} values of 5–10 μM independent of the p53-state of the tumor cells, induced cell death independent of caspase activation, and led to dysfunctional autophagy by the formation of p62-oligomers/aggregates [59]. It decreased the interaction of Hsp70 with APAF1, p53, and the co-chaperones Hsp40, CHIP, and BAG-1M and suppressed NF-κB signaling and activity. In vivo, compound **9a** (40 mg/kg, i.p., every five days for 30 days) blocked Myc-based lymphopathogenesis and led to prolonged survival in Eμ-Myc transgenic mice. Compound **9a** exhibited considerable antiproliferative activity against acute leukemia (AML and ALL) cells (IC_{50} = 2.5–12.7 μM), induced apoptosis in these cells by caspase activation, and led to a degradation of Akt and ERK1/2 [60]. In addition, compound **9a** sensitized acute leukemia cells to treatment with cytarabine (an antimetabolite), 17-AAG (a Hsp90 inhibitor), vorinostat (a HDAC inhibitor), and sorafenib (a RTK inhibitor). In primary effusion lymphoma (PEL), compound **9a** exerted cytotoxic effects on BC3 and BCBL1 cells by apoptosis and another cell death mechanism, which was associated with immunogenic activity such as activation of dendritic cells [61]. Compound **9a** increased lysosome permeabilization and cathepsin D release in PEL cells, accompanied by Bid cleavage, outer mitochondrial depolarization, and AIF translocation to the nucleus. Moreover, compound **9a** sensitized PEL cells to bortezomib treatment. The combination of compound **9a** with DNA-targeting platinum complexes such as cisplatin and oxaliplatin also revealed synergy effects in HT-29 colon carcinoma and PC-3 prostate carcinoma cells [62]. In prostate cancer cells, compound **9a** also increased the antitumor effects of hyperthermia (HT, 43 °C), best when given immediately before HT started, which was accompanied by upregulation of p21 and suppression of c-Myc and cyclin D1 [63]. The combination of compound **9a** (100 μg in 50 μL) and HT (43 °C for 1 h, twice on days 0 and 4), led to significant PC-3 prostate carcinoma xenograft growth inhibition. Its modified analog compound **9b** (Figure 2, PES-Cl) was antiproliferative against a panel of BRAF-V600E mutant melanoma (IC_{50} values between 2–5 μM, while inactive against melanocytes) and showed higher antiproliferative activity than compound **9a** against SkBr3 breast carcinoma, FaDu head and neck squamous cell carcinoma, and H1299 lung adenocarcinoma cells [64]. The cytotoxic activity of compound **9b** is based on apoptosis induction (caspase activation) and inhibition of autophagy (p62 accumulation), while the HeLa cell cycle was arrested in the G2-M phase by compounds **9a** and **9b** associated with cyclin B1 degradation. At doses of 20 mg/kg (i.p., once per week), compound **9b** led to a much higher survival rate of Eμ-Myc mice (71.4% survival) than compound **9a** (35% survival) after 210 days. Compound **9b** also induced apoptosis in A375 melanoma cells, accompanied by Her2 degradation in these cells [65]. In contrast to compounds **1** and **3c**, only compound **9b** led to G2-M arrest in H1299 and A375 cells based on cyclin B1 degradation.

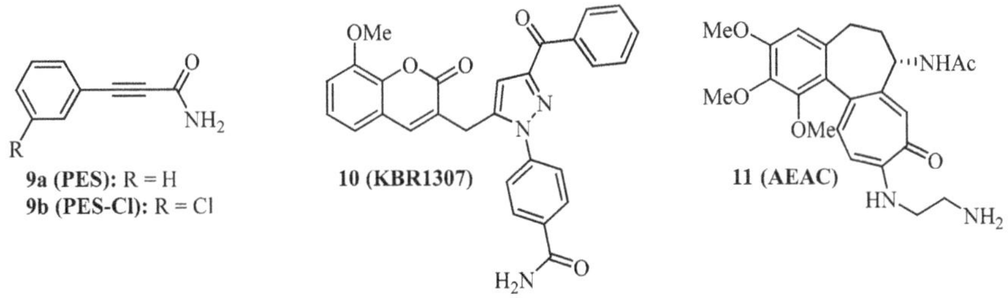

Figure 2. Structures of SBD-interacting Hsp70 inhibitors **9–11**.

Compound **9a** does not interact with Hsp70 when no nucleotide is bound to the protein. The coumarin-pyrazole hybrid, compound **10** (Figure 2, KBR1307) was designed, which binds Hsp70 in the presence and absence of nucleotides [66]. Both compounds **9a** and **10** reduce the activity of Hsp70 ATPase significantly, but compound **10** was more active against MCF-7 cells than compound **9a**. Similar coumarin-thiazole hybrids were described by the same group before as binders to the C-terminus of Hsp70 with activity against DLD-1 colon carcinoma and HepG2 hepatoma cells [67].

The screening of the InterBioScreen compound library for Hsp70 inhibitors revealed that the semi-synthetic colchicine derivative, compound **11** (Figure 2, *N*-aminoethylaminocolchicine, AEAC), interferes with substrate binding and refolding functions of Hsp70, based on a nanomolar affinity for Hsp70 (K_D = 149 nM) [68]. Although the antiproliferative and cytotoxic activities of compound **11** are low, it sensitized C6 rat glioblastoma and B16 mouse melanoma cells to doxorubicin treatment. The combination of compound **11** (2 mg/kg) and doxorubicin (1 mg/kg) inhibited in vivo B16 tumor growth by 71% and increased the lifespan of treated mice by ca. 15 days when compared with untreated mice.

The structures of SBD-interacting compounds are shown in Figure 2. Table 2 summarizes the anticancer effects of the described SBD-domain interacting Hsp70 inhibitors.

Table 2. SBD-interacting Hsp70 inhibitors and their effects on cancer model(s).

Compound	Cancer Model(s)	Effects
9a (PES)	Colon, breast, prostate and pancreas carcinoma, osteosarcoma, lymphoma, acute leukemia	Antiprolierative independent from p53-state, caspase activation, dysfunctional autophagy, prolonged survival of Eμ-Myc mice, NF-κB suppression, degradation of Akt and ERK1/2, immunogenic activity, sensitization of cancers to drugs and hyperthermia [59–63]
9b (PES-Cl)	BRAF-V600E mutant melanoma, SkBr3 breast carcinoma, FaDu head and neck squamous cell carcinoma, H1299 lung adenocarcinoma, lymphoma	More antiproliferative than **9a**, apoptosis induction, autophagy inhibition, G2-M phase arrest, degradation of cyclin B1 and Her2, prolonged survival of Eμ-Myc mice [64,65]
10 (KBR1307)	MCF-7 breast carcinoma	More antiproliferative than **9a**, binds Hsp70 in absence of nucleotide [66]
11 (AEAC)	C6 rat glioblastoma, B16 mouse melanoma	Increased doxorubicin activity in vitro and in vivo, tumor growth inhibition (71%) and prolonged survival in B16 mice [68]

2.2. Grp78 Inhibitors

The specific targeting of organelle-specific Hsp70 isoform proteins such as Grp78 (ER) and mortalin (mitochondria) has become a valuable strategy to combat cancer. Grp78 is the "master protein" of UPR (unfolded protein response) and redirects misfolded polypeptides for degradation or refolding [14]. Some of the already described Hsp70 inhibitors also inhibit Grp78. For instance, compound **1** is likewise a Grp78 inhibitor. In OSA cells, compound **1** inhibited Grp78 in addition to Hsp70, followed by antiproliferative and proapoptotic effects [29]. In MCF-7 and MDA-MB-231 breast cancer cells, Grp78 was upregulated upon tamoxifen treatment, leading to resistance; however, Grp78-inhibitory **1** enhanced apoptosis induction by tamoxifen in these cells, accompanied by suppression of tamoxifen-induced phosphor-GSK-3β, which is a downstream factor of Akt signaling [69]. The quinoline analog **12** (Figure 3) of compound **1** was slightly more active against and selective for Grp78 (K_D = 0.6 μM for Grp78, 0.3 μM for Hsp70) than compound **1** (K_D = 0.8 μM for Grp78, 0.1 μM for Hsp70), however, compound **12** showed no antiproliferative activity against HCT-116 colon cancer cells in contrast to compound **1** (GI_{50} = 5.0 μM) [70]. The absence of antiproliferative activity in compound **12** was explained by the detrimental physicochemical properties of this compound, which need to be improved in order to obtain a valuable adenosine-based Grp78-selective inhibitor in the future.

The thiazole benzenesulfonamide **13** (Figure 3, HA15) was identified as a Grp78 inhibitor that enhanced ER stress associated with autophagy and apoptosis, leading to cell death in melanoma cells, in particular in BRAF-mutant cells, at a concentration of 10 μM [71]. In vivo, compound **13** (0.7 mg/mouse/day for 2 weeks, i.p.) inhibited A375 melanoma growth without causing side effects such as mouse weight loss and liver damage. Grp78 was found to be upregulated in lung cancers, and treatment with compound **13** led to antiproliferative effects on A549 lung cancer cells as well as to apoptosis induction, autophagy, and increased ER stress [72]. The GRp78 inhibitor **13** suppressed KRAS expression and revealed antiproliferative and pro-apoptotic activities in various KRAS-mutant cancer cell lines (A427 lung adenocarcinoma, H460 non-small cell lung carcinoma, HCT-116 and LS180 colon carcinomas, PANC-1 and CFPAC-1 pancreatic ductal adenocarcinomas), and induced apoptosis in A427, HCT-116, and PANC-1 cells by caspase activation [73]. Compound **13** also suppressed cell proliferation and steroidogenesis in adrenocortical carcinoma (ACC) cells, and showed synergy effects in combination with the approved drug mitotane, which is also an activator of ER stress [74].

In carboplatin-resistant canine osteosarcoma cells (HMPOS-2.5R and HMPOS-10R), compound **13** and the atypical Grp78 inhibitor **14** (Figure 3, the celecoxib derivative OSU-03012, which binds directly to the Grp78 ATPase domain) still showed considerable activity (EC_{50} = 1.9–3.5 μM for **13** and 5.3–8.5 μM for **14**), while the Hsp70 inhibitor **1** was distinctly less active against these resistant cells (EC_{50} = 25–30 μM) than against the parent HMPOS cells (EC_{50} = 1.8 μM) [75,76]. Compound **14** reduced the expression of Bag2, and treatment with compound **14**, especially in combination with the PDE5 inhibitor sildenafil, formed toxic autophagosomes that inducing cell death in GBM5 and GBM12 glioblastoma cells [77]. Similar effects were observed for the multi-kinase inhibitor sorafenib, which is an approved anticancer drug and also able to bind to the N-terminal domain of Grp78.

Analogously to compound **13**, the hydroxyquinoline derivative **15** (Figure 3, YUM70) was able to downregulate KRAS, leading to antiproliferative and pro-apoptotic effects on KRAS-mutant cancer cells [73]. Initially, compound **15** was identified as an antiproliferative and caspase-dependent pro-apoptotic compound in pancreas cancer cells (IC_{50} = 2.8 μM for MiaPaCa-2, 4.5 μM for PANC-1, and 9.6 μM for BxPC-3 cells) based on Grp78 inhibition (by binding to the SBD) followed by ER stress via eIF2α phosphorylation as well as AT4 and CHOP activation [78]. Synergy effects of compound **15** on MiaPaCa-2 cells were observed in combination with vorinostat or topotecan. In the MiaPaCa-2 xenograft model, compound **15** (30 mg/kg, 5 days a week for 7 weeks, i.p.) inhibited tumor growth and caused weight loss in treated mice.

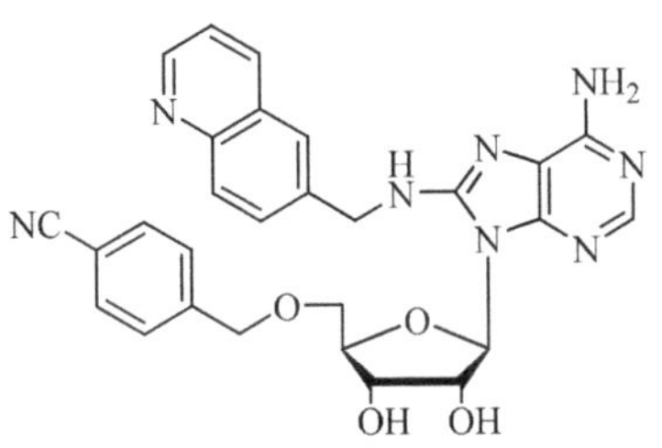

12

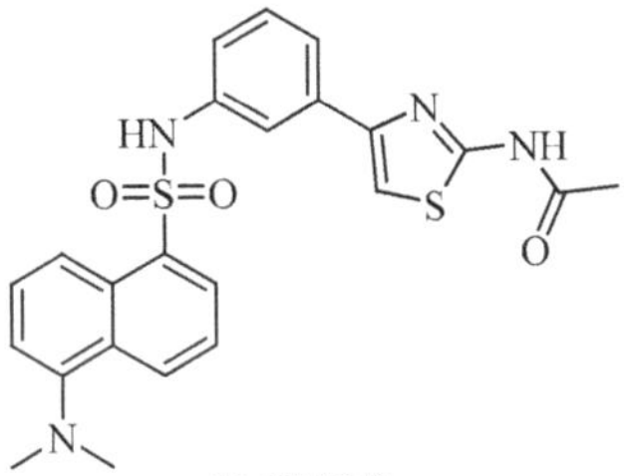

13 (HA15)

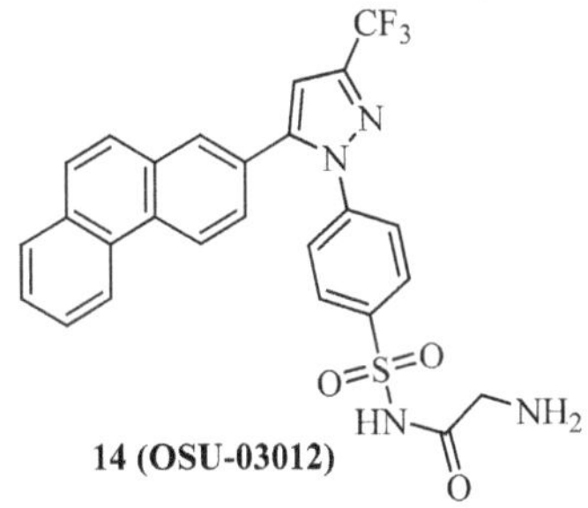

14 (OSU-03012)

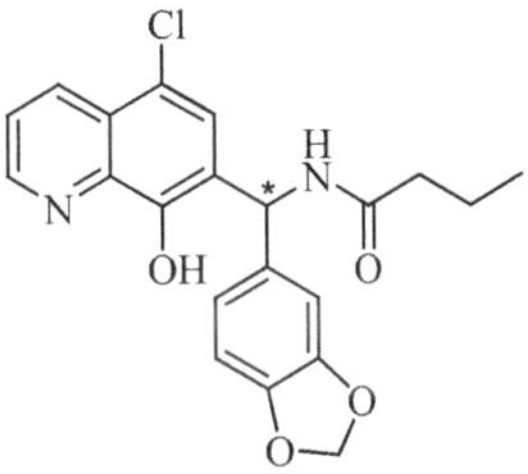

15 (YUM70)

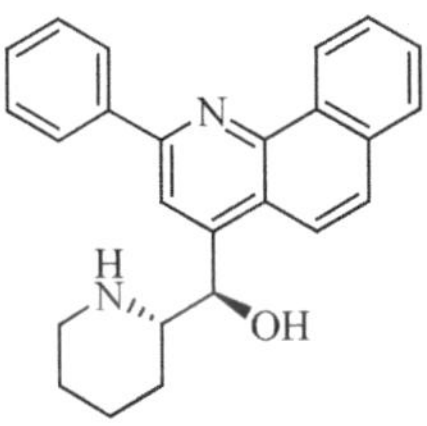

16 (HM01)

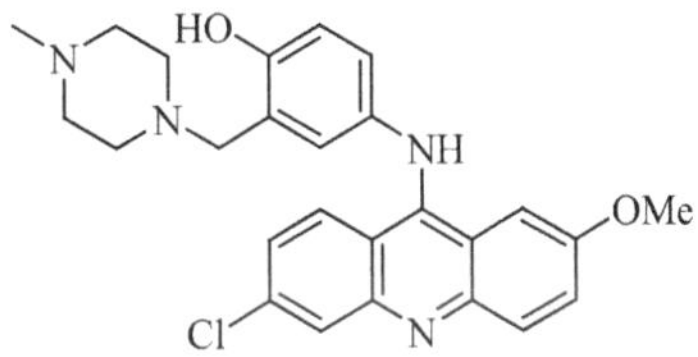

17 (HM03)

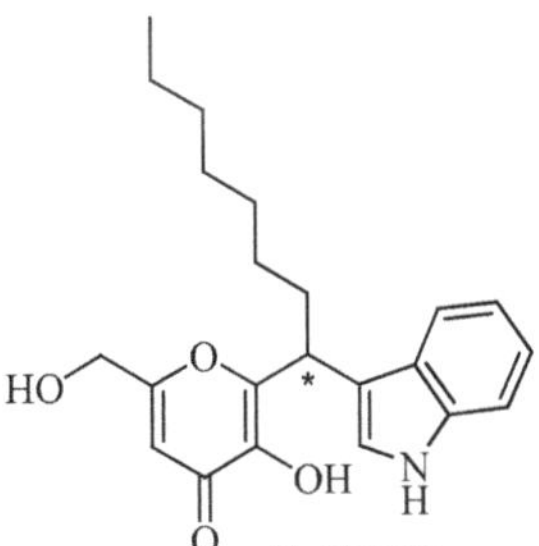

18 (IKM5)

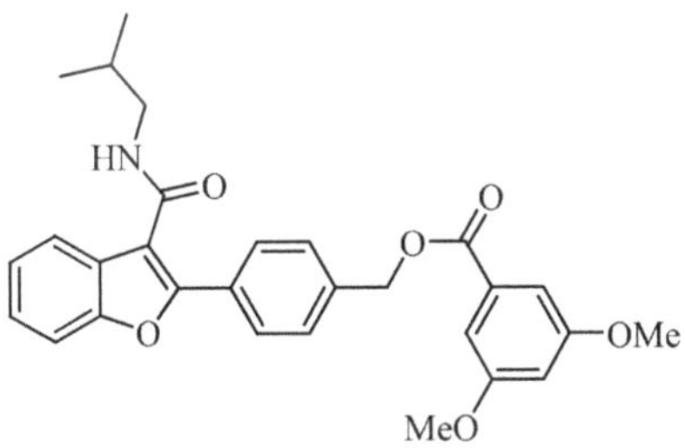

19 (FL5)

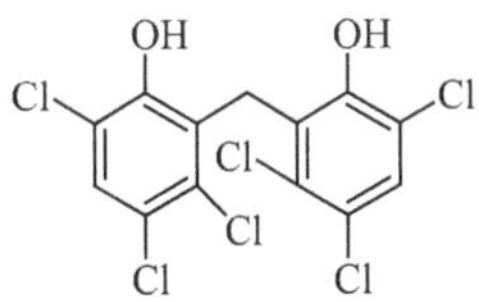

20 (Hexachlorophene)

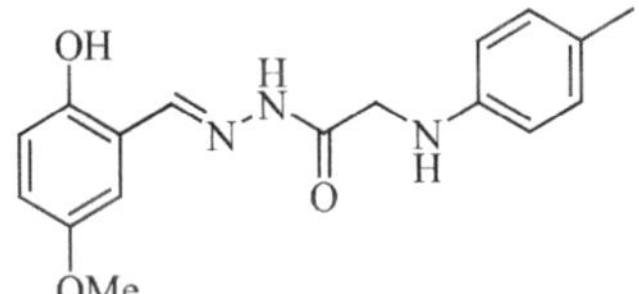

21 (VH1019)

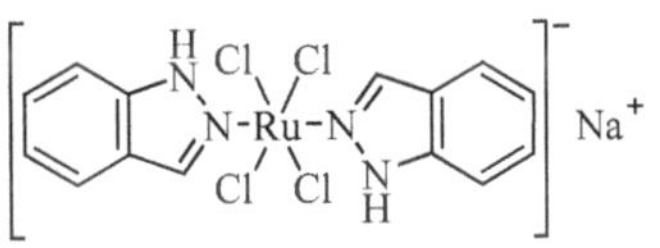

22 (KP1339)

Figure 3. Structures of Grp78 inhibitors **12–22**. *: chiral center or asymmetric carbon atom.

The *N*-heteroaromatic compounds **16** (Figure 3, HM01) and **17** (Figure 3, HM03) were identified as Grp78 inhibitor hits from the virtual screening of an NCI diversity set, and they showed considerable affinity to the substrate-binding channel of Grp78 [79]. Both compounds exerted moderate antiproliferative activity against HCT-116 colon cancer cells (IC_{50} between 10 and 25 μM) but may serve as lead compounds for the design of more potent Grp78 inhibitors.

The indolylkojyl derivative **18** (Figure 3, IKM5) was prepared by a simple three-component reaction and identified as a potent Grp78 inhibitor (K_i = 1.4 μM), which showed high antiproliferative activities against a panel of breast cancer cell lines (IC_{50} = 0.15 μM for MCF-7, 0.21 μM for MDA-MB-231, 0.54 μM for MDA-MB-468, and 3.5 μM for BT474 cells) [80]. Compound **18** induced TIMP-1 by blocking its interaction with Grp78, suppressed EMT markers such as MMP-2, Twist1, and vimentin, and upregulated the Par-4 tumor suppressor, which controls NF-κB signaling. Compound **18** increased the activity of doxorubicin against invasive breast cancer cells. In vivo, compound **18** (30 mg/kg, b.w.) inhibited 4T1 breast tumor growth by 79.2% and lung metastasis formation by 84.5%.

The benzofuran derivative **19** (Figure 3, FL5) is a strong binder and stabilizer of Grp78 (T_m increase > 2 °C), which was associated with its anticancer and antiangiogenic activities against renal cell carcinoma (RCC) cells (10 μM of **19** led to 50% cell death) and HUVECs (EC_{50} = 1.5 μM), while it was inactive against mouse fibroblasts [81]. Docking studies showed that compound **19** does not interfere with the ATPase activity of Grp78.

A high-throughput substrate binding assay led to the identification of the disinfectant hexachlorophene (Figure 3, **20**) as a Grp78 inhibitor binding to the SBD [82]. Compound **20** was cytotoxic against HCT-116 colon carcinoma cells (CC_{50} = 3.4 μM) and induced apoptosis and autophagy, as well as an unfolded protein response associated with upregulated ATF4, XBP1s, and CHOP.

In silico methods for the design of new Grp78 inhibitors led to the identification of **21** (Figure 3, VH1019) with antiproliferative activity against MCF-7 breast cancer cells (IC_{50} = 12.7 μM) [83]. It is noteworthy that compound **21** mimicked ATP in its binding to Grp78.

In addition to organic compounds, metal complexes can be a valuable source of Grp78 inhibitors. The anionic ruthenium complex **22** (Figure 3, KP1339, BOLD-100) is an efficient Grp78 suppressor that has already demonstrated promising anticancer activity in phase 1 clinical trials, both as compound **22** and in its benzindazolium salt form, KP1019 [84]. A phase 1b/2 clinical trial with the metallodrug compound **22** for patients suffering from gastrointestinal cancer is ongoing (NCT04421820). Mechanistic studies revealed apoptosis induction in sensitive cancer cell lines by activation of caspase-8, which was associated with disruption of the ER upon suppression of key chaperones, leading to the degradation of vital proteins [85]. In contrast, cancer cells with a low response to complex compound **22**-induced G2 cell cycle arrest may serve as a general hint at cancers resistant to Grp78 inhibitors. Interactions with the ribosomal proteins RPL10 and RPL24 and with the transcription factor GTF2I were identified in HCT-116 colon carcinoma cells and associated with ribosomal disturbance and ER stress induction [86]. Grp78 is upregulated in asbestos-associated pleural mesothelioma, and, thus, the activity of **22** against this cancer was evaluated. Complex compound **22** was cytotoxic against mesothelioma cells (EC_{50} = 71 μM for REN and 90 μM for MM98 cells), inhibited REN cell colony formation at a concentration of 100 μM, induced apoptosis in REN cells by activation of caspase-3/7 and caspase-8, increased ROS formation and cytosolic Ca^{2+} levels, and suppressed Grp78 expression in REN cells [87]. CHOP and XPB1 expression were upregulated by compound **22** in REN cells. Sensitive HCT-116 colon carcinoma (IC_{50} = 76 μM) and Capan1 pancreatic carcinoma cells (IC_{50} = 40 μM) were used to study cell-based resistance to treatment with compound **22** [88]. Increased glucose uptake and upregulated glycolysis were observed in sensitive cancer cells upon treatment, but especially in the resistant cell lines HCTR and CapanR obtained from HCT-116 and Capan1 cells, respectively, upon exposure to compound **22**. However, this specific mechanism made the resistant tumor cells highly vulnerable to the treatment with the glycolysis inhibitor and ER stress inducer 2-dexyglucose, which led to synergy effects of combinations of compound **22** with 2-deoxyglucose in resistant HCTR cells.

The structures of the described Grp78 inhibitors are shown in Figure 3, and their activities are summarized in Table 3.

Table 3. Grp78 inhibitors and their effects on cancer model(s).

Grp78 Inhibitor	Cancer Model(s)	Effects
1 (VER-155008)	Osteosarcoma, MCF-7 and MDA-MB-231 breast cancer	Antiproliferative, apoptosis induction, suppression of tamoxifen-induced phosphor-GSK-3β [30,69]
12	HCT-116 colon carcinoma	More selective for Grp78 than **1**, no antiproliferative activity [70]
13 (HA15)	Melanoma (BRAF-mutant), A549 NSCLC, KRAS-mutant cancer, adrenocortical carcinoma	ER stress, induction of apoptosis and autophagy, in vivo inhibition of A375 melanoma growth, suppression of KRAS and steroidogenesis, synergy with mitotane [71–74]
14 (OSU-03012)	Carboplatin-resistant canine osteosarcoma (HMPOS-2.5R and HMPOS-10R), GBM5 and GBM12 glioblastoma	Antiproliferative, Bag2 suppression, formation of toxic autophagosomes [75–77]
15 (YUM70)	KRAS-mutant cancer, pancreatic cancer (Mia-PaCa-2, PANC-1, BxPC-3)	Antiproliferative, ER stress, eIF2α phosphorylation, AT4 and CHOP activation, synergy with vorinostat and topotecan, in vivo MiaPaCa-2 tumor growth inhibition [73,78]
16 (HM01), **17** (HM03)	HCT-116 colon carcinoma	Antiproliferative [79]
18 (IKM5)	Breast cancer (MCF-7, MDA-MB-231, MDA-MB-468, BT474, 4T1)	Antiproliferative, suppression of MMP-2, Twist1 and vimentin, induction of TIMP-1 and Par-4, in vivo inhibition of breast tumor growth and lung metastasis formation [80]
19 (FL5)	Renal cell carcinoma, HUVECs	Cell death, anti-angiogenic [81]
20 (Hexachlorophene)	HCT-116 colon carcinoma	Cytotoxic, induction of apoptosis and autophagy, upregulated ATF4, XBP1s, and CHOP [82]
21 (VH1019)	MCF-7 breast carcinoma	ATP-mimic, antiproliferative [83]
22 (KP1339/BOLD-100)	Miscellaneous, HCT-116 colon carcinoma, REN pleural mesothelioma, Capan1 pancreatic carcinoma	Apoptosis induction and ER disruption in **22**-sensitive cells, G2 cell cycle arrest in **22**-resistant cells, binding to ribosomal proteins, ER stress, cytotoxic, ROS formation, induction of CHOP and XPB1, glycolysis upregulation, synergy with 2-deoxyglucose [85–88]

2.3. Mortalin Inhibitors

Mortalin is the mitochondrial Hsp70 isoform with an N-terminal mitochondrial localization motif and has become an important target for cancer therapy [15]. Due to its preferred localization in mitochondria, mortalin is involved in the regulation of mitochondrial metabolism and of key tumor factors such as p53, PI3K/AKT, Raf/MEK/ERK, and JAK/STAT pathways [15,89]. The triphenylphosphonium (TPP) moiety was described as a mitochondria-targeting device, and, thus, compound **2b** was conjugated with TPP in order to obtain the mortalin-targeting apoptozole conjugate compound **23** (Figure 4) [40]. Compound **23** showed higher antiproliferative activity against a panel of 20 cancer cell lines (IC_{50} = 0.5–1.5 μM) than compound **2b** and induced caspase-dependent apoptosis in HeLa cells by blocking the interaction of p53 with mortalin in the mitochondria, followed by Bak-mediated mitochondrial outer membrane permeabilization.

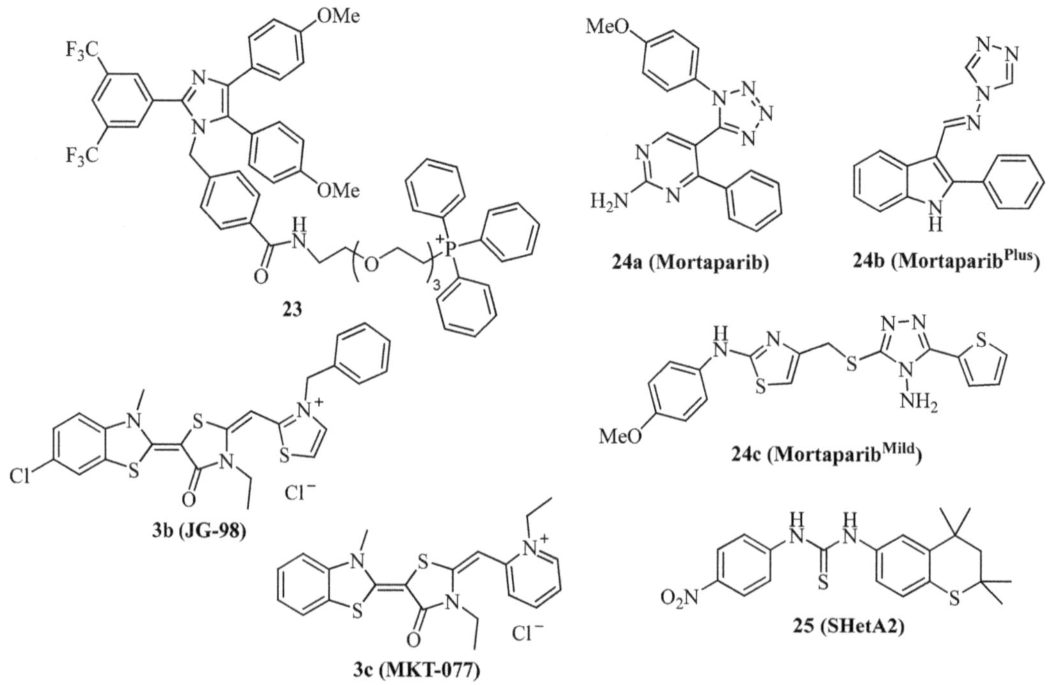

Figure 4. Structures of mortalin inhibitors **3b**, **3c**, **23–25**.

Cationic Hsp70 inhibitors such as compound **3b** (Figure 4, JG-98) were identified as mortalin inhibitors with pronounced activity against proteasome inhibitor-resistant MM cells associated with 55S mitoribosome degradation [90]. Compound **3c** (Figure 4, MKT-077) overlaps with the p53-binding region of mortalin, releasing active p53 followed by upregulation of p21 in treated cancer cells [91,92]. In addition, compound **3c** sensitized K562 leukemia cells to complement-mediated lysis by inhibition of mortalin and its interaction with the C9 complement protein [93]. Compound **3c** was identified already in the 1990s as a mitochondria-targeting anticancer compound with preference for ras-associated cancers, and entered clinical trials after promising in vivo results [94–97]. However, complications such as renal toxicity prevented the approval of compound **3c** as an anticancer drug [98].

The tetrazole derivative **24a** (Figure 4, mortaparib) is a relatively new dual mortalin and PARP1 inhibitor that was developed in 2019 [99]. Compound **24a** led to the activation and nuclear accumulation of p53 by inhibiting mortalin. In addition, PARP1 was downregulated by compound **24a,** followed by increased double-strand breaks and apoptosis induction in HeLa cervix carcinoma and SKOV-3 ovarian cancer cells. In vivo, compound **24a** (20 mg/kg i.p.) was well tolerated, inhibited the tumor growth of SKOV-3 xenografts, and suppressed the formation of metastases in the lungs and kidneys. Based on these promising anticancer activities of compound **24a**, the same group developed further mortaparib derivatives. The 1,2,4-triazole **24b** (Figure 4, mortaparib^{Plus}) blocked the interaction of p53 with mortalin, leading to p53 activation and suppression of PARP1 and CARF in HCT-116 cells [100]. Compound **24b** induced apoptosis and activated p21 in HCT-116 (p53 wildtype) and DLD-1 (p53^{S241F}) colon carcinoma cells in p53-dependent and –independent ways. In contrast to colon cancer cells, in breast cancer cells, compound **24b** upregulated only p21 in the p53-wildtype MCF-7 cells, while T47D cells (p53^{L194F}) treated with compound **24b** showed no p21 changes but activation of PARP1, albeit compound **24b** inhibited the mortalin-p53 interaction in both cell lines [101]. More recently, compound **24c** (Figure 4, mortaparib^{mild}) was identified as an inhibitor of PARP1 and mortalin-p53 interaction in HCT-116 cells, however, at higher concentrations than compounds **24a** and **24c**, thus leading to the attribute "mild" for compound **24c** [102].

The thiochromane derivative **25** (Figure 4, SHetA2) is an orally bioavailable mortalin inhibitor that blocks the interaction of mortalin with p53 in ovarian cancer cells [103]. Due to its promising chemo-preventive and selective anticancer activities, including apoptosis induction, compound **25** has entered clinical phase 1 studies for patients with advanced/recurrent cervical, endometrial, or ovarian cancer [15]. Since mutant p53 led to drug resistance upon treatment with compound **25**, strategies to overcome this resistance are sought. A promising strategy is the combination of compound **25** with the p53 reactivator PRIMA-1^{Met}, which was studied in a panel of ovarian cancer cell lines [104]. PRIMA-1^{Met} reduced resistance to compound **25** and exhibited synergy effects in combination with compound **25** in p53-mutant and p53-wildtype ovarian cancer cells, accompanied by caspase activation, increased ROS formation, and reduced ATP. The combination of compound **25** (60 mg/kg by gavage every day for 2 weeks, then every second day for 3 weeks) and PRIMA-1^{Met} (10 mg/kg, i.p. every other day) inhibited MESOV tumor growth (tumor free rate of 67%) in an additive way without toxicity to the liver and kidneys.

The structures of the described small-molecule mortalin inhibitors are shown in Figure 4. Their anticancer activities are summarized in Table 4.

Table 4. Mortalin inhibitors and their anticancer effects.

Mortalin Inhibitor	Cancer Model(s)	Effects
3b (JG-98)	Multiple myeloma	Antiproliferative, 55S mitoribosome degradation [90]
3c (MKT-077)	Miscellaneous, ras-induced cancer, K562 leukemia	Antiproliferative, mitochondria accumulation, activation of p53 and p21, inhibition of mortalin-C9 [91–97]
23	Miscellaneous, HeLa cervix carcinoma	Antiproliferative, p53 and Bak activation [40]
24a (Mortaparib)	HeLa cervix and SKOV-3 ovarian carcinoma	Dual mortalin and PARP1 inhibitor, p53 activation, apoptosis induction, in vivo inhibition of SKOV-3 tumor growth and metastases [99]
24b (Mortaparib^{Plus})	HCT-116 and DLD-1 colon carcinoma	Inhibition of mortalin-p53 and PARP1, CARF-1 suppression, induction of apoptosis and p21 [100]
24c (Mortaparib^{Mild})	HCT-116 colon carcinoma	Inhibition of mortalin-p53 and PARP1 [102]
25 (SHetA2)	Ovarian cancer	Inhibition of mortalin-p53, synergy with p53-reactivator PRIMA-1^{Met}, caspase-activation, increased ROS formation, reduced ATP, in vivo inhibition of MESOV tumor growth, clinical phase 1 studies (advanced/recurrent cervical, endometrial and ovarian cancer) [15,103,104]

3. Modulators of Co-Chaperone Hsp40

The co-chaperones of the Hsp40 family are vital for the ATPase activity of Hsp70 proteins, and modulation of Hsp40 activity has tremendous effects on the Hsp70 network of protein interaction and integrity. The natural Hsp70 ATPase stimulator 15-deoxyspergualin served as a lead compound for the design of the uracil derivative NSC-630668-R/1, which inhibits endogenous and Hsp40-induced ATPase activity of Hsp70 [105,106]. Further synthetic derivatives were prepared using the straightforward Biginelli multicomponent reaction, leading to the identification of **26a** (Figure 5, MAL3-101), which specifically blocks Hsp70 ATPase in an allosteric way by inhibition of the Hsp40 co-chaperone protein TAg [107]. In this way, the strategy to target the Hsp70/Hsp40 axis as a possible treatment for cancer diseases was established. In contrast, the synthetic small-molecule dihydropyrimidine compounds **26b** (Figure 5, 115-3b) and **26c** (Figure 5, 115-7c), which lack the amide side chain of **26a**, stimulated protein folding by the bacterial DnaK (Hsp70) chaperone in the presence of the DnaJ (Hsp40) and GrpE co-chaperones, while compound **26a** inhibited protein folding in this assay as expected with an EC_{50} value of 3.2 μM [108,109].

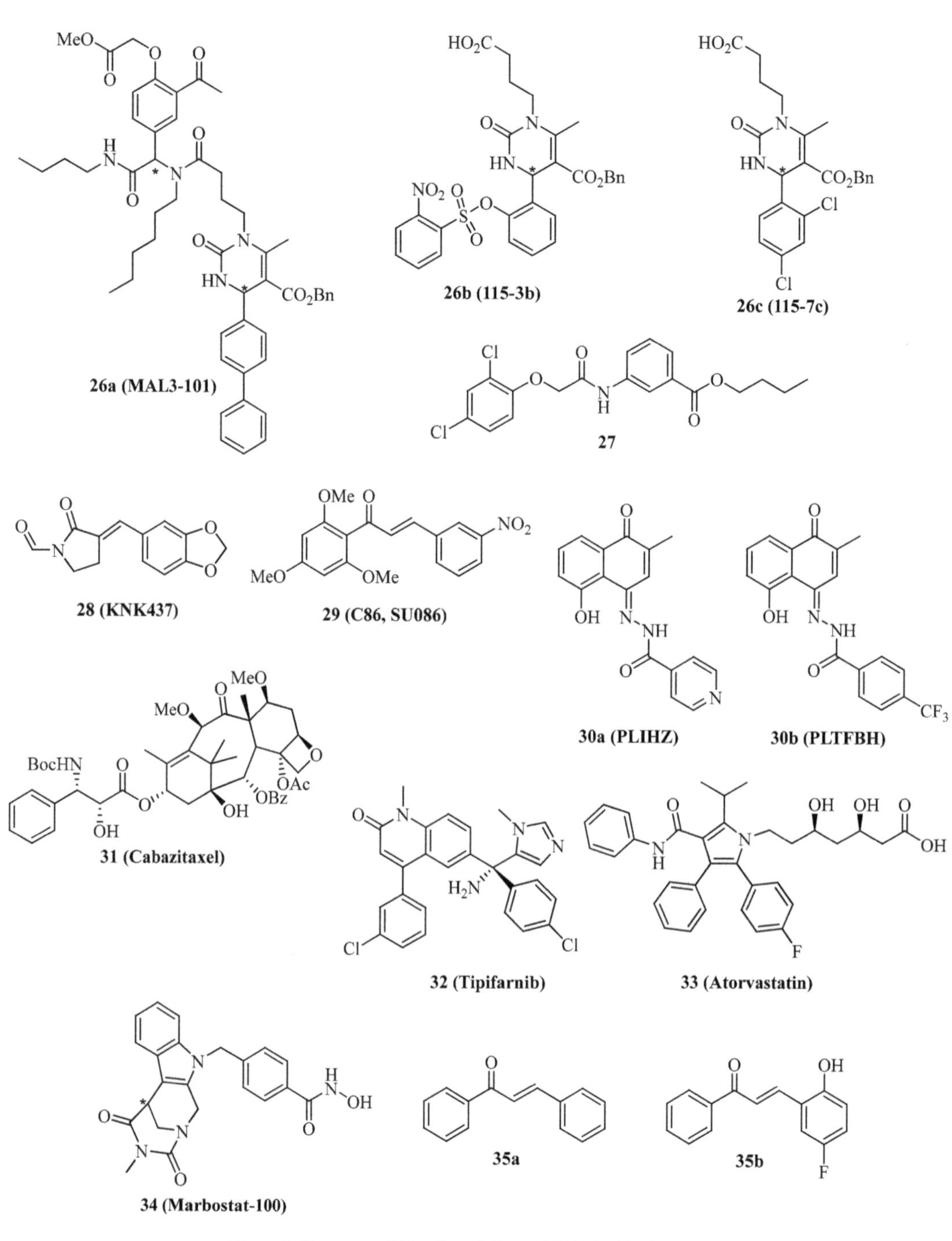

Figure 5. Structures of Hsp40 modulators **26–35**. *: chiral center or asymmetric carbon atom.

Compound **26a** showed various anticancer properties, e.g., antiproliferative activity against Merkel cell carcinoma (MCC) accompanied by apoptosis induction [110]. In addition, compound **26a** (40 mg/kg i.p. every other day) inhibited WaGa MCC growth in mice, again associated with apoptosis induction in vivo. Moreover, compound **26a** (10 μM) showed antiproliferative activity against T24 and SW780 MIBC cells, and the combination of compound **26a** with the Hsp70 inhibitor **1** or with the Hsp90 inhibitor STA-9090 led to synergistic antiproliferative activity against MIBC cells, which were resistant to compound **26a** alone [111]. RMS13 rhabdomyosarcoma cells were also sensitive to treatment with **26a** based on the induction of UPR and apoptosis; however, dose-escalation led to the isolation of a resistant RMS13-R cell line with upregulated autophagy and an ER-associated degradation pathway [112]. Inhibition of autophagy restored the sensitivity to **26a** in the RMS13-R cells.

The phenoxy-*N*-arylacetamide **27** (Figure 5) was found to inhibit Hsp70 by direct binding to the co-chaperone DnaJ (IC_{50} = 0.13 μM) [113].

In 2000, the synthetic lactam derivative **28** (Figure 5, KNK437) was identified as a suppressor of heat-induced Hsp70 and Hsp40 in COLO 320DM human colon cancer cells, whose effect was more pronounced than for the known natural Hsp inhibitor quercetin [114]. In vivo, compound **28** (200 mg/kg i.p.) showed synergy effects in combination with heat treatment (44 °C) in mice bearing SCC VII squamous cell carcinomas [115]. At a non-toxic concentration of 300 μM, compound **28** inhibited colony formation of p53-mutant SAS/mp53 human squamous cell carcinoma cells at a temperature of 42 °C [116]. This inhibitory effect of compound **28** in combination with heat was stronger than in the p53-wildtype SAS/neo cells and accompanied by the induction of apoptosis at a reduced concentration (100 μM). Compound **28** (100 μM) also inhibited heat-induced (44 °C) transcription-activating histone H3-Lys4 methylation both in thermoresistant HSC4 and in thermosensitive KB oral squamous cell carcinoma cells, which was associated with Hsp70 suppression by compound **28** [117]. In immortalized Cos-1 cells, 100 μM of compound **28** inhibited mTOR and S6K phosphorylation and suppressed mTORC1 activity, leading to apoptosis induction [118]. Compound **28** (50 μM) sensitized MDA-MB-231 breast cancer cells to ionizing radiation by mechanisms independent from Hsp suppression [119]. Instead, hypoxia-related AKT and HIF-1α survival pathways were inhibited by compound **28,** which were interesting off-targets of this compound. The Hsp40 protein DNAJA1 was upregulated in colorectal cancer cells, but treatment with compound **28** strongly inhibited the level of the Hsp40 protein DNAJA1 in SW480 (IC_{50} = 24.7 μM), RKO (IC_{50} = 25.5 μM), LOVO (IC_{50} = 56.0 μM) and SW620 cells (IC_{50} = 48.3 μM), while no effects on Hsp70 and Hsp90 were found [120]. DNAJA1-overexpressing SW480 and SW620 cells exhibited much faster and stronger tumor growth in vivo, which was inhibited by treatment with compound **28** (20 mg/kg i.p. for 20 days). The combination of 5-FU/L-OHP with compound **28** (20 mg/kg i.p.) efficiently suppressed liver metastasis formation in DNAJA1-overexpressing SW480 colorectal tumors. Mechanistically, compound **28** suppressed CDC45 and upregulated ubiquitin in DNAJA1-overexpressing SW480 cells as a consequence of its strong DNAJA1 downregulating activity.

The chalcone compound **29** (Figure 5, C86/SU086) was initially identified as a synthetic xanthohumol analog with antiproliferative activity against HeLa cervix carcinoma cells (IC_{50} = 1.4 μM), pro-apoptotic activity by caspase-3 activation, and thioredoxin reductase inhibitory activity (IC_{50} = 3.5 μM) followed by induction of ROS formation in treated HeLa cells [121]. The effects of compound **29** on castration-resistant prostate cancer (CRPC) were studied in more detail, revealing strong antiproliferative activity against 22Rv1 CRPC cells (IC_{50} between 1 and 2.5 μM) and a suppression of FL-AR/ARv7 signaling based on enhanced degradation of FL-AR and ARv7 in 22Rv1 CRPC cells treated with compound **29** (10 μM) [122]. Experiments using biotinylated 29 confirmed a direct binding as a pan-Hsp40/DNAJ inhibitor, which interacts with DNAJA, DNAJB, and DNAJC proteins likely via the J domains of the Hsp40 proteins. In addition, compound **29** is bound to Hsp40 in complexes with AR and Arv7 and with Hsp70/Bag3/CHIP. Similar effects were observed for the Hsp70 inhibitor **3b** (JG98, IC_{50} = 0.4–0.5 μM for 22Rv1 cells, destabilization of FL-AR/ARv7 proteins). Both compounds **29** (15 mg/kg, i.v. 3 x per week) and **3f** (8 mg/kg, i.p. every other day) inhibited the growth of 22Rv1 CRPC xenografts in mice; however, the combination of both drug candidates surpassed the antitumor activity of the single compounds. More recently, compound **29** was also identified as a Hsp90 inhibitor in prostate cancer cells, which interferes with tumor cell glycolysis [123]. In C4-2 prostate cancer xenografts, compound **29** (50 mg/kg/day i.p.) inhibited tumor growth similar to the approved prostate cancer drugs enzalutamide (10 mg/kg/day p.o.) and abiraterone (200 mg/kg/day p.o.). Yet, the combination of compound **29** with either enzalutamide or abiraterone led to synergistic and additive tumor growth inhibitory effects, respectively.

The semi-synthetic hydrazones **30a** (Figure 5, PLIHZ) and **30b** (Figure 5, PLTFBH), which were made of the natural naphthoquinone plumbagin, were identified as Hsp40 inhibitors by molecular docking [124]. Compounds **30a** and **30b** revealed antiproliferative activities against HN31 pharyngeal squamous cell carcinoma cells (IC_{50} = 1.2 μM for **30a** and 0.6 μM for **30b**) and suppressed DNAJA1 and conformational mutant p53 levels in HN31 cells, accompanied by suppression of tumor cell migration based on downregulation of active Cdc42 and Rac. No effects were observed for wild-type p53 or DNA-contact mutant p53 proteins, indicating a selective action against conformational mutant p53. In addition to DNAJA1, compound **30b** showed distinct inhibitory activities against DNAJA2, DNAJA3, DNAJB1, DNAJB12, and DNAJC3.

The semi-synthetic 3rd generation taxane derivative cabazitaxel (Figure 5, **31**) was able to inhibit LNCaP and PC-3 prostate cancer cell proliferation and suppress Hsp40, HOP, and AR in prostate cancer cells at very low doses (0.3 nM) [125].

The inhibition of farnesylation of HDJ-2 Hsp40 proteins has become a reasonable marker for the clinical outcome of anticancer active farnesyl transferase inhibitors [126–128]. Farnesyl transferase inhibitors such as tipifarnib (Figure 5, **32**) entered clinical trials for various cancer diseases because they inhibit the farnesylation of Ras proteins, leading to their inactivation [129,130]. The radio-sensitizing effect on SF763 glioblastoma cells treated with compound **32** was associated with a suppression of the radio-induced translocation of HDJ-2 by compound **32** based on reduced levels of farnesylated HDJ-2 [131]. In addition, compound **32** inhibited U87 and SF763 glioblastoma cell proliferation (IC_{50} = 3.1 μM for U87 and 1.9 μM for SF763) and led to G2/M arrest in these cells based on p21 induction. The HMG-CoA reductase inhibitory statin drug **33** (Figure 5, atorvastatin), which is applied for the treatment of cardiovascular disease based on its cholesterol depletion activity, was also studied for its inhibition of DNAJA1 farnesylation in pancreatic cancer cells expressing wild-type or mutant p53 proteins [132]. Compound **33** induced apoptosis, upregulated p21, and degraded mutant p53 (R172H) and cyclin D1 in PO3 cells based on the suppression of farnesylated DNAJA1. In addition, compound **33** suppressed mutant p53 in SU 86.86 (G245S), BXPC-3 (Y220C), and Pan 10.05 (I255N) pancreatic cancer cells, while mutant p53 levels in MIA-PaCa-2 (R248W) and PANC-1 (R273H) were not affected by compound **33**. Nuclear translocation of mutant p53 was inhibited in PO-3 cells by compound **33**, which also suppressed PO-3 cell migration and invasion.

Selective inhibition of cytoplasmic HDAC6 by the HDAC6 inhibitor **34** (Figure 5, marbostat) led to MYC degradation and apoptosis in MYC-overexpressing B-cell lymphoma cells. It was shown that HDAC6 inhibition by compound **34** led to hyperacetylation of tubulin followed by enhanced binding of DNAJA3 to hyperacetylated tubulin in the cytoplasm of B-cell lymphoma cells, which led to enhanced Myc degradation [133].

In contrast to the Hsp40-inhibiting chalcone **29**, the trans-chalcone **35a** (50 μM) and its 5-fluoro-2-hydroxy derivative **35b** (10 μM) activated Hsp40 and p53 expression in U2OS osteosarcoma cells, accompanied by suppression of CRM1 (Figure 5) [134].

The structures of the described small-molecule Hsp40 modulators are shown in Figure 5. Their anticancer activities are summarized in Table 5.

Table 5. Hsp40 modulators and their anticancer effects.

Hsp40 Modulator	Cancer Model(s)	Effects
26a (MAL3-101)	Merkel cell carcinoma, muscle invasive bladder cancer, RMS13 rhabdomyosarcoma	Antiproliferative, apoptosis induction, in vivo MCC growth inhibition, synergy with **1** and STA9090, UPR induction [110–112]
28 (KNK437)	Squamous cell carcinoma (KB, SCC VII and SAS/mp53), immortalized Cos-1, MDA-MB-231 breast carcinoma, colon carcinoma (COLO 320DM, SW480, SW620, RKO, LOVO)	Suppression of Hsp70 and Hsp40, in vivo inhibition of SCC VII squamous cell carcinoma at 44 °C, apoptosis induction and colony formation inhibition at 42 °C, inhibition of heat-induced H3-Lys4 methylation, suppression of AKT and HIF-1α pathways, selective inhibition of DNAJA1, in vivo inhibition of DNAJA1-overexpressing SW480 and SW620 tumors, suppressed CDC45, upregulated ubiquitin, in vivo suppressed liver metastasis formation with 5-FU/L-OHP [113–120]
29 (C86/SU086)	HeLa cervix carcinoma, 22Rv1 and C4-2 prostate cancer	Antiproliferative, pan-Hsp40/DNAJ inhibition, apoptosis induction, ROS formation, inhibition of thioredoxin reductase, degradation of FL-AR and ARv7, Hsp90 inhibition, in vivo 22Rv1 and C4-2 prostate tumor growth inhibition [121–123]
30a (PLIHZ), **30b** (PLTFBH)	HN31 pharyngeal squamous cell carcinoma	Antiproliferative, suppression of DNAJA1, mutant p53, Cdc42 and Rac [124]
31 (Cabazitaxel)	LNCaP and PC-3 prostate cancer	Antiproliferative, suppression of Hsp40, HOP and AR [125]
32 (Tipifarnib)	SF763 and U87 glioblastoma	Farnesyltransferase inhibition, reduction of farnesylated HDJ-2, radio-sensitizing, antiproliferative, p21 induction, G2/M arrest [131]
33 (Atrovastatin)	Pancreatic carcinoma (PO3, SU 86.86, BXPC-3, Pan 10.05)	HMG-CoA reductase inhibition, suppression of DNAJA1 farnesylation, induction of apoptosis and p21, degradation of mutant p53 (blocked nuclear transport), inhibition of migration [132]
34 (Marbostat-100)	MYC-overexpressing B-cell lymphoma	HDAC6 inhibition, apoptosis induction, MYC-degradation, tubulin hyperacetylation, relocation of DNAJA3 to acetyltubulin [133]
35a, **35b**	U2OS osteosarcoma	Activation of Hsp40 and p53, suppression of CRM1 [134]

4. Discussion

The design and development of inhibitors of the Hsp70-Hsp40 axis is a prospering field of research. Numerous Hsp70 inhibitors were described, and the number of Hsp40 inhibitors and modulators is also growing. Hsp70 is essential for the survival of proliferating cancer cells, where it is often overexpressed, while it is more or less dispensable for non-transformed cells, making it an excellent anticancer drug target for tumor-selective drug candidates. Hsp70 is composed of two distinct domains, a 40 kDa N-terminal nucleotide-binding domain (NBD) that regulates client association and a 25 kDa C-terminal substrate-binding domain (SBD). Hsp70 is found in the cytosol and possesses multiple cellular functions. Together with Hsp90, it acts as a multi chaperone complex to modulate cellularly-protective heat stress responses. Organelle-specific Hsp70 isoforms such as Grp78 in the endoplasmic reticulum or mortalin in the mitochondria were also identified as promising anticancer drug targets for Hsp70 inhibitors [14,15]. Hsp40 is another member of the Hsp family and functions as a client protein of Hsp70. It is a small co-chaperone of Hsp70 that prevents the aggregation of unfolded polypeptides and proteins before transferring them to Hsp70 [17]. Hsp40 also binds to the nucleotide-binding domain of Hsp70, thereby increasing the ATPase activity of Hsp70. Moreover, Hsp70 is also involved in modulating receptor tyrosine kinase (RTK) signaling pathway activity, such as the Ras/Raf-MAPK or the AKT pathway. Synthetic Hsp70 inhibitors can interact with the N-terminal NBD or the C-terminal SBD of Hsp70. Some inhibitors also interfere with Hsp40 or act on organelle-specific Hsp70 isoforms such as Grp78 (ER) or mortalin (mitochondria), leading to pronounced anticancer effects such as apoptosis, cell cycle arrest, or cellular senescence (Figure 6) [14,15].

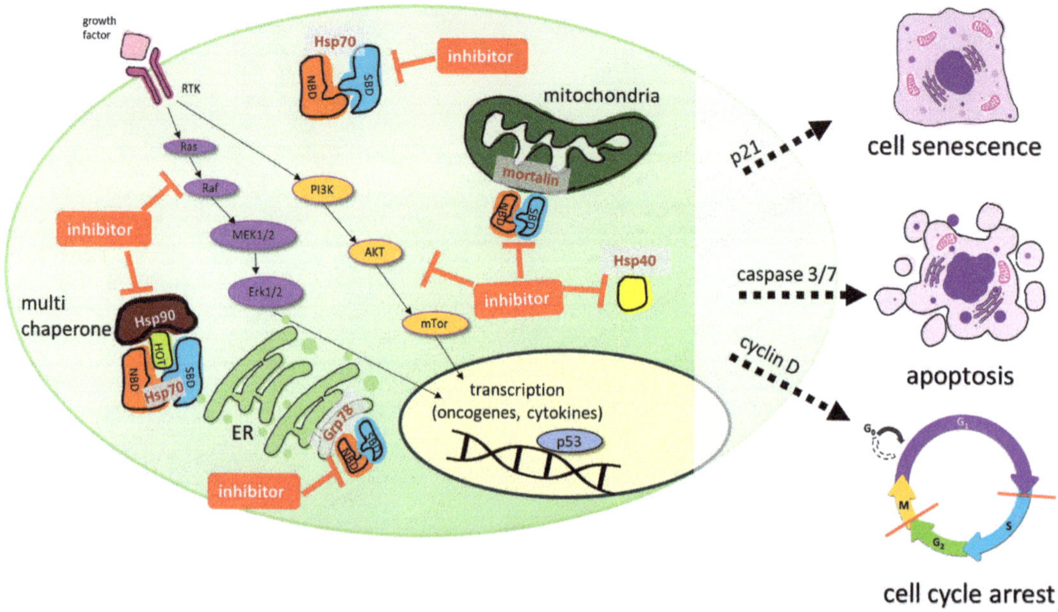

Figure 6. Cellular effects of Hsp70 isoforms and their inhibition by synthetic small molecules. Hsp70 is composed of two distinct domains, a 40 kDa N-terminal nucleotide-binding domain (NBD) that regulates client association and a 25 kDa C-terminal substrate-binding domain (SBD). Hsp70 is found in the cytosol with multicellular functions. Together with Hsp90 it acts as a multi-chaperone complex to modulate heat stress responses. Organelle-specific Hsp70 isoforms such as Grp78 in the endoplasmic reticulum or mortalin in mitochondria were identified as promising anticancer drug targets. Inhibition of these multicellular processes by (semi-)synthetic inhibitors (red lines) leads to cell senescence, apoptosis and cell cycle arrest in various cancers.

These effects are often synergistic in combination with well-established Hsp90 inhibitors, and have the potential to overcome the drawbacks of single Hsp90 inhibition in cancer therapy. A conjugate of 17-AAG with apoptozole (**2c**) designed as a dual Hsp70/Hsp90 inhibitor was superior to Hsp70 inhibitor **2b** in terms of anticancer activity [39,40]. Synthetic efforts led to promising compounds with intrinsic dual or multimodal activities, such as dual mortalin/PARP1 inhibition, which are pointing the way to more efficient anticancer drug candidates [99–101].

Synthetic conjugate strategies also led to subcellular isoform targeting by attachment of a mitochondria-specific triphenylphosphinium moiety to the apoptozole scaffold in the selective mortalin inhibitor **23** [40]. It is noteworthy that mitochondria-targeting cationic dyes such as rhodacyanines (**3a–f**) and methylene blue (**8**) were identified as Hsp70 inhibitors, which were initially developed for other applications than cancer therapy. Methylene blue was the first synthetic drug applied for the treatment of malaria and the lead compound for the development of tricyclic anti-depressant drugs [135]. Meanwhile, several rhodacyanine derivatives have also been investigated as antiprotozoal agents [136]. Thus, the repurposing of drugs appears promising and is not a one-way road. As far as drug repurposing is concerned, the disinfectant hexachlorophene (**20**) and the spasmolytic drug pinaverium bromide (**4**) are further interesting examples of Hsp70 inhibitory activity [50,82]. In contrast to the cationic mitochondria-targeting compounds with preference for mortalin binding, the anionic ruthenium complex **22** (KP1339) is an efficient Grp78 suppressor [85–88]. However, the exact anticancer mechanisms of action of such pleiotropic ruthenium complexes are only partially understood. In terms of Grp78 isoform selectivity, the modification of compounds **1** through **12** already showed a simple way to improve selectivity for Grp78 [70].

From a chemical point of view, the (semi-)synthetic Hsp70 and Hsp40 inhibitors described in this review are structurally heterogenic and comprise various compound classes. While natural Hsp70 and Hsp40 inhibitors were largely excluded from this review for clarity's sake, some promising chemically modified natural products (**11**, **30a**, **30b**) and synthetic surrogates (**26a**, **29**) were mentioned. In terms of Hsp40 protein targeting, structurally related synthetic compounds can have opposite effects on their target. While **26a** inhibited Hsp40, its analogs compounds, **26b** and **26c**, activated Hsp40 [107–112]. In addition, chalcone **29** inhibited Hsp40, while chalcones **35a** and **35b** activated Hsp40 [121–123,134]. Interestingly, in the Chalcone case, both effects led to anticancer activity. Posttranslational modifications of Hsp40 (farnesylation) and its binding partners (acetylated tubulin) were also described as targetable Hsp40-modulatory mechanisms [125,131–133]. In this way, potential combination partners such as farnesyltransferase inhibitors and HDAC6 inhibitors emerged as modulators of the Hsp70-Hsp40 axis, which might be considered for future in vivo experiments.

As mentioned above, the combination of Hsp70 and/or Hsp40 inhibitors with various approved anticancer drugs revealed promising antitumor effects both in vitro and in vivo, suggesting suitable therapy regimens for new clinical trials. The Hsp70 inhibitors **3c** and **22** (in its benzindazolium form, KP1019) already underwent early-stage clinical trials decades ago, even before their Hsp70 inhibitory activity was discovered [84,94]. Meanwhile, compound **22** experienced a revival and is currently in phase 1b/2 clinical trials (NCT04421820) in combination with FOLFOX (folinic acid, 5-fluorouracil, and oxaliplatin) for the treatment of advanced gastrointestinal tumors. However, with various new inhibitors and optimized analogs being available now, more valuable outcomes can be expected for Hsp70 and Hsp40 inhibitors from future clinical studies when designed and conducted properly.

Hsp40 is another member of the Hsp protein family and functions as a client protein of Hsp70. It is a small co-chaperone of Hsp70 that prevents the aggregation of unfolded polypeptides and proteins before transferring them to Hsp70. Hsp40 also binds to the nucleotide-binding domain (NBD) of Hsp70, thereby increasing the ATPase activity of Hsp70. Moreover, Hsp70 is also involved in modulating receptor tyrosine kinase (RTK) signaling pathway activity, such as the Ras/Raf-MAPK or the AKT pathway.

Synthetic Hsp70 inhibitors can act at the N-terminal (NBD) or substrate-binding C-terminal (SBD) binding domains of Hsp70. Some inhibitors also interfere with Hsp40 or act on organelle specific Hsp isoforms such as Grp78 (ER) or mortalin (mitochondria), leading to pronounced anticancer effects such as apoptosis, cell cycle arrest, or cellular senescence.

Author Contributions: Writing—original draft preparation, B.B.; writing—review and editing, B.N. and M.H. All authors have read and agreed to the published version of the manuscript.

Funding: This research received no external funding.

Institutional Review Board Statement: Not applicable.

Informed Consent Statement: Not applicable.

Data Availability Statement: Not applicable.

Conflicts of Interest: The authors declare no conflict of interest.

References

1. Santoro, M.G. Heat shock factors and the control of stress response. *Biochem. Pharmacol.* **2000**, *59*, 55–63. [CrossRef] [PubMed]
2. Ritossa, F. A new puffing pattern induced by temperature shock and DNP in drosophila. *Experientia* **1962**, *18*, 571–573. [CrossRef]
3. Tissiéres, A.; Mitchell, H.K.; Tracy, U.M. Protein synthesis in salivary glands of *Drosophila melanogaster*: Relation to chromosome puffs. *J. Mol. Biol.* **1974**, *84*, 389–398. [CrossRef] [PubMed]

4. Hasday, J.D.; Singh, I.S. Fever and the heat shock response: Distinct, partially overlapping processes. *Cell Stress Chaperones* **2000**, *5*, 471–480. [CrossRef] [PubMed]
5. Rodina, A.; Wang, T.; Yan, P.; Gomes, E.D.; Dunphy, M.P.S.; Pillarsetty, N.; Koren, J.; Gerecitano, J.F.; Taldone, T.; Zong, H.; et al. The epichaperome is an integrated chaperome network that facilitates tumour survival. *Nature* **2016**, *538*, 397–401. [CrossRef]
6. Macario, A.J.L.; Conway de Macario, E. Chaperone Proteins and Chaperonopathies. In *Stress: Physiology, Biochemistry, and Pathology Handbook of Stress Series*; Fink, G., Ed.; Academic Press: Cambridge, MA, USA, 2019; Volume 3, pp. 135–152.
7. Kampinga, H.H.; Hageman, J.; Vos, V.J.; Kubota, H.; Tanguay, R.M.; Bruford, E.A.; Cheetham, M.E.; Chen, B.; Hightower, L.E. Guidelines for the nomenclature of the human heat shock proteins. *Cell Stress Chaperones* **2009**, *14*, 105–111. [CrossRef] [PubMed]
8. Li, L.; Wang, L.; You, Q.-D.; Xu, X.-L. Heat shock protein 90 inhibitors: An update on achievements, challenges, and future directions. *J. Med. Chem.* **2020**, *63*, 1798–1822. [CrossRef]
9. Birbo, B.; Madu, E.E.; Madu, C.O.; Jain, A.; Lu, Y. Role of Hsp90 in cancer. *Int. J. Mol. Sci.* **2021**, *22*, 10317. [CrossRef]
10. Rosenzweig, R.; Nillegoda, N.B.; Mayer, M.P.; Bukau, B. The Hsp70 chaperone network. *Nat. Rev. Mol. Cell Biol.* **2019**, *20*, 665–680. [CrossRef]
11. Albakova, Z.; Armeev, G.A.; Kanevsky, L.M.; Kovalenko, E.I.; Sapozhnikov, A.M. Hsp70 multi-functionality in cancer. *Cells* **2020**, *9*, 587. [CrossRef]
12. Fu, X.; Liu, J.; DiSanto, M.E.; Zhang, X. Heat shock protein 70 and 90 family in prostate cancer. *Life* **2022**, *12*, 1489. [CrossRef] [PubMed]
13. Yang, S.; Xiao, H.; Cao, L. Recent advance in heat shock proteins in cancer diagnosis, prognosis, metabolism and treatment. *Biomed. Pharmacother.* **2021**, *142*, 112074. [CrossRef] [PubMed]
14. Elfiky, A.A.; Baghdady, A.M.; Ali, S.A.; Ahmed, M.I. GRP78 targeting: Hitting two birds with a stone. *Life Sci.* **2020**, *260*, 118317. [CrossRef] [PubMed]
15. Elwakeel, A. Abrogating the interaction between p53 and mortalin (Grp75/HSPA9/mtHsp70) for cancer therapy: The story so far. *Front. Cell Dev. Biol.* **2022**, *10*, 879632. [CrossRef] [PubMed]
16. Albakova, Z.; Siam, M.K.S.; Sacitharan, P.K.; Ziganshin, R.H.; Ryazantsev, D.Y.; Sapozhnikov, A.M. Extracellular heat shock proteins and cancer: New perspectives. *Transl. Oncol.* **2021**, *14*, 100995. [CrossRef]
17. Li, J.; Qian, X.; Sha, B. Heat shock protein 40: Structural studies and their functional implications. *Protein Pept. Lett.* **2009**, *16*, 606–612. [CrossRef]
18. Liu, Q.; Liang, C.; Zhou, L. Structural and functional analysis of the Hsp70/Hsp40 chaperone system. *Protein Sci.* **2020**, *29*, 378–390. [CrossRef]
19. Williamson, D.S.; Borgognoni, J.; Clay, A.; Daniels, Z.; Dokurno, P.; Drysdale, M.J.; Foloppe, N.; Francis, G.L.; Graham, C.J.; Howes, R.; et al. Novel adenosine-derived inhibitors of 70 kDa heat shock protein, discovered through structure-based design. *J. Med. Chem.* **2009**, *52*, 1510–1513. [CrossRef]
20. Massey, A.J.; Williamson, D.S.; Browne, H.; Murray, J.B.; Dokurno, P.; Shaw, T.; Macias, A.T.; Daniels, Z.; GEoffroy, S.; Dopson, M.; et al. A novel, small molecule inhibitor of Hsc70/Hsp70 potentiates Hsp90 inhibitor induced apoptosis. *Cancer Chemother. Pharmacol.* **2010**, *66*, 535–545. [CrossRef] [PubMed]
21. Wen, W.; Liu, W.; Shao, Y.; Chen, L. VER-155008, a small molecule inhibitor of HSP70 with potent antio-cancer activity on lung cancer cell lines. *Exp. Biol. Med.* **2014**, *239*, 638–645. [CrossRef]
22. Sakai, K.; Inoue, M.; Mikami, S.; Nishimura, H.; Kuwabara, Y.; Kojima, A.; Toda, M.; Ogawa-Kobayashi, Y.; Kikuchi, S.; Hirata, Y.; et al. Functional inhibition of heat shock protein 70 by VER-155008 suppresses pleural mesothelioma cell proliferation via an autophagy mechanism. *Thoracic Cancer* **2021**, *12*, 491–503. [CrossRef] [PubMed]
23. Kita, K.; Shiota, M.; Tanaka, M.; Otsuka, A.; Matsumoto, M.; Kato, M.; Tamada, S.; Iwao, H.; Miura, K.; Nakatani, T.; et al. Heat shock protein 70 inhibitors suppress androgen receptor expression in LNCAP95 prostate cancer cells. *Cancer Sci.* **2017**, *108*, 1820–1827. [CrossRef]
24. Brünnert, D.; Langer, C.; Zimmermann, L.; Bargou, R.C.; Burchardt, M.; Chatterjee, M.; Stope, M.B. The heat shock protein 70 inhibitor VER155008 suppresses the expression of HSP27, HOP and HSP90β and the androgen receptor, induces apoptosis, and attenuates prostate cancer cell growth. *J. Cell. Biochem.* **2020**, *121*, 407–417. [CrossRef] [PubMed]
25. Yu, B.; Yang, H.; Zhang, X.; Li, H. Visualizing and quantifying the effect of the inhibition of HSP70 on breast cancer cells based on laser scanning microscopy. *Technol. Cancer Res. Treat.* **2018**, *17*, 1–7. [CrossRef] [PubMed]
26. Kim, S.H.; Kang, J.G.; Kim, C.S.; Ihm, S.-H.; Choi, M.G.; Yoo, H.J.; Lee, S.J. The hsp70 inhibitor VER155008 induces paraptosis requiring de novo protein synthesis in anaplastic thyroid carcinoma cells. *Biochem. Biophys. Res. Commun.* **2014**, *454*, 36–41. [CrossRef] [PubMed]
27. Shervington, L.; Patil, H.; Shervington, A. Could the anti-chaperone VER155008 replace temozolomide for glioma treatment. *J. Cancer* **2015**, *6*, 786–794. [CrossRef]
28. Cavanaugh, A.; Juengst, B.; Sheridan, K.; Danella, J.F.; Williams, H. Combined inhibition of heat shock proteins 90 and 70 leads to simultaneous degradation of the oncogenic signaling proteins involved in muscle invasive bladder cancer. *Oncotarget* **2015**, *6*, 39821–39838. [CrossRef]
29. Kim, S.H.; Kang, J.G.; Kim, C.S.; Ihm, S.-H.; Choi, M.G.; Yoo, H.J.; Lee, S.J. Hsp70 inhibition potentiates radicicol-induced cell death in anaplastic thyroid carcinoma cells. *Anticancer Res.* **2014**, *34*, 4829–4838.

30. Asling, J.; Morrison, J.; Mutsaers, A.J. Targeting HSP70 and GRP78 in canine osteosarcoma cells in combination with doxorubicin chemotherapy. *Cell Stress Chaperones* **2016**, *21*, 1065–1076. [CrossRef]
31. Sojka, D.R.; Hastorek, S.; Vydra, N.; Toma-Jonik, A.; Wieczorek, A.; Gogler-Piglowska, A.; Scieglinska, D. Inhibition of the heat shock protein A (HSPA) family potentiates the anticancer effects of manumycin A. *Cells* **2021**, *10*, 1418. [CrossRef]
32. Tang, X.; Tan, L.; Shi, K.; Peng, J.; Xiao, Y.; Li, W.; Chen, L.; Yang, Q.; Qian, Z. Gold nanorods together with HSP inhibitor-VER-155008 micelles for colon cancer mild-temperature photothermal therapy. *Acta Pharmaceut. Sin.* **2018**, *8*, 587–601. [CrossRef] [PubMed]
33. Huang, L.; Wang, Y.; Bai, J.; Yang, Y.; Wang, F.; Feng, Y.; Zhang, R.; Li, F.; Zhang, P.; Lv, N.; et al. Blockade of HSP70 by VER-155008 synergistically enhances bortezomib-induced cytotoxicity in multiple myeloma. *Cell Stress Chaperones* **2020**, *25*, 357–367. [CrossRef] [PubMed]
34. Wada, N.; Kawano, Y.; Fujiwara, S.; Kikukawa, Y.; Okuno, Y.; Tasaki, M.; Ueda, M.; Ando, Y.; Yoshinaga, K.; Ri, M.; et al. Shikonin, dually functions as a proteasome inhibitor and a necroptosis inducer in multiple myeloma cells. *Int. J. Oncol.* **2015**, *46*, 963–972. [CrossRef] [PubMed]
35. Reikyam, H.; Nepstad, I.; Sulen, A.; Gjertsen, B.T.; Hatfield, K.J.; Bruserud, O. Increased antileukemic effects in human acute myeloid leukemia by combining HSP70 and HSP90 inhibitors. *Exp. Opin. Investig. Drugs* **2013**, *22*, 551–563. [CrossRef]
36. Kul, P.; Tuncbilek, M.; Ergul, M.; Tunoglu, E.N.Y.; Tutar, Y. A novel 6,8,9.trisubstituted purine analogue drives breast cancer luminal A subtype MCF-7 to apoptosis and senescence through Hsp70 inhibition. *Anticancer Agents Med. Chem.* **2023**, *23*, 585–598. [CrossRef]
37. Williams, D.R.; Ko, S.-K.; Park, S.; Lee, M.-R.; Shin, I. An apoptosis-inducing small molecule that binds to heat shock protein 70. *Angew. Chem. Int. Ed.* **2008**, *47*, 7466–7469. [CrossRef]
38. Ko, S.-K.; Kim, J.; Na, D.C.; Park, S.; Park, S.-H.; Hyun, J.Y.; Baek, K.-H.; Kim, N.D.; Kim, N.-K.; Park, Y.N.; et al. A small molecule inhibitor of ATPase activity of HSP70 induces apoptosis and has antitumor activities. *Chem. Biol.* **2015**, *22*, 391–403. [CrossRef]
39. Park, S.-H.; Kim, W.-J.; Li, H.; Seo, W.; Park, S.-H.; Kim, H.; Shin, S.C.; Zuiderweg, E.R.P.; Kim, E.E.; Sim, T.; et al. Anti-leukemia activity of a Hsp70 inhibitor and its hybrid molecules. *Sci. Rep.* **2017**, *7*, 3537. [CrossRef]
40. Park, S.-H.; Baek, K.-H.; Shin, I.; Shin, I. Subcellular Hsp70 inhibitors promote cancer cell death via different mechanisms. *Cell Chem. Biol.* **2018**, *25*, 1242–1254. [CrossRef]
41. Colvin, T.A.; Gabai, V.L.; Calderwood, S.K.; Li, H.; Gummuluru, S.; Matchuk, O.N.; Smirnova, S.G.; Orlova, N.V.; Zamulaeva, I.A.; Garcia-Marcos, M.; et al. Hsp70-Bag3 interactions regulate cancer-related signaling networks. *Cancer Res.* **2014**, *74*, 4731–4740. [CrossRef]
42. Antonietti, P.; Linder, B.; Hehlgans, S.; Mildenberger, I.C.; Burger, M.C.; Fulda, S.; Steinbach, J.P.; Gessler, F.; Rödel, F.; Mittelbronn, M.; et al. Interference with the HSF1/HSP70/BAG3 pathway primes glioma cells to matrix detachment and BH3 mimetic-induced apoptosis. *Mol. Cancer Ther.* **2016**, *16*, 156–168. [CrossRef] [PubMed]
43. Das, C.K.; Linder, B.; Bonn, F.; Rothweiler, F.; Dikic, I.; Michaelis, M.; Cinatl, J.; Mandal, M.; Kögel, D. BAG3 overexpression and cytoprotective autophagy mediate apoptosis resistance in chemoresistant breast cancer cells. *Neoplasia* **2018**, *20*, 263–279. [CrossRef] [PubMed]
44. Li, X.; Srinivasan, S.R.; Connarn, J.; Ahmad, A.; Young, Z.T.; Kabza, A.M.; Zuiderweg, E.R.P.; Sun, D.; Gestwicki, J.E. Analogues of the allosteric heat shock protein 70 (Hsp70) inhibitor, MKT-077, as anti-cancer agents. *ACS Med. Chem. Lett.* **2013**, *4*, 1042–1047. [CrossRef] [PubMed]
45. Li, X.; Colvin, T.; Rauch, J.N.; Acosta-Alvear, D.; Kampmann, M.; Dunyak, B.; Hann, B.; Aftab, B.T.; Murnane, M.; Cho, M.; et al. Validation of the Hsp70-Bag3 protein-protein interaction as a potential therapeutic target in cancer. *Mol. Cancer Ther.* **2015**, *14*, 642–648. [CrossRef]
46. Yaglom, J.A.; Wang, Y.; Li, A.; Li, Z.; Monti, S.; Alexandrov, I.; Lu, X.; Sherman, M.Y. Cancer cell responses to Hsp70 inhibitor JG-98: Comparison with Hsp90 inhibitors and finding synergistic drug combinations. *Sci. Rep.* **2018**, *8*, 3010. [CrossRef]
47. Gabai, V.L.; Yaglom, J.A.; Wang, Y.; Meng, L.; Shao, H.; Kim, G.; Colvin, T.; Gestwicki, J.; Sherman, M.Y. Anticancer effects of targeting Hsp70 in tumor stromal cells. *Cancer Res.* **2016**, *76*, 5926–5932. [CrossRef]
48. Chang, C.-S.; Kumar, V.; Lee, D.-Y.; Chen, Y.; Wu, Y.-C.; Gao, J.-Y.; Chu, P.-C. Development of novel rhodacyanine-based heat shock 70 inhibitors. *Curr. Med. Chem.* **2021**, *28*, 5431–5446. [CrossRef]
49. Shao, H.; Li, X.; Moses, M.A.; Gilbert, L.A.; Kalyanaraman, C.; Young, Z.T.; Chernova, M.; Journey, S.N.; Weissman, J.S.; Hann, B.; et al. Exploration of benzothiazole rhodacyanines as allosteric inhibitors of protein-protein interactions with heat shock protein 70 (Hsp70). *J. Med. Chem.* **2018**, *61*, 6163–6177. [CrossRef]
50. Dublang, L.; Underhaug, J.; Flydal, M.I.; Velasco-Carneros, L.; Maréchal, J.-D.; Moro, F.; Boyano, M.D.; Martinez, A.; Muga, A. Inhibition of the humN Hac/0 system by small ligands as a potential anticancer approach. *Cancers* **2021**, *13*, 2936. [CrossRef]
51. Song, T.; Guo, Y.; Xue, Z.; Guo, Z.; Wang, Z.; Lin, D.; Zhang, H.; Pan, H.; Zhang, X.; Yin, F.; et al. Small-molecule inhibitor targeting the Hsp70-Bim protein-protein interaction in CML cells overcomes BCR-ABL-independent TKI resistance. *Leukemia* **2021**, *35*, 2862–2874. [CrossRef]
52. Wang, Z.; Song, T.; Guo, Z.; Uwituze, L.B.; Guo, Y.; Zhang, H.; Wang, H.; Zhang, X.; Pan, H.; Ji, T.; et al. A novel Hsp70 inhibitor specifically targeting the cancer-related Hsp70-Bim protein-protein interaction. *Eur. J. Med. Chem.* **2021**, *220*, 113452. [CrossRef]

53. Howe, M.K.; Bodoor, K.; Carlson, D.A.; Hughes, P.F.; Alwarawrah, Y.; Loiselle, D.R.; Jaeger, A.M.; Darr, D.B.; Jordan, J.L.; Hunter, L.M.; et al. Identification of an allosteric small-molecule inhibitor selective for the inducible form of heat shock protein 70. *Chem. Biol.* **2014**, *21*, 1648–1659. [CrossRef]
54. Kang, Y.; Taldone, T.; Patel, H.J.; Patel, P.D.; Rodina, A.; Gozman, A.; Maharaj, R.; Clement, C.C.; Patel, M.R.; Brodsky, J.L.; et al. Heat shock protein 70 inhibitors. 1. 2,5′-Thiopyrimidine and 5-(phenylthio)pyrimidine acrylamides as irreversible binders to an allosteric site on heat shock protein 70. *J. Med. Chem.* **2014**, *57*, 1188–1207. [CrossRef]
55. Jinwal, U.K.; Miyata, Y.; Koren III, J.; Jones, J.R.; Trotter, J.H.; Chang, L.; O'Leary, J.; Morgan, D.; Lee, D.C.; Shults, C.L.; et al. Chemical manipulation of Hsp70 ATPase activity regulates Tau stability. *J. Neurosci.* **2009**, *29*, 12079–12088. [CrossRef]
56. Wang, A.M.; Morishima, Y.; Clapp, K.M.; Peng, H.-M.; Pratt, W.B.; Gestwicki, J.E.; Osawa, Y.; Lieberman, A.P. Inhibition of Hsp70 by methylene blue affects signaling protein function and ubiquitination and modulates polyglutamine protein degradation. *J. Biol. Chem.* **2010**, *285*, 15714–15723. [CrossRef] [PubMed]
57. Davis, A.L.; Cabello, M.C.; Qiao, S.; Azimian, S.; Wondrak, G.T. Phenotypic identification of the redox dye methylene blue as an antagonist of heat shock response gene expression in metastatic melanoma cells. *Int. J. Mol. Sci.* **2013**, *14*, 4185–4202. [CrossRef] [PubMed]
58. Sanchala, D.; Bhatt, L.K.; Pethe, P.; Shelat, R.; Kulkarni, Y.A. Anticancer activity of methylene blue via inhibition of heat shock protein 70. *Biomed. Pharmacother.* **2018**, *107*, 1037–1045. [CrossRef] [PubMed]
59. Leu, J.I.; Pimkina, J.; Frank, A.; Murphy, M.E.; George, D.L. A small molecule inhibitor of inducible heat shock protein 70. *Mol. Cell* **2009**, *36*, 15–27. [CrossRef]
60. Kaiser, M.; Kühnl, A.; Reins, J.; Fischer, S.; Ortiz-Tanchez, J.; Schlee, C.; Mochmann, L.H.; Heesch, S.; Benlasfer, O.; Hofmann, W.-K.; et al. Antileukemic activity of the HSP70 inhibitor pifithrin-μ in acute leukemia. *Blood Cancer J.* **2011**, *1*, e28. [CrossRef]
61. Granato, M.; Lacconi, V.; Peddis, M.; Lotti, L.V.; Di Renzo, L.; Gonnella, R.; Santarelli, R.; Trivedi, P.; Frati, L.; D'Orazi, G.; et al. HSP70 inhibition by 2-phenylethynesulfonamide induces lysosomal cathepsin D release and immunogenic cell death in primary effusion lymphoma. *Cell Death Dis.* **2013**, *4*, e730. [CrossRef]
62. McKeon, A.M.; Egan, A.; Chandanshive, J.; McMahon, H.; Griffith, D.M. Novel improved synthesis of HSP70 inhibitor, Pifithrin-μ. In vitro synergy quantification of Pifithrin-μ combined with Pt drugs in prostate and colorectal cancer cells. *Molecules* **2016**, *21*, 949. [CrossRef] [PubMed]
63. Sekihara, K.; Harashima, N.; Tongu, M.; Tamaki, Y.; Uchida, N.; Inomata, T.; Harada, M. Pifithrin-μ, and inhibitor of heat-shock protein 70, can increase the antitumor effects of hyperthermia against human prostate cancer cells. *PLoS ONE* **2013**, *8*, e78772. [CrossRef] [PubMed]
64. Balaburski, G.M.; Leu, J.I.; Beeharry, N.; Hayik, S.; Andrake, M.D.; Zhang, G.; Herlyn, M.; Villanueva, J.; Dunbrack, R.L., Jr.; Yen, T.; et al. A modified HSP70 inhibitor shows broad activity as an anticancer agent. *Mol. Cancer Ther.* **2013**, *11*, 219–229. [CrossRef] [PubMed]
65. Budina-Kolomets, A.; Balaburski, G.M.; Bondar, A.; Beeharry, N.; Yen, T.; Murphy, M.E. Comparison of the activity of three different HSP70 inhibitors on apoptosis, cell cycle arrest, autophagy inhibition, and HSP90 inhibition. *Cancer Biol. Ther.* **2014**, *15*, 194–199. [CrossRef]
66. Coskun, K.A.; Koca, I.; Gümüs, M.; Tutar, Y. Designing specific HSP70 substrate binding domain inhibitor for perturbing protein folding pathways to inhibit cancer mechanism. *Anticancer Agents Med. Chem.* **2021**, *21*, 1472–1480. [CrossRef]
67. Kocam, I.; Gümüs, M.; Özgür, A.; Disli, A.; Tutar, Y. A noverl approach to inhibit heat shock response as anticancer strategy by coumarine compounds containing thiazole skeleton. *Anticancer Agents Med. Chem.* **2015**, *15*, 916–930. [CrossRef]
68. Lazarev, V.F.; Sverchinsky, D.V.; Mikhaylova, E.R.; Semenyuk, P.I.; Komarova, E.Y.; Niskanen, S.A.; Nikotina, A.D.; Burakov, A.V.; Kartsev, V.G.; Guzhova, I.V.; et al. Sensitizing tumor cells to conventional drugs: HSP70 chaperone inhibitors, their selection and application in cancer models. *Cell Death Dis.* **2018**, *9*, 41. [CrossRef]
69. Pujari, R.; Jose, J.; Bhavnani, V.; Kumar, N.; Shastry, P.; Pal, J.K. Tamoxifen-induced cytotoxicity in breast cancer cells is mediated by glucose-regulated protein 78 (GRP78) via AKT (Thr308) regulation. *Int. J. Biochem. Cell Biol.* **2016**, *77*, 57–67. [CrossRef]
70. Macias, A.T.; Williamson, D.S.; Allen, N.; Borgognoni, J.; Clay, A.; Daniels, Z.; Dokurno, P.; Drysdale, M.J.; Francis, G.L.; Graham, C.J.; et al. Adenosine-derived inhibitors of 78 kDa glucose regulated protein (Grp78) ATPase: Insights into isoform selectivity. *J. Med. Chem.* **2011**, *54*, 4034–4041. [CrossRef]
71. Cerezo, M.; Lehraki, A.; Millet, A.; Rouaud, F.; Plaisant, M.; Jaune, E.; Botton, T.; Ronco, C.; Abbe, P.; Amdouni, H.; et al. Compounds triggering ER stress exert anti-melanoma effects and overcome BRAF inhibitor resistance. *Cancer Cell* **2016**, *29*, 805–819. [CrossRef]
72. Wu, J.; Wu, Y.; Lian, X. Targeted inhibition of GRP78 by HA15 promotes apoptosis of lung cancer cells accompanied by ER stress and autophagy. *Biol. Open* **2020**, *9*, bio053298. [CrossRef]
73. Ha, D.P.; Huang, B.; Wang, H.; Rangel, D.F.; Van Krieken, R.; Liu, Z.; Samanta, S.; Neamati, N.; Lee, A.S. Targeting GRP78 suppresses oncogenic KRAS protein expression andreduced viability of cancer cells bearing various *KRAS* mutations. *Neoplasia* **2022**, *33*, 100837. [CrossRef] [PubMed]
74. Ruggiero, C.; Doghman-Bouguerra, M.; Ronco, C.; Benhida, R.; Rocchi, S.; Lalli, E. The GRP78/BiP inhibitor HA15 synergizes with mitotane action against adrenocortical carcinoma cells through convergent activation of ER stress pathways. *Mol. Cell. Endocrinol.* **2018**, *474*, 57–64. [CrossRef] [PubMed]

75. Mattos, D.R.; Weinman, M.A.; Wan, X.; Goodall, C.P.; Serrill, J.D.; McPhail, K.L.; Milovancev, M.; Bracha, S.; Ishmael, J.E. Canine osteosarcoma cells exhibit basal accumulation of multiple chaperone proteins and are sensitive to small molecule inhibitors of GRP78 and heat shock protein function. *Cell Stress Chaperones* **2022**, *27*, 223–239. [CrossRef] [PubMed]
76. Bhattacharjee, R.; Devi, A.; Mishra, S. Molecular docking and molecular dynamics studies reveal structural basis pof inhibition and selectivity of inhibitors ECGC and OSU-03012 towards glucose regulated protein-78 (GRP78) overexpressed in glioblastoma. *J. Mol. Model.* **2015**, *21*, 272. [CrossRef] [PubMed]
77. Booth, L.; Shuch, B.; Albers, T.; Roberts, J.L.; Tavallai, M.; Proniuk, S.; Zukiwski, A.; Wang, D.; Chen, C.-S.; Bottaro, D.; et al. Multi-kinase inhibitors can associate with heat shock proteins through their NH_2-termini by which they suppress chaperone function. *Oncotarget* **2016**, *7*, 12975–12996. [CrossRef] [PubMed]
78. Samanta, S.; Yang, S.; Debnath, B.; Xue, D.; Kuang, Y.; Ramkumar, K.; Lee, A.S.; Ljungman, M.; Neamati, N. The hydroxyquinoline analogue YUM70 inhibits GRP78 to induce ER stress-mediated apoptosis in pancreatic cancer. *Cancer Res.* **2021**, *81*, 1883–1895. [CrossRef]
79. Huang, M.; Li, Z.; Li, D.; Walker, S.; Greenan, C.; Kennedy, R. Structure-based design of HSPA5 inhibitors: From peptide to small molecule inhibitors. *Bioorg. Med. Chem. Lett.* **2013**, *23*, 3044–3050. [CrossRef]
80. Nayak, D.; Katoch, A.; Sharma, D.; Faheem, M.M.; Chakraborty, S.; Sahu, P.K.; Chikan, N.A.; Amin, H.; Gupta, A.P.; Gandhi, S.G.; et al. Indolylkojyl methane analogue IKM5 potentially inhibits invasion of breast cancer cells via attenuation of GRP78. *Breast Cancer Res. Treat.* **2019**, *177*, 307–323. [CrossRef]
81. Qiao, Y.; Dsouza, C.; Matthews, A.A.; Jin, Y.; He, W.; Bao, J.; Jiang, F.; Chandna, R.; Ge, R.; Fu, L. Discovery of small molecules targeting GRP78 for antiangiogenic and anticancer therapy. *Eur. J. Med. Chem.* **2020**, *193*, 112228. [CrossRef]
82. Ambrose, A.J.; Zerio, C.J.; Sivinski, J.; Schmidlin, C.J.; Shi, T.; Ross, A.B.; Widrick, K.J.; Johnson, S.M.; Zhang, D.D.; Chapman, E. A high throughput substrate binding assay reveals hexachlorophene as an inhibitor of the ER-resident HSP70 chaperone GRP78. *Bioorg. Med. Chem. Lett.* **2019**, *29*, 1689–1693. [CrossRef] [PubMed]
83. Viswanath, A.N.I.; Lim, J.W.; Seo, S.H.; Lee, J.Y.; Lim, S.M.; Pae, A.N. GRP78-targeted in-silico virtual screening of novel anticancer agents. *Chem. Biol. Drug. Des.* **2018**, *92*, 1555–1566. [CrossRef] [PubMed]
84. Hartinger, C.G.; Jakupec, M.A.; Zorbas-Seifried, S.; Groessl, M.; Egger, A.; Berger, W.; Zorbas, H.; Dyson, P.J.; Keppler, B.K. KP1019, a new redox-active anticancer agent—Preclinical development and results of a clinical phase I study in tumor patients. *Chem. Biodivers.* **2008**, *5*, 2140–2155. [CrossRef] [PubMed]
85. Schoenhacker-Alte, B.; Mohr, T.; Pirker, C.; Kryezu, K.; Kuhn, P.-S.; Buck, A.; Hofmann, T.; Gerner, C.; Hermann, G.; Koellensperger, G.; et al. Sensitivity towards the Grp78 inhibitor KP1339/IT-139 is characterized by apoptosis induction via caspase 8 upon disruption or ER homeostasis. *Cancer Lett.* **2017**, *404*, 79–88. [CrossRef]
86. Neuditschko, B.; Legin, A.A.; Baier, D.; Schintlmeister, A.; Reipert, S.; Wagner, M.; Keppler, B.K.; Berger, W.; Meier-Menches, S.M.; Gerner, C. Interaction with ribosomal proteins accompanies stress induction of the anticancer metallodrug BOLD-100/KP1339 in the endoplasmic reticulum. *Angew. Chem. Int. Ed.* **2021**, *60*, 5063–5068. [CrossRef]
87. Ranzato, E.; Bonsignore, G.; Martinotti, S. ER stress response and induction of apoptosis in malignant pleural mesothelioma: The Achilles heel targeted by the anticancer ruthenium drug BOLD-100. *Cancers* **2022**, *14*, 4126. [CrossRef]
88. Baier, D.; Schoenhacker-Alte, B.; Rusz, M.; Pirker, C.; Mohr, T.; Mendrina, T.; Kirchhofer, D.; Meier-Menches, S.M.; Hohenwallner, K.; Schaier, M.; et al. The anticancer ruthenium compound BOLD-100 targets glycolysis and generates a metabolic vulnerability towards glucose deprivation. *Pharmaceutics* **2022**, *14*, 238. [CrossRef]
89. Yoon, A.-R.; Wadhwa, R.; Kaul, S.C.; Yun, C.-O. Why is mortalin a potential therapeutic target for cancer? *Front. Cell Dev. Biol.* **2022**, *10*, 914540. [CrossRef]
90. Ferguson, I.D.; Tin, Y.-H.T.; Lam, C.; Shao, H.; Tharp, K.M.; Hale, M.; Kasap, C.; Mariano, M.C.; Kishishita, A.; Escobar, B.P.; et al. Allosteric HSP70 inhibitors perturb mitochondrial proteostasis and overcome proteasome inhibitor resistance in multiple myeloma. *Cell Chem. Biol.* **2022**, *29*, 1288–1302. [CrossRef]
91. Kaul, S.C.; Reddel, R.R.; Mitsui, Y.; Wadhwa, R. An N-terminal region of Mot-2 binds to P53 in vitro. *Neoplasia* **2001**, *3*, 110–114. [CrossRef]
92. Wadhwa, R.; Sugihara, T.; Yoshida, A.; Nomura, H.; Reddel, R.R.; Simpson, R.; Maruta, H.; Kaul, S.C. Selective toxicity of MKT-077 to cancer cells is mediated by its binding to the Hsp70 family protein Mot-2 and reactivation of p53 function. *Cancer Res.* **2000**, *60*, 6818–6821.
93. Pilzer, D.; Saar, M.; Koya, K.; Fishelson, Z. Mortalin inhibitors sensitize K562 leukemia cells to complement-dependent cytotoxicity. *Int. J. Cancer* **2010**, *126*, 1428–1435. [CrossRef]
94. Koya, K.; Li, Y.; Wang, H.; Ukai, T.; Tatsuda, N.; Kawakami, M.; Shishido, T.; Chen, L.B. MKT-077, a novel rhodacyanine dye in clinical trials, exhibits anticarcinoma activity in preclinical studies based on selective mitochondrial accumulation. *Cancer Res.* **1996**, *56*, 538–543.
95. Modica-Napolitano, J.S.; Koya, K.; Weisberg, E.; Brunelli, B.T.; Li, Y.; Chen, L.B. Selective damage to carcinoma mitochondria by the rhodacyanine MKT-007. *Cancer Res.* **1996**, *56*, 544–550.
96. Chiba, Y.; Kubota, T.; Watanabe, M.; Matsuzaki, S.W.; Otani, Y.; Teramoto, T.; Matsumoto, Y.; Koya, K.; Kitajima, M. MKT-077, localized lipophilic cation: Antitumor activity against human tumor xenografts serially transplanted into nude mice. *Anticancer Res.* **1998**, *18*, 1047–1052. [PubMed]

97. Tikoo, A.; Shakri, R.; Connolly, L.; Hirokawa, Y.; Shishido, T.; Bowers, B.; Ye, L.H.; Kohama, K.; Simpson, R.J.; Maruta, H. Treatment of ras-induced cancers, by the F-actin-bundling drug MKT-077. *Cancer J.* **2000**, *6*, 162–168. [PubMed]
98. Propper, D.J.; Braybrooke, J.P.; Taylor, D.J.; Lodi, R.; Styles, P.; Cramer, J.A.; Collins, W.C.; Levitt, N.C.; Talbot, D.C.; Ganesan, T.S.; et al. Phase I trial of the selective mitochondrial toxin MKT077 in chemo-resistant solid tumors. *Ann. Oncol.* **1999**, *10*, 923–927. [CrossRef] [PubMed]
99. Putri, J.F.; Bhargava, P.; Dhanjal, J.K.; Yaguchi, T.; Sundar, D.; Kaul, S.C.; Wadhwa, R. Mortaparib, a novel dual inhibitor of mortalin and PARP1, is a potential drug candidate for ovarian and cervical cancers. *J. Exp. Clin. Cancer Res.* **2019**, *28*, 499. [CrossRef] [PubMed]
100. Sari, A.N.; Elwakeel, A.; Dhanjal, J.K.; Kumar, V.; Sundar, D.; Kaul, S.C.; Wadhwa, R. Identification and characterization of mortaparib^{Plus}—A novel triazole derivative that targets mortalin-p53 interaction and inhibits cancer-cell proliferation by wild-type p53-dependent and -independent mechanisms. *Cancers* **2021**, *13*, 835. [CrossRef]
101. Elwakeel, A.; Sari, A.N.; Dhanjal, J.K.; Meidinna, H.N.; Sundar, D.; Kaul, S.C.; Wadhwa, R. Mutant p53^{L194F} harboring luminal-A breast cancer cells are refractory to apoptosis and cell cycle arrest in response to mortaparib^{Plus}, a multimodal small molecule inhibitor. *Cancers* **2021**, *13*, 3043. [CrossRef] [PubMed]
102. Meidinna, H.N.; Shefrin, S.; Sari, A.N.; Zhang, H.; Dhanjal, J.K.; Kaul, S.C.; Sundar, D.; Wadhwa, R. Identification of a new member of mortaparib class of inhibitors that target mortalin and PARP1. *Front. Cell Dev. Biol.* **2022**, *10*, 918970. [CrossRef] [PubMed]
103. Benbrook, D.M.; Nammalwar, B.; Long, A.; Matsumoto, H.; Singh, A.; Bunce, R.A.; Berlin, K.D. SHetA2 interference with mortalin binding to p66shc and p53 identified using drug-conjugated magnetic microspheres. *Investig. New Drugs* **2013**, *32*, 412–423. [CrossRef] [PubMed]
104. Ramraj, S.K.; Elayapillai, S.P.; Pelikan, R.C.; Zhao, Y.D.; Isingizwe, Z.R.; Kennedy, A.L.; Lightfoot, S.A.; Benbrook, D.M. Novel ovarian cancer maintenance therapy targeted at mortalin and mutant p53. *Int. J. Cancer* **2020**, *147*, 1086–1097. [CrossRef] [PubMed]
105. Brodsky, J.L. Selectivity of the molecular chaperone-specific immunosuppressive agent 15-deoxyspergualin: Modulation of Hsc70 ATPase activity without compromising DnaJ chaperone interactions. *Biochem. Pharmacol.* **1999**, *57*, 877–880. [CrossRef] [PubMed]
106. Fewell, S.W.; Day, B.W.; Brodsky, J.L. Identification of an inhibitor of hsc70-mediated protein translocation and ATP hydrolysis. *J. Biol. Chem.* **2001**, *276*, 910–914. [CrossRef] [PubMed]
107. Fewell, S.W.; Smith, C.M.; Lyon, M.A.; Dumitrescu, T.P.; Wipf, P.; Day, B.W.; Brodsky, J.L. Small molecule modulators of endogenous and co-chaperone-stimulated Hsp70 ATPase activity. *J. Biol. Chem.* **2004**, *279*, 51131–51140. [CrossRef] [PubMed]
108. Wisén, S.; Gestwicki, J.E. Identification of small molecules that modify the protein folding activity of heat shock protein 70. *Anal. Biochem.* **2008**, *374*, 371–377. [CrossRef]
109. Wisén, S.; Bertelsen, E.B.; Thompson, A.D.; Patury, S.; Ung, P.; Chang, L.; Evans, C.G.; Walter, G.M.; Wipf, P.; Carlson, H.A.; et al. Binding of a small molecule at a protein-protein interface regulates the chaperone activity of Hsp70-Hsp40. *ACS Chem. Biol.* **2010**, *5*, 611–622. [CrossRef]
110. Adam, C.; Baeurle, A.; Brodsky, J.L.; Wipf, P.; Schrama, D.; Becker, J.C.; Houben, R. The Hsp70 modulator MAL3-101 inhibits Merkel cell carcinoma. *PLoS ONE* **2014**, *9*, e92041. [CrossRef]
111. Prince, T.; Ackerman, A.; Cavanaugh, A.; Schreiter, B.; Juengst, B.; Andolino, C.; Danella, J.; Chernin, M.; Williams, H. Dual targeting of HSP70 does not induce the heat shock response and synergistically reduces viability in muscle invasive bladder cancer. *Oncotarget* **2018**, *9*, 32702–32717. [CrossRef]
112. Sannino, S.; Guerriero, C.J.; Sabnis, A.J.; Stolz, D.B.; Wallace, C.T.; Wipf, P.; Watkins, S.C.; Bivona, T.G.; Brodsky, J.L. Compensatory increases of select proteostasis networks, after Hsp70 inhibition in cancer cells. *J. Cell Sci.* **2018**, *131*, jcs217760. [CrossRef] [PubMed]
113. Cassel, J.A.; Ilyin, S.; McDonnell, M.E.; Reitz, A.B. Novel inhibitors of heat shock protein Hsp70-mediated luciferase refolding that bind to DnaJ. *Bioorg. Med. Chem.* **2012**, *20*, 3609–3614. [CrossRef] [PubMed]
114. Yokota, S.; Kitahara, M.; Nagata, K. Benzylidene lactam compound, KNK437, a novel inhibitor of acquisition of thermotolerance and heat shock protein induction in human colon carcinoma cells. *Cancer Res.* **2000**, *60*, 2942–2948. [PubMed]
115. Koishi, M.; Yokota, S.; Mae, T.; Nishimura, S.; Kanamori, S.; Horii, N.; Shibuya, K.; Sasai, K.; Hiraoka, M. The effects of KNK437, a novel inhibitor of heat shock protein synthesis, on the acquisition of thermotolerance in a murine transplantable tumor in vivo. *Clin. Cancer Res.* **2001**, *7*, 215–219. [PubMed]
116. Ohnishi, K.; Takahashi, A.; Yokota, S.; Ohnishi, T. Effects of a heat shock protein inhibitor KNK437 on heat sensitivity and heat tolerance in human squamous cell carcinoma cell lines differing in p53 status. *Int. J. Radiat. Biol.* **2004**, *80*, 607–614. [CrossRef]
117. Matsuda, K.; Nakagawa, S.-Y.; Nakano, T.; Asaumi, J.-I.; Jagetia, G.C.; Kawasaki, S. Effects of KNK437 on heat-induced methylation of histone H3 in human oral squamous cell carcinoma cells. *Int. J. Hyperthermia* **2006**, *22*, 729–735. [CrossRef]
118. Inoue, H.; Uyama, T.; Hayashi, J.; Watanabe, A.; Kobayashi, K.-I.; Tadokoro, T.; Yamamoto, Y. *N*-Formyl-3,4-methylenedioxy-benzylidene-γ-butyrolactam, KNK437 induces caspase-3 activation through inhibition of mTORC1 activity in Cos-1 cells. *Biochem. Biophys. Res. Commun.* **2010**, *395*, 56–60. [CrossRef]
119. Oommen, D.; Prise, K.M. KNK437, abrogates hypoxia-induced radioresistance by dual targeting of the AKT and HIF-1α survival pathways. *Biochem. Biophys. Res. Commun.* **2012**, *421*, 538–543. [CrossRef]
120. Yang, S.; Ren, X.; Liang, Y.; Yan, Y.; Zhou, Y.; Hu, J.; Wang, Z.; Song, F.; Wang, F.; Liao, W.; et al. KNK437 restricts the growth and metastasis of colorectal cancer via targeting DNAJA1/CDC45 axis. *Oncogene* **2020**, *39*, 249–261. [CrossRef]

121. Zhang, B.; Duan, D.; Ge, C.; Yao, J.; Liu, Y.; Li, X.; Fang, J. Synthesis of xanthohumol analogues and discovery of potent thiredoxin reductase inhibitor as potential anticancer agent. *J. Med. Chem.* **2015**, *58*, 1795–1805. [CrossRef]
122. Moses, M.A.; Kim, Y.S.; Rivera-Marquez, G.M.; Oshima, N.; Watson, M.J.; Beebe, K.E.; Wells, C.; Lee, S.; Zuehlke, A.D.; Shao, H.; et al. Targeting the Hsp40/Hsp70 chaperone axis as a novel strategy to treat castration-resistant prostate cancer. *Cancer Res.* **2018**, *78*, 4022–4035. [CrossRef] [PubMed]
123. Rice, M.A.; Kumar, V.; Tailor, D.; Garcia-Marques, F.J.; Hsu, E.-C.; Liu, S.; Bermudez, A.; Kanchustambham, V.; Shankar, V.; Inde, Z.; et al. SU086, an inhibitor of HSP90, impairs glycolysis and represents a treatment strategy for advanced prostate cancer. *Cell Rep. Med.* **2022**, *3*, 100502. [CrossRef] [PubMed]
124. Alalem, M.; Bhosale, M.; Ranjan, A.; Yamamoto, S.; Kaida, A.; Nishikawa, S.; Parrales, A.; Farooki, S.; Anant, S.; Padhye, S.; et al. Mutant p53 depletion by novel inhibitors for HSP40/J-domain proteins derived from the natural compound plumbagin. *Cancers* **2022**, *14*, 4187. [CrossRef] [PubMed]
125. Rottach, A.-M.; Ahrend, H.; Martin, B.; Walther, B.; Zimmermann, U.; Burchardt, M.; Stope, M.B. Cabazitaxel inhibits prostate cancer cell growth by inhibition of androgen receptor and heat shock protein expression. *World J. Urol.* **2019**, *37*, 2137–2145. [CrossRef]
126. Adjei, A.A.; Davis, J.N.; Erlichman, C.; Svingen, P.A.; Kaufmann, S.H. Comparison of potential markers of farnesyltransferase inhibition. *Clin. Cancer Res.* **2000**, *6*, 2318–2325.
127. Karp, J.E.; Lancet, J.E.; Kaufmann, S.H.; End, D.W.; Wright, J.J.; Bol, K.; Horak, I.; Tidwell, M.L.; Liesveld, J.; Kottke, T.J.; et al. Clinical and biologic activity of the farnesyltransferase inhibitor R115777 in adults with refractory and relapsed acute leukemias: A phase 1 clinical-laboratory correlative trial. *Blood* **2001**, *97*, 3361–3369. [CrossRef]
128. Cohen, S.J.; Ho, L.; Ranganathan, S.; Abbruzzese, J.L.; Alpaugh, R.K.; Beard, M.; Lewis, N.L.; McLaughlin, S.; Rogatko, A.; Perez-Ruixo, J.J.; et al. Phase II and pharmacodynamics study of the farnesyltransferase inhibitor R115777 as initial therapy in patients with metastatic pancreatic adenocarcinoma. *J. Clin. Oncol.* **2003**, *21*, 1301–1306. [CrossRef]
129. Wang, J.; Yao, X.; Huang, J. New tricks for human farnesyltransferase inhibitor: Cancer and beyond. *Med. Chem. Commun.* **2017**, *8*, 841–854. [CrossRef]
130. Marchwicka, A.; Kaminska, D.; Monirialamdari, M.; Blazewska, K.M.; Gendaszewska-Darmach, E. Protein prenyltransferases and their inhibitors: Structural and functional characterization. *Int. J. Mol. Sci.* **2022**, *23*, 5424. [CrossRef]
131. Wang, C.-C.; Liao, Y.-P.; Mischel, P.S.; Iwamoto, K.S.; Cacalano, N.A.; McBride, W.H. HDJ-2 as a target for radiosensitization of glioblastoma multiforme cells by the farnesyltransferase inhibitor R115777 and the role of the p53/p21 pathway. *Cancer Res.* **2006**, *66*, 6756–6762. [CrossRef]
132. Xu, D.; Tong, X.; Sun, L.; Li, H.; Jones, R.D.; Liao, J.; Yang, G.-Y. Inhibition of mutant Kras and p53-driven pancreatic carcinogenesis by atorvastatin: Mainly via targeting of the farnesylated DNAJA1 in chaperoning mutant p53. *Mol. Carcinogen.* **2019**, *58*, 2052–2064. [CrossRef] [PubMed]
133. Winkler, R.; Mägdefrau, A.-S.; Piskor, E.-M.; Kleemann, M.; Beyer, M.; Linke, K.; Hansen, L.; Schaffer, A.-M.; Hoffmann, M.E.; Poepsel, S.; et al. Targeting the MYC interaction network in B-cell lymphoma via histone deacetylase 6 inhibition. *Oncogene* **2022**, *41*, 4560–4572. [CrossRef] [PubMed]
134. Silva, G.; Marins, M.; Chaichanasak, N.; Yoon, Y.; Fachin, A.L.; Pinhanelli, V.C.; Regasini, L.O.; dos Santos, M.B.; Ayusso, G.M.; de Carvalho Marques, B.; et al. Trans-chalcone increases p53 activity via DNAJB1/HSP40 induction and CRM1 inhibition. *PLoS ONE* **2018**, *13*, e0202263. [CrossRef] [PubMed]
135. Schirmer, R.H.; Coulibaly, B.; Stich, A.; Scheiwein, M.; Merkle, H.; Eubel, J.; Becker, H.; Müller, O.; Zich, T.; Schiek, W.; et al. Methylene blue as an antimalarial agent. *Redox Rep.* **2003**, *8*, 272–275. [CrossRef]
136. Takasu, K. π-Delocalized lipophilic cations as new candidates for antimalarial, antitrypanosomal and antileishmanial agents: Synthesis, evaluation of antiprotozoal potency, and insight into their action mechanisms. *Chem. Pharm. Bull.* **2016**, *64*, 656–667. [CrossRef]

International Journal of
Molecular Sciences

Article

eIF4A1 Is a Prognostic Marker and Actionable Target in Human Hepatocellular Carcinoma

Sara M. Steinmann [1,†], Anabel Sánchez-Martín [1,2,†], Elisabeth Tanzer [1], Antonio Cigliano [3], Giovanni M. Pes [3], Maria M. Simile [3], Laurent Desaubry [4], Jose J.G. Marin [2], Matthias Evert [1] and Diego F. Calvisi [3,*]

1 Experimental Tumor Pathology, Institute of Pathology, University Regensburg, Franz-Josef-Strauss-Allee 11, 93053 Regensburg, Germany
2 Experimental Hepatology and Drug Targeting (HEVEPHARM) Group, University of Salamanca, IBSAL, 37007 Salamanca, Spain
3 Department of Medicine, Surgery, and Pharmacy, University of Sassari, 07100 Sassari, Italy
4 Therapeutic Innovation Laboratory, UMR7200, CNRS/University of Strasbourg, CEDEX, 67401 Illkirch, France
* Correspondence: calvisid@uniss.it; Tel.: +39-079-228353
† These authors contributed equally to this work.

Abstract: Hepatocellular carcinoma (HCC) is a primary liver tumor with high lethality and increasing incidence worldwide. While tumor resection or liver transplantation is effective in the early stages of the disease, the therapeutic options for advanced HCC remain limited and the benefits are temporary. Thus, novel therapeutic targets and more efficacious treatments against this deadly cancer are urgently needed. Here, we investigated the pathogenetic and therapeutic role of eukaryotic initiation factor 4A1 (eIF4A1) in this tumor type. We observed consistent eIF4A1 upregulation in HCC lesions compared with non-tumorous surrounding liver tissues. In addition, *eIF4A1* levels were negatively correlated with the prognosis of HCC patients. In HCC lines, the exposure to various eIF4A inhibitors triggered a remarkable decline in proliferation and augmented apoptosis, paralleled by the inhibition of several oncogenic pathways. Significantly, anti-growth effects were achieved at nanomolar concentrations of the eIF4A1 inhibitors and were further increased by the simultaneous administration of the pan mTOR inhibitor, Rapalink-1. In conclusion, our results highlight the pathogenetic relevance of eIF4A1 in HCC and recommend further evaluation of the potential usefulness of pharmacological combinations based on eIF4A and mTOR inhibitors in treating this aggressive tumor.

Keywords: hepatocellular carcinoma; eIF4A1; translation inhibitors; targeted therapies

Citation: Steinmann, S.M.; Sánchez-Martín, A.; Tanzer, E.; Cigliano, A.; Pes, G.M.; Simile, M.M.; Desaubry, L.; Marin, J.J.; Evert, M.; Calvisi, D.F. eIF4A1 Is a Prognostic Marker and Actionable Target in Human Hepatocellular Carcinoma. *Int. J. Mol. Sci.* **2023**, *24*, 2055. https://doi.org/10.3390/ijms24032055

Academic Editor: Bernhard Biersack

Received: 10 December 2022
Revised: 17 January 2023
Accepted: 18 January 2023
Published: 20 January 2023

1. Introduction

Liver cancer is one of the most frequent malignancies and has poor prognosis, ranked as the third leading cause of cancer-related deaths worldwide and thus remains a global health issue [1]. The incidence of liver cancer is increasing, and the World Health Organization has estimated that the number of deaths from this cancer will reach one million annually in 2030 [2]. Hepatocellular carcinoma (HCC) is the most common type of primary liver cancer, accounting for ~90% of cases. HCC commonly develops in the setting of underlying liver diseases such as cirrhosis or chronic liver inflammation. Although hepatitis B or C virus infection and alcohol abuse remain major HCC risk factors, the number of HCC cases related to non-alcoholic fatty liver disease (NAFLD) or associated with metabolic syndrome or diabetes mellitus is rapidly growing, and this condition might become the leading cause of HCC in Western countries in the near future [3,4]. Surgical resection, radiofrequency ablation, transarterial chemoembolization, and liver transplantation are potentially curative treatments for early-stage HCC lesions [5,6]. However, most patients with HCC are diagnosed in an advanced stage of the disease when they are no longer good candidates for curative strategies. Palliative systemic therapy is the only therapeutic option for these patients. Several tyrosine multikinase inhibitors, including sorafenib, lenvatinib, regorafenib, and cabozantinib, are approved as the first- or second-line treatments for unresectable advanced HCC. However, only a slight increase in overall survival has been reached in these patients [7–10]. Immune-based therapies are also emerging for the treatment of advanced HCC. The combination of azetolizumab (anti-PD-L1) with bevacizumab (anti-VEGF) has shown superior efficacy compared to sorafenib, and it is now recommended as the standard first-line treatment for advanced HCC [6,11]. Nevertheless, the benefits of immunotherapy are still modest, and many patients remain non-responders. Therefore, developing novel therapeutic strategies is imperative to improving the outcome of HCC patients.

Targeted inhibition of translation machinery is emerging as a promising cancer treatment approach. Protein synthesis is a highly controlled process that plays a major role in gene expression regulation. However, this mechanism is frequently dysregulated in human malignancies [12]. Studies have shown that altered mRNA translation leads to tumorigenesis and cancer progression by selectively enhancing the synthesis of proteins involved in cell proliferation, activation of invasion and metastasis, and other neoplastic characteristics-related processes [13]. Translation initiation is the rate-limiting step; it is regulated by multiple eukaryotic initiation factors (eIFs) [14,15]. The eIF4F complex, which includes eIF4E, eIF4G, and eIF4A proteins, is essential for cap-dependent ribosome recruitment and translation initiation. Thus, eIF4E binds to the m^7G-cap structure at the mRNA '5′ end, whereas eIF4G acts as a scaffold protein recruiting eIF4E and eIF4A. The latter, the only component of the eIF4F complex with enzymatic activity, is an ATP-dependent DEAD-box RNA helicase that unwinds the mRNA secondary structures in the 5′ untranslated regions (5′-UTRs) to enable ribosome scanning. The helicase eIF4A is required to efficiently translate mRNAs with long and highly structured 5′-UTRs, including those with G-quadruplex structures [16]. These complex structures characterize the 5-UTRs of many oncogenes considered eIF4A-dependent genes, such as KRAS, BLC2, NOTCH1, CDK6, and CCND1 [17,18].

There are two eIF4A isoforms involved in mRNA translation, eIF4A1 and eIF4A2, which share ~90% homology in their amino acid sequence [19]. Despite their similarity, eIF4A1, but not eIF4A2, is essential to cell viability. In addition, eIF4A2 cannot compensate for eIF4A1 dysfunction, suggesting that both isoforms have distinct biological properties [20,21]. Their expression also varies between different types of tissues, with eIF4A1 being the most abundant paralog [22]. The overexpression of the eIF4A1 isoform occurs in a wide variety of malignancies, where it correlates with metastasis and poor prognoses [23–25]. The c-Myc protooncogene upregulates EIF4A1 transcription [26]; in addition, the levels of free functional eIF4A1 protein are increased by the degradation of the tumor suppressor programmed cell death 4 (PDCD4), which is itself regulated by mammalian target of rapamycin (mTOR) signaling [27,28]. Although little is known about eIF4A2 in cancer, recent studies have revealed the upregulation of this isoform in colorectal cancer [29] and paclitaxel-resistant breast cancer [30]. High eIF4A2 levels are also associated with poor prognosis in colorectal and esophageal squamous cell carcinomas [29,31].

Consequently, the characteristics of eIF4A isoforms make them attractive drug targets for anti-tumor therapy. Rocaglates, also known as flavaglines, are a group of small molecules harboring a common cyclopean[b]benzofuran core that suppresses cap-dependent translation by inhibiting eIF4A activity. More than one hundred natural rocaglates have been isolated from *Aglaia* species, including silvestrol and rocaglamide. In addition, numerous synthetic derivatives have recently been developed to improve their potency and bioavailability [32–34]. The rocaglates stabilize the non-specific binding of eIF4A to RNA, which prevents its incorporation into the eIF4F complex [35,36], and they target both eIF4A isoforms [37]. Silvestrol, one of the best-studied drugs of this family, has demonstrated potent anti-tumor activity both in vitro and in vivo [38,39], and the antiproliferative potency of other active rocaglates derivatives, such as the synthetic compound CR-1-31-B, has also been reported in several tumors [40–43]. Similarly, the synthetic flavagline FL3 exhibited anticancer properties in various experimental cancer models [44]. Furthermore, the synthetic eIF4A inhibitor, Zotatifin, is currently being evaluated in a phase I/II clinical trial for solid tumors [45].

In this study, we investigated the pathogenetic role of eIF4A1 in human HCC and its potential usefulness as a novel molecular target for the development of novel pharmacological strategies.

2. Results

2.1. *eIF4A1 Is Significantly Overexpressed in Human Hepatocellular Carcinoma and Correlates with a Worse Outcome*

First, *eIF4A1* mRNA levels in human cancers were investigated using data obtained from The Cancer Genome Atlas (TCGA) and the UALCAN (http://ualcan.path.uab.edu/; accessed on 24 November 2022) analysis tool. *eIF4A1* mRNA expression was elevated in various cancer types, including colon adenocarcinoma (COAD), glioblastoma multiforme (GBM), esophageal carcinoma (ESCA), and kidney renal clear cell carcinoma (KIRC), among others (Figure 1A). In HCC (normal tissue: n = 50; tumor tissue: n = 371), *eIF4A1* was significantly overexpressed (Figure 1A; LIHC: liver hepatocellular carcinoma; Figure 1B; P = 8.871×10^{-10}) with comparable incidences for women (P = 2.566×10^{-10}) and men (P = 2.155×10^{-10}; Figure 1C) and elevated expression for patients aged between 21 and 80 years (Figure 1D; 21–40 years: P = 1.463×10^{-06}; 41–60 years: P = 2.841×10^{-08}; 61–80 years: P = 7.070×10^{-09}) compared to that in normal tissue. Maximal *eIF4A1* expression was detected in tumors of young patients between 21 and 40 years of age (Figure 1D; 21–40 vs. 61–80 years: P = 2.678×10^{-02}). Examination of multiple clinicopathological features revealed that HCC patients with high *eIF4A1* expression had poorer prognoses. The survival time of patients with high *eIF4A1* expression was significantly shorter than that of patients with low or medium *eIF4A1* levels (Figure 1E; P = 0.027). Furthermore, the TCGA data showed that in comparison to normal tissue, *eIF4A1* expression was increased in all tumor grades, with grades 3 and 4 tumors revealing even higher levels than grades 1 and 2 tumors (Figure 1F; Normal vs. Grade 1: P = 2.359×10^{-03}; Normal vs. Grade 2: P = 3.212×10^{-09}; Normal vs. Grade 3: P = 4.581×10^{-09}; Normal vs. Grade 4: P = 6.219×10^{-03}; Grade 1 vs. Grade 3: P = 5.378×10^{-04}; Grade 1 vs. Grade 4: P = 3.380×10^{-02}; Grade 2 vs. Grade 3: P = 2.050×10^{-02}; Grade 2 vs. Grade 4: P = 2.020×10^{-02}). The analysis of HCC cancer stages (displayed significantly increased *eIF4A1* expression in stage 1 to 3 tumors (Figure 1G; Normal vs. Stage 1: P = 1.878×10^{-08}; Normal vs. Stage 2: P = 3.364×10^{-07}; Normal vs. Stage 3: P = 1.058×10^{-06}; Stage 1 vs. Stage 3: P = 2.366×10^{-02}). Importantly, high *eIF4A1* expression was found in N0 and N1 nodal metastasis status (Figure 1H; Normal vs. N0: P = 4.421×10^{-12}; Normal vs. N1: P = 1.023×10^{-04}), with a clear tendency toward enhanced *eIF4A1* expression in N1 specimens.

To confirm the findings in the TCGA dataset analysis, the eIF4A1 levels in 47 human HCC samples and paired normal tissues collected at the University Regensburg were evaluated. In compliance with the TCGA data set, western blot analysis revealed that eIF4A1 protein levels were significantly increased in HCC specimens compared to those in paired non-tumorous tissues (Figure 2; *** p = 0.0002; Wilcoxon test; Supplementary Figure S1). In addition, immunoblotting showed augmented protein expression of the eukaryotic initiation factor (eIF) family member eIF4A2 (* p = 0.0249; Wilcoxon test) and the translation initiator and rocaglate target DEAD-box protein 3 DDX3 [46] (** p = 0.0088; Wilcoxon test) in HCC lesions (Figure 2A,B; Supplementary Figure S1).

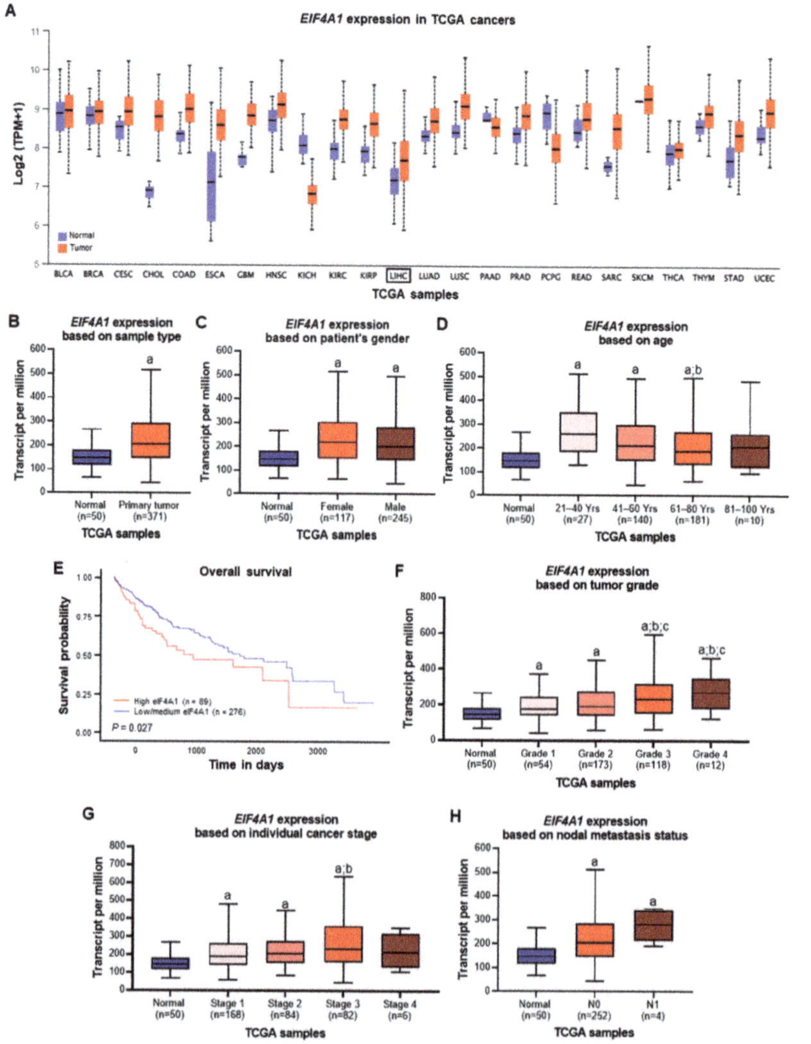

Figure 1. Biostatistical analysis from TGCA data of the *eIF4A1* gene in various tumor types and human hepatocellular carcinoma (HCC). Please refer to the main text for detailed explanations. (**A**) mRNA levels of *eIF4A1* in various tumor types, including liver hepatocellular carcinoma (LIHC). (**B**) mRNA levels of *eIF4A1* in non-tumorous (normal) and HCC (primary tumor) specimens. (**C**) mRNA levels of *eIF4A1* in HCC specimens based on gender. (**D**) mRNA levels of *eIF4A1* in HCC specimens based on age (Yrs: Years). (**E**) Kaplan-Meier curves in HCC patients based on *eIF4A1* mRNA levels. (**F**) mRNA levels of *eIF4A1* in HCC specimens based on tumor grade (grade 1: well-differentiated; grade 2: moderately differentiated; grade 3: poorly differentiated; grade 4: undifferentiated). (**G**) mRNA levels of *eIF4A1* in HCC specimens based on the individual cancer stage (stage 1: cancers localized to one part of the body; stage 2 and 3: locally advanced cancers; stage 4: spread to other organs and often metastasized cancers). (**H**) mRNA levels of *eIF4A1* in HCC specimens based on nodal metastasis status (N0: no evidence of cancer in regional lymph nodes; N1: cancer has spread to a single lymph node near the liver).

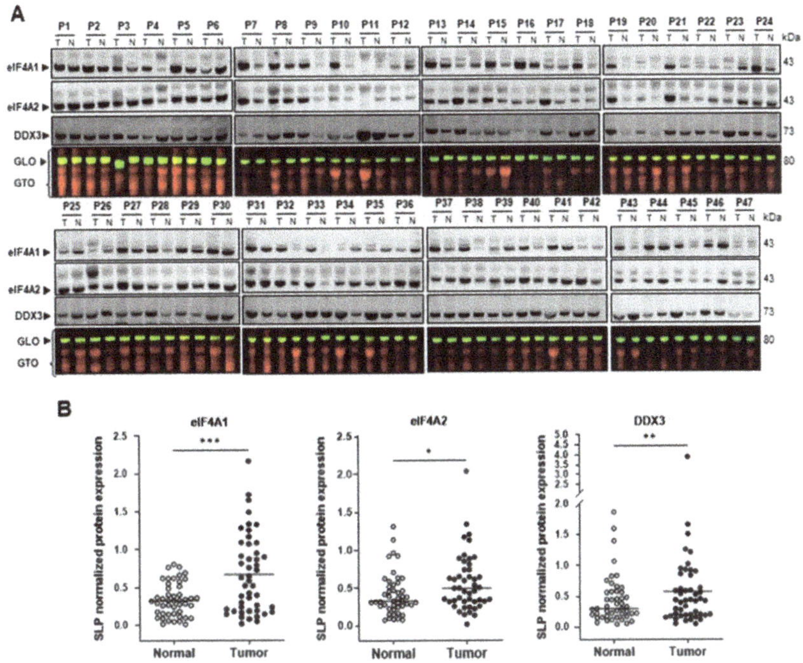

Figure 2. eIF4A1 is upregulated in human hepatocellular carcinoma (HCC; n = 47). (**A**) Levels of eIF4A1, eIF4A2, and DDX3 proteins in human HCC samples, as detected by western blotting using the SPL method. (**B**) Quantification of western blot band intensities. Dots representing the individual protein expression of one patient and their medians are shown (* $p < 0.5$; ** $p < 0.1$; *** $p < 0.01$; Wilcoxon matched-pairs signed rank test). Abbreviations: P, patient; T, tumor; N, non-tumor; GLO: Gel loading control (green); GTO: Total protein (red).

Remarkably, when assessing the prognostic relevance of the three proteins, eIF4A1 levels were inversely associated with patient survival time ($p < 0.0001$; Figure 3A; Supplementary Materials). In contrast, levels of eIF4A2 did not separate HCC patients based on survival length (P = 0.741; Figure 3B; Supplementary Materials). Moreover, DDX3 displayed a trend toward more prolonged survival without reaching significance (P = 0.142; Figure 3C; Supplementary Materials).

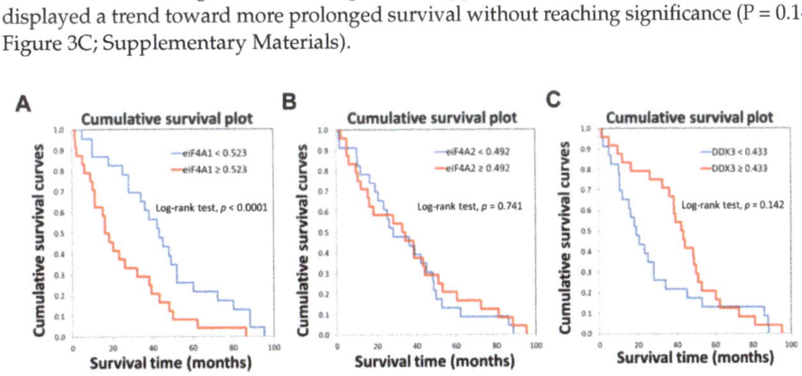

Figure 3. Levels of eIF4A1 protein negatively correlate with the survival of patients with hepatocellular carcinoma (HCC) (n = 47). Kaplan-Meyer curves in HCC patients for eIF4A1 (**A**), eIF4A2 (**B**), and DDX3 (**C**) proteins.

Subsequently, we evaluated the levels of eIF4A1 in a large collection of formalin-fixed, paraffin-embedded HCC samples (n = 356) by immunohistochemistry (Figure 4). In agreement with western blot data, increased immunoreactivity for eIF4A1 was detected ubiquitously in HCC lesions compared to that in the matching non-neoplastic liver tissues (Figure 4, first lower panel). In contrast, non-tumorous epithelial cells exhibited faint or absent eIF4A1 staining (Figure 4, second lower panel). HCC lesions displayed either homogeneous or scattered upregulation of eIF4A1 (Figure 4, third to fifth lower panels).

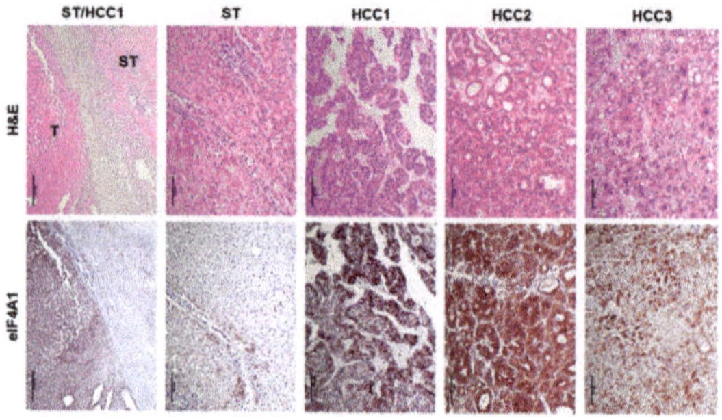

Figure 4. Representative immunohistochemistry patterns of eIF4A1 protein in human hepatocellular carcinoma (HCC; n = 356). The first left panels show an example of human HCC (denominated HCC1) at low magnification. Note the enhanced immunoreactivity for eIF4A1 in the tumorous part (T) compared with the neighboring non-tumorous surrounding tissue (ST), which exhibits faint/absent eIF4A1 staining. The second and third left panels depict higher magnifications for ST and HCC1. A second tumor (HCC2), characterized by intense, homogeneous eIF4A1 immunoreactivity, is shown in the fourth panel from the left. Finally, a third tumor (HCC3) in the right panels displays heterogeneous eIF4A1 immunolabeling. Abbreviation: H&E, hematoxylin and eosin staining. Original magnifications: 40× in the first left panels, 200× in all the other panels. Scale bar: 500 μm in the first left panels, 100 μm in all the other panels.

Overall, the present data indicated that eIF4A1 is highly expressed in primary HCC lesions and associated with a poor outcome. These findings suggest the usefulness of eIF4A1 as a prognostic biomarker in human HCC.

2.2. Targeting eIF4A1 with Rocaglates Inhibits Tumor Cell Growth in HCC Cell Lines

Next, to evaluate the therapeutic potential of targeting eIF4A1 activity in HCC, we challenged human HCC cell lines in vitro with five distinct rocaglate derivatives, namely CR-1-31-B, FL3, Rocaglamide, Silvestrol, and Zotatifin. These drugs promote the binding affinity of eIF4A1 to mRNA species containing polypurine sequence motifs, sequentially impeding the translation initiation during protein synthesis by blocking the formation of the heterotrimeric eIF4F cap-binding complex. Prior to pharmacological treatments, we assessed the endogenous eIF4A1, eIF4A2, and DDX3 protein levels using western blot analysis in various human HCC cell lines. eIF4A1, eIF4A2, and DDX3 proteins were expressed in all HCC cell lines tested (Figure 5A,B; Supplementary Figure S2).

Subsequently, to assess the cytotoxic potential of eIF4A inhibitors, the HLE, HLF, and PLC/PRF/5 tumor cell lines were randomly selected and treated with a range of concentrations between 0 and 200 nM per eIF4A inhibitor, and colorimetric MTT assays were performed 48 h after starting compound exposure. Principally, cellular metabolic activity was measured to indicate cell viability, proliferation, and cytotoxicity. Overall, all five rocaglates efficiently reduced tumor cell viability in the three cell lines in a dose-dependent manner (Figure 6A–C). Notably, in HCC cell lines, the calculated half-maximal inhibitory concentration (IC50) values for the five drugs resided in the low nanomolar range (Figure 6D).

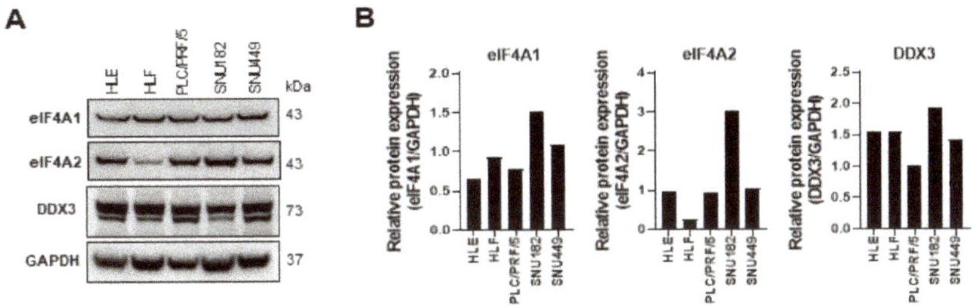

Figure 5. Human hepatocellular carcinoma cell lines express eIF4A1, eIF4A2, and DDX3 proteins. (**A**) The levels of the three proteins were assessed in HLE, HLF, PLF/PRF/5, SNU182, and SNU449 cell lines by western blot analysis. GAPDH was used as a loading control, and protein band intensities of eIF4A1, eIF4A2, and DDX3 were normalized to GAPDH levels (**B**).

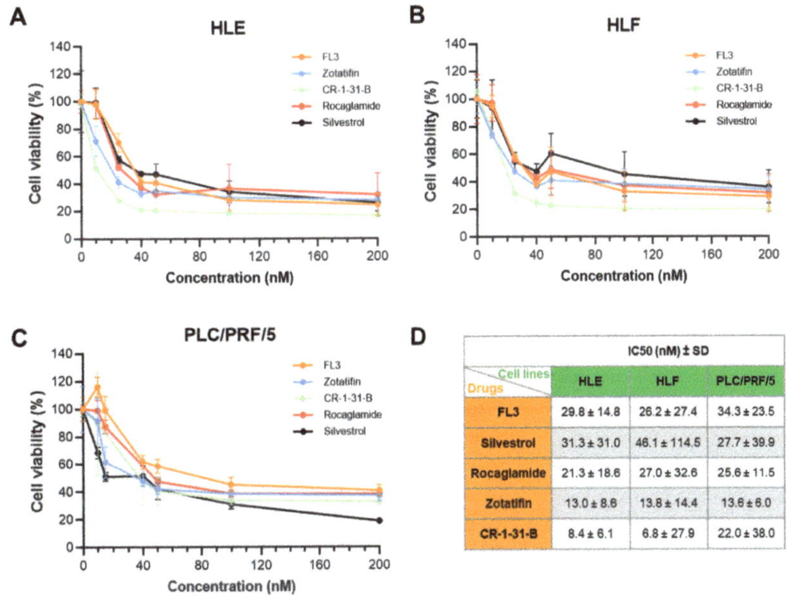

Drugs \ Cell lines	HLE	HLF	PLC/PRF/5
	IC50 (nM) ± SD		
FL3	29.8 ± 14.8	26.2 ± 27.4	34.3 ± 23.5
Silvestrol	31.3 ± 31.0	46.1 ± 114.5	27.7 ± 39.9
Rocaglamide	21.3 ± 18.6	27.0 ± 32.6	25.6 ± 11.5
Zotatifin	13.0 ± 8.6	13.8 ± 14.4	13.6 ± 6.0
CR-1-31-B	8.4 ± 6.1	6.8 ± 27.9	22.0 ± 38.0

Figure 6. Cell viability of human hepatocellular carcinoma cell lines exposed to eIF4A inhibitors, as assessed by MTT assay. The eIF4A1 inhibitors, FL3, Zotatifin, CR-1-31-B, Rocaglamide, and Silvestrol, were administered to HLE (**A**), HLF (**B**), and PLC/PRF/5 (**C**) cell lines. Similar data were obtained in the SNU449 cell line (not shown). Data of two independent experiments with n = 4 technical replicates are represented as the percentage of DMSO-treated cells ± SD. A summary of the IC50 ± SD of the five drugs in the three cell lines is depicted in (**D**).

To further elucidate the mechanisms of action of eIF4A inhibitors in HCC cell lines, we selected the two most effective drugs, i.e., CR-1-31-B and Zotatifin, for the following experiments. Zotatifin was also selected because, to date, it is the only eIF4A inhibitor in clinical trials. When evaluating cell proliferation, we found that incubation with CR-1-31-B and Zotatifin induced a marked reduction in BrdU incorporation in both HLE and HLF cell lines, with a slightly more pronounced anti-growth effect in the case of CR-1-31-B (Figure 7A,B).

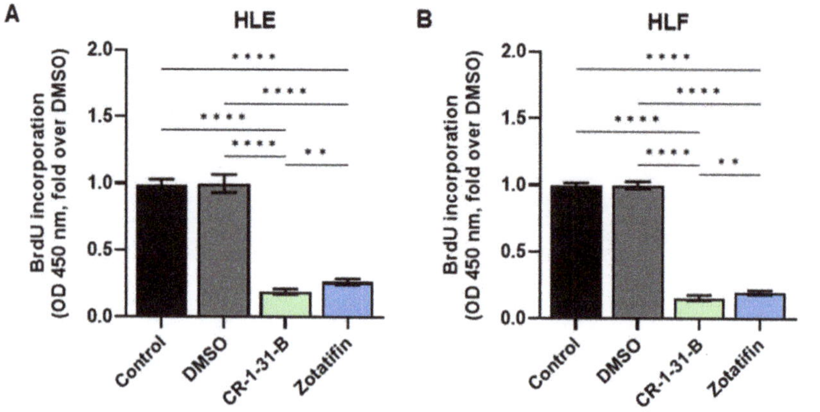

Figure 7. Effects of the eIF4A inhibitors, CR-1-31-B and Zotatifin, on the proliferation of hepatocellular carcinoma cell line monolayer cultures indicated by BrdU incorporation. BrdU incorporation assay was conducted on (**A**) HLE and (**B**) HLF cells treated for 48 h with 10 nM CR-1-31-B and 20 nM Zotatifin. Untreated cells and cells treated with DMSO served as controls. The fold-changes over DMSO of optical densities (OD) at 450 nm are presented. All results are expressed as mean ± SD of three independent experiments in triplicate. For statistical analysis, Tukey's multiple comparisons test was performed (** $p < 0.1$; **** $p < 0.001$).

Subsequently, apoptosis was determined in the same cell lines treated with the two eIF4A inhibitors. Both CR-1-31-B and Zotatifin induced higher apoptotic cell death in the cell lines than was observed in untreated and DMSO-treated cells (Figure 8A,B).

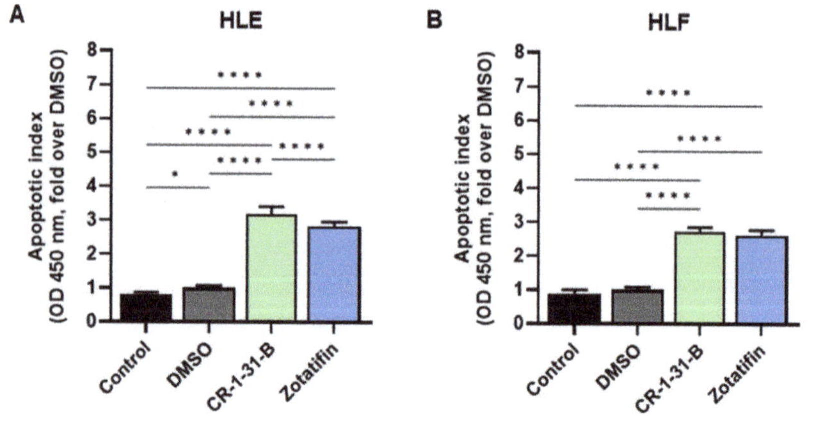

Figure 8. Effects of the eIF4A inhibitors, CR-1-31-B and Zotatifin, on the apoptosis of hepatocellular carcinoma cell line monolayer cultures. Cell death assay was conducted on (**A**) HLE and (**B**) HLF cells treated for 48 h with 10 nM CR-1-31-B and 20 nM Zotatifin. Untreated cells and cells treated with DMSO served as controls. The fold-changes over DMSO of optical densities (OD) at 405 nm are presented. All results are expressed as mean ± SD of three independent experiments in triplicate. For statistical analysis, Tukey's multiple comparisons test was performed (* $p < 0.05$; **** $p < 0.001$).

Next, we assessed the effects of CR-1-31-B and Zotatifin on some of the most relevant oncogenic pathways in hepatocarcinogenesis by western blot analysis (Figure 9; Supplementary Figures S3 and S4). Both drugs did not affect eIF4A1 and DDX3 protein levels, supporting the hypothesis that they act on eIF4A1 and DDX3 activity without affecting their expression. In contrast, similarly to what has been described in other tumor types, the treatment of HCC cells with these drugs led to eIF4A2 upregulation, which is considered a compensatory response to marked eIF4A1 suppression. Nonetheless, it has been shown that eIF4A2 induction cannot functionally compensate for eIF4A1 loss or inactivation [20,21]. Regarding oncogenic pathways, both CR-1-31-B and Zotatifin administration resulted in the downregulation of activated/phosphorylated (p-)STAT3, AKT, ERK1/2, and total SKP2 proteins in a dose-dependent manner. In addition, levels of the Hippo pathway effector protein TAZ were decreased, whereas the effects on the ortholog YAP were inconsistent. Furthermore, the levels of prohibitin 1 and 2 (PHB1 and 2), which are supposed to be critical targets of rocaglates [47], were decreased in CR-1-31-B- and Zotatifin-treated cells. In contrast, levels of mTOR effectors, such as activated/phosphorylated (p-)RPS6 and phosphorylated/inactivated (p-)4EBP1, were either increased or unchanged following CR-1-B-31 and Zotatifin administration.

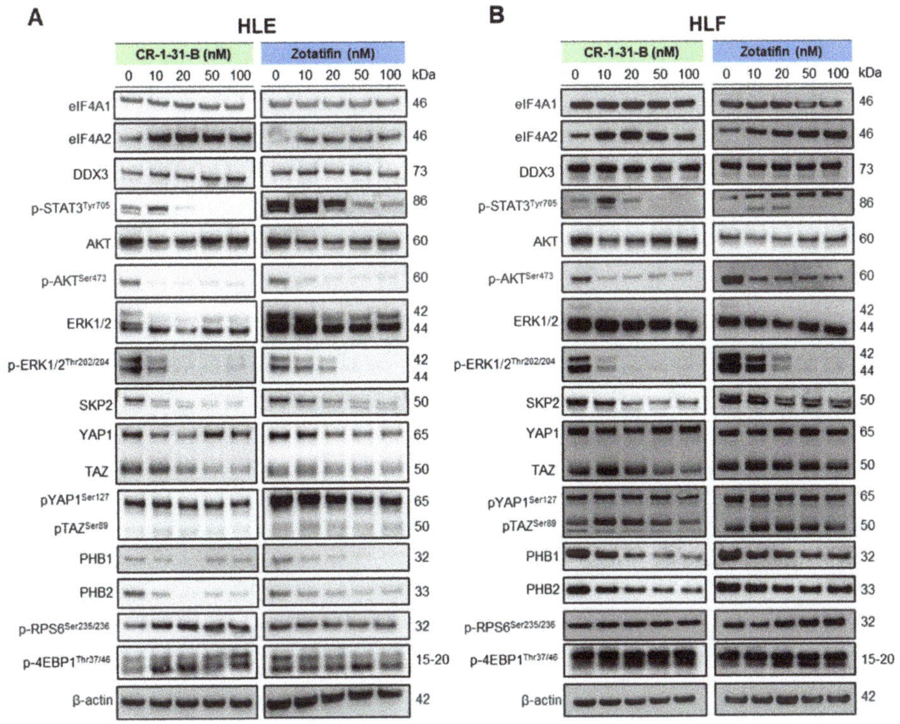

Figure 9. Effects of the eIF4A inhibitors, CR-1-31-B and Zotatifin, in HCC cell lines on the levels of elements of oncogenic pathways involved in hepatocarcinogenesis. Western blot analysis was used to assess the levels of several effectors of oncogenic cascades in HLE (A) and HLF (B) cell lines exposed to increasing concentrations of CR-1-31-B and Zotatifin inhibitors (0, 10, 20, 50, and 100 nM). Cells were treated for 48 h, and western blot analysis was conducted at this time point.

2.3. *Targeting the mTOR Pathway Synergizes with eIF4A Inhibitors to Restrain HCC Cell Growth In Vitro*

Inspired by the findings that eIF4A inhibitors did not reduce the activation of mTOR effectors, we tested whether mTOR suppression could synergize with eIF4A inhibitors to hamper HCC cell growth. For this purpose, the HLE and HLF cell lines were treated with the pan mTOR inhibitor, Rapalink-1 [48], either alone or in combination with CR-1-31-B or Zotatifin. Rapalink-1 was chosen over common mTORC1 inhibitors, such as Rapamycin or Rapamycin homologs, because the latter drugs effectively inhibit RPS6 but not 4EBP1 [49].

When assessing HLE and HLF metabolic activity with the MTT assay, we discovered that Rapalink-1 inhibited HCC cell growth at nanomolar concentrations (IC50: 20 nM). Significantly, the combination of Rapalink-1 with CR-1-31-B or Zotatifin triggered an even more potent, synergistic reduction of HLE and HLF metabolic activity than that induced by treatments alone (Figure 10).

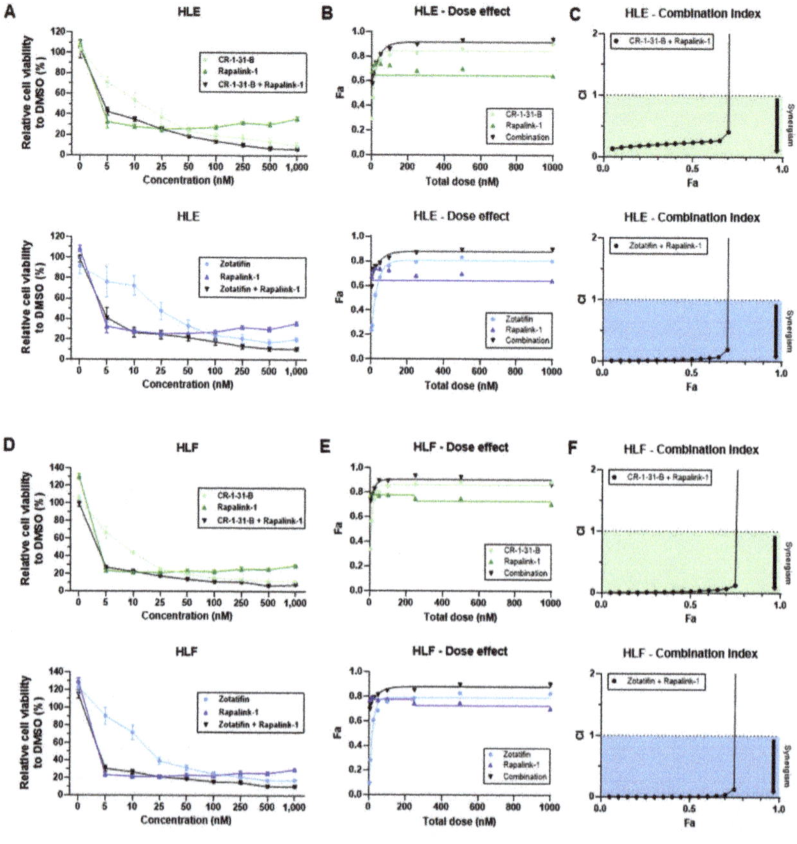

Figure 10. Highly synergistic cytotoxicity of combining the pan mTOR inhibitor, Rapalink-1, with CR-1-31-B and Zotatifin in hepatocarcinoma cell lines. (**A**) HLE and (**D**) HLF cells were exposed to various concentrations of compounds, alone or in combination, for 48 h. Cell proliferation based on the metabolic activity was assessed using the MTT assay and expressed as fold-change over DMSO control. The mean ± SEM of three independent experiments performed in technical triplicates are shown. (**B**,**E**) Dose effect curves of obtained fractions of dead cells (Fa) were used to determine the combination index (CI) (**C**,**F**) and synergism of drug combinations using CompuSyn software (https://www.combosyn.com, accessed on 1 November 2022) based on the Chou and Talalay method [50]. CI < 1 suggests synergism.

Subsequently, we assessed proliferation and apoptosis in the same cell lines using Rapalink-1 alone and in combination with CR-1-B-R1 or Zotatifin. Again, the combined treatments resulted in a significant reduction in proliferation (Figure 11A) and a robust increase in apoptosis (Figure 11B) in the two cell lines.

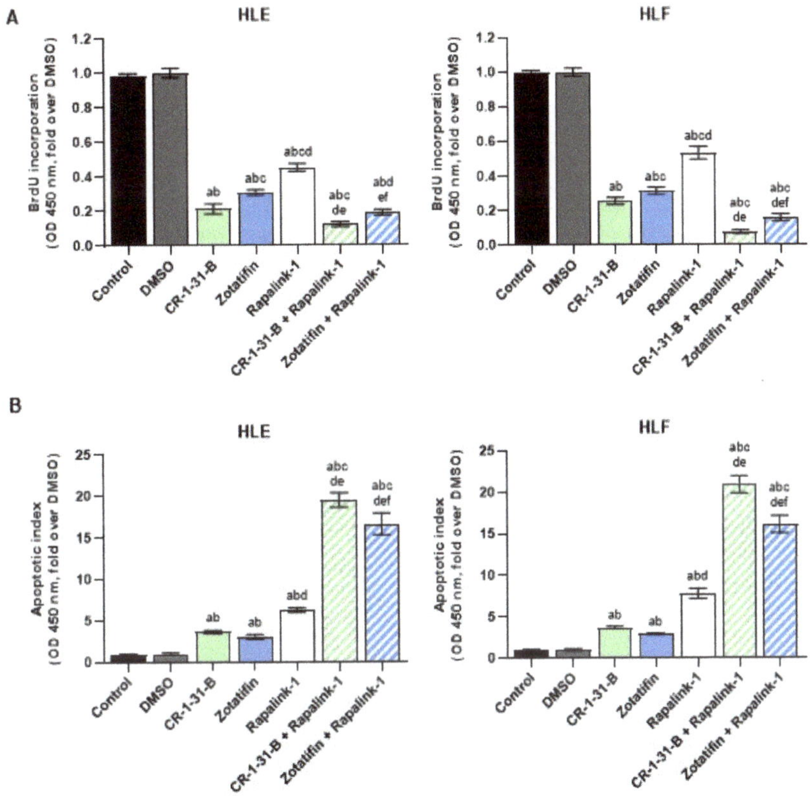

Figure 11. Combining the pan mTOR inhibitor, Rapalink-1, with CR-1-31-B and Zotatifin results in synergistic anti-growth activity in hepatocarcinoma cell lines. (**A**) HLE and HLF cells were exposed to the three drugs, alone or in combination, for 48 h and proliferation was assessed. (**B**) HLE and HLF cells were exposed to the three drugs, alone or in combination, for 48 h and apoptosis was determined. For the experiments in (**A**,**B**), the drugs were administered at their IC50 concentrations. All results are expressed as mean ± SD of three independent experiments in triplicate. For statistical analysis, Tukey's multiple comparisons test was performed. Lowercase letters are used to denote statistical significance (a–f: $p > 0.0001$; a, vs. Control; b, vs. DMSO; c, vs. CR-1-31-B, d, vs. Zotatifin; e, vs. Rapalink-1; f, CR-1-31-B + Rapalink-1).

Finally, western blot analysis confirmed that the mTOR inhibitor, Rapalink-1, alone and in combination with the eIF4A inhibitors, CR-1-31-B or Zotatifin, fully inhibited p-RPS6 and p-4EBP1 levels. In addition, Rapalink-1 alone resulted in the downregulation of eIF4A1, eIF4A2, and DDX3, whereas inconsistent results were obtained for these proteins when Rapalink-1 was combined with CR-1-31-B or Zotatifin (Figure 12; Supplementary Figure S5). These data indicated that targeting the eIF4A family and mTOR oncogenic pathway might be a potent anti-tumor treatment strategy in HCC cell lines (Figure 12; Supplementary Figure S5).

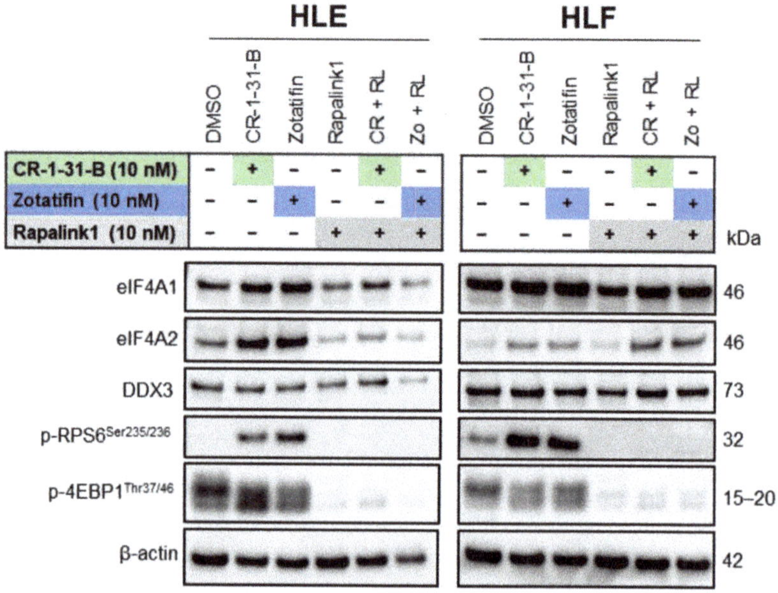

Figure 12. Complete inactivation of mTOR effectors after synergistic eIF4A and mTOR inhibition using CR-1-31-B and Zotatifin in combination with Rapalink-1 in HCC cell lines. Western blot analysis in HLE and HLF cell lines challenged with CR-1-31-B (10 nM) and Zotatifin (10 nM) and Rapalink-1 (10 nM) inhibitors, alone or in combination (1:1). Cells were treated for 48 h, and western blot analysis was conducted at this time point.

3. Discussion

HCC, the most common liver primary cancer, is characterized by its late diagnosis, increasing incidence worldwide, and limited treatment options [51]. Despite the most recent and effective therapeutic strategies [11,52–54], the prognosis for advanced HCC remains poor [55]. Therefore, the identification of novel molecular targets and development of more efficacious therapies are imperative to improving the life expectancy of HCC patients.

Cumulating evidence indicates that translation dysregulation is critical to tumor development and progression [56]. Therefore, targeting translation initiation components might be a potential strategy for therapeutic interventions against cancer [57]. In the present study, we investigated in silico and in vitro the pathogenetic and therapeutic relevance of eIF4A1, a pivotal player in initiating cap-dependent protein translation, in HCC [19,57]. Our results strongly suggest a pro-tumorigenic function of eIF4A1. Specifically, we show that HCC patients with high levels of eIF4A1 mRNA and protein carry poor outcomes. Thus, eIF4A1 might represent a novel prognostic marker in this disease. Accordingly, a negative prognostic role of eIF4A1 has been previously detected in other tumor types, such as lung adenocarcinoma [58], clear cell renal carcinoma [59], and gastric cancer [23], among many others.

Furthermore, we revealed that targeting eIF4A1 using rocaglates induced strong growth suppression of human HCC cells by inhibiting proliferation and triggering apoptosis. Notably, the growth inhibitory activity of these compounds on HCC cells occurred at low nanomolar concentrations, implying their potency, at least in vitro. These data, confirming previous findings obtained in different tumor types, support the targeting of eIF4A1 as a promising therapeutic strategy against HCC [57].

At the molecular level, we discovered that various oncoproteins belonging to the JAK/STAT, ERK/MAPK, AKT, and Hippo/TAZ pathways are markedly downregulated by eIF4A inhibitors. On the other hand, we revealed that treatment with these compounds did not affect the activity of members of the mTOR signaling cascade. Strikingly, exposure of HCC cells to combinations of eIF4A inhibitors and the mTOR inhibitor, Rapalink-1, a third-generation mTOR inhibitor that links rapamycin and MLN0128 [48], resulted in synergistic cytostatic effects. Indeed, this treatment led to higher suppression of proliferation and induction of massive apoptosis compared to that induced by incubation with one of the compounds alone. To the best of our knowledge, this is the first report showing the synergistic effects of eIF4A and mTOR inhibitors against the growth of HCC cells. Furthermore, preliminary data from our laboratory indicate that this drug combination is also synergistic in other tumor entities in vitro, suggesting that cancer cells might be highly vulnerable to eIF4A and mTOR pathway inhibition.

Moreover, this is the first study to demonstrate in the setting of HCC that Rapalink-1-induced inhibition of mTOR signaling can be achieved through suppression of two key downstream substrates, p-RPS6$^{Ser235/236}$ and p-4EBP1$^{Thr37/46}$, even in the presence of eIF4A inhibitors. Hence, we suggest that a combinatory therapy with eIF4A inhibitors and Rapalink-1, a drug that was originally developed to overcome mTOR resistance mutations, might serve as a refinement strategy for systemic management of HCC by simultaneously and effectively targeting both mTOR signaling and eIF4A-dependent translation. Overall, the present data indicate that eIF4A is a potential prognostic biomarker in HCC patients and a promising therapeutic target in this deadly tumor type. In particular, eIF4A inhibitors displayed formidable cytostatic activity against HCC cells in vitro, further augmented in combination with an mTOR inhibitor. Although the antineoplastic effectiveness of these drugs, alone and in combination, requires validation in vivo, the present data suggest that targeting translation initiation can provide an innovative and valuable strategy in the development of novel treatments against human HCC.

4. Materials and Methods

4.1. Human Tissue Specimens

Forty-seven human HCC tumor tissue samples and paired surrounding non-tumorous tissues were collected at the Institute of Pathology at the University of Regensburg (Regensburg, Germany). Patient features are summarized in Supplementary Table S1. HCC tumors were divided between shorter survival/poorer outcome (HCCP; $n = 25$) and longer survival/better outcome (HCCB; $n = 22$), characterized by <3 and >3 years' survival following partial liver resection, respectively. The study was conducted according to the guidelines of the Declaration of Helsinki and approved by the Clinical Research Ethics Committee of the Medical University of Regensburg (protocol code 17-1015-101; 4 July 2018).

4.2. Cell Lines and Reagents

The human HCC cell lines HLE (JCRB0404, Xenotech, Kansas City, MO, USA), HLF (JCRB0405, Tebu-bio GmbH, Offenbach, Germany), PLC/PRF/5 (300315, CLS Cell lines Service GmbH, Eppelheim, Germany), SNU182 (CRL-2235, LGC Standards GmbH), and SNU449 (CRL-2234, LGC Standards GmbH) were cultured at 37 °C in a 5% CO_2 humidified atmosphere. HLE, HLF, and PLC/PRF/5 cell lines were maintained in Dulbecco's Modified Eagle Medium (DMEM, high glucose, Anprotec, Bruckberg, Germany), and SNU182 and SNU449 cell lines were maintained in RPMI 1640 (Anprotec) supplemented with 10% (*v/v*) FCS (Anprotec, Bruckberg, Germany), 2 mM L-glutamine, 10 mM HEPES, 1 mM sodium pyruvate, and 1% penicillin/streptomycin solution (all from Anprotec). Mycoplasma-free status for all cell lines was monitored regularly using the PCR Mycoplasma Test Kit I (PK-CA91-1096, PromoCell, Heidelberg, Germany), and cell line authentication was performed by Cell Lines Service (Eppelheim, Germany). CR-1-31-B (HY-136453, MedChemExpress, Cas. No. 1352914-52-3), FL3 [60], Rapalink-1 (HY-111373, MedChemExpress, Cas. No. 1887095-82-0), Rocaglamide (HY-19356, MedChemExpress, Cas. No. 84573-16-0), Silvestrol (HY-13251, MCE, Cas. No. 697235-38-4), and Zotatifin (HY-112163, MedChemExpress, Cas. No. 2098191-53-6) were used for in vitro experiments. Stock solutions (1 and 10 mM) were prepared in dimethylsulfoxide (DMSO) and aliquots were stored at −20 °C.

4.3. MTT Viability, Proliferation, and Apoptosis Assays

Cells were seeded in flat-bottom 96 well plates at a density of 10 x 10^4 cells per well. After an overnight attachment period, cells were exposed to various concentrations of CR-1-31-B, FL3, Rocaglamide, Silvestrol, Zotatifin, and Rapalink-1 alone or CR-1-31-B-Rapalink-1 and Zotatifin-Rapalink-1 combinations for 48 h. Cells treated with DMSO and wells containing only culture medium served as negative and background controls, respectively.

For the MTT assay, following treatment, 10 µL of 5 mg/mL methyl-thiazolyl-diphenyl-tetrazolium bromide (MTT) solution was added per 96 well and incubated at 37 °C and under 5% CO_2 for 2 h. The medium was fully aspirated, and 100 µL of 100% (*v/v*) DMSO was added per well to dissolve the formazan crystals. After 15 min of shaking at room temperature, the absorbances were measured at 570 and 630 nm using the FLUORstar Omega multiplate reader and MARS data analysis software (both from BMG Labtech, Ortenberg, Germany). All experiments were performed in triplicate or quadruplicate and repeated at least twice. The average absorbance of the DMSO-treated cells was defined as 100% cell viability. IC50 values and standard deviations were calculated using GraphPad Prism software. Average cell viability values of three independent MTT assay experiments were employed to determine dose-effect curves (Fa = fraction affected) and the combination indices of two drugs using CompuSyn PC software version 1 (ComboSyn Inc., Paramus, NJ, USA).

Cell proliferation was evaluated in the cell lines at the 48-h time point using the BrdU Cell Proliferation Assay Kit (Cell Signaling Technology, Danvers, MA, USA). Briefly, following drug treatment, cells were incubated with 1× bromodeoxyuridine (BrdU) for 2 h and fixed for 30 min at room temperature. The fixing solution was discarded, and cells were incubated with the BrdU mouse detection antibody for 1 h at room temperature. After washing, cells were stained with HRP-conjugated anti-mouse secondary antibody for 30 min at room temperature and washed again. After incubation with TMB substrate solution for a further 30 min at room temperature, stop solution was added, and the absorbance was measured at 450 nm. The results are expressed as fold-change over DMSO control.

Apoptosis was determined in the HCC cell lines using the Cell Death Detection Elisa plus Kit (Roche Molecular Biochemicals, Indianapolis, IN, USA), following the manufacturer's instructions. All cell line experiments were repeated at least three times in triplicate.

4.4. Protein Extraction, Western Blot Analysis, and Smart Protein Layers (SPL) Approach

Human HCC tumor and non-tumorous liver tissues were homogenized in 100–250 µL of T-PER^{TM} lysis buffer (78510, Thermo Fisher Scientific) containing 1x Halt^{TM} Protease and Phosphatase Inhibitor Cocktail (78443, Thermo Scientific) in the Bullet Blender Storm 24 (Next Advance, Troy, NY, USA) for 3 min at speed 8, placed in the rotator mixer for 30–60 min at +4 °C and sonicated. Lysis of pellets from the cell lines was performed in the same lysis buffer for 1 h on ice. Protein concentrations were determined using the colorimetric BioRad Protein Assay Dye Reagent Concentrate (500-0006, Bio-Rad, Hercules, CA, USA) with bovine serum albumin (BSA) as the standard.

For western blot analysis, protein lysates were denatured in 1× LDS Bolt^{TM} sample buffer (B0007) and 1× Bolt^{TM} reducing agent (B00009, both from Invitrogen) by boiling for 5 min at 95 °C. Samples with 10 µL of protein per lane were separated by SDS-PAGE on Bolt™ 4–12% Bis-Tris Mini Protein Gels (NW04125BOX) in 1× Bolt™ MES SDS Running Buffer (B000202, both from Invitrogen) and transferred onto nitrocellulose membranes (iBlot^{TM} 2 Mini/Regular Transfer Stacks, IB23001/2) by electroblotting running program P0 in the iBlot™ 2 Gel Transfer Device (IB21001, both from Thermo Fisher Scientific). Membranes were blocked in EveryBlot Blocking Buffer (12010020, Bio-Rad) and probed at +4°C overnight with specific antibodies: eIF4A1 (1:1000; ab31217, Abcam, Cambridge, UK); eIF4A2 (1:1000; ab31218, Abcam); DDX3 (1:1000; A300-474A, Bethyl Laboratories, Montgomery, USA); p-RPS6$^{Ser235//236}$ (4856), p-4EBP1^{Thr3747} (2885), p-STAT3^{Tyr705} (9145), AKT (4691),), ERK1/2 (4695), p-ERK1/2$^{Thr202/204}$ (4370), SKP2 (2652), YAP/TAZ (8418), p-YAP/TAZ (13008), PHB1 (2426), PHB2 (14085) (all 1:1000; Cell Signaling Technology); p-AKT^{Ser473} (1:1000; 66444-1-IG, Proteintech (Rosemont, IL, USA); and β-actin (1:5000; ab20272, Abcam). Horseradish peroxidase-conjugated secondary antibody (1:20,000) was incubated for 1 h at room temperature, and blots were imaged with Clarity Max ECL Western Blotting Substrate (1705062S, Bio-Rad) using ChemiDoc MP (17001402, Bio-Rad). Band volumes were quantified with Image Lab™ software version 6.0.1 (Bio-Rad) and normalized to GAPDH or β-actin.

High patient-to-patient variation of reference protein expression, such as GAPDH and β-actin, impedes proper quantification of western blot data. Hence, for accurate target protein quantification of human HCC samples, western blotting was performed using the Smart Protein Layers (SPL) Kit Red from NH DyeAgnostics (PR913, Halle, Germany) according to the manufacturer's instructions. The SPL technology is based on three components: (i) red-fluorescent Smart Label (SMA label) reagent for labeling and visualization of total protein on gels and blots, (ii) bi-fluorescent sample-specific standard Smartalyzer L (SMA basic L) for normalization, standardization, and quantification of total protein per sample, and (iii) a calibrator (CAL) for comparison of fluorescence and target protein signals derived from different gels. Briefly, before gel electrophoresis, every protein sample was labeled with 2 μL of SMA basic L and 1 μL of Smart label working solution. Samples with of 10 μg of protein per sample, 12 μL of calibrator (CAL; 12.5, 25, and 80 kDa), and 5 μL of Biotinylated Protein Ladder (SeeBlue™ Plus 2, Thermo Fisher Scientific) were applied to the gels. After conventional gel electrophoresis, the gel fluorescence of SMA basic L (signal: band at 80 kDa; channel: Alexa Fluor 488; gel loading control = GLO) and SMA label (signal: whole lane; channel: Cy5; gel total protein = GTO) were imaged. Following blotting, the blot fluorescence of the SMA label (signal: whole lane; channel: Cy5; blot total protein = BTO) and the blot chemiluminescent signal of the target protein (signal: band; chemiluminescence; band target protein = BTA) were detected. Finally, the normalization of gel load (based on SMA basic; GLO), normalization of Smart Label (based on SMA label), and normalization of target protein signal (based on total protein) were conducted using the SPL-normalization template following the manufacturer's protocol. Results were expressed as SPL normalized target volume. Visualization and quantification were performed using the Chemidoc MP Imaging System and Image Lab™ software version 6.0.1 (both from Bio-Rad).

4.5. Immunohistochemistry

Human liver specimens were harvested and fixed in 10% formalin overnight at 4 °C and embedded in paraffin. Hematoxylin and eosin (Thermo Fisher Scientific, Waltham, MA, USA) staining was conducted using a standard protocol on liver sections. Specifically, antigen retrieval was performed in 10 mM sodium citrate buffer (pH 6.0) by heating in a microwave on high for 10 min, followed by a 20-min cool down at room temperature. After blocking with 5% goat serum and the Avidin-Biotin blocking kit (Vector Laboratories, Burlingame, CA, USA), the slides were incubated with the anti-eIF4A1 primary antibody (1:100; ab31217, Abcam) overnight at 4 °C. Slides were then subjected to 3% hydrogen peroxide for 10 min to quench endogenous peroxidase activity and, subsequently, the biotin-conjugated secondary antibody was applied at a 1:500 dilution for 30 min at room temperature. Reaction detection was achieved using the Vectastain ABC-Elite Peroxidase Kit (Vector Laboratories, # PK-6100) with ImmPACT DAB (Vector Laboratories, SK-4105) as the chromogen. Slides were counterstained with Mayer's hematoxylin.

4.6. Statistical Analysis

GraphPad Prism version 9.3.1 software (GraphPad Software Inc., La Jolla, CA, USA) and IBM SPSS version 26 software (IBM Corp., Armonk, NY, USA) were employed to evaluate statistical significance. The p values for TCGA data were obtained from the UALCAN analysis tool. The non-parametric Wilcoxon signed-rank test was used for paired sample comparison. Kaplan-Meier curves were evaluated using the log-rank test. For calculation of IC50s, data were transformed to log2, normalized, and non-linear regression (log)inhibitor vs. response–variable slope (four parameters) was performed. For the transformation of SEM to SD, SEM was multiplied by the square root of the sample size n. Tukey's multiple comparison test was applied for multiple comparisons. Two-tailed values of * $p < 0.5$, ** $p < 0.1$, *** $p < 0.01$, and **** $p < 0.001$ were considered significant. Lowercase letters denoted statistical significance, as stated in the associated figure legends. All data are expressed as the mean ± SD or SEM.

5. Conclusions

Our results, obtained in silico and in vitro, constitute the initial step in an innovative direction to develop pharmacological strategies to treat liver cancer. This is particularly valuable owing to the lack of available tools for efficaciously treating advanced non-surgically resectable HCC. The fact that eIF4A1 was consistently upregulated in HCC in comparison with adjacent non-tumor liver tissue and the magnitude of this change was negatively correlated with the prognosis of HCC patients supports a role of eIF4A1 in HCC malignancy. This led us to evaluate the impact of eIF4A1 inhibitors on HCC cell viability. Interestingly, even at low nanomolar concentrations, these inhibitors triggered a remarkable reduction in proliferation and augmented apoptosis, paralleled by the inhibition of several oncogenic pathways. Moreover, the cytostatic effect of eIF4A1 inhibitors was further increased by the pan mTOR inhibitor, Rapalink-1.

During the last decade, considerable efforts have been made to characterize the molecular basis for the high refractoriness of HCC to currently available first- and second-line drugs [61]. This, together with the identification of new sensitizing targets, will permit us to overcome the current marked limitations in treating these patients, which in clinical practice means that they are subjected to treatments of low efficacy but with harmful side effects. In addition, unsatisfactory management delays the selection of alternative therapeutic options while allowing a Darwinian selection of the most resistant clones, which complicates the situation due to the development of cross-resistance.

In summary, the results of the present study highlight the pathogenic relevance of eIF4A1 in HCC and constitute an encouraging advance in developing new sensitizing strategies. Thus, the usefulness of combining eIF4A1 and mTOR inhibitors with currently used chemotherapeutic agents deserves further investigation.

Supplementary Materials: The following supporting information can be downloaded at: https://www.mdpi.com/article/10.3390/ijms24032055/s1.

Author Contributions: Individual contributions for this research article: Conceptualization, D.F.C.; methodology, A.S.-M., S.M.S., M.M.S., A.C., L.D. and E.T.; validation, D.F.C.; formal analysis, A.S.-M. and S.M.S.; investigation, A.S.-M., D.F.C. and S.M.S.; resources, D.F.C.; data curation, D.F.C.; writing—original draft preparation, A.S.-M., D.F.C. and S.M.S.; writing—review and editing, D.F.C., S.M.S., A.S.-M. and J.J.M.; visualization, A.S.-M., D.F.C. and S.M.S.; supervision, D.F.C., J.J.M. and G.M.P.; project administration, D.F.C.; funding acquisition, M.E. and D.F.C. All authors have read and agreed to the published version of the manuscript.

Funding: DFC was supported by a grant from the Italian Association Against Cancer (AIRC) (grant number IG19175) and the Fondo di Ateneo per la Ricerca 2019 from the University of Sassari (Sassari, Italy). The funders had no role in the study design, data collection and analysis, publication decision, or manuscript preparation.

Institutional Review Board Statement: The study was conducted according to the guidelines of the Declaration of Helsinki, and approved by the Clinical Research Ethics Committee of the Medical University of Regensburg (protocol code 17-1015-101; 4 July 2018).

Informed Consent Statement: Informed consent was obtained from all subjects involved in the study.

Data Availability Statement: Not applicable.

Acknowledgments: We thank Manfred Meyer and Ingrid Winkel (Institute of Pathology, University Regensburg, Regensburg, Germany) for their excellent technical support.

Conflicts of Interest: The authors declare no conflict of interest.

References

1. Sung, H.; Ferlay, J.; Siegel, R.L.; Laversanne, M.; Soerjomataram, I.; Jemal, A.; Bray, F. Global Cancer Statistics 2020: GLOBOCAN Estimates of Incidence and Mortality Worldwide for 36 Cancers in 185 Countries. *CA Cancer J. Clin.* **2021**, *71*, 209–249. [CrossRef] [PubMed]
2. Sleeman, K.E.; de Brito, M.; Etkind, S.; Nkhoma, K.; Guo, P.; Higginson, I.J.; Gomes, B.; Harding, R. The escalating global burden of serious health-related suffering: Projections to 2060 by world regions, age groups, and health conditions. *Lancet Glob. Health* **2019**, *7*, e883–e892. [CrossRef] [PubMed]
3. Akinyemiju, T.; Abera, S.; Ahmed, M.; Alam, N.; Alemayohu, M.A.; Allen, C.; Al-Raddadi, R.; Alvis-Guzman, N.; Amoako, Y.; Artaman, A.; et al. The Burden of Primary Liver Cancer and Underlying Etiologies From 1990 to 2015 at the Global, Regional, and National Level: Results from the Global Burden of Disease Study 2015. *JAMA Oncol.* **2017**, *3*, 1683–1691. [CrossRef]
4. Younossi, Z.; Stepanova, M.; Ong, J.P.; Jacobson, I.M.; Bugianesi, E.; Duseja, A.; Eguchi, Y.; Wong, V.W.; Negro, F.; Yilmaz, Y.; et al. Nonalcoholic Steatohepatitis Is the Fastest Growing Cause of Hepatocellular Carcinoma in Liver Transplant Candidates. *Clin. Gastroenterol. Hepatol.* **2019**, *17*, 748–755.e3. [CrossRef] [PubMed]
5. Grigorie, R.; Alexandrescu, S.; Smira, G.; Ionescu, M.; Hrehoreţ, D.; Braşoveanu, V.; Dima, S.; Ciurea, S.; Boeţi, P.; Dudus, I.; et al. Curative Intent Treatment of Hepatocellular Carcinoma-844 Cases Treated in a General Surgery and Liver Transplantation Center. *Chirurgia* **2017**, *112*, 289–300. [CrossRef]
6. Vogel, A.; Martinelli, E. Updated treatment recommendations for hepatocellular carcinoma (HCC) from the ESMO Clinical Practice Guidelines. *Ann. Oncol.* **2021**, *32*, 801–805. [CrossRef]
7. Marin, J.J.G.; Romero, M.R.; Herraez, E.; Asensio, M.; Ortiz-Rivero, S.; Sanchez-Martin, A.; Fabris, L.; Briz, O. Mechanisms of Pharmacoresistance in Hepatocellular Carcinoma: New Drugs but Old Problems. *Semin. Liver Dis.* **2022**, *42*, 87–103. [CrossRef]
8. Llovet, J.M.; Ricci, S.; Mazzaferro, V.; Hilgard, P.; Gane, E.; Blanc, J.-F.; de Oliveira, A.C.; Santoro, A.; Raoul, J.-L.; Forner, A.; et al. Sorafenib in advanced hepatocellular carcinoma. *N. Engl. J. Med.* **2008**, *359*, 378–390. [CrossRef]
9. Bruix, J.; Qin, S.; Merle, P.; Granito, A.; Huang, Y.-H.; Bodoky, G.; Pracht, M.; Yokosuka, O.; Rosmorduc, O.; Breder, V.; et al. Regorafenib for patients with hepatocellular carcinoma who progressed on sorafenib treatment (RESORCE): A randomised, double-blind, placebo-controlled, phase 3 trial. *Lancet* **2017**, *389*, 56–66. [CrossRef]
10. Abou-Alfa, G.K.; Meyer, T.; Cheng, A.-L.; El-Khoueiry, A.B.; Rimassa, L.; Ryoo, B.-Y.; Cicin, I.; Merle, P.; Chen, Y.; Park, J.-W.; et al. Cabozantinib in Patients with Advanced and Progressing Hepatocellular Carcinoma. *N. Engl. J. Med.* **2018**, *379*, 54–63. [CrossRef]
11. Finn, R.S.; Qin, S.; Ikeda, M.; Galle, P.R.; Ducreux, M.; Kim, T.-Y.; Kudo, M.; Breder, V.; Merle, P.; Kaseb, A.O.; et al. Atezolizumab plus Bevacizumab in Unresectable Hepatocellular Carcinoma. *N. Engl. J. Med.* **2020**, *382*, 1894–1905. [CrossRef] [PubMed]
12. Bhat, M.; Robichaud, N.; Hulea, L.; Sonenberg, N.; Pelletier, J.; Topisirovic, I. Targeting the translation machinery in cancer. *Nat. Rev. Drug Discov.* **2015**, *14*, 261–278. [CrossRef] [PubMed]
13. Silvera, D.; Formenti, S.C.; Schneider, R.J. Translational control in cancer. *Nat. Rev. Cancer* **2010**, *10*, 254–266. [CrossRef] [PubMed]
14. Sonenberg, N.; Hinnebusch, A.G. Regulation of translation initiation in eukaryotes: Mechanisms and biological targets. *Cell* **2009**, *136*, 731–745. [CrossRef] [PubMed]
15. Jackson, R.J.; Hellen, C.U.T.; Pestova, T.V. The mechanism of eukaryotic translation initiation and principles of its regulation. *Nat. Rev. Mol. Cell Biol.* **2010**, *11*, 113–127. [CrossRef]
16. Wolfe, A.L.; Singh, K.; Zhong, Y.; Drewe, P.; Rajasekhar, V.K.; Sanghvi, V.R.; Mavrakis, K.J.; Jiang, M.; Roderick, J.E.; van der Meulen, J.; et al. RNA G-quadruplexes cause eIF4A-dependent oncogene translation in cancer. *Nature* **2014**, *513*, 65–70. [CrossRef]
17. Singh, K.; Lin, J.; Lecomte, N.; Mohan, P.; Gokce, A.; Sanghvi, V.R.; Jiang, M.; Grbovic-Huezo, O.; Burčul, A.; Stark, S.G.; et al. Targeting eIF4A-Dependent Translation of KRAS Signaling Molecules. *Cancer Res.* **2021**, *81*, 2002–2014. [CrossRef]
18. Rubio, C.A.; Weisburd, B.; Holderfield, M.; Arias, C.; Fang, E.; DeRisi, J.L.; Fanidi, A. Transcriptome-wide characterization of the eIF4A signature highlights plasticity in translation regulation. *Genome Biol.* **2014**, *15*, 476. [CrossRef]
19. Li, Q.; Imataka, H.; Morino, S.; Rogers, G.W.; Richter-Cook, N.J.; Merrick, W.C.; Sonenberg, N. Eukaryotic translation initiation factor 4AIII (eIF4AIII) is functionally distinct from eIF4AI and eIF4AII. *Mol. Cell. Biol.* **1999**, *19*, 7336–7346. [CrossRef]
20. Lu, W.-T.; Wilczynska, A.; Smith, E.; Bushell, M. The diverse roles of the eIF4A family: You are the company you keep. *Biochem. Soc. Trans.* **2014**, *42*, 166–172. [CrossRef]
21. Galicia-Vázquez, G.; Cencic, R.; Robert, F.; Agenor, A.Q.; Pelletier, J. A cellular response linking eIF4AI activity to eIF4AII transcription. *RNA* **2012**, *18*, 1373–1384. [CrossRef] [PubMed]
22. Nielsen, P.J.; Trachsel, H. The mouse protein synthesis initiation factor 4A gene family includes two related functional genes which are differentially expressed. *EMBO J.* **1988**, *7*, 2097–2105. [CrossRef] [PubMed]
23. Gao, C.; Guo, X.; Xue, A.; Ruan, Y.; Wang, H.; Gao, X. High intratumoral expression of eIF4A1 promotes epithelial-to-mesenchymal transition and predicts unfavorable prognosis in gastric cancer. *Acta Biochim. Biophys. Sin.* **2020**, *52*, 310–319. [CrossRef] [PubMed]
24. Liang, S.; Zhou, Y.; Chen, Y.; Ke, G.; Wen, H.; Wu, X. Decreased expression of EIF4A1 after preoperative brachytherapy predicts better tumor-specific survival in cervical cancer. *Int. J. Gynecol. Cancer* **2014**, *24*, 908–915. [CrossRef]

25. Modelska, A.; Turro, E.; Russell, R.; Beaton, J.; Sbarrato, T.; Spriggs, K.; Miller, J.; Gräf, S.; Provenzano, E.; Blows, F.; et al. The malignant phenotype in breast cancer is driven by eIF4A1-mediated changes in the translational landscape. *Cell Death Dis.* **2015**, *6*, e1603. [CrossRef]
26. Lin, C.-J.; Cencic, R.; Mills, J.R.; Robert, F.; Pelletier, J. c-Myc and eIF4F are components of a feedforward loop that links transcription and translation. *Cancer Res.* **2008**, *68*, 5326–5334. [CrossRef]
27. Dorrello, N.V.; Peschiaroli, A.; Guardavaccaro, D.; Colburn, N.H.; Sherman, N.E.; Pagano, M. S6K1- and betaTRCP-mediated degradation of PDCD4 promotes protein translation and cell growth. *Science* **2006**, *314*, 467–471. [CrossRef]
28. Roux, P.P.; Topisirovic, I. Signaling Pathways Involved in the Regulation of mRNA Translation. *Mol. Cell. Biol.* **2018**, *38*, e00070-18. [CrossRef]
29. Chen, Z.-H.; Qi, J.-J.; Wu, Q.-N.; Lu, J.-H.; Liu, Z.-X.; Wang, Y.; Hu, P.-S.; Li, T.; Lin, J.-F.; Wu, X.-Y.; et al. Eukaryotic initiation factor 4A2 promotes experimental metastasis and oxaliplatin resistance in colorectal cancer. *J. Exp. Clin. Cancer Res.* **2019**, *38*, 196. [CrossRef]
30. Liu, M.; Gong, C.; Xu, R.; Chen, Y.; Wang, X. MicroRNA-5195-3p enhances the chemosensitivity of triple-negative breast cancer to paclitaxel by downregulating EIF4A2. *Cell. Mol. Biol. Lett.* **2019**, *24*, 47. [CrossRef]
31. Lyu, S.; Lu, J.; Chen, W.; Huang, W.; Huang, H.; Xi, S.; Yan, S. High expression of eIF4A2 is associated with a poor prognosis in esophageal squamous cell carcinoma. *Oncol. Lett.* **2020**, *20*, 177. [CrossRef] [PubMed]
32. Kim, S.; Salim, A.A.; Swanson, S.M.; Kinghorn, A.D. Potential of cyclopentabbenzofurans from Aglaia species in cancer chemotherapy. *Anticancer. Agents Med. Chem.* **2006**, *6*, 319–345. [CrossRef] [PubMed]
33. Pan, L.; Woodard, J.L.; Lucas, D.M.; Fuchs, J.R.; Kinghorn, A.D. Rocaglamide, silvestrol and structurally related bioactive compounds from Aglaia species. *Nat. Prod. Rep.* **2014**, *31*, 924–939. [CrossRef] [PubMed]
34. Ebada, S.S.; Lajkiewicz, N.; Porco, J.A.; Li-Weber, M.; Proksch, P. Chemistry and biology of rocaglamides (= flavaglines) and related derivatives from aglaia species (meliaceae). *Prog. Chem. Org. Nat. Prod.* **2011**, *94*, 1–58. [CrossRef] [PubMed]
35. Iwasaki, S.; Floor, S.N.; Ingolia, N.T. Rocaglates convert DEAD-box protein eIF4A into a sequence-selective translational repressor. *Nature* **2016**, *534*, 558–561. [CrossRef] [PubMed]
36. Cencic, R.; Carrier, M.; Galicia-Vázquez, G.; Bordeleau, M.-E.; Sukarieh, R.; Bourdeau, A.; Brem, B.; Teodoro, J.G.; Greger, H.; Tremblay, M.L.; et al. Antitumor Activity and Mechanism of Action of the Cyclopentabbenzofuran, Silvestrol. *PLoS ONE* **2009**, *4*, e5223. [CrossRef] [PubMed]
37. Chu, J.; Zhang, W.; Cencic, R.; Devine, W.G.; Beglov, D.; Henkel, T.; Brown, L.E.; Vajda, S.; Porco, J.A.; Pelletier, J. Amidino-Rocaglates: A Potent Class of eIF4A Inhibitors. *Cell Chem. Biol.* **2019**, *26*, 1586–1593.e3. [CrossRef]
38. Bordeleau, M.-E.; Robert, F.; Gerard, B.; Lindqvist, L.; Chen, S.M.H.; Wendel, H.-G.; Brem, B.; Greger, H.; Lowe, S.W.; Porco, J.A.; et al. Therapeutic suppression of translation initiation modulates chemosensitivity in a mouse lymphoma model. *J. Clin. Investig.* **2008**, *118*, 2651–2660. [CrossRef]
39. Chen, W.-L.; Pan, L.; Kinghorn, A.D.; Swanson, S.M.; Burdette, J.E. Silvestrol induces early autophagy and apoptosis in human melanoma cells. *BMC Cancer* **2016**, *16*, 17. [CrossRef]
40. Zhang, X.; Bi, C.; Lu, T.; Zhang, W.; Yue, T.; Wang, C.; Tian, T.; Zhang, X.; Huang, Y.; Lunning, M.; et al. Targeting translation initiation by synthetic rocaglates for treating MYC-driven lymphomas. *Leukemia* **2020**, *34*, 138–150. [CrossRef]
41. Cao, Y.; He, Y.; Yang, L.; Luan, Z. Targeting eIF4A using rocaglate CR-1-31B sensitizes gallbladder cancer cells to TRAIL-mediated apoptosis through the translational downregulation of c-FLIP. *Oncol. Rep.* **2021**, *45*, 230–238. [CrossRef] [PubMed]
42. Manier, S.; Huynh, D.; Shen, Y.J.; Zhou, J.; Yusufzai, T.; Salem, K.Z.; Ebright, R.Y.; Shi, J.; Park, J.; Glavey, S.V.; et al. Inhibiting the oncogenic translation program is an effective therapeutic strategy in multiple myeloma. *Sci. Transl. Med.* **2017**, *9*, eaal2668. [CrossRef] [PubMed]
43. Chan, K.; Robert, F.; Oertlin, C.; Kapeller-Libermann, D.; Avizonis, D.; Gutierrez, J.; Handly-Santana, A.; Doubrovin, M.; Park, J.; Schoepfer, C.; et al. eIF4A supports an oncogenic translation program in pancreatic ductal adenocarcinoma. *Nat. Commun.* **2019**, *10*, 5151. [CrossRef] [PubMed]
44. Wang, D.; Tabti, R.; Elderwish, S.; Abou-Hamdan, H.; Djehal, A.; Yu, P.; Yurugi, H.; Rajalingam, K.; Nebigil, C.G.; Désaubry, L. Prohibitin ligands: A growing armamentarium to tackle cancers, osteoporosis, inflammatory, cardiac and neurological diseases. *Cell. Mol. Life Sci.* **2020**, *77*, 3525–3546. [CrossRef]
45. National Library of Medicine. A Phase 1-2 Dose-Escalation and Cohort-Expansion Study of Intravenous Zotatifin (eFT226) in Subjects with Selected Advanced Solid Tumor Malignancies. Clinicaltrials.gov. 2019. Available online: https://clinicaltrials.gov/ct2/show/NCT04092673 (accessed on 1 December 2022).
46. Chen, M.; Asanuma, M.; Takahashi, M.; Shichino, Y.; Mito, M.; Fujiwara, K.; Saito, H.; Floor, S.N.; Ingolia, N.T.; Sodeoka, M.; et al. Dual targeting of DDX3 and eIF4A by the translation inhibitor rocaglamide A. *Cell Chem. Biol.* **2021**, *28*, 475–486.e8. [CrossRef]
47. Nebigil, C.G.; Moog, C.; Vagner, S.; Benkirane-Jessel, N.; Smith, D.R.; Désaubry, L. Flavaglines as natural products targeting eIF4A and prohibitins: From traditional Chinese medicine to antiviral activity against coronaviruses. *Eur. J. Med. Chem.* **2020**, *203*, 112653. [CrossRef]
48. Rodrik-Outmezguine, V.S.; Okaniwa, M.; Yao, Z.; Novotny, C.J.; McWhirter, C.; Banaji, A.; Won, H.; Wong, W.; Berger, M.; de Stanchina, E.; et al. Overcoming mTOR resistance mutations with a new-generation mTOR inhibitor. *Nature* **2016**, *534*, 272–276. [CrossRef]

49. Lu, X.; Paliogiannis, P.; Calvisi, D.F.; Chen, X. Role of the Mammalian Target of Rapamycin Pathway in Liver Cancer: From Molecular Genetics to Targeted Therapies. *Hepatology* **2021**, *73* (Suppl. 1), 49–61. [CrossRef]
50. Chou, T.C.; Talalay, P. Quantitative analysis of dose-effect relationships: The combined effects of multiple drugs or enzyme inhibitors. *Adv. Enzyme Regul.* **1984**, *22*, 27–55. [CrossRef]
51. Llovet, J.M.; Kelley, R.K.; Villanueva, A.; Singal, A.G.; Pikarsky, E.; Roayaie, S.; Lencioni, R.; Koike, K.; Zucman-Rossi, J.; Finn, R.S. Hepatocellular carcinoma. *Nat. Rev. Dis. Primers* **2021**, *7*, 6. [CrossRef]
52. D'Alessio, A.; Cammarota, A.; Prete, M.G.; Pressiani, T.; Rimassa, L. The evolving treatment paradigm of advanced hepatocellular carcinoma: Putting all the pieces back together. *Curr. Opin. Oncol.* **2021**, *33*, 386–394. [CrossRef] [PubMed]
53. Garuti, F.; Neri, A.; Avanzato, F.; Gramenzi, A.; Rampoldi, D.; Rucci, P.; Farinati, F.; Giannini, E.G.; Piscaglia, F.; Rapaccini, G.L.; et al. The changing scenario of hepatocellular carcinoma in Italy: An update. *Liver Int.* **2021**, *41*, 585–597. [CrossRef] [PubMed]
54. Facciorusso, A.; Abd El Aziz, M.A.; Sacco, R. Efficacy of Regorafenib in Hepatocellular Carcinoma Patients: A Systematic Review and Meta-Analysis. *Cancers* **2019**, *12*, 36. [CrossRef] [PubMed]
55. Finn, R.S.; Zhu, A.X. Evolution of Systemic Therapy for Hepatocellular Carcinoma. *Hepatology* **2021**, *73* (Suppl. S1), 150–157. [CrossRef] [PubMed]
56. Xue, C.; Gu, X.; Li, G.; Bao, Z.; Li, L. Expression and Functional Roles of Eukaryotic Initiation Factor 4A Family Proteins in Human Cancers. *Front. Cell Dev. Biol.* **2021**, *9*, 711965. [CrossRef]
57. Jiang, S.-L.; Mo, J.-L.; Peng, J.; Lei, L.; Yin, J.-Y.; Zhou, H.-H.; Liu, Z.-Q.; Hong, W.-X. Targeting translation regulators improves cancer therapy. *Genomics* **2021**, *113*, 1247–1256. [CrossRef]
58. Wu, K.-L.; Huang, Y.-C.; Wu, Y.-Y.; Chang, C.-Y.; Chang, Y.-Y.; Chiang, H.-H.; Liu, L.-X.; Tsai, Y.-M.; Hung, J.-Y. Characterization of the Oncogenic Potential of Eukaryotic Initiation Factor 4A1 in Lung Adenocarcinoma via Cell Cycle Regulation and Immune Microenvironment Reprogramming. *Biology* **2022**, *11*, 975. [CrossRef]
59. Zhang, L.-L.; Chang, W.; He, S.-B.; Zhang, B.; Ma, G.; Shang, P.-F.; Yue, Z.-J. High expression of eIF4A1 predicts unfavorable prognosis in clear cell renal cell carcinoma. *Mol. Cell. Probes* **2022**, *65*, 101845. [CrossRef]
60. Thuaud, F.; Bernard, Y.; Türkeri, G.; Dirr, R.; Aubert, G.; Cresteil, T.; Baguet, A.; Tomasetto, C.; Svitkin, Y.; Sonenberg, N.; et al. Synthetic analogue of rocaglaol displays a potent and selective cytotoxicity in cancer cells: Involvement of apoptosis inducing factor and caspase-12. *J. Med. Chem.* **2009**, *52*, 5176–5187. [CrossRef]
61. Marin, J.J.G.; Briz, O.; Herraez, E.; Lozano, E.; Asensio, M.; Di Giacomo, S.; Romero, M.R.; Osorio-Padilla, L.M.; Santos-Llamas, A.I.; Serrano, M.A.; et al. Molecular bases of the poor response of liver cancer to chemotherapy. *Clin. Res. Hepatol. Gastroenterol.* **2018**, *42*, 182–192. [CrossRef]

International Journal of *Molecular Sciences*

Article

Pterostilbene-Mediated Inhibition of Cell Proliferation and Cell Death Induction in Amelanotic and Melanotic Melanoma

Joanna Wawszczyk [1,*,†], Katarzyna Jesse [2,†] and Małgorzata Kapral [1,*]

[1] Department of Biochemistry, Faculty of Pharmaceutical Sciences in Sosnowiec, Medical University of Silesia in Katowice, Jedności 8, 41-200 Sosnowiec, Poland
[2] Silesian Park of Medical Technology Kardio-Med Silesia, M. Curie-Skłodowskiej 10C, 41-800 Zabrze, Poland
* Correspondence: jwawszczyk@sum.edu.pl (J.W.); mkapral@sum.edu.pl (M.K.); Tel.: +48-32-364-12-61 (J.W.); +48-32-364-12-61 (M.K.)
† These authors contributed equally to this work.

Abstract: Melanoma is one of the fastest-growing cancers worldwide. Treatment of advanced melanoma is very difficult; therefore, there is growing interest in the identification of new therapeutic agents. Pterostilbene is a natural stilbene that has been found to have several pharmacological activities. The aim of this study was to evaluate the influence of pterostilbene on the proliferation and apoptosis of human melanoma cells. Proliferation of pterostilbene-treated amelanotic (C32) and melanotic (A2058) melanoma cells was determined by BRDU assay. Flow cytometric analyses were used to determine cell cycle progression, and further molecular investigations were performed using real-time RT-qPCR. The expression of the p21 protein and the DNA fragmentation assay were determined by the ELISA method. The results revealed that pterostilbene reduced the proliferation of both amelanotic and melanotic melanoma cells. Pterostilbene induced apoptosis in amelanotic C32 melanoma cells, and this effect was mediated by an increase in the expression of the *BAX*, *CASP9*, and *CASP9* genes; induction of caspase 3 activity; and DNA degradation. Pterostilbene did not affect the activation of apoptosis in the A2058 cell line. It may be concluded that pterostilbene has anticancer potential against human melanoma cells; however, more studies are still needed to fully elucidate the effects of pterostilbene on amelanotic and melanotic melanoma cells.

Keywords: pterostilbene; melanoma; anticancer activity; apoptosis; proliferation

Citation: Wawszczyk, J.; Jesse, K.; Kapral, M. Pterostilbene-Mediated Inhibition of Cell Proliferation and Cell Death Induction in Amelanotic and Melanotic Melanoma. *Int. J. Mol. Sci.* **2023**, *24*, 1115. https://doi.org/10.3390/ijms24021115

Academic Editor: Bernhard Biersack

Received: 13 December 2022
Revised: 3 January 2023
Accepted: 3 January 2023
Published: 6 January 2023

1. Introduction

Malignant melanoma originating from melanocytes is one of the fastest-growing and extremely heterogeneous cancers. The occurrence of this cancer continues to increase in most white populations around the world [1]. The genetic and biochemical heterogeneity of this aggressive cancer with high metastatic potential and limited response to chemotherapeutic agents is associated, among other occurrences, with high frequency of mutations such as those in the *BRAF*, *NRAS*, and *C-KIT* genes [2]. Among all melanomas, about 2–8% are recognized as amelanotic melanoma [3]. The amelanotic subtype is a rare form of melanoma lacking melanin that often results in a delayed diagnosis. Late diagnosis, delayed treatment, and more aggressive pathological characteristics contribute to a higher risk of death and recurrence [4,5]. The molecular heterogeneity of melanoma is an obstacle to accurate diagnosis and effective treatment. Malignant melanoma cells have developed many molecular mechanisms leading to protection of abnormal melanocytes against death, including the activation of molecular pathways involved in the regulation of melanoma cell survival, as well as accumulation of gene mutations that promote the survival of altered melanocytes, which consequently lead to tumor promotion [6].

One of the main therapeutic targets for the treatment of melanoma is the induction of cell death, e.g., apoptosis [7]. Melanoma cells avoid programmed cell death through the dysregulation of the balance between proapoptotic and antiapoptotic proteins, as well as caspase activity. One of the most promising strategies is the development of new drugs or therapies targeting proteins of the BCL family, caspases, or other mediators involved in apoptosis [8]. Numerous efforts are being put toward the development of cancer therapeutic strategies with the use of plant-derived compounds [9,10]. Natural compounds, particularly polyphenols, have been proposed to be more potent anticancer drugs than synthetic drugs due to their lower adverse effects, high precision, and secure mode of action. Numerous studies have reported that polyphenols could be considered as a possible therapeutic option in the treatment of cancer cells in the future [11–13].

Pterostilbene (trans-3,5-dimethoxy-4-hydroxystilbene, PTB) (Figure 1a) is a naturally occurring polyphenol that has been revealed to exhibit a variety of pharmacological properties, including the desired antioxidant, anti-inflammatory, and anticancer activity [14]. It has been shown to inhibit the cell cycle and induce both apoptotic and nonapoptotic cell death in several types of cancer cells, such as breast [15], liver [16], or lung cancer [17].

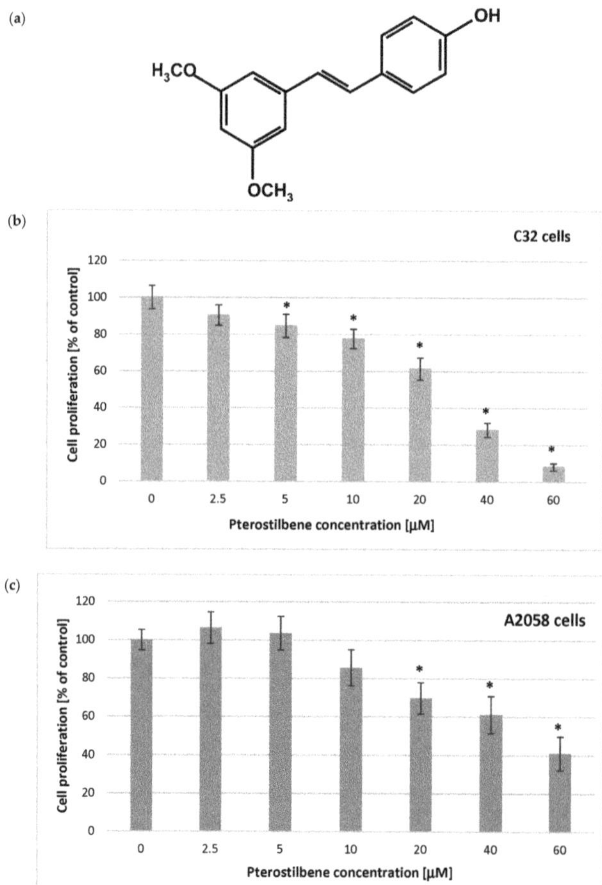

Figure 1. Chemical structures of pterostilbene. (**a**) Influence of pterostilbene on the proliferation of (**b**) C32 and (**c**) A2058 cells after 48 h of treatment. The results are expressed as a percentage of the untreated control (means ± SD; * $p < 0.05$ vs. control).

The broad spectrum of biological activity of pterostilbene suggests that this compound may have a preventive and therapeutic effect on malignant melanoma. To date, there are only a few reports on the biological activity of pterostilbene in melanoma cells and the mechanism by which pterostilbene can exert potential anticancer effects on skin cancers [18]. Most anticancer drugs exert their biological activity by inhibition of the cell cycle and induction of programmed cell death, such as apoptosis, autophagy, and necroptosis [7,8]. Therefore, the objective of this study was to evaluate the influence of pterostilbene in a wide range of concentrations on the proliferation and apoptosis of human amelanotic and melanotic melanoma cells in vitro.

2. Results

2.1. Antiproliferation Activity of Pterostilbene on C32 and A2058 Melanoma Cells

The design of the experiments focused first on investigating the response of pterostilbene-treated melanoma cells. The 5-bromo-2′-deoxyuridine (BrdU) incorporation assay was used to monitor cellular proliferative activity in response to PTB treatment of melanoma cells. Cells were treated with pterostilbene at increasing concentrations (2.5–60 μM) for 48 h. The obtained results have shown that pterostilbene decreased the incorporation of BrdU into newly synthesized DNA after 48 h in a concentration-dependent pattern (Figure 1). Additionally, this depended on the cell type. A substantial reduction in the proliferative activity of C32 cells was found after incubation with pterostilbene at concentrations $\geq$ 5 μM. In A2058 melanotic melanoma cells, a significant decrease in the DNA synthesis level was achieved after treatment with higher concentrations of PTB ($\geq$20 μM). No relevant inhibition of cell proliferation was observed in cultures incubated with 2.5 μM PTB. The IC50 values of 21.45 μM and 42.70 μM for C32 and A2058 cells, respectively, reflected their different levels of sensitivity to pterostilbene. Taken together, these experiments revealed that C32 amelanotic cells are more sensitive to pterostilbene than A2058 melanotic cells.

2.2. The Influence of Pterostilbene on Melanoma Cell Cycle

The cell cycle distribution of C32 and A2058 melanoma cells was assessed by flow cytometry following PTB treatment at concentrations of 20, 40, and 60 μM for 72 h (Figure 2). It was observed that PTB, in a dose-independent manner, markedly decreased the amount of G1/G0 cells and induced cell cycle arrest at the S phase in amelanotic C32 cells. Furthermore, pterostilbene, at concentrations of 40 and 60 μM, significantly reduced the number of C32 cells in the G2/M phase and increased the sub-G1 fraction of cells (Figure 2a). In A2058 melanotic cells, we observed that treatment with 40 and 60 μM PTB significantly altered the percentage of G1/G0 phase cells compared to the control. Exposure of cells to pterostilbene at all concentrations resulted in a significant increase in the population of cells in the sub-G1 phase, suggesting that it induced cancer cell death (Figure 2b).

2.3. The Impact of Pterostilbene on Transcriptional Activity of Genes Encoding the Cell Cycle-Regulating Proteins

The *CCND1* and *CDKN1A* genes encode key cell cycle proteins, i.e., cyclin CD1 and $p21^{Waf1/Cip1}$, respectively. Cyclin D1 determines the cell's transition from the G1 phase to the S phase of the cell cycle, and the p21 protein acts as an inhibitor of cyclin-dependent kinases, resulting in the course of cell cycle inhibition. Therefore, we determined the expression of *CCND1* and *CDKN1A* mRNAs in melanoma cells treated with 20, 40, and 60 μM. Exposure of amelanotic C32 melanoma cells to pterostilbene at all concentrations for 12 h resulted in down-expression of *CCND1* mRNA compared to untreated cells (Figure 3a). However, the obtained results demonstrated that PTB had no statistically important influence on this gene expression in melanotic A2058 cells (Figure 3b).

(**a**)

(**b**)

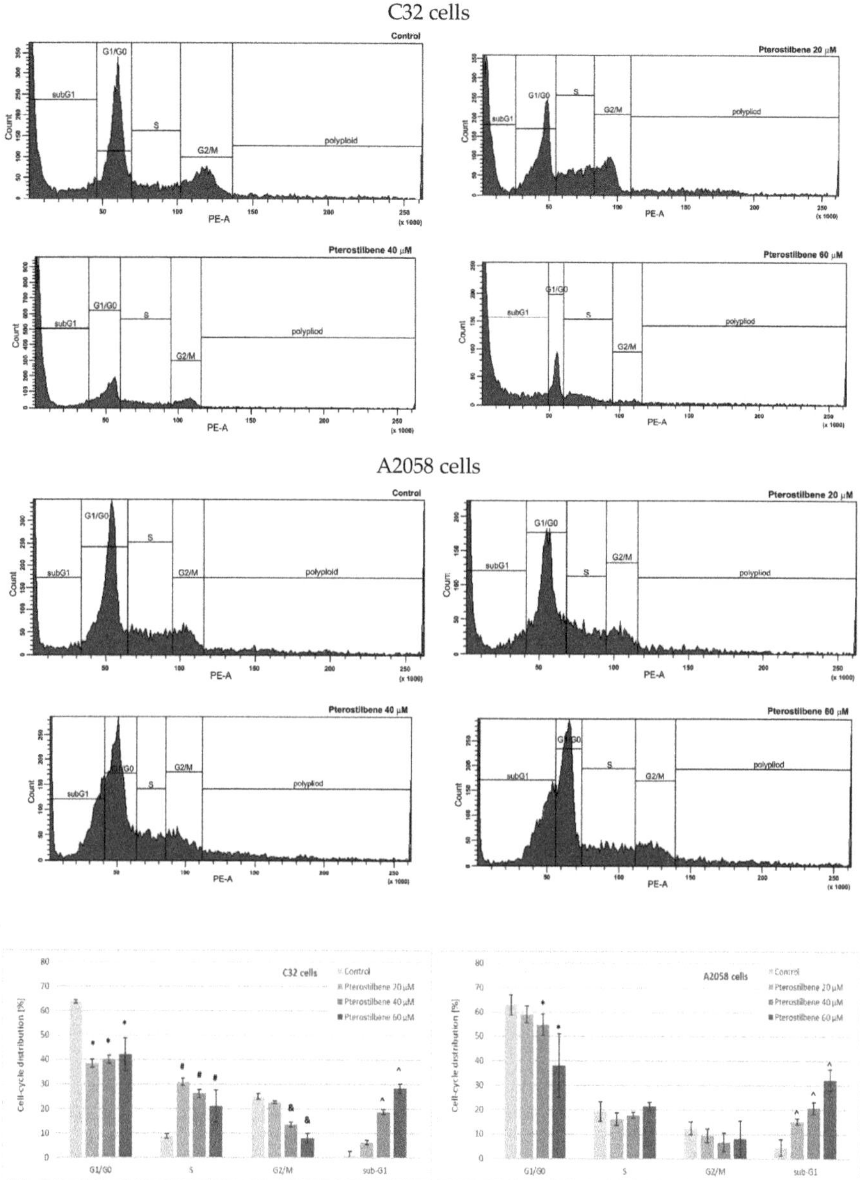

Figure 2. The effect of pterostilbene on the C32 and A2058 cell cycle after 72 h. The cells were labeled with propidium iodide for DNA contents and analyzed by flow cytometry. (**a**) Representative histograms of cell cycle analysis. (**b**) Cell cycle distribution. The data indicate the percentage of cells in each phase of the cell cycle. (means ± SD; * (G1/G0), # (S phase), and & (G2/M), ^ (subG1), $p < 0.05$ vs. control).

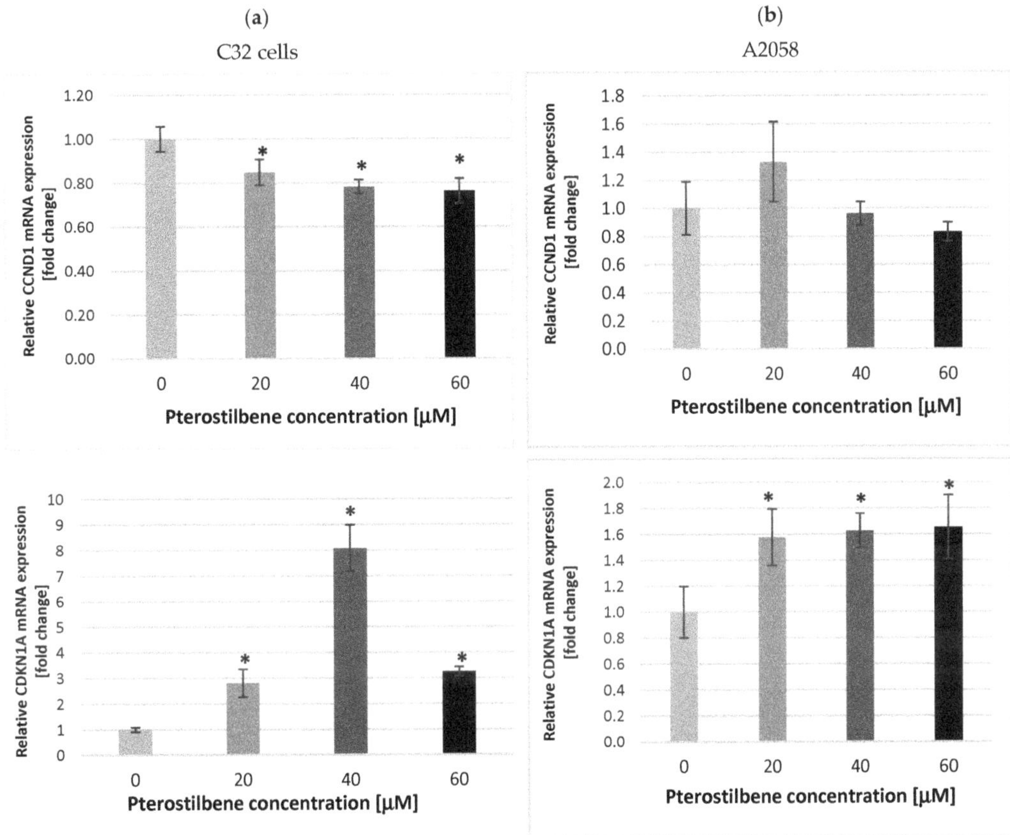

Figure 3. Expression of *CCND1* and *CDKN1A* mRNAs in (**a**) C32 and (**b**) A2058 melanoma cells treated with 20, 40, and 60 μM pterostilbene for 12 h. The results are presented as mean ± SD; * $p < 0.05$ vs. the control.

Compared to the control, an increase in the *CDKN1A* mRNA expression was detected in both melanoma cell lines treated with PTB for 12 h. However, in the C32 culture, a stronger effect of PTB action than on A2058 was observed (Figure 3). The increase in the transcriptional activity of the *CDKN1A* gene in both amelanotic and melanotic melanoma cells under the influence of pterostilbene may result in cell cycle arrest and a decrease in the proliferation potential of melanoma cells.

2.4. The Effect of Pterostilbene on the p21$^{Waf1/Cip1}$ Protein Level

To establish whether the increase in *CDKN1A* mRNA expression corresponded with p21$^{Waf1/Cip1}$ protein level at the next step of the study, the influence of pterostilbene on the concentration of p21 protein in melanoma cells was evaluated (Figure 4). Amelanotic C32 cells characterized higher p21 protein levels than melanotic A2058 cells. The obtained results revealed that PTB had no effect on the p21 concentration in C32 cultures. Exposure of A2058 cells to PTB at all concentrations for 24 h resulted in up-expression of the p21 protein compared to untreated cells. These findings indicated a pterostilbene effect on the *CDKN1A* gene in melanotic A2058 cells, both at the mRNA and protein levels.

(a)

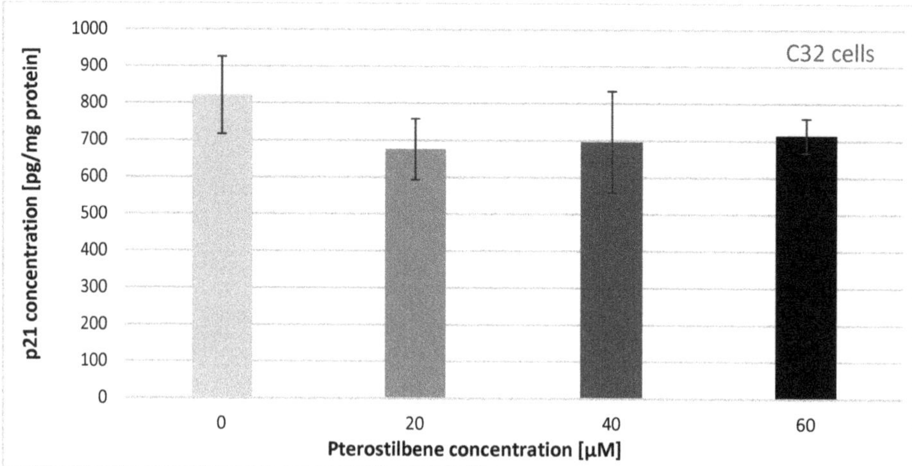

(b)

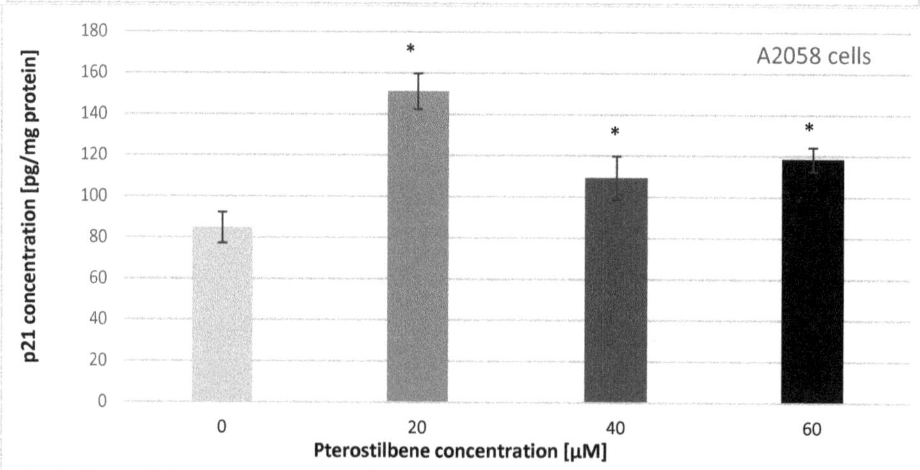

Figure 4. Effect of pterostilbene at concentrations of 20, 40, and 60 µM on the p21 protein in (**a**) C32 and (**b**) A2058 melanoma cells at 24 h. The results are presented as mean ± SD; * $p < 0.05$ vs. control.

2.5. The Influence of Pterostilbene on Melanoma Cell Apoptosis

To further examine whether the mechanism of action of pterostilbene in amelanotic and melanotic melanoma cells involves the induction of apoptosis, the influence of pterostilbene on the expression of genes encoding apoptosis-related proteins (BAX, caspases 3 and 9), caspase 3 activity, and DNA fragmentation levels was evaluated. The transcriptional activity of the examined genes was analyzed in control cells and cells treated with different concentrations of pterostilbene for 12 h using quantitative RT-PCR (Figure 5). BAX belongs to the BCL2 protein family and functions as an apoptotic activator that determines the survival or death of cells. The experimental data revealed that pterostilbene was found, at all concentrations, to significantly increase the expression of the *BAX* gene in amelanotic C32 cells in relation to the controls (Figure 5a). However, treatment of melanotic cells with pterostilbene negatively regulated *BAX* mRNA expression (Figure 5b).

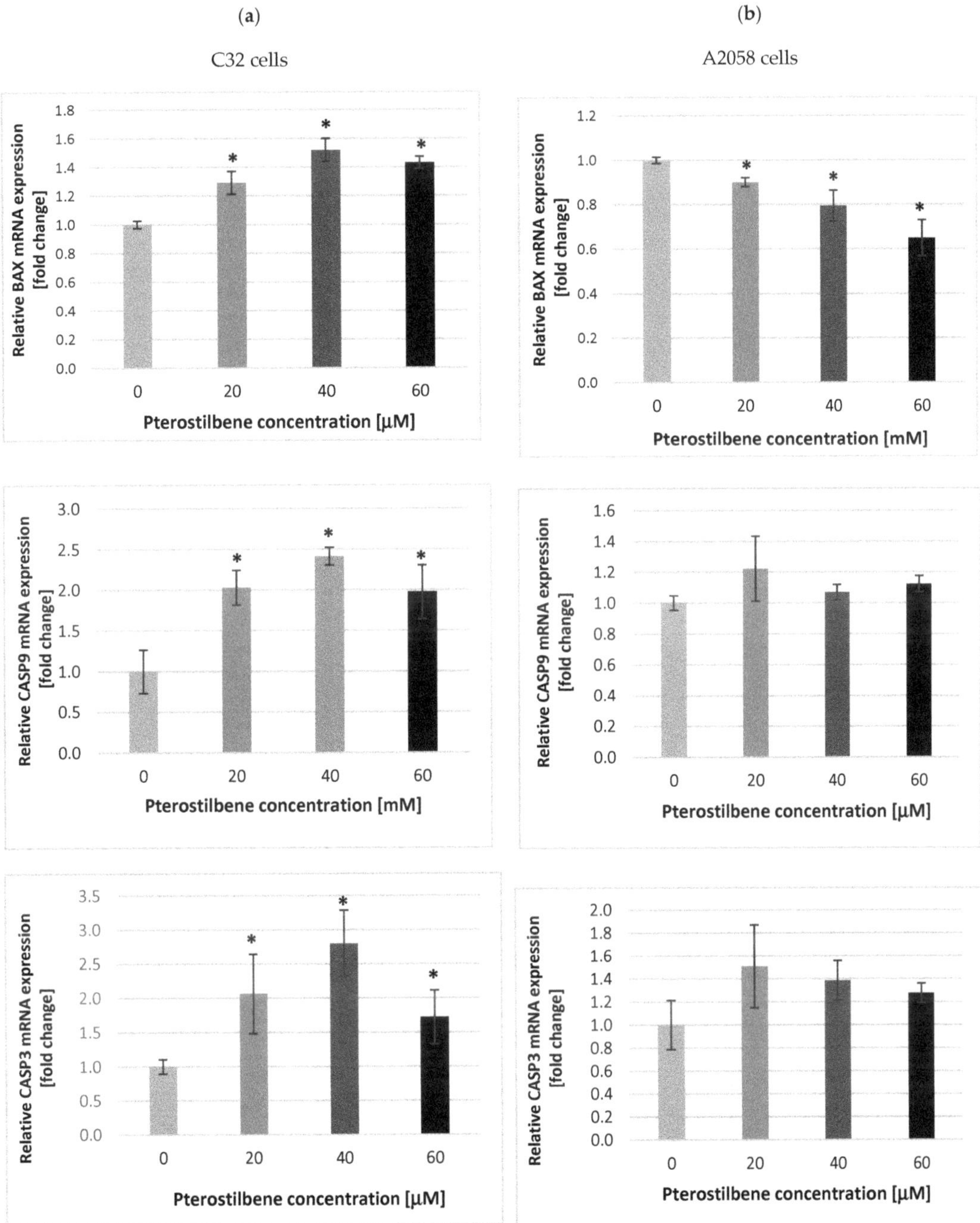

Figure 5. Expression of the *BAX, CASP9* and *CASP3* genes in (**a**) C32 and (**b**) A2058 cells after 12 h of incubation with pterostilbene. The results are presented as mean ± SD; * $p < 0.05$ vs. control.

In the next step, the transcription level of the genes encoding initiator caspase 9 (*CASP9*) and the executive caspase 3 (*CASP3*) were analyzed. Compared to the control, the expression of the *CASP9* and *CASP3* mRNAs in amelanotic C32 cells showed that pterostilbene, at all concentrations, significantly increased the transcriptional activity of both genes by approximately two-fold. However, the strongest effect was observed in cells exposed to PTB at a concentration of 40 µM (Figure 5a). The results indicated that pterostilbene had no influence on the expression of either *CASP9* or *CASP3* mRNAs in A2058 cells (Figure 5b).

Caspase 3 is an enzyme that plays a crucial role in the execution phase of cell apoptosis. To determine the influence of pterostilbene on the induction of caspase 3 activity, cells were incubated with pterostilbene at concentrations of 20, 40, and 60 µM for 48 and 72 h. The effect of pterostilbene on caspase 3 activity in melanoma cells is presented in Figure 6. Pterostilbene, at all concentrations, statistically increased caspase 3 activity in amelanotic C32 cells after 48 and 72 h. The strongest effect was observed in cells exposed to pterostilbene at a concentration of 40 µM at both time points. On the contrary, pterostilbene did not influence caspase 3 activity in melanotic A2058 cells, which may indicate the inability of pterostilbene to induce apoptosis of these cells. Thus, these data suggest that active caspase 3 could be involved in the induction of apoptosis by pterostilbene in amelanotic C32 cells, but not in melanotic A2058 cells.

(**a**)

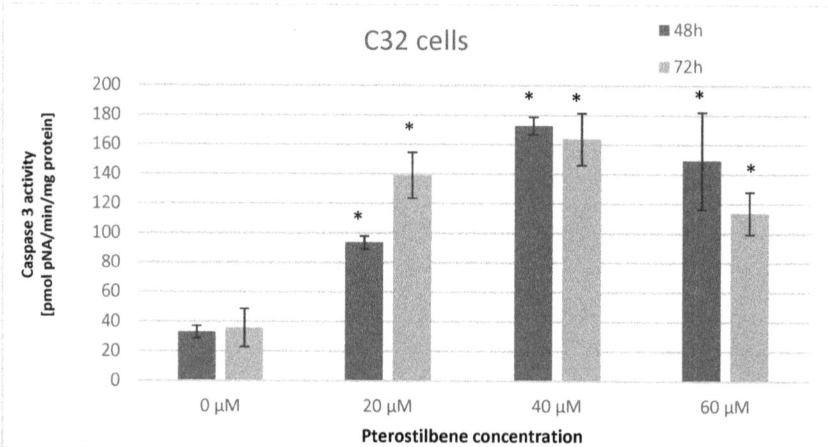

(**b**)

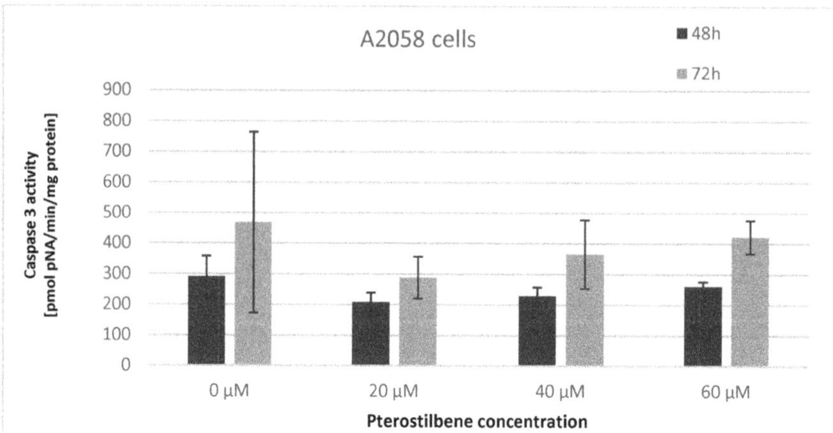

Figure 6. Effect of pterostilbene on caspase 3 activity in (**a**) C32 and (**b**) A2058 cells after 48 and 72 h of incubation. The results are presented as mean ± SD; * $p < 0.05$ vs. control.

To confirm the apoptotic activity of pterostilbene in melanoma cells, a Cell Death Detection ELISA^{PLUS} assay was performed. Cultures C32 and A2058 were incubated with pterostilbene for 24 and 48 h. After 24 h, statistically significant 6.7-fold, 7.6-fold, and 6.6-fold increases in enrichment factors were observed in cells treated with 20, 40, and 60 μM pterostilbene, respectively. Longer incubation (72 h) of C32 with this stilbene caused a similar effect on released nucleosomes (Figure 7). Our results demonstrated that pterostilbene did not influence the formation of nucleosomes in A2058 cells.

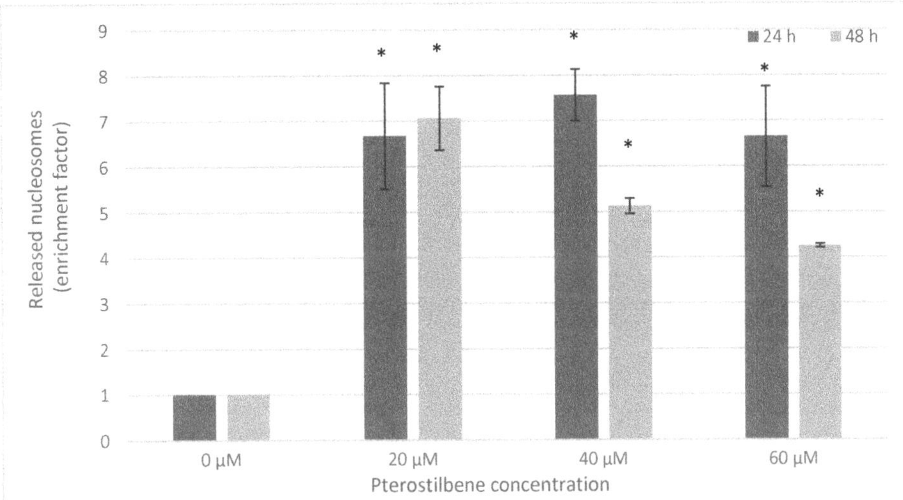

Figure 7. Effect of pterostilbene on the induction of apoptosis in C32 cells evaluated by the mean DNA fragmentation after 24 h and 48 h of incubation. Data are expressed as enrichment factors (means ± SD; * $p < 0.05$ vs. control).

These findings may suggest that pterostilbene induces caspase-dependent apoptotic death of amelanotic C32 cells, but does not stimulate this pathway in melanotic A2058 melanoma cells. The different effects of this stilbene on melanoma cells may depend on melanin pigmentation of cells and the difference in the origin of the tested cells.

3. Discussion

Melanoma is a serious medical problem. Relatively high mortality and a growing number of newly diagnosed cases result in a growing need for research on melanoma. Treatment of advanced melanoma is supported by chemotherapy, immunotherapy, or radiotherapy, but its anticancer effects are unsatisfactory [19]. Therefore, novel and more effective treatment strategies need to be developed.

Phytochemicals are natural compounds that have gained attention as promising chemopreventive and chemotherapeutic agents due to studies that have demonstrated their ability to prevent the development of skin cancer [20,21]. Different phytochemicals perform various functions, including inducing the death of cancer cells by arresting the cell cycle and inhibiting angiogenesis or metastasis [22]. Among the different classes of phytochemicals, polyphenols are promising in the treatment of melanoma [20]. One of the most studied polyphenols is resveratrol. Several preclinical studies have revealed that resveratrol is active against melanoma; it has been shown to decrease the growth of amelanotic and melanotic melanoma cells [23]. Furthermore, it has been shown to induce the death of melanoma cells through caspase-dependent and caspase-independent pathways [24,25]. Despite its promising anticancer role, resveratrol has unfavorable pharmacodynamics due to its high metabolism leading to a reduced concentration in the human body [26]. Therefore, much focus has shifted towards other polyphenols with biological activity. Pterostilbene is a naturally occurring polyphenol found in blueberries and grapes that shows a higher level of bioavailability than resveratrol [27]. It has gained increasing attention due to its role in the prevention of different diseases, such as cardiovascular diseases, neurological disorders, and metabolic diseases [28–30]. Pterostilbene has previously been described to have promising anticancer activity against various types of cancer by influencing cancer cell proliferation and death, among other effects [31]. It has been reported to arrest cell-cycle progression at the G1/G0 phase in breast cancer cells by up-regulation of p21 and cyclin D1 suppression [32]. Furthermore, the antiproliferative and cell death-inducing activity of pterostilbene has been shown against pancreatic [33], colon [34], lung [35], and ovarian [36] cancer cells in vitro and in vivo. Despite reports on the anticancer properties of pterostilbene, data on its activity against melanoma cells are still limited. Therefore, it is important to investigate the effect of pterostilbene on melanoma cells.

Since evaluation of cancer cell proliferation and death is a key component in the discovery and development of antineoplastic drugs in the current study, changes in the growth of two different human melanoma cells, melanotic A2058 and amelanotic C32, were studied after incubation with pterostilbene, as was the ability of this stilbene to induce cell death. In more detail, the results of our study revealed a significant dose-dependent reduction in the proliferation rate of both types of melanoma cells after incubation with pterostilbene. The study by Bennloch et al. [37] also revealed that pterostilbene, at low concentrations (1–5 μM), did not alter the growth nor the viability of melanoma cells (A2058, MeWo, and MelJuso) after 72 h of incubation. The time- and dose-dependent growth-inhibitory effect of pterostilbene was observed at concentrations of 10–100 μM against A375 melanoma cells, as well as lung cancer cells (A549), colon cancer cells (HT29), and breast cancer cells, in studies carried out by Mena et al. [38]. The obtained value of pterostilbene's half-maximal inhibitory concentrations for A549 cells was 14.7 μM, and was lower than that for breast cancer cells (44 μM) and colon cancer cells (60.3 μM). Due to the different sensitivities of the cells studied to pterostilbene, the authors suggested that the reduction in the number of tumor cells depends on differences in the cell lines assayed.

Cell cycle analysis was performed to examine the mechanism underlying the effect of pterostilbene on melanoma cells. The analysis performed herein indicated that pterostilbene caused changes in the cell cycle profile of both C32 and A2058 cells; however, the influence appeared to be different depending on the type of cells. Pterostilbene decreased the number of cells in the G1/G0 and G2/M phases, and also induced cell cycle arrest at the S phase in amelanotic C32 cells. Treatment of melanotic cells with pterostilbene also reduced the percentage of cells in the G1/G0 phase, but the effect was weaker than in amelanotic cells. In both cell lines, an increase in cell population in the sub-G1 phase was observed after exposure to pterostilbene. Therefore, these results indicate the ability of pterostilbene to induce the arrest of the melanoma cell cycle and the induction of cell death. Although many studies have revealed the influence of pterostilbene on the cancer cell cycle and changes in its distribution in various phases, so far only single studies have been published on its effect on melanoma cells. It has been found that the mechanism of action of pterostilbene varies in different cells. Pterostilbene led to inhibition of the cell cycle in the S phase with a concomitant decrease in the number of cells in the G2/M phase in a SMMC-7721 hepatocellular carcinoma [39]. The results of these studies are consistent with the data obtained in the present study for C32 melanoma cells. The accumulation of cells in the S phase of the cell cycle may result from the activation of DNA repair processes. In lymphoma cells, pterostilbene was found to increase the level of CHEK2, a protein kinase known as an important mediator of the DNA damage checkpoint for phosphorylate proteins involved in DNA repair and cell cycle arrest [40,41]. Chen et al. [42] showed a pterostilbene-induced increase in a marker protein of double-stranded DNA breaks contributing to cells' genomic instability in H929R myeloma cells. They suggested that pterostilbene could induce DNA damage leading to cell cycle arrest in the S phase. In oral cancer cells SAS and OECM-1, pterostilbene caused both S phase cycle arrest and increased the percentage of the cell population in the G1/G0, which may indicate that arrest of cell division was inhibited in the G1 phase [43]. Pterostilbene also induced cell cycle arrest in the G1 phase in HL-40 leukemia. However, in the case of these cells, it also caused a decrease in the number of cells in the S and G2/M phases [44]. The effect on pterostilbene on melanoma cells was only analyzed by Mena et al. [38]. They revealed that a high concentration of pterostilbene (75 μM) induced the portion of A375 cells in the G1 phase. However, pterostilbene, at lower concentrations, reduced the number of A375 melanoma cells in the G1 phase with an increase in the population of cells in the S phase. These results are in line with the findings of the current study. The presented studies suggest that the effect of pterostilbene on the cell cycle varies depending on the concentration used, the origin of the cells, and their invasiveness. Therefore, further experiments should be conducted to determine the exact mechanism of pterostilbene activity in melanoma cells.

To clarify the mechanism of pterostilbene activity against colon cancer cells, we also evaluated changes in the expression of the cell cycle regulatory genes *CCND1* and *CDKN1A*. Molecular analysis of human melanoma reveals that cell cycle regulators are frequently mutated. The frequency of mutations in the *CCND1* gene encoding cyclin D1 depends on the melanoma subtype and ranges from 19 to 80%. Moreover, CCND1/cyclin D1 up-regulation favors the growth and development of the primary tumor; therefore, it is considered as the proto-oncogene [45]. The *CDKN1A* gene encodes the protein p21, which acts as an inhibitor of cyclin-dependent kinases (CDKs). It not only promotes cell cycle arrest, but also interacts with the proliferating cell nuclear antigen (PCNA), leading to the arrest of DNA synthesis and the initiation of repair processes [46]. In melanoma cells, disturbances in *CDKN1A* expression are frequent, and usually lead to their uncontrolled proliferation [47]. Cyclin D and the p21 protein participate in the regulation of the cell cycle at the G1/S restriction point. Our data indicated that treatment of C32 amelanotic melanoma cells with pterostilbene significantly decreased gene expression in C32, suggesting disruption of the uncontrolled progression of the cell cycle of these cells. Surprisingly, pterostilbene did not significantly affect *CCND1* transcriptional activity in A2058 melanotic melanoma cells; therefore, more studies are required to determine the antiproliferative mechanism of pterostilbene against these cells. Interestingly, there is a lack of RB protein that is regulated by cyclin D in A2058 cells. The loss of the functional RB protein was associated with the failure of cells to arrest the cell cycle at the G1/S checkpoint [48]. Until now, the expression of cyclin D in pterostilbene-treated melanoma cells has not been evaluated. However, published studies have shown the ability of pterostilbene to down-regulate cyclin D in several types of cancer cells, such as breast cancer [32] and colon cancer [36] cells. Furthermore, pterostilbene increased the expression of the *CDKN1A* gene that encodes the p21 protein in amelanotic and melanotic cells. However, changes in *CDKN1A* mRNA were accompanied by up-regulation of p21 expression only in melanotic cells. Based on data from the literature, increased expression of p21 blocks the transition of cells from the G1 to the S phase. Similar results were obtained in the present study. Incubation of A2058 cells with pterostilbene resulted in an increase in the p21 protein; however, it did not influence the number of cells in phase S. In C32 cells, pterostilbene did not affect the concentration of the p21 protein and did not increase the number of cells in S phase [49]. In this context, it can be suggested that one of the possible mechanisms of the biological activity of pterostilbene is related to the inhibition of proliferation. The arrest cell cycle in the G1 phase is an opportunity for cells to go through the repair mechanism or follow the apoptosis pathway.

As observed, pterostilbene revealed a strong growth-inhibitory effect. Incubation of melanoma cells with pterostilbene caused a reduction in DNA synthesis and cell cycle arrest, as well as a significant decrease in the number of cells, which can only be explained when cell death is induced. Pterostilbene has been shown to activate cancer cell death through different mechanisms such as apoptosis, autophagy, and necrosis; however, apoptosis is assumed to be the main type of pterostilbene-induced cell death [50–52]. Apoptosis plays a crucial role in the elimination of damaged and abnormal cells, and dysregulation in the apoptotic cell death machinery is a hallmark of a variety of cancer cells [53]. Melanoma cells are characterized by a low sensitivity to apoptosis associated with high expression of antiapoptotic proteins of the BCL-2 family [7], loss of activity of Apaf-1 factor [54] and up-regulation of the PI3K/AKT/mTOR pathway [55]. Therefore, apoptosis-targeted drugs are promising in melanoma therapy. To determine the mechanism of pterostilbene-induced cell death in the current study, the possible impact of pterostilbene on *BAX*, *CASP9*, and *CASP3* gene expression was evaluated. We found that pterostilbene significantly increased the expression of *BAX*, *CASP9*, and *CASP3* in amelanotic C32 cells, but not in melanotic A2058 cells. Apoptosis is tightly regulated by effector molecules such as executioner caspase 3, which is responsible for the breakdown of the nucleus and other cellular compartments during apoptosis [56]. The current study revealed that pterostilbene activated caspase 3 in C32 cells, but not in A2058 cells. To confirm the pro-apoptotic influence of pterostilbene, we investigated nucleosome formation, a marker of apoptotic death. The results showed that pterostilbene significantly increased nucleosome occurrence in amelanotic C32 cells, but had no effect in melanotic A2058 cells, indicating different mechanisms of cell death-inducing activity of pterostilbene in C32 and A2058 melanoma cells. Based on the current results, we confirmed that pterostilbene activated caspase-dependent apoptosis in human melanoma amelanotic C32 cells, as well as caspase-independent cell death in melanotic A2058. This implies that pterostilbene could induce cell death through various molecular pathways. Despite limited reports on the pro-apoptotic effects of pterostilbene on melanoma cells, numerous studies have shown multidirectional mechanisms of pterostilbene-induced cell death in breast, gastric, prostate, colon, and lung cancer cells [57–61]. Pterostilbene has been shown to cause depolarization of mitochondrial membranes, release of cytochrome C [59,62], up-regulation of the expression of pro-apoptotic proteins, increases in caspase activity, and inhibition of the expression of anti-apoptotic proteins [63]. Research evidence suggests that pterostilbene, dependent on the cell type, can induce different mechanisms of cell death. Studies by Mena et al. [38] showed that pterostilbene significantly induced caspase 3 activity in melanoma A375 cells and lung cancer A549 cells; however, it did not have this effect in colon HT-29 cancer cells or breast MCF-7 cancer cells. Studies carried out by Ferrer et al. [64] found that the pro-apoptotic properties of pterostilbene in B16M-F10 melanoma cells were indicated by an increase in expression of the *BAX* gene and inhibition of expression of the *BCL-2* gene. Pterostilbene has also been reported to induce the formation of apoptotic bodies and increase the activity of caspases 3 and 7 in SK-MEL-2 melanoma cells [52]. Furthermore, pterostilbene has also been reported to induce necrosis, cancer cell autophagy [50], and permeabilization of the lysosomal membrane [38].

Overviewing all of our data together, pterostilbene demonstrates a potent anticancer influence against melanoma cells in vitro thought cyclin D1, p21, and caspase 3, causing cell cycle arrest as well as both caspase-dependent and caspase-independent cell death (Figure 8).

Furthermore, our results showed significant differences between the studied C32 amelanotic and A2058 melanotic cells. As observed, a more pronounced effect of pterostilbene was detected against amelanotic cells than against A2058 cells. Moreover, in amelanotic cells, it induced caspase-dependent apoptotic cell death.

Despite significant progress in our understanding of melanoma, its heterogeneity is one of the major obstacles to clinical efficiency of anticancer drugs. Melanoma cells are known for their high plasticity and ability to switch back and forth between different melanoma cell states [65]. Heterogeneity can be observed at both the genetic and biological levels, even within melanoma cell lines. Biological heterogeneity includes switching between mostly melanotic proliferative and invasive mesenchymal-like states of cells. Different phenotypes may also have different sensitivities to a drug. Moreover, some melanoma cells that are responsive to drugs can evolve into a drug-tolerant state [65,66]. The consequences of the highly heterogeneous nature of melanoma are reflected in the clinical presentation of the therapeutic response, and represent a challenge for identifying effective treatment strategies. The sensitivity of melanoma cells to targeted inhibitors and immunotherapies has been shown to be associated with different levels of cancer cell differentiation (differentiated phenotypes exhibit higher sensitivity to BRAF inhibitors) [67,68].

Furthermore, cells with innate sensitivities to BRAF inhibitors may become drug-resistant as a result of BRAF inhibition [65]. Recently published studies have shown that even isogenic melanoma cells can undertake different, independent trajectories between drug-responsive and drug-tolerant states, each of them characterized by unique metabolic and signaling networks [66]. New methods, such as single-cell functional proteomics, allow the identification of signaling pathways activated after BRAF inhibition and before the appearance of drug-resistant phenotypes. That cell-to-cell differences seem critical to therapeutic response to single-agent therapies, and simultaneously, targeting these distinct pathways is essential to fully prevent drug tolerance. Combination therapy using drugs targeting those pathways with BRAF inhibition may be able to halt the adaptive transition [67,68]. Therefore, it is important to incorporate intratumor heterogeneity and the expected evolutionary trajectories to drug tolerance into the design of a new drug combination for melanoma treatment. Studies also showed the relationship between cellular de-differentiation and metabolic reprogramming. Therefore, it seems interesting to evaluate whether the metabolic differences between cancer cells can be used to resensitize them to drugs. Differentiated melanoma cell lines have been shown to be more sensitive to fatty acid synthesis inhibitors, while differentiated ones are more sensitive to lipid monounsaturation [69]. Understanding the role of heterogeneity in the aggressiveness of melanoma and resistance to treatment could allow us to determine more effective types of therapy; develop new therapeutics, including targeted polypharmacology; and achieve better outcomes for patients [70].

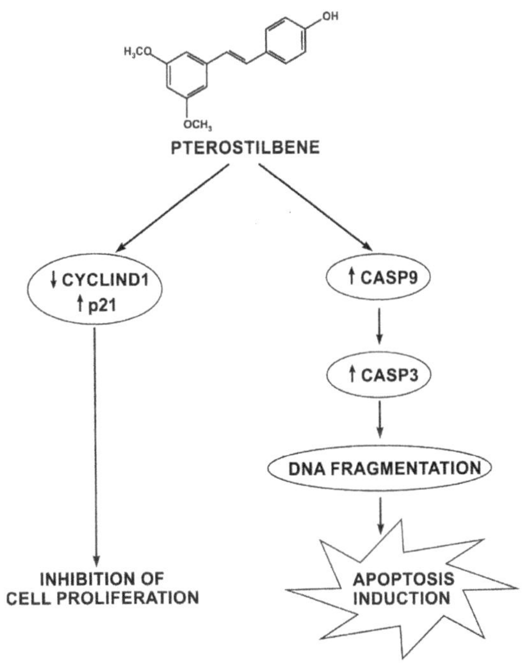

Figure 8. The potential mechanism of pterostilbene action on melanoma cells.

The presented results suggest that pterostilbene is more effective towards the amelanotic C32 cell line. Recognition of the underlying pathways affected by pterostilbene in melanoma cells, as well as the relationship between the mechanism of action of pterostilbene and its ability to synthesize melanin and the mutation profile, seems especially important due to the high heterogeneity of melanoma. It is also essential due to the high aggressiveness of amelanotic melanoma and diagnoses at a more advanced clinical stage than pigmented melanomas. The question still remains as to how to explain the different sensitivities of C32 and A2058 cells to pterostilbene. The melanin content of the C32 and A2058 cells varies. Melanin plays a crucial role in preventing damage caused by free radicals, so melanoma cells with a low melanin content are significantly less resistant to reactive oxygen species damage than cells with higher melanin content [71]. Further investigation of the activity of pterostilbene against amelanotic and melanotic melanoma cells is clearly warranted.

4. Materials and Methods

4.1. Cell Lines and Cell Culture

The A2058 melanotic and C32 amelanotic human melanoma lines were obtained from the American Type Culture Collection (ATCC, Rockvile, MD, USA). Both melanoma cell lines are highly tumorigenic in nude mice with the BRAFV600E mutation, but differ in their ability to synthesize melanin. Cells were routinely cultured in MEM medium (Sigma Aldrich, St. Louis, MO, USA) supplemented with 10% fetal bovine serum (BioWest, Nualillé, France), 100 U/mL penicillin, and 100 µg/mL streptomycin (Sigma Aldrich) in a humidified atmosphere containing 5% CO_2 and 95% air at 37 °C. Cells were treated with pterostilbene solutions, and the experiment proceeded as described below.

4.2. Preparation of Pterostilbene Solution

The stock solution of pterostilbene (Sigma Aldrich) was prepared by dissolving it in dimethylsulfoxide (DMSO, Sigma Aldrich). Subsequently, pterostilbene solution was diluted in a sterile culture medium to the desired concentration directly prior to use. The final concentration of DMSO was 0.1%.

4.3. Cell Proliferation Analysis

To evaluate the effect of pterostilbene on melanoma cell proliferation, a 5-bromo-2′-deoxyuridine (BrdU) colorimetric enzyme-linked immunosorbent assay kit (ELISA) (Roche, Mannheim, Germany) was used. Cells were seeded in 96-well plates at a density of 8×10^3 in 200 µL medium, followed by overnight incubation. Subsequently, the medium was replaced with a fresh one containing pterostilbene (2.5; 5; 10; 15; 20; 40; 60 µM) and treated for 48 h. The BrdU solution was added to the media for the last 4 h of incubation. After removal of the labeling media, cells were fixed and DNA was denatured with FixDenat solution for 20 min. Immune complexes were formed using a peroxidase-conjugated antibody. The incorporation of BrdU into DNA was determined by absorbance measuring at $\lambda = 450$ nm (with reference $\lambda = 690$ nm) using a microplate spectrophotometer Labtech LT-5000c. The growth of treated cells was expressed as a percentage of untreated control cells. The pterostilbene concentration that reduced proliferation by 50% (IC50) compared to control was determined by fitting a four-parameter logistic model (Hill equation) to the experimental data using computer curve fitting software (GraphPad Prism version 9, San Diego, CA, USA).

4.4. Cell Cycle Analysis

To evaluate the distribution of the cell cycle, A2058 cells (1.5×10^5 cells) and C32 cells (5×10^5 cells) were seeded in 21.5 cm^2 dishes (Nunc International, Rochester, NY, USA) and grown for 24 h. Subsequently, cell cultures were treated with PTB at concentrations of 0, 20, 40, and 60 µM. After 72 h of incubation, cells were harvested by trypsinization, washed in PBS buffer, and fixed in cold 70% ethanol at −20 °C overnight. Cell pellets were incubated with RNaseA (final concentration, 200 µg/mL) in PBS buffer for 1 h at 37 °C in the dark and then stained with a solution of propidium iodide (final concentration, 10 µg/mL) (Sigma Aldrich). The DNA content and cell cycle distribution of cells were analyzed by a BD FACS Aria II flow cytometer and BD FACSDiva software (BD Biosciences, San Jose, CA, USA).

4.5. DNA Fragmentation Assay

To demonstrate the pro-apoptotic potential of pterostilbene "Cell Death Detection, ELISA^{PLUS}" (Roche) was used. This method reveals the occurrence of nuclear DNA fragmentation by the use of antibodies against DNA and histones. Melanoma cells were seeded in 96-well plates at a density of 7.5×10^3 cells/200 µL and cultured for 24 h. Then, cells were treated with pterostilbene at concentrations of 20, 40, and 60 µM for 24 h or 48 h. According to the manufacturer's instructions, cells were lysed and centrifuged to produce a nucleosome-containing supernatant. The enrichment of mono- and oligonucleosomes released into the cytoplasm of cell lysates was detected by biotinylated anti-histone- and peroxidase-coupled anti-DNA-Ab. DNA fragmentation was expressed as the enrichment of histone-associated mono- and oligonucleosomes released into the cytoplasm. Spectrophotometric results were measured at a wavelength of 405 nm. The enrichment factor used as the parameter of apoptosis was calculated using the formula of absorbance of sample cells/absorbance of control cells to estimate the fold increase of DNA fragmentation in treated samples with reference to the control.

4.6. Total RNA Extraction and Quantitative Real-Time RT-PCR (RT-qPCR)

To evaluate the expression of the *CCND1, CDKN1A, CASP3, CASP9*, and *BAX* genes, cells were seeded at a density of 8×10^5 onto 21.5 cm^2 culture plates (Nunc International, Rochester, NY, USA) and grown for 48 h. Then, PTB, at concentrations of 20, 40, and 60 µM, was added to the cell cultures for 12 h. Total RNA was extracted from cells with the use of TRI REAGENT (Zymo Research, Irvine, CA, USA) according to the manufacturer's instructions. RNA concentration and purity were detected using the Shimadzu UV-1800 spectrophotometer (Shimadzu, Kyoto, Japan). Samples showing a ratio of Abs 260/280 nm between 1.8 and 2.0 were only used for experiments. Detection of the expression of the examined genes was carried out using a RT-qPCR technique with SYBR Green chemistry (SensiFastTM SYBR Green No-ROX One-Step) (Bioline, Meridian Bioscience, Cincinnati, OH, USA) and CFX Connect Real-Time PCR Detection System (Bio-Rad, Hercules, CA, USA). Aliquots (0.1 µg) of total cellular RNA were applied to one-step RT-qPCR at 20 µL reaction volume. Oligonucleotide primers specific for *CASP3* and *CASP9* mRNAs were synthesized in Oligo.pl at the Institute of Biochemistry and Biophysics of the Polish Academy of Sciences (Warsaw, Poland). The primers for *CCND1, CDKN1A*, and *BAX* were commercially available (Sigma-Aldrich). The characteristics of the primers are presented in Table 1. The thermal profile for RT-qPCR was as follows: 45 °C for 10 min for reverse transcription and 95 °C for 2 min, followed by 45 cycles at 95 °C for 5 s, 60 °C for 10 s, and 72 °C for 5 s for amplification. Each gene analysis was performed in triplicate. The mRNA copy numbers of the examined genes were determined on the basis of the commercially available standard of β-actin (TaqMan DNA Template Reagent Kit, Invitrogen, Waltham, MA, USA) and recalculated per µg of total RNA. The expression levels of all genes in cultured cells were expressed as a fold change relative to the corresponding controls. The fold change > 1 and <1 were set as an increased and a decreased expression of the target gene, respectively.

Table 1. Characteristics of the primers used in the experiment.

Gene	Forward Primer (5′-3′)	Reverse Primer (5′-3′)
CCND1	GCCTCTAAGATGAAGGAGAC	CCATTTGCAGCAGCTC
CDKN1A	AGGGATTTCTTCTGTTCAGG	GACAAAGTCGAAGTTCCATC
CASP3	GGCCTGCCGTGGTACAGAACTGG	AGCGACTGGATGAACCAGGAGCCA
CASP9	GACCGGAAACACCCAGACCAGTGGA	GCAGTGGCCACAGGGCTCCAT
BAX	TCTGAGCAGATCATGAAGAC	TCCATGTTACTGTCCAGTTC

4.7. Detection of the $p21^{Waf1/Cip1}$ Protein Level

To determine the concentration of the $p21^{Waf1/Cip1}$ protein, melanoma cells were seeded at a density of 4×10^6 onto 56.7 cm^2 culture dishes and cultured for 24 h. One day after plating, the cultures were exposed to pterostilbene at concentrations of 20, 40, and 60 μM for 24 h. Afterwards, cells were washed with ice-cold PBS, scrapped from the dishes, centrifuged, and lysed on ice in a cell extraction buffer. Expression of the p21 protein was determined with commercially available ELISA kits (Invitrogen) following the manufacturer's instructions. The absorbance was measured using the Labtech LT-5000 multiplate reader (Labtech International) at λ = 450 nm. The concentration of p21 was evaluated on the basis of the standard curve generated under identical conditions. The results obtained were normalized to the total protein content in the cells, as measured by the Bradford assay (Sigma-Aldrich).

4.8. Caspase 3 Activity

Detection of caspase 3 activity was performed with the commercially available "Colorimetric Caspase 3 Assay Kit" (Sigma Aldrich) based on the hydrolysis of the substrate, Ac-DEVD-pNA (acetyl-Asp-Glu-Val-Asp-pNA), by active caspase 3. Cells were seeded (2.5×10^6) onto 56.7 cm^2 dishes and cultured for 24 h. Then, cells were treated with pterostilbene at concentrations of 20, 40, and 60 μM for 48 h or 72 h. Subsequently, cells were scraped from the dishes, lysed, and centrifuged. Caspase-3 activity in cell lysates was determined according to the manufacturer's instructions and normalized to the total content of cellular proteins, which was determined by the Bradford method.

4.9. Statistical Analysis

Statistical analysis was performed with the use of Statistica PL ver. 12.0 Software (StatSoft Polska, Cracow, Poland). One-way analysis of variance (ANOVA) with Tukey's post hoc was used to evaluate significance in the examined group. All data expressed as means ± SD were representative of at least three independent experiments. Values of $p < 0.05$ were considered statistically significant.

5. Conclusions

Based on the present study, it may be concluded that pterostilbene exhibits antiproliferative activity in human melanoma cells and induces both caspase-dependent and caspase-independent melanoma cell death. The molecular mechanism of pterostilbene includes the regulation of cyklinD1 and p21 expression, activation of caspase 3, and induction of DNA fragmentation. Furthermore, the results showed significant differences between the studied C32 amelanotic and A2058 melanotic melanoma cells. As observed, a more pronounced effect of pterostilbene was detected against amelanotic. Further research is warranted to fully elucidate the effects of pterostilbene on amelanotic and melanotic melanoma cells.

Author Contributions: Conceptualization, J.W., K.J. and M.K.; methodology J.W. and K.J.; software, J.W., K.J. and M.K.; validation, J.W. and K.J.; investigation, J.W. and K.J.; resources, J.W. and K.J.; data curation, K.J.; writing—original draft preparation, J.W. and K.J.; writing—review and editing, J.W. and M.K.; supervision, M.K.; project administration, M.K. and J.W.; funding acquisition, M.K. and J.W. All authors have read and agreed to the published version of the manuscript.

Funding: This research was funded by grants no. PCN-1-101/K/0/F and PCN-1-007/N/2/F from the Medical University of Silesia, Katowice, Poland.

Institutional Review Board Statement: Not applicable.

Informed Consent Statement: Not applicable.

Data Availability Statement: Data are contained within this article.

Acknowledgments: The authors would like to especially thank Monika Paul-Samojedny (Department of Medical Genetics, School of Pharmaceutical Sciences in Sosnowiec, Medical University of Silesia, Katowice, Poland) for her contribution to performing the flow cytometry experiment.

Conflicts of Interest: The authors declare no conflict of interest.

References

1. von Schuckmann, L.A.; Hughes, M.C.B.; Ghiasvand, R.; Malt, M.; van der Pols, J.C.; Beesley, V.L.; Khosrotehrani, K.; Smithers, B.M.; Green, A.C. Risk of Melanoma Recurrence After Diagnosis of a High-Risk Primary Tumor. *JAMA Dermatol.* **2019**, *155*, 688. [CrossRef] [PubMed]
2. Lopes, J.; Rodrigues, C.M.P.; Gaspar, M.M.; Reis, C.P. Melanoma Management: From Epidemiology to Treatment and Latest Advances. *Cancers* **2022**, *14*, 4652. [CrossRef] [PubMed]
3. Moreau, J.F.; Weissfeld, J.L.; Ferris, L.K. Characteristics and Survival of Patients with Invasive Amelanotic Melanoma in the USA. *Melanoma Res.* **2013**, *23*, 408–413. [CrossRef]
4. Ricci, F.; Fania, L.; Paradisi, A.; Di Lella, G.; Pallotta, S.; Sobrino, L.; Panebianco, A.; Annessi, G.; Abeni, D. Delayed Melanoma Diagnosis in the COVID-19 Era: Increased Breslow Thickness in Primary Melanomas Seen after the COVID-19 Lockdown. *J. Eur. Acad. Dermatol. Venereol.* **2020**, *34*, e778–e779. [CrossRef] [PubMed]
5. Chacón, M.; Pfluger, Y.; Angel, M.; Waisberg, F.; Enrico, D. Uncommon Subtypes of Malignant Melanomas: A Review Based on Clinical and Molecular Perspectives. *Cancers* **2020**, *12*, 2362. [CrossRef]
6. Broussard, L.; Howland, A.; Ryu, S.; Song, K.; Norris, D.; Armstrong, C.A.; Song, P.I. Melanoma Cell Death Mechanisms. *Chonnam Med. J.* **2018**, *54*, 135. [CrossRef] [PubMed]
7. Mattia, G.; Puglisi, R.; Ascione, B.; Malorni, W.; Carè, A.; Matarrese, P. Cell death-based treatments of melanoma:conventional treatments and new therapeutic strategies. *Cell Death Dis.* **2018**, *9*, 112. [CrossRef]
8. Hamilton, C.; Fox, J.P.; Longley, D.B.; Higgins, C.A. Therapeutics Targeting the Core Apoptotic Machinery. *Cancers* **2021**, *13*, 2618. [CrossRef]
9. Forni, C.; Rossi, M.; Borromeo, I.; Feriotto, G.; Platamone, G.; Tabolacci, C.; Mischiati, C.; Beninati, S. Flavonoids: A Myth or a Reality for Cancer Therapy? *Molecules* **2021**, *26*, 3583. [CrossRef]
10. Miyata, Y.; Shida, Y.; Hakariya, T.; Sakai, H. Anti-Cancer Effects of Green Tea Polyphenols Against Prostate Cancer. *Molecules* **2019**, *24*, 193. [CrossRef]
11. Cháirez-Ramírez, M.H.; de la Cruz-López, K.G.; García-Carrancá, A. Polyphenols as Antitumor Agents Targeting Key Players in Cancer-Driving Signaling Pathways. *Front. Pharmacol.* **2021**, *12*, 710304. [CrossRef]
12. Hazafa, A.; Iqbal, M.O.; Javaid, U.; Tareen, M.B.K.; Amna, D.; Ramzan, A.; Piracha, S.; Naeem, M. Inhibitory Effect of Polyphenols (Phenolic Acids, Lignans, and Stilbenes) on Cancer by Regulating Signal Transduction Pathways: A Review. *Clin. Transl. Oncol.* **2022**, *24*, 432–445. [CrossRef]
13. Maleki Dana, P.; Sadoughi, F.; Asemi, Z.; Yousefi, B. The Role of Polyphenols in Overcoming Cancer Drug Resistance: A Comprehensive Review. *Cell. Mol. Biol. Lett.* **2022**, *27*, 10. [CrossRef] [PubMed]
14. Ma, Z.; Zhang, X.; Xu, L.; Liu, D.; Di, S.; Li, W.; Zhang, J.; Zhang, H.; Li, X.; Han, J.; et al. Pterostilbene: Mechanisms of Its Action as Oncostatic Agent in Cell Models and in Vivo Studies. *Pharmacol. Res.* **2019**, *145*, 104265. [CrossRef]
15. Elsherbini, A.M.; Sheweita, S.A.; Sultan, A.S. Pterostilbene as a Phytochemical Compound Induces Signaling Pathways Involved in the Apoptosis and Death of Mutant P53-Breast Cancer Cell Lines. *Nutr. Cancer* **2021**, *73*, 1976–1984. [CrossRef] [PubMed]
16. Yu, C.L.; Yang, S.F.; Hung, T.W.; Lin, C.L.; Hsieh, Y.H.; Chiou, H.L. Inhibition of eIF2α dephosphorylation accelerates pterostilbene-induced cell death in human hepatocellular carcinoma cells in an ER stress and autophagy-dependent manner. *Cell Death Dis.* **2019**, *10*, 418. [CrossRef] [PubMed]
17. Ma, Z.; Yang, Y.; Di, S.; Feng, X.; Liu, D.; Jiang, S.; Hu, W.; Qin, Z.; Li, Y.; Lv, J.; et al. Pterostilbene Exerts Anticancer Activity on Non-Small-Cell Lung Cancer via Activating Endoplasmic Reticulum Stress. *Sci. Rep.* **2017**, *7*, 8091. [CrossRef] [PubMed]
18. Surien, O.; Masre, S.F.; Basri, D.F.; Ghazali, A.R. Chemopreventive Effects of Oral Pterostilbene in Multistage Carcinogenesis of Skin Squamous Cell Carcinoma Mouse Model Induced by DMBA/TPA. *Biomedicines* **2022**, *10*, 2743. [CrossRef]
19. Garbe, C.; Peris, K.; Hauschild, A.; Saiag, P.; Middleton, M.; Bastholt, L.; Grob, J.-J.; Malvehy, J.; Newton-Bishop, J.; Stratigos, A.J.; et al. Diagnosis and Treatment of Melanoma. European Consensus-Based Interdisciplinary Guideline—Update 2016. *Eur. J. Cancer* **2016**, *63*, 201–217. [CrossRef]
20. Iqbal, J.; Abbasi, B.A.; Ahmad, R.; Batool, R.; Mahmood, T.; Ali, B.; Khalil, A.T.; Kanwal, S.; Afzal Shah, S.; Alam, M.M.; et al. Potential Phytochemicals in the Fight against Skin Cancer: Current Landscape and Future Perspectives. *Biomed. Pharmacother.* **2019**, *109*, 1381–1393. [CrossRef]
21. Strickland, L.R.; Pal, H.C.; Elmets, C.A.; Afaq, F. Targeting Drivers of Melanoma with Synthetic Small Molecules and Phytochemicals. *Cancer Lett.* **2015**, *359*, 20–35. [CrossRef]
22. Ranjan, A.; Ramachandran, S.; Gupta, N.; Kaushik, I.; Wright, S.; Srivastava, S.; Das, H.; Srivastava, S.; Prasad, S.; Srivastava, S.K. Role of Phytochemicals in Cancer Prevention. *Int. J. Mol. Sci.* **2019**, *20*, 4981. [CrossRef] [PubMed]
23. Niles, R.M.; McFarland, M.; Weimer, M.B.; Redkar, A.; Fu, Y.-M.; Meadows, G.G. Resveratrol Is a Potent Inducer of Apoptosis in Human Melanoma Cells. *Cancer Lett.* **2003**, *190*, 157–163. [CrossRef] [PubMed]

24. Wang, M.; Yu, T.; Zhu, C.; Sun, H.; Qiu, Y.; Zhu, X.; Li, J. Resveratrol Triggers Protective Autophagy Through the Ceramide/Akt/MTOR Pathway in Melanoma B16 Cells. *Nutr. Cancer* **2014**, *66*, 435–440. [CrossRef] [PubMed]
25. Yang, H.-Z.; Zhang, J.; Zeng, J.; Liu, S.; Zhou, F.; Zhang, F.; Giampieri, F.; Cianciosi, D.; Forbes-Hernandez, T.Y.; Ansary, J.; et al. Resveratrol Inhibits the Proliferation of Melanoma Cells by Modulating Cell Cycle. *Int. J. Food Sci. Nutr.* **2020**, *71*, 84–93. [CrossRef] [PubMed]
26. Pannu, N.; Bhatnagar, A. Resveratrol: From Enhanced Biosynthesis and Bioavailability to Multitargeting Chronic Diseases. *Biomed. Pharmacother.* **2019**, *109*, 2237–2251. [CrossRef]
27. Chakraborty, A.; Gupta, N.; Ghosh, K.; Roy, P. In Vitro Evaluation of the Cytotoxic, Anti-Proliferative and Anti-Oxidant Properties of Pterostilbene Isolated from Pterocarpus Marsupium. *Toxicol. Vitr.* **2010**, *24*, 1215–1228. [CrossRef]
28. Pari, L.; Satheesh, M.A. Effect of Pterostilbene on Hepatic Key Enzymes of Glucose Metabolism in Streptozotocin- and Nicotinamide-Induced Diabetic Rats. *Life Sci.* **2006**, *79*, 641–645. [CrossRef]
29. Rimando, A.M.; Nagmani, R.; Feller, D.R.; Yokoyama, W. Pterostilbene, a New Agonist for the Peroxisome Proliferator-Activated Receptor α-Isoform, Lowers Plasma Lipoproteins and Cholesterol in Hypercholesterolemic Hamsters. *J. Agric. Food Chem.* **2005**, *53*, 3403–3407. [CrossRef]
30. Li, Y.-R.; Li, S.; Lin, C.-C. Effect of Resveratrol and Pterostilbene on Aging and Longevity: Effect of Resveratrol and Pterostilbene on Aging and Longevity. *BioFactors* **2018**, *44*, 69–82. [CrossRef]
31. Estrela, J.M.; Ortega, A.; Mena, S.; Rodriguez, M.L.; Asensi, M. Pterostilbene: Biomedical Applications. *Crit. Rev. Clin. Lab. Sci.* **2013**, *50*, 65–78. [CrossRef] [PubMed]
32. Wakimoto, R.; Ono, M.; Takeshima, M.; Higuchi, T.; Nakano, S. Differential Anticancer Activity of Pterostilbene Against Three Subtypes of Human Breast Cancer Cells. *Anticancer Res.* **2017**, *37*, 6153–6159. [CrossRef] [PubMed]
33. McCormack, D.E.; Mannal, P.; McDonald, D.; Tighe, S.; Hanson, J.; McFadden, D. Genomic Analysis of Pterostilbene Predicts Its Antiproliferative Effects Against Pancreatic Cancer In Vitro and In Vivo. *J. Gastrointest. Surg.* **2012**, *16*, 1136–1143. [CrossRef]
34. Paul, S.; Rimando, A.M.; Lee, H.J.; Ji, Y.; Reddy, B.S.; Suh, N. Anti-Inflammatory Action of Pterostilbene Is Mediated through the P38 Mitogen-Activated Protein Kinase Pathway in Colon Cancer Cells. *Cancer Prev. Res.* **2009**, *2*, 650–657. [CrossRef]
35. Wang, Z.; Wang, T.; Chen, X.; Cheng, J.; Wang, L. Pterostilbene Regulates Cell Proliferation and Apoptosis in Non-small-cell Lung Cancer via Targeting COX-2. *Biotechnol. Appl. Biochem.* **2022**. [CrossRef] [PubMed]
36. Wen, W.; Lowe, G.; Roberts, C.; Finlay, J.; Han, E.; Glackin, C.; Dellinger, T. Pterostilbene Suppresses Ovarian Cancer Growth via Induction of Apoptosis and Blockade of Cell Cycle Progression Involving Inhibition of the STAT3 Pathway. *Int. J. Mol. Sci.* **2018**, *19*, 1983. [CrossRef]
37. Benlloch, M.; Obrador, E.; Valles, S.L.; Rodriguez, M.L.; Sirerol, J.A.; Alcácer, J.; Pellicer, J.A.; Salvador, R.; Cerdá, C.; Sáez, G.T.; et al. Pterostilbene Decreases the Antioxidant Defenses of Aggressive Cancer Cells In Vivo: A Physiological Glucocorticoids- and Nrf2-Dependent Mechanism. *Antioxid. Redox Signal.* **2016**, *24*, 974–990. [CrossRef]
38. Mena, S.; Rodríguez, M.L.; Ponsoda, X.; Estrela, J.M.; Jäättela, M.; Ortega, A.L. Pterostilbene-Induced Tumor Cytotoxicity: A Lysosomal Membrane Permeabilization-Dependent Mechanism. *PLoS ONE* **2012**, *7*, e44524. [CrossRef]
39. Qian, Y.; Liu, Z.; Zhang, Z.; Levenson, A.; Li, K. Pterostilbene Increases PTEN Expression through the Targeted Downregulation of MicroRNA-19a in Hepatocellular Carcinoma. *Mol. Med. Rep.* **2018**, *17*, 5193–5201. [CrossRef]
40. Kong, Y.; Chen, G.; Xu, Z.; Yang, G.; Li, B.; Wu, X.; Xiao, W.; Xie, B.; Hu, L.; Sun, X.; et al. Pterostilbene Induces Apoptosis and Cell Cycle Arrest in Diffuse Large B-Cell Lymphoma Cells. *Sci. Rep.* **2016**, *6*, 37417. [CrossRef]
41. Dai, B.; Zhao, X.F.; Mazan-Mamczarz, K.; Hagner, P.; Corl, S.; Bahassi, E.M.; Lu, S.; Stambrook, P.J.; Shapiro, P.; Gartenhaus, R.B. Functional and Molecular Interactions between ERK and CHK2 in Diffuse Large B-Cell Lymphoma. *Nat. Commun.* **2011**, *2*, 402. [CrossRef] [PubMed]
42. Chen, G.; Xu, Z.; Chang, G.; Hou, J.; Hu, L.; Zhang, Y.; Yu, D.; Li, B.; Chang, S.; Xie, Y.; et al. The Blueberry Component Pterostilbene Has Potent Anti-Myeloma Activity in Bortezomib-Resistant Cells. *Oncol. Rep.* **2017**, *38*, 488–496. [CrossRef] [PubMed]
43. Ko, C.-P.; Lin, C.-W.; Chen, M.-K.; Yang, S.-F.; Chiou, H.-L.; Hsieh, M.-J. Pterostilbene Induce Autophagy on Human Oral Cancer Cells through Modulation of Akt and Mitogen-Activated Protein Kinase Pathway. *Oral Oncol.* **2015**, *51*, 593–601. [CrossRef] [PubMed]
44. Siedlecka-Kroplewska, K.; Jozwik, A.; Boguslawski, W.; Wozniak, M.; Zauszkiewicz-Pawlak, A.; Spodnik, J.H.; Rychlowski, M.; Kmiec, Z. Pterostilbene Induces Accumulation of Autophagic Vacuoles Followed by Cell Death in HL60 Human Leukemia Cells. *J. Physiol. Pharmacol. Off. J. Pol. Physiol. Soc.* **2013**, *64*, 545–556.
45. González-Ruiz, L.; González-Moles, M.Á.; González-Ruiz, I.; Ruiz-Ávila, I.; Ramos-García, P. Prognostic and Clinicopathological Significance of CCND1/Cyclin D1 Upregulation in Melanomas: A Systematic Review and Comprehensive Meta-Analysis. *Cancers* **2021**, *13*, 1314. [CrossRef]
46. Karimian, A.; Ahmadi, Y.; Yousefi, B. Multiple Functions of P21 in Cell Cycle, Apoptosis and Transcriptional Regulation after DNA Damage. *DNA Repair* **2016**, *42*, 63–71. [CrossRef] [PubMed]
47. Piepkorn, M. Melanoma Genetics: An Update with Focus on the CDKN2A(P16)/ARF Tumor Suppressors. *J. Am. Acad. Dermatol.* **2000**, *42*, 705–726. [CrossRef] [PubMed]

48. Xing, F.; Persaud, Y.; Pratilas, C.A.; Taylor, B.S.; Janakiraman, M.; She, Q.-B.; Gallardo, H.; Liu, C.; Merghoub, T.; Hefter, B.; et al. Concurrent Loss of the PTEN and RB1 Tumor Suppressors Attenuates RAF Dependence in Melanomas Harboring V600EBRAF. *Oncogene* **2012**, *31*, 446–457. [CrossRef]
49. Abbas, T.; Dutta, A. P21 in Cancer: Intricate Networks and Multiple Activities. *Nat. Rev. Cancer* **2009**, *9*, 400–414. [CrossRef]
50. Chen, R.-J.; Lee, Y.-H.; Yeh, Y.-L.; Wu, W.-S.; Ho, C.-T.; Li, C.-Y.; Wang, B.-J.; Wang, Y.-J. Autophagy-Inducing Effect of Pterostilbene: A Prospective Therapeutic/Preventive Option for Skin Diseases. *J. Food Drug Anal.* **2017**, *25*, 125–133. [CrossRef]
51. Chen, R.-J.; Kuo, H.-C.; Cheng, L.-H.; Lee, Y.-H.; Chang, W.-T.; Wang, B.-J.; Wang, Y.-J.; Cheng, H.-C. Apoptotic and Nonapoptotic Activities of Pterostilbene against Cancer. *Int. J. Mol. Sci.* **2018**, *19*, 287. [CrossRef]
52. Schneider, J.G.; Alosi, J.A.; McDonald, D.E.; McFadden, D.W. Effects of Pterostilbene on Melanoma Alone and in Synergy with Inositol Hexaphosphate. *Am. J. Surg.* **2009**, *198*, 679–684. [CrossRef] [PubMed]
53. Pfeffer, C.; Singh, A. Apoptosis: A Target for Anticancer Therapy. *Int. J. Mol. Sci.* **2018**, *19*, 448. [CrossRef] [PubMed]
54. Soengas, M.S.; Capodieci, P.; Polsky, D.; Mora, J.; Esteller, M.; Opitz-Araya, X.; McCombie, R.; Herman, J.G.; Gerald, W.L.; Lazebnik, Y.A.; et al. Inactivation of the Apoptosis Effector Apaf-1 in Malignant Melanoma. *Nature* **2001**, *409*, 207–211. [CrossRef]
55. Robertson, G.P. Functional and Therapeutic Significance of Akt Deregulation in Malignant Melanoma. *Cancer Metastasis Rev.* **2005**, *24*, 273–285. [CrossRef] [PubMed]
56. Prokhorova, E.A.; Kopeina, G.S.; Lavrik, I.N.; Zhivotovsky, B. Apoptosis Regulation by Subcellular Relocation of Caspases. *Sci. Rep.* **2018**, *8*, 12199. [CrossRef]
57. Nutakul, W.; Sobers, H.S.; Qiu, P.; Dong, P.; Decker, E.A.; McClements, D.J.; Xiao, H. Inhibitory Effects of Resveratrol and Pterostilbene on Human Colon Cancer Cells: A Side-by-Side Comparison. *J. Agric. Food Chem.* **2011**, *59*, 10964–10970. [CrossRef] [PubMed]
58. Moon, D.; McCormack, D.; McDonald, D.; McFadden, D. Pterostilbene Induces Mitochondrially Derived Apoptosis in Breast Cancer Cells in Vitro. *J. Surg. Res.* **2013**, *180*, 208–215. [CrossRef] [PubMed]
59. Pan, M.-H.; Chang, Y.-H.; Badmaev, V.; Nagabhushanam, K.; Ho, C.-T. Pterostilbene Induces Apoptosis and Cell Cycle Arrest in Human Gastric Carcinoma Cells. *J. Agric. Food Chem.* **2007**, *55*, 7777–7785. [CrossRef]
60. Lin, C.-S.; Lin, W.-S.; Lin, C.-L.; Kao, C.-H. Carvedilol Use Is Associated with Reduced Cancer Risk: A Nationwide Population-Based Cohort Study. *Int. J. Cardiol.* **2015**, *184*, 9–13. [CrossRef]
61. Lin, V.C.-H.; Tsai, Y.-C.; Lin, J.-N.; Fan, L.-L.; Pan, M.-H.; Ho, C.-T.; Wu, J.-Y.; Way, T.-D. Activation of AMPK by Pterostilbene Suppresses Lipogenesis and Cell-Cycle Progression in P53 Positive and Negative Human Prostate Cancer Cells. *J. Agric. Food Chem.* **2012**, *60*, 6399–6407. [CrossRef]
62. Pan, M.-H.; Lin, Y.-T.; Lin, C.-L.; Wei, C.-S.; Ho, C.-T.; Chen, W.-J. Suppression of Heregulin- β 1/HER2-Modulated Invasive and Aggressive Phenotype of Breast Carcinoma by Pterostilbene via Inhibition of Matrix Metalloproteinase-9, P38 Kinase Cascade and Akt Activation. *Evid. Based Complement. Altern. Med.* **2011**, *2011*, 562187. [CrossRef]
63. Chen, R.-J.; Ho, C.-T.; Wang, Y.-J. Pterostilbene Induces Autophagy and Apoptosis in Sensitive and Chemoresistant Human Bladder Cancer Cells. *Mol. Nutr. Food Res.* **2010**, *54*, 1819–1832. [CrossRef] [PubMed]
64. Ferrer, P.; Asensi, M.; Segarra, R.; Ortega, A.; Benlloch, M.; Obrador, E.; Varea, M.T.; Asensio, G.; Jordá, L.; Estrela, J.M. Association between Pterostilbene and Quercetin Inhibits Metastatic Activity of B16 Melanoma. *Neoplasia* **2005**, *7*, 37–47. [CrossRef] [PubMed]
65. Su, Y.; Bintz, M.; Yang, Y.; Robert, L.; Ng, A.H.C.; Liu, V.; Ribas, A.; Heath, J.R.; Wei, W. Phenotypic heterogeneity and evolution of melanoma cells associated with targeted therapy resistance. *PLoS Comput. Biol.* **2019**, *15*, e1007034. [CrossRef] [PubMed]
66. Su, Y.; Ko, M.E.; Cheng, H.; Zhu, R.; Xue, M.; Wang, J.; Lee, J.W.; Frankiw, L.; Xu, A.; Wong, S.; et al. Multi-Omic Single-Cell Snapshots Reveal Multiple Independent Trajectories to Drug Tolerance in a Melanoma Cell Line. *Nat. Commun.* **2020**, *11*, 2345. [CrossRef] [PubMed]
67. Su, Y.; Shi, Q.; Wei, W. Single cell proteomics in biomedicine: High-dimensional data acquisition, visualization, and analysis. *Proteomics* **2017**, *17*, 10. [CrossRef]
68. Su, Y.; Wei, W.; Robert, L.; Xue, M.; Tsoi, J.; Garcia-Diaz, A.; Homet Moreno, B.; Kim, J.; Ng, R.H.; Lee, J.W.; et al. Single-cell analysis resolves the cell state transition and signaling dynamics associated with melanoma drug-induced resistance. *Proc. Natl. Acad. Sci. USA* **2017**, *114*, 13679–13684. [CrossRef]
69. Du, J.; Su, Y.; Qian, C.; Yuan, D.; Miao, K.; Lee, D.; Ng, A.H.C.; Wijker, R.S.; Ribas, A.; Levine, R.D.; et al. Raman-guided subcellular pharmaco-metabolomics for metastatic melanoma cells. *Nat. Commun.* **2020**, *11*, 4830. [CrossRef]
70. Hachey, S.J.; Boiko, A.D. Therapeutic implications of melanoma heterogeneity. *Exp. Dermatol.* **2016**, *25*, 497–500. [CrossRef]
71. Nawrot-Hadzik, I.; Choromańska, A.; Abel, R.; Preissner, R.; Saczko, J.; Matkowski, A.; Hadzik, J. Cytotoxic Effect of Vanicosides A and B from Reynoutria Sachalinensis against Melanotic and Amelanotic Melanoma Cell Lines and in Silico Evaluation for Inhibition of BRAFV600E and MEK1. *Int. J. Mol. Sci.* **2020**, *21*, 4611. [CrossRef] [PubMed]

International Journal of
Molecular Sciences

Article

Discovery of Simple Diacylhydrazine-Functionalized Cinnamic Acid Derivatives as Potential Microtubule Stabilizers

Xiang Zhou [1,*,†], Yi-Hong Fu [2,†], Ya-Yu Zou [3], Jiao Meng [1], Gui-Ping Ou-Yang [3], Qiang-Sheng Ge [3] and Zhen-Chao Wang [1,3,*]

1 State Key Laboratory Breeding Base of Green Pesticide and Agricultural Bioengineering, Key Laboratory of Green Pesticide and Agricultural Bioengineering, Ministry of Education, Center for R & D of Fine Chemicals, Guizhou University, Guiyang 550025, China
2 Science and Technology College, Hubei University of Arts and Science, Xiangyang 441025, China
3 College of Pharmacy, Guizhou University, Guiyang 550025, China
* Correspondence: xiangzhou@gzu.edu.cn or zhoux1534@163.com (X.Z.); zcwang@gzu.edu.cn or wzc.4884@163.com (Z.-C.W.); Tel./Fax: +86-851-8830-8717 (Z.-C.W.)
† These authors contributed equally to this work.

Abstract: To develop novel microtubule-binding agents for cancer therapy, an array of *N*-cinnamoyl-*N′*-(substituted)acryloyl hydrazide derivatives were facilely synthesized through a two-step process. Initially, the antiproliferative activity of these title compounds was explored against A549, 98 PC-3 and HepG2 cancer cell lines. Notably, compound $\mathbf{I_{23}}$ exhibited the best antiproliferative activity against three cancer lines with IC_{50} values ranging from 3.36 to 5.99 μM and concurrently afforded a lower cytotoxicity towards the NRK-52E cells. Anticancer mechanism investigations suggested that the highly bioactive compound $\mathbf{I_{23}}$ could potentially promote the protofilament assembly of tubulin, thus eventually leading to the stagnation of the G2/M phase cell cycle of HepG2 cells. Moreover, compound $\mathbf{I_{23}}$ also disrupted cancer cell migration and significantly induced HepG2 cells apoptosis in a dosage-dependent manner. Additionally, the in silico analysis indicated that compound $\mathbf{I_{23}}$ exhibited an acceptable pharmacokinetic profile. Overall, these easily prepared *N*-cinnamoyl-*N′*-(substituted)acryloyl hydrazide derivatives could serve as potential microtubule-interacting agents, probably as novel microtubule-stabilizers.

Keywords: cinnamic acid derivatives; antiproliferative activity; microtubule; tubulin stabilizer

Citation: Zhou, X.; Fu, Y.-H.; Zou, Y.-Y.; Meng, J.; Ou-Yang, G.-P.; Ge, Q.-S.; Wang, Z.-C. Discovery of Simple Diacylhydrazine-Functionalized Cinnamic Acid Derivatives as Potential Microtubule Stabilizers. *Int. J. Mol. Sci.* **2022**, *23*, 12365. https://doi.org/10.3390/ijms232012365

Academic Editor: Bernhard Biersack

Received: 3 September 2022
Accepted: 5 October 2022
Published: 15 October 2022

1. Introduction

Cancer remains the chief and ever-expanding culprit in human mortality that prompts significant concerns in every country [1]. As the International Agency for Research on Cancer (IARC) reported, this intractable disease had been emerging in approximately 18.1 million new cases and thus leading to 9.6 million cancer deaths in 2018 [2,3]. Except for the use of radiotherapy and surgery, chemotherapy is still an effective approach to treat cancer in view of its fast-acting performance against adversaries, the practical applicability and the available effectiveness, but sometimes certain pharmacologically induced side effects coexist [4–7]. Therefore, exploring and developing a highly effective therapeutic approach to selectively eliminate cancer cells should be actively pursued.

To date, cell cycle modulators or inhibitors, which can arrest uncontrollable tumor growth, are considered as a kind of hopeful antiproliferative agent. Among these modulators, microtubule-interacting agents that target microtubule and subsequently affect multiple cellular processes (e.g., mitosis, cell division, and intracellular transportation) have become a promising group of anticancer agents that have been introduced to the market for cancer therapy over 50 years [8–15]. To date, there have been three key binding sites in tubulin involving paclitaxel (Taxol), colchicine, and vinca alkaloid. These agents have also been divided into two categories: microtubule destabilizing agents and microtubule-stabilizing agents [16–22]. Particularly, microtubule-stabilizing agents (MSAs) were considered promising for clinical cancer treatment [23]. For instance, Taxol, the first diterpene isolated from the western yew, had been verified with excellent antiproliferative activity and was used in the clinic. Subsequently, docetaxel, obtained from the semisynthetic derivatives of Taxol, was also approved in 1996 by the Food and Drug Administration (FDA) for clinical treatment [24,25]. Moreover, epothilones and laulimalide, as natural microtubule-stabilizing agents that could promote tubulin self-assembly into microtubules, were also commercialized [26]. However, the wide application of these compounds was restricted due to some urgent problems, including high toxicity, limited sources, complex isolation processes, and the already discovered drug resistance [22]. To solve this issue, some chemically synthesized small molecules were persistently explored and actively developed as microtubule stabilizers, such as GS-164, Synstab A, 4′-methoxy-2-styrylchromone, and compound A (Figure 1) [22,27–30]. It is notable that *α*,*β*-unsaturated ketones, as the privileged chemical scaffolds, frequently appeared in the microtubule modulators, exemplified by curcumin and its derivatives that could inhibit tubulin self-assembly by interacting with the unique binding site of tubulin [22,31–33]. Meanwhile, these similar substrates have also been exploited as the correlative olefin polymer materials in the field of advanced materials science [34]. Herein, to discover some potential anti-cancer agents targeting tubulin based on the chemical modifications of natural ingredients of *α*,*β*-unsaturated carbonyl compounds, a series of facilely synthetic *N*-cinnamoyl-*N*′-(substituted)acryloyl hydrazide derivatives from raw material cinnamic acid were designed, synthesized, and screened for their anticancer activity in vitro (Figure 1). We expected that these designed frameworks would have the ability to disrupt tubulin assembly, which should be investigated by fluorescence imaging, tubulin polymerization assay and TEM imaging.

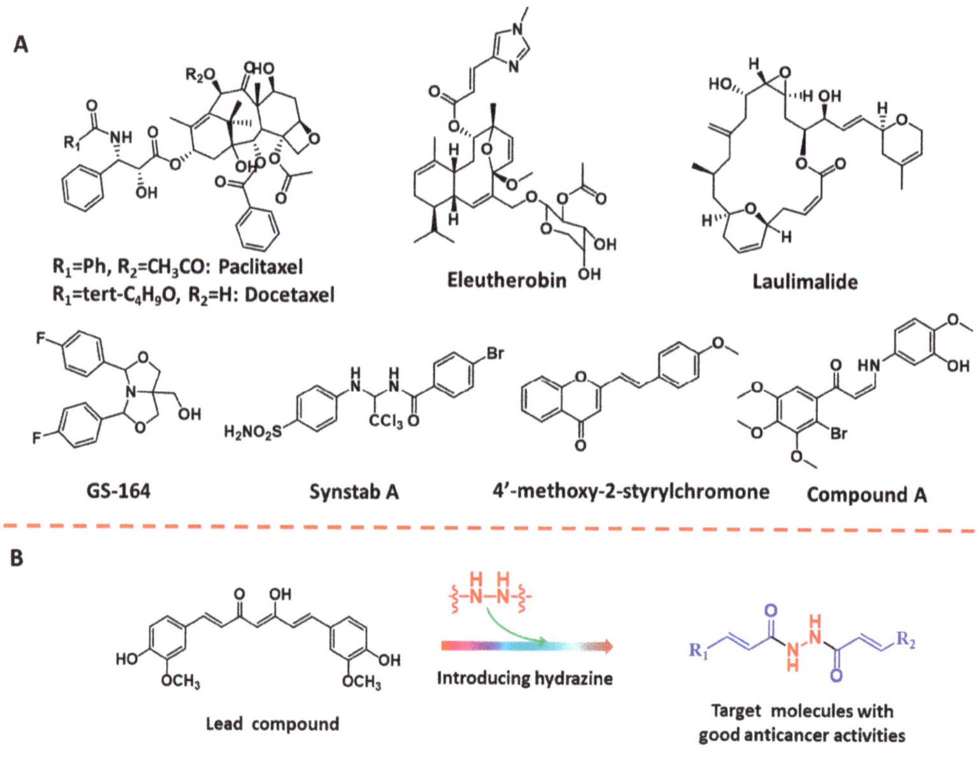

Figure 1. Chemical structures of some microtubule-targeting agents (**A**). Design concepts of target molecules (**B**).

2. Results

2.1. Chemistry

To effectively obtain the target compounds, a facile synthetic route was designed as depicted in Scheme 1. Briefly, the cinnamic acid was reacted with 60% hydrazine hydrate, EDCI, and HOBt under basic conditions to give the intermediate cinnamohydrazide [35]. Then, the corresponding cinnamic acid analogue was reacted with cinnamohydrazide, EDCI, and HOBt and subsequently yielded the precipitates. The final target compounds **I_1–I_{22}** were afforded by being filtered and washed by CH_2Cl_2, respectively. Similarly, compounds **I_{23}–I_{26}** were synthesized according to the same reaction condition by changing to different starting substrates. Finally, these structures were confirmed by NMR and HRMS analysis.

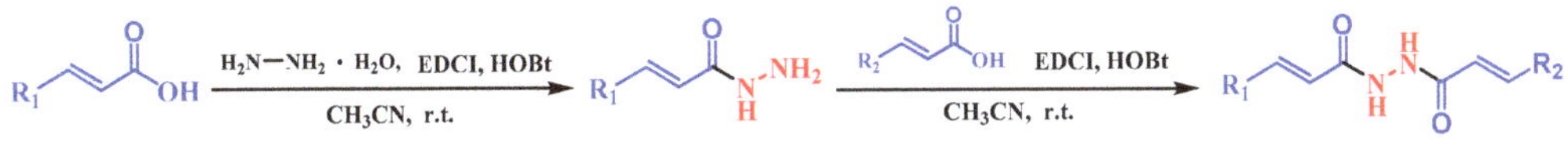

I_1: R_1= phenyl, R_2=phenyl
I_2: R_1= phenyl, R_2=2-NO_2-phenyl
I_3: R_1= phenyl, R_2=2-Cl-phenyl
I_4: R_1= phenyl, R_2=3-NO_2-phenyl
I_5: R_1= phenyl, R_2=3-OH-phenyl
I_6: R_1= phenyl, R_2=3-OCH_3-phenyl
I_7: R_1= phenyl, R_2=4-NO_2-phenyl
I_8: R_1= phenyl, R_2=4-Cl-phenyl
I_9: R_1= phenyl, R_2=4-F-phenyl
I_{10}: R_1= phenyl, R_2=4-CH_3-phenyl
I_{11}: R_1= phenyl, R_2=4-i-Pr-phenyl
I_{12}: R_1= phenyl, R_2=4-OCH_3-phenyl
I_{13}: R_1= phenyl, R_2=4-OC_2H_5-phenyl
I_{14}: R_1= phenyl, R_2=4-$N(CH_3)_2$-phenyl
I_{15}: R_1= phenyl, R_2=-3-OCH_3-4-OH-phenyl
I_{16}: R_1= phenyl, R_2=cyclopropyl
I_{17}: R_1= phenyl, R_2=furanyl
I_{18}: R_1= phenyl, R_2=thiophenyl
I_{19}: R_1= phenyl, R_2=4-CH_3-thiazolyl
I_{20}: R_1= phenyl, R_2=cyclohexyl
I_{21}: R_1= phenyl, R_2=3-cyclohexen-1-yl
I_{22}: R_1= phenyl, R_2=3-pyridine-1-yl
I_{23}: R_1= 4-NO_2-phenyl, R_2=3-OCH_3-phenyl
I_{24}: R_1= 4-NO_2-phenyl, R_2=4-OCH_3-phenyl
I_{25}: R_1= 4-OCH_3-phenyl, R_2=3-OCH_3-phenyl
I_{26}: R_1= 3-OCH_3-phenyl, R_2=3-OCH_3--phenyl

Scheme 1. Synthetic route for the target molecules **I_1–I_{26}**.

2.2. The Antiproliferative Activity of Title Compounds

The antiproliferative activity of all the synthesized target compounds against A549, PC-3 and HepG2 cancer cells were evaluated using the MTT assay and expressed by IC_{50} values. As shown in Table 1, most of target compounds exerted certain antiproliferative activity against A549, PC-3, and HepG2. Among them, compounds **I_5**, **I_7** and **I_{23}** displayed good antiproliferative activity (IC_{50} = 5.09 μM, 4.17 μM, 5.99 μM, respectively) against A549, which were significantly better than curcumin (IC_{50} = 69.6 μM) and similar to gefitinib (IC_{50} = 5.47 μM). In addition, compounds **I_1**, **I_9** and **I_{23}** revealed excellent anti-PC-3 activity with IC_{50} values of 5.95 μM, 4.68 μM and 4.17 μM, respectively, which were equipotent to the positive drug gifitinib (5.99 μM). Moreover, compounds **I_9**, **I_{14}** and **I_{23}** exhibited outstanding antiproliferative activity (IC_{50} = 4.75 μM, 3.42 μM, 3.36 μM, respectively) against HepG2, which were better than curcumin, gefitinib and colchicine (IC_{50} = 40.2, 31.5 and 5.46 μM, respectively). Particularly, compound **I_{23}** displayed comprehensive antiproliferative activity with IC_{50} values of 5.99, 4.17 and 3.36 μM against A549, PC-3 and HepG2, respectively. It was superior to those of the curcumin and gefitinib. The structure–activity relationship (SAR) of target compounds against HepG2 cancer cells was further summarized as follows: (1) a bulky electron-donating group at the *meta*-position was unfavorable to the bioactivity, such as **I_4** (3-NO_2-Ph, 6.32 μM) > **I_5** (3-OH-Ph, 8.69 μM) > **I_6** (3-OCH_3-Ph, 46.8 μM); (2) a bulky electron-donating group at the *para*-position showed good bioactivity, illustrated by **I_{14}** (4-$N(CH_3)_2$-Ph, 3.42 μM) > **I_{13}** (4-OC_2H_5-Ph, 15.4 μM) > **I_{12}** (4-OCH_3-Ph, 21.1 μM) > **I_{11}** (4-i-Pr-Ph, 18.6 μM) and **I_{10}** (4-CH_3-Ph, 23.6 μM); (3) a bulky electron-withdrawing group at the *para*-position was unfavorable to the bioactivity, such as **I_9** (4-F-Ph, 4.75 μM) > **I_8** (4-Cl-Ph, 6.98 μM) > **I_9** (4-NO_2-Ph, 7.63 μM); (4) as R_1 was 4-NO_2, a 3-OCH_3 located at another benzene ring was beneficial to bioactivity, **I_{23}** (3-OCH_3, 3.36 μM) and **I_{24}** (4-OCH_3, 8.08 μM). Finally, a detailed SAR analysis of target compounds is depicted in Figure 2.

Table 1. The antiproliferative activity of all designed compounds against cancer cell lines [a].

No.	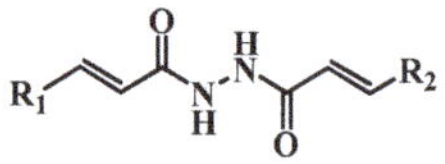		IC_{50} (μM) [b]		
	R_1	R_2	A549	PC-3	HepG2
I_1	-phenyl	-phenyl	26.7 ± 0.43	5.95 ± 1.52	8.17 ± 0.46
I_2	-phenyl	2-NO_2-phenyl	7.55 ± 1.22	8.08 ± 2.24	9.33 ± 0.95
I_3	-phenyl	2-Cl-phenyl	6.48 ± 0.41	7.38 ± 0.24	9.53 ± 1.74
I_4	-phenyl	3-NO_2-phenyl	6.05 ± 0.08	15.8 ± 1.67	6.32 ± 1.00
I_5	-phenyl	3-OH-phenyl	5.09 ± 0.08	6.68 ± 0.96	8.69 ± 1.04
I_6	-phenyl	3-OCH_3-phenyl	41.0 ± 4.47	160 ± 10.9	46.8 ± 0.14
I_7	-phenyl	4-NO_2-phenyl	4.17 ± 0.93	8.79 ± 3.19	7.63 ± 1.75
I_8	-phenyl	4-Cl-phenyl	9.57 ± 1.92	63.2 ± 3.86	6.98 ± 1.30
I_9	-phenyl	4-F-phenyl	24.9 ± 1.54	4.68 ± 0.69	4.75 ± 0.36
I_{10}	-phenyl	4-CH_3-phenyl	19.5 ± 4.95	70.3 ± 9.88	23.6 ± 2.67
I_{11}	-phenyl	4-i-Pr-phenyl	43.0 ± 7.30	19.2 ± 6.57	18.6 ± 3.90
I_{12}	-phenyl	4-OCH_3-phenyl	10.9 ± 0.25	39.4 ± 3.54	21.1 ± 4.07
I_{13}	-phenyl	4-OC_2H_5-phenyl	15.0 ± 4.90	22.5 ± 2.64	15.4 ± 0.93
I_{14}	-phenyl	4-$(CH_3)_2N$-phenyl	12.8 ± 0.60	7.81 ± 1.69	3.42 ± 0.49
I_{15}	-phenyl	3-OCH_3-4-OH-phenyl	35.8 ± 0.59	>300	31.1 ± 7.66
I_{16}	-phenyl	cyclopropyl	16.4 ± 0.98	80.5 ± 5.43	78.5 ± 1.39
I_{17}	-phenyl	furanyl	13.0 ± 2.40	36.0 ± 0.70	34.5 ± 0.28
I_{18}	-phenyl	thiophenyl	13.8 ± 0.29	15.9 ± 1.14	21.1 ± 3.76
I_{19}	-phenyl	4-CH_3-5-thiazolyl	26.6 ± 5.08	96.4 ± 8.36	89.3 ± 8.89
I_{20}	-phenyl	cyclohexyl	16.3 ± 3.25	21.8 ± 3.15	>300
I_{21}	-phenyl	3-cyclohexen-1-yl	8.27 ± 0.80	21.6 ± 0.89	20.4 ± 0.86
I_{22}	-phenyl	3-pyridine-1-yl	60.2 ± 2.37	159 ± 5.88	43.3 ± 1.10
I_{23}	4-NO_2-phenyl	3-OCH_3-phenyl	5.99 ± 1.07	4.17 ± 0.57	3.36 ± 0.80
I_{24}	4-NO_2-phenyl	4-OCH_3-phenyl	5.17 ± 0.41	15.7 ± 2.98	8.08 ± 2.25
I_{25}	4-OCH_3-phenyl	3-OCH_3-phenyl	144 ± 8.95	142 ± 6.79	80.0 ± 4.04
I_{26}	3-OCH_3-phenyl	3-OCH_3-phenyl	11.0 ± 1.72	21.0 ± 2.32	14.5 ± 2.56
Curcumin			69.6 ± 1.18	73.4 ± 4.99	40.2 ± 1.74
Colchicine			1.35 ± 0.08	2.12 ± 0.29	5.46 ± 1.76
Gefitinib			5.47 ± 1.06	5.99 ± 1.47	31.5 ± 10.2

[a] The anti-proliferation activities of all target compounds against cancer cells were determined by MTT assay after 48 h. The data represent the mean of triplicate determinations. [b] IC_{50} values are indicated as the mean ± SD (standard error) of at least three independent experiments.

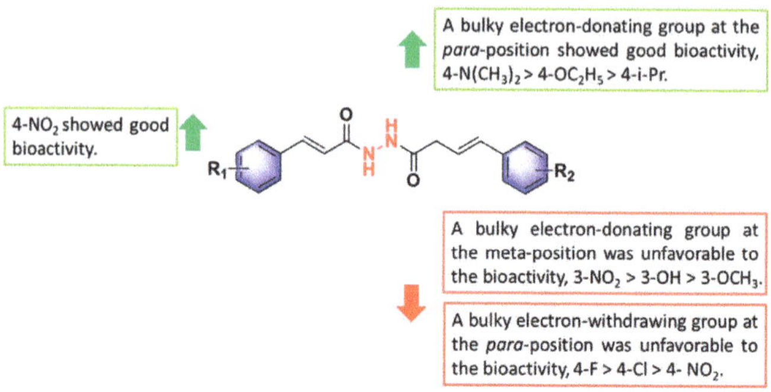

Figure 2. Description diagram of structure-activity relationship (SAR) analysis.

To evaluate the selectivity of target compounds against cancer cell lines and normal cell lines, certain target compounds were evaluated by using the NRK-52E cell line. As Table 2 illustrates, compounds $\mathbf{I_5}$ and $\mathbf{I_7}$ showed good cytotoxicity against A549, PC-3 and HepG2 cell lines, but exhibited weaker cytotoxic towards NRK-52E cells. Meanwhile, compound $\mathbf{I_{14}}$ revealed lower selectivity towards three cell lines and NRK-52E cells. Specially, compared with colchicine and gefitinib, compound $\mathbf{I_{23}}$ afforded a relatively lower cytotoxicity towards normal NRK-52E cells, but exhibited the best cytotoxicity against three cell lines. Given these obtained results, a possible disrupting effect of compound $\mathbf{I_{23}}$ on microtubule was further investigated.

Table 2. Cytotoxicity and selectivity of title compounds $\mathbf{I_5}$, $\mathbf{I_7}$, $\mathbf{I_{14}}$ and $\mathbf{I_{23}}$ [a].

No.	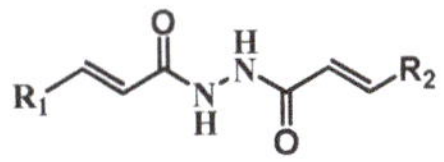		IC_{50} (μM) [b]			
	R_1	R_2	A549	PC-3	HepG2	NRK-52E
I_5	phenyl	3-OH-phenyl	5.09 ± 0.08	6.68 ± 0.96	8.69 ± 1.04	50.0 ± 4.40
I_7	phenyl	4-NO_2-phenyl	4.17 ± 0.93	8.79 ± 3.19	7.63 ± 1.75	21.0 ± 1.75
I_{14}	phenyl	4-$(CH_3)_2N$-phenyl	12.8 ± 0.60	7.81 ± 1.69	3.42 ± 0.93	7.00 ± 0.30
I_{23}	4-NO_2-phenyl	3-OCH_3-phenyl	5.99 ± 1.07	4.17 ± 0.57	3.36 ± 0.80	6.99 ± 1.60
Curcumin			69.6 ± 1.18	73.4 ± 4.99	40.2 ± 1.74	>300
Colchicine			1.35 ± 0.08	2.12 ± 0.29	5.46 ± 1.76	3.32 ± 0.81
Gefitinib			5.47 ± 1.06	5.99 ± 1.47	31.5 ± 10.2	21.0 ± 2.10

[a] The anti-proliferation activities of all target compounds against cancer cells were determined by MTT assay after 48 h. The data represent the mean of triplicate determinations. [b] IC_{50} values are indicated as the mean ± SD (standard error) of at least three independent experiments.

2.3. *Immunofluorescence Staining of Tubulin*

The disrupting effect of microtubule triggered by compound $\mathbf{I_{23}}$ in living cells was disclosed using the immunofluorescence staining assay. Primarily, the HepG2 cells were treated with 0, 3, and 6 μM of compound $\mathbf{I_{23}}$ for 24 h before cell imaging. Clearly, the agminated microtubule network and wrinkled nucleus were observed in cells after incubating with compound $\mathbf{I_{23}}$, thereby leading to the morphological change in HepG2 cells (Figure 3). By contrast, the cellular microtubule network was well assembled and arranged normally in the control group. This result revealed that compound $\mathbf{I_{23}}$ might be capable of leading to cell cycle disorder through disturbing microtubule assembly and targeting tubulin.

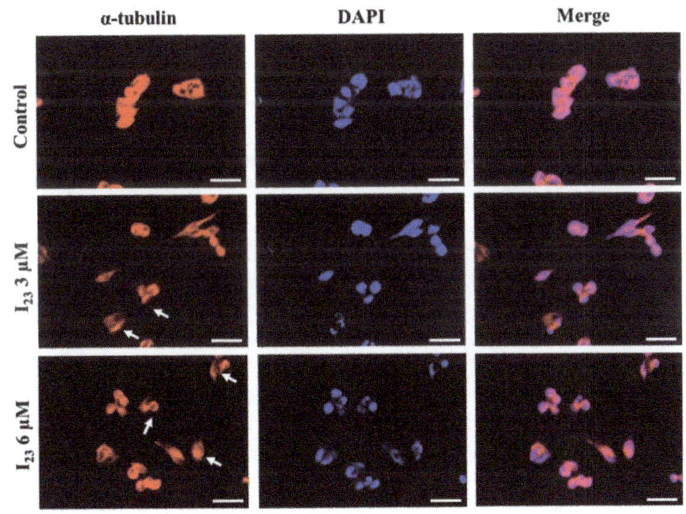

Figure 3. Effects of compound $\mathbf{I_{23}}$ on the cellular microtubule network were visualized by immunofluorescence assay. HepG2 cells were treated with vehicle control 0.1% DMSO, 3 μM and 6 μM $\mathbf{I_{23}}$ for 24 h. Then, cells were fixed and stained with anti-α-tubulin antibody (red) and counterstained with DAPI (blue). Detection of the fixed and stained cells was performed using fluorescence microscope. Scale bars are 50 μm.

2.4. Effects of Compound $\mathbf{I_{23}}$ on Tubulin Polymerization

Tubulin polymerization assay [36–38] triggered by compound $\mathbf{I_{23}}$ was carried out to verify the interaction mode. Meanwhile, paclitaxel, served as a known microtubule stabilizer that could promote the protofilament assembly. As shown in Figure 4, similar to paclitaxel, compound $\mathbf{I_{23}}$ could stabilize tubulin assembly and promote protofilament assembly in a dosage-dependent manner, which was consistent with the outcome from previous reports [38,39], indicating that the potential microtubule-stabilizer might be developed.

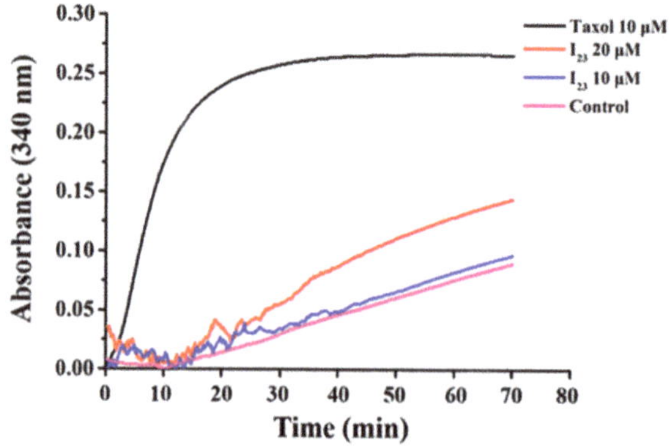

Figure 4. Progress of tubulin polymerization in the presence of 10 μM and 20 μM compound $\mathbf{I_{23}}$. Taxol (10 μM) was used as positive polymerization control, whereas untreated tubulin was used as negative control.

2.5. Tubulin Polymerization Affected by Compound $\mathbf{I_{23}}$ via TEM

Transmission electron microscopy (TEM) was employed to directly visualize the influence of tubulin polymerization stimulated by compound $\mathbf{I_{23}}$. Evidently, without the addition of compound $\mathbf{I_{23}}$, the observed microtubules presented uniform fibrous nanostructures, suggesting the spontaneous formation of microtubules occurred in general tubulin buffer solution (Figure 5A). However, the spontaneous assembly of α/β tubulin heterodimers was disturbed by the existence of 20 μM compound $\mathbf{I_{23}}$, subsequently affording more large nonlinear disorganized aggregations (Figure 5B). These microscopic investigation results manifested that $\mathbf{I_{23}}$ could promote the microtubular aggregation.

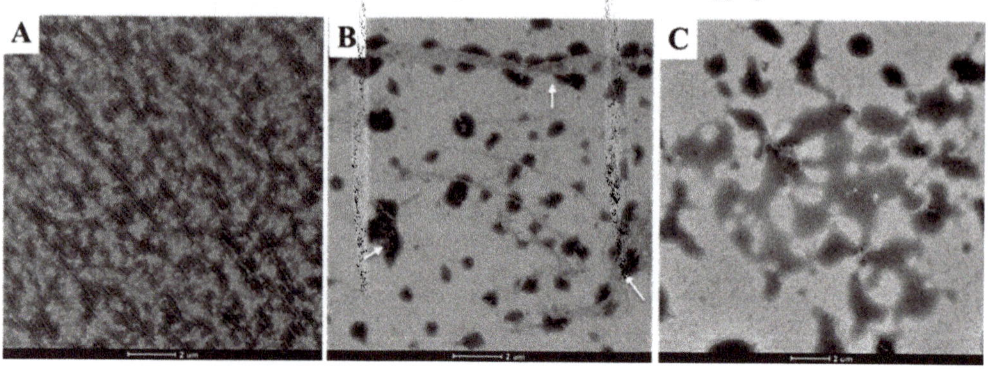

Figure 5. Effect of different concentrations of compound $\mathbf{I_{23}}$ on tubulin polymerization: (**A**) 4 mg/mL tubulin, (**B**) 20 μM $\mathbf{I_{23}}$ + 4 mg/mL tubulin, (**C**) 20 μM $\mathbf{I_{23}}$ without tubulin. Scale bars are 2 μm.

2.6. Cell Cycle Analysis

Microtubules play a critical role during the eukaryotic cell division [40]. Based on the above results, compound $\mathbf{I_{23}}$ could clearly affect the tubulin polymerizations, and this event probably led to the stagnation of cell cycle. To investigate this effect, flow cytometry analysis on HepG2 cell cycle arrest was performed as Figure 6A–C, and the corresponding results were displayed as Figure 6D. Interestingly, the G2/M pattern was arrested after treatment with compound **I$_{23}$** (24 μM), thus providing the relevant percentages from 17.68% (0 μM) to 20.56% (Figure 6), indicating the designed compound could cause cell cycle disorder through targeting tubulin and the subsequent disturbance on microtubule assembly.

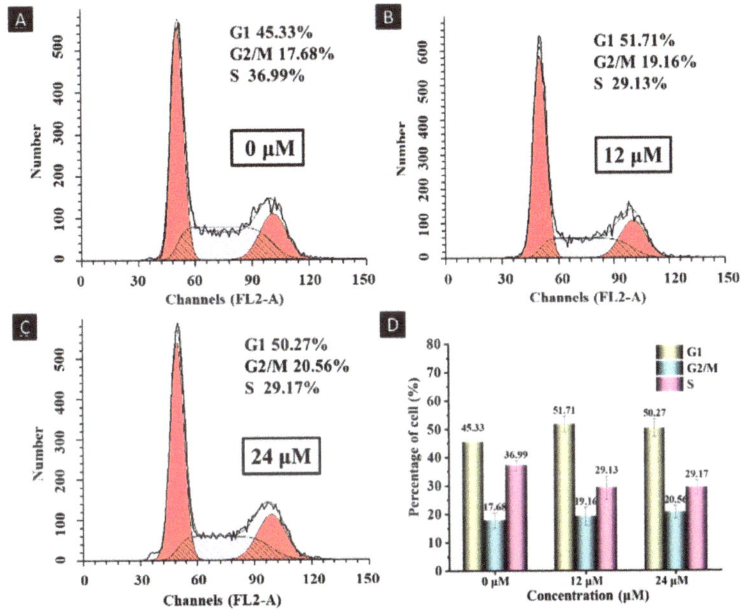

Figure 6. Compound **I$_{23}$** affected the cell cycle distribution in HepG2 cancer cells. Untreated (Control) and cells treated with compound **I$_{23}$** at 24 μM concentration for 24 h.

2.7. Molecular Docking of Compound **I$_{23}$** *with Tubulin*

To better understand the possible interaction mode and binding sites between compound **I$_{23}$** and tubulin, the related software Sybyl X 2.0, PyMOL and Discovery Studio (DS) 2020 were exploited. Meanwhile, the reported 3D crystal structure of tubulin (PDB code: 5syf) was used by the removal of the ligand molecule paclitaxel. Docking study (Figure 7) displayed that compound **I$_{23}$** could embed in the active pocket around with the amino acid residues, including Leu227, Leu217, His229, Leu230, Ala233, Phe272, Pro360, Arg320, and Ser374. Apart from conventional hydrogen bonds and carbon hydrogen bonds, other non-covalent interactions including π-donor hydrogen bonds, π-alkyl, etc., also were very crucial donations for the interaction [41–43]. In detail, the Ser374 residue could form strong hydrogen bond interactions with compound **I$_{23}$**, thus affording a distance of 2.2 Å located in the S9–S10 stabilizing loop of tubulin, which was the binding site of Taxol. The correlative Ala233, Leu217, Leu230, Phe272, Pro360, and Arg320 residues could form π-alkyl interactions. Simultaneously, π-π stacked interaction was also observed between His229 residue and compound **I$_{23}$**. These results indicated that a strong interaction occurred between the designed compound and tubulin, thereby disturbing the normal assembly of tubulin, which was consistent with the aforementioned tubulin polymerization assays.

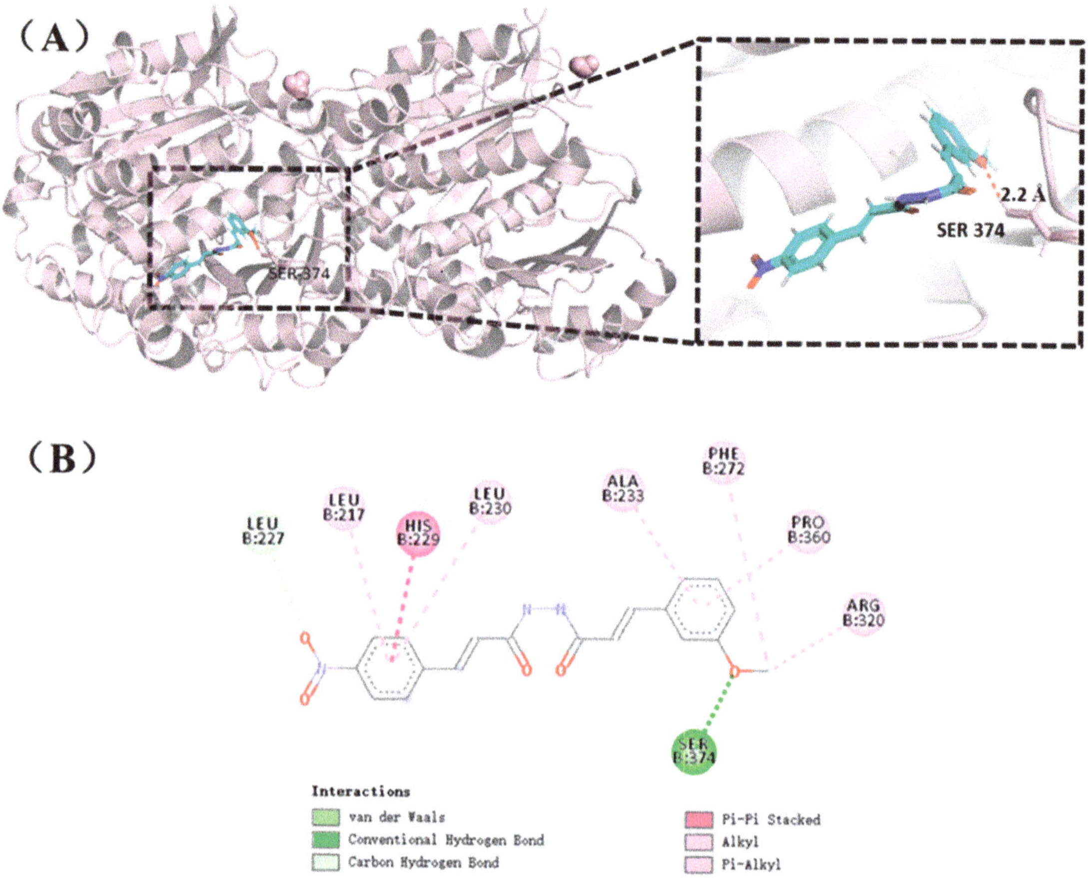

Figure 7. Predicted binding modes of compound $\mathbf{I_{23}}$ with tubulin. Binding of $\mathbf{I_{23}}$ into the active site of tubulin was performed by PyMol software (**A**) and the 2D diagram of binding modes was shown through Discovery Studio 4.5 software (**B**).

2.8. Effects on Cell Migration of Compound $\mathbf{I_{23}}$

Because the targeting-tubulin agents were validated, having the ability to interfere with cell migration [44,45], the cell migration effect triggered by compound $\mathbf{I_{23}}$ was investigated by the calculation of average migration rates of scratched A549 cells monolayer, which was a widely used model. As depicted in Figure 8, a 12.5% migration rate was found in the control after 12 h, while the decreased migration rates reached to 8.90%, 6.10%, and 3.98% after treatment of 3 μM $\mathbf{I_{23}}$, 6 μM $\mathbf{I_{23}}$, and 6 μM gefitinib, respectively. After incubation for another 12 h, the migration rate of A549 cells that coexisted with compound $\mathbf{I_{23}}$ changed to 13.2% (blank control reached to 24.4%). This outcome indicated that compound $\mathbf{I_{23}}$ could significantly attenuate the migration of A549 cells in a dosage-dependent manner, which could be a lead structure for the exploration of novel microtubule-stabilizing agents.

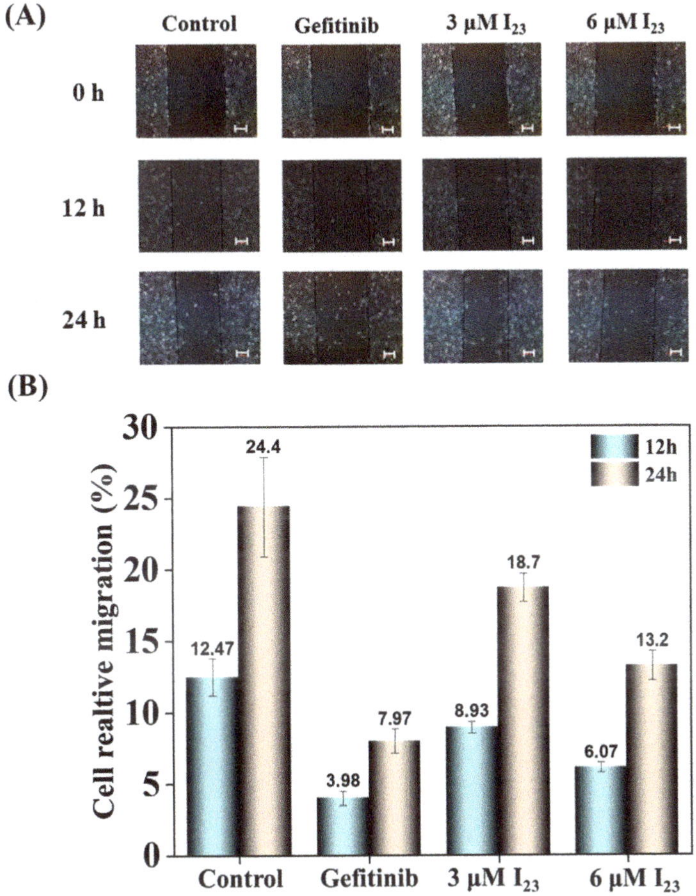

Figure 8. Cell migration assay showed that compound $\mathbf{I_{23}}$ inhibited the migration of A549 cells. (**A**) The cancer cells were incubated with 0 μM, 3 μM, 6 μM compound $\mathbf{I_{23}}$, the 6 μM gefitinib was served as the positive control. Scale bars are 100 μm (**B**) the cell relative migration was calculated using the software.

2.9. Apoptosis Effects of HepG2 Cells Caused by Compound $\mathbf{I_{23}}$

To date, many reports have demonstrated that anti-mitotic cancer agents could also cause cell death through inducing apoptosis [46,47]. Thus, the apoptotic behavior of HepG2 cells triggered by compound $\mathbf{I_{23}}$ should be tested since an appreciable antiproliferative activity and substantial disturbance on microtubule assembly were achieved. As shown in Figure 9, apoptosis effects were observed after incubating cells with 0 μM, 3 μM and 6 μM of compound $\mathbf{I_{23}}$, respectively. Particularly, as the dose was up to 6 μM, a strong apoptotic phenomenon happened, indicating that compound $\mathbf{I_{23}}$ acting as a potential microtubule-stabilizer could slightly induce HepG2 cancer cells apoptosis.

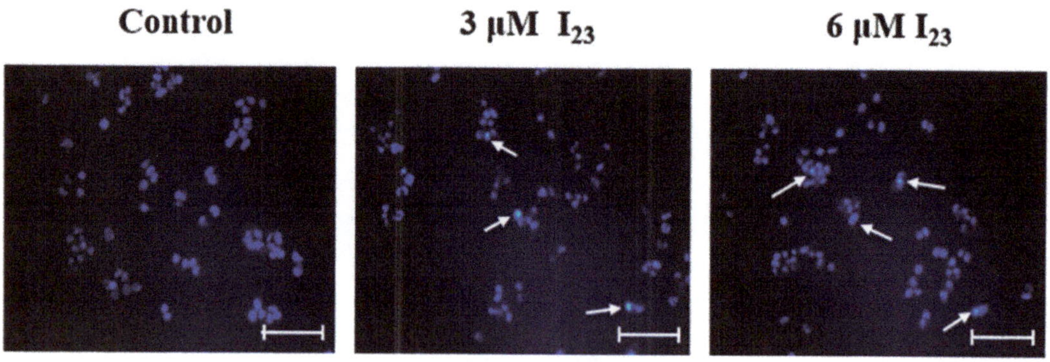

Figure 9. The cells were incubated with 0 μM, 3 μM and 6 μM compound $\mathbf{I_{23}}$ for 24 h, subsequently stained with Hoechst 33258 (10 μg/mL) and visualized by fluorescence microscopy. Scale bars are 100 μm.

2.10. In Silico Drug-likeness Evaluation

Finally, the early evaluation of the lead compound for its potential is a critical step in drug development. Therefore, to assess whether compound $\mathbf{I_{23}}$ has the potential as the promising lead compound, ADMETlab 2.0 software was used to obtain more information on the pharmacokinetic profile, including ADMET and drug-likeness properties (Figure 10). The corresponding physicochemical properties, ADMET, and drug-likeness properties of compound $\mathbf{I_{23}}$ are presented in Figure 11 and Table 3, respectively. Notably, the outcomes illustrated that compound $\mathbf{I_{23}}$ exhibited acceptable physicochemical properties, ADMET, and drug-likeness properties. For instance, compound $\mathbf{I_{23}}$ possessed favorable physicochemical properties: molecular weight = 369.13, logS = −4.140, logP = 2.821, and log D = 3.559. Meanwhile, ADMET and drug-likeness properties of compound $\mathbf{I_{23}}$ were provided as follows: (1) compound $\mathbf{I_{23}}$ had appreciable absorption potency: for example, compound $\mathbf{I_{23}}$ was active in both human intestinal absorption (HIA) and had 20% bioavailability (F20%); (2) compound $\mathbf{I_{23}}$ was active in blood–brain barrier (BBB) penetration; (3) compound $\mathbf{I_{23}}$ possessed acceptable safety profiles (e.g., hERG blockers, eye corrosion and respiratory toxicity), and displayed some satisfactory metrics on metabolism (See support information Table S1) and excretion potency; and (4) more interestingly, compound $\mathbf{I_{23}}$ met all drug-likeness properties including Lipinski rule, Pfzer rule, Golden triangle, and GSK rule. Based on the above-mentioned outcomes, compound $\mathbf{I_{23}}$ displayed an acceptable pharmacokinetic profile, making it a promising lead compound to excavate and discover novel microtubule stabilizers.

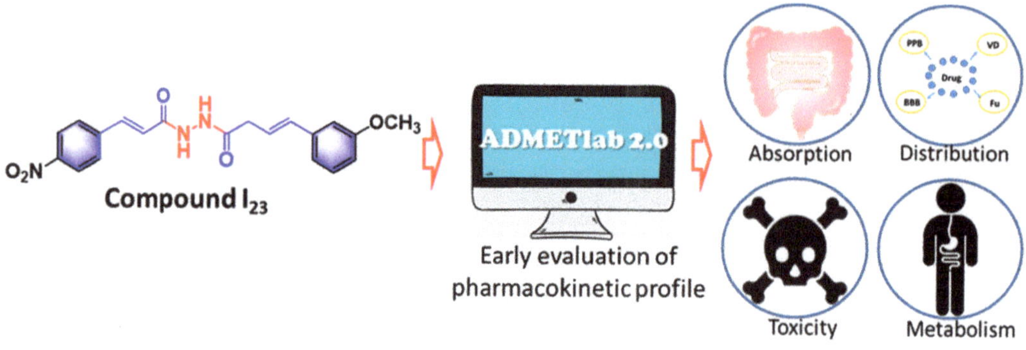

Figure 10. The predicted model of pharmacokinetic profile for compound $\mathbf{I_{23}}$.

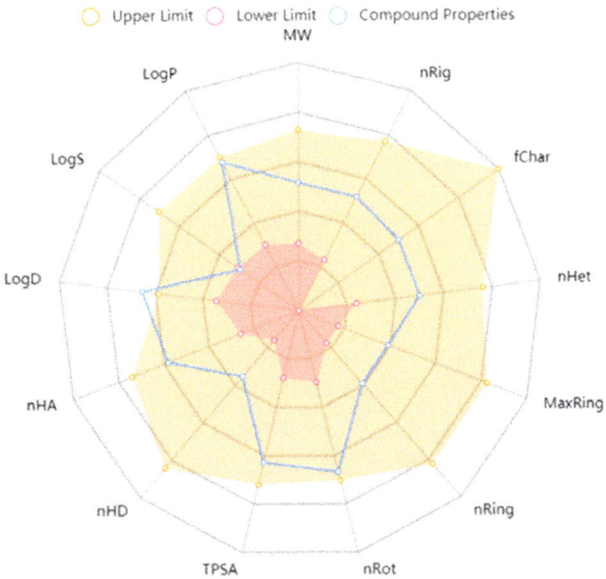

Figure 11. The physicochemical properties of compound $\mathbf{I_{23}}$.

Table 3. ADMET and drug-likeness properties of compound $\mathbf{I_{23}}$ through the online prediction tool ADMETlab 2.0.

Property	Value	Decision
Absorption		
Caco-2 Permeability	−4.82	Excellent
Madin–Darby canine kidney cells (MDCK) permeability	0.000185	Excellent
P-glycoprotein (Pgp)-inhibitor	0.984	Bad
P-glycoprotein (Pgp)-substrate	0.002	Excellent
Human intestinal absorption (HIA)	0.021	Excellent
20% bioavailability (F20%)	0.002	Excellent
Distribution		
Plasma protein binding (PPB)	95.57%	Bad
Volume distribution (VD)	0.285	Excellent
Blood–brain barrier (BBB) penetration	0.038	Excellent
The fraction unbound in plasma (Fu)	1.635%	Bad
Excretion		
Clearance	7.066	Excellent
The half-life ($T_{1/2}$)	0.404	
Toxicity		
hERG blockers	0.329	Medium
Rat oral acute toxicity	0.195	Excellent
Maximum recommended daily dose (FDAMDD)	0.449	Medium
Skin sensitization	0.916	Bad
Eye corrosion	0.004	Excellent
Eye irritation	0.305	Medium
Respiratory toxicity	0.145	Excellent
Drug-likeness		
Lipinski rule	Accepted	Excellent
Pfzer rule	Accepted	Excellent
Golden triangle	Accepted	Excellent
GSK rule	Accepted	Excellent

3. Experimental Section

3.1. Instruments and Chemicals

NMR spectra were performed using the JEOL-ECX500 instrument (Akishima, Japan) or Bruker Biospin AG-400 instrument (Bruker Optics, Ettlingen, Germany) using DMSO-d_6 as the solvent and tetramethylsilane as the internal standard. HRMS spectra were obtained on Waters Xevo G2-S QTOF MS (Waters MS Technologies, Manchester, UK). Immunofluorescence staining assay was performed by the tubulin-tracker red kit (Beyotime Institute of Biotechnology, Shanghai, China) and observed using a Nikon ECLIPSE Ti—S fluorescent microscope (Nikon, Tokyo, Japan). The tubulin polymerization assay was performed by the tubulin polymerization assay kit (cytoskeleton, #BK004P) and recorded by Cytation™ 5 multi-mode readers (BioTek Instruments, Inc., Winooski, VT, USA). The trans-cinnamic acid and analogues were purchased from Aladdin Industrial Inc. (Shanghai, China). The FACSCalibur™ flow cytometer (Becton Dickinson Immunocytometry Systems, San Jose, CA, USA) was employed to analyze the cell cycle arrest.

3.2. Pharmacology

3.2.1. Cell Culture

An A549 (human non-small cell lung cancer cell line) cell line was purchased from the Shanghai Cell Bank of the Chinese Academy of Sciences; PC-3 (human prostate cancer cell line) cell line was donated by the Key Laboratory of Natural Product Chemistry of the Chinese Academy of Sciences of Guizhou Province; HepG2 (human liver cancer cell line) and NRK 52E (normal rat kidney cell line) cell line was donated by Guizhou Medical University; all cell lines were kept by the laboratory. Cells were maintained in RPMI 1640 or DMEM complete medium.

3.2.2. MTT Assay

The antiproliferative activity of target compounds against A549, PC-3, HepG2, and NRK-52E were determined by MTT assay, as described in previous articles [48–50]. The assay was recorded at 490 nm by an Infinite® M200 PRO multimode microplate (Tecan, Männedorf, Switzerland). Curcumin, gefitinib, and colchicine were used as positive controls.

3.2.3. Immunofluorescence Staining Pattern

According to the previous methods [51,52], compound solutions of various concentrations were added to a 6-well plate in which HepG2 cells had been seeded for 12 h. After incubation for 24 h and washed with phosphate-buffered saline (PBS, 10 mM, pH 7.3), each well was treated with 4% formaldehyde solution for 15 min and washed with 0.1% Triton X-100 of phosphate-buffered saline (PBS, 10 mM, pH 7.3). After that, the fixed cells were incubated with diluted tubulin-tracker red for 40 min in a dark environment, followed by washing three times by 0.1% Triton X-100 of phosphate-buffered saline (PBS, 10 mM, pH 7.3). Finally, each well was added DAPI (2 μg/mL). Finally, the results were observed using a Nikon ECLIPSE Ti-S fluorescent microscope (Nikon Co., Tokyo, Japan).

3.2.4. Tubulin Polymerization Assay In Vitro

The tubulin polymerization assay in vitro was determined by HTS-Tubulin Polymerization Assay Kit (#BK004P, Cytoskeleton, Inc., Denver, CO, USA) as described in previous articles [36–38,53,54]. The plate was pre-warmed at 37 °C, and the reaction assay contained 100 μL 4 mg/mL tubulin in G-PEM buffer. Then, 10 μL (10×) compounds solution or 100 μM paclitaxel solution as a control. Finally, the polymerization was carried out at 37 °C and recorded at 340 nm each 10 s for 70 min using Cytation™ 5 multi-mode readers.

3.2.5. Tubulin Polymerization Affected by Target Compounds via TEM

Briefly, according to the protocol of tubulin polymerization assay [38,39], the 100 μL centrifuge tubes were pre-warmed at 37 °C, and the reaction assay contained 20 μL 4 mg/mL tubulin in G-PEM buffer. Then, 2 μL (10×) compounds solution was added. Finally, the assay was carried out at 37 °C for 30 min. The formed microtubules were transferred to Formvar-caron-coated copper grids, negatively stained with 1% phosphotungstic acid, and visualized under a transmission electron microscope.

3.2.6. Cell Cycle Analysis

The cells in the 6-well plates were carefully collected 24 h after dosing. After the cells were washed with pre-cooled PBS, the pre-cooled 70% ethanol solution was added for fixation (4 °C overnight). After that, the cells were washed with PBS, and incubated with 1 mg/mL RNase A for 30 min at 37 °C). Then, 20 mg/mL PI staining solution was added and incubated in the dark for 30 min at 4 °C. Finally, the cell arrest was analyzed by the FACSCalibur™ flow cytometer (Becton Dickinson Immunocytometry Systems, San Jose, CA, USA) as previously described [40].

3.2.7. Computational Docking Studies

The docking study was performed by Sybyl X 2.0 and the tubulin protein (PDB: 5syf) was downloaded from RCSB Protein Data Bank (www.rcsb.org, accessed on 1 October 2020.). The protein and all ligands were prepared by minimization with the CHARMM force field. Molecular docking was carried out using Sybyl X 2.0 protocol without constraint. All bound water and ligands were eliminated from the protein and the polar hydrogen was added to the proteins. The docking results were performed by PyMol software and Discovery Studio (DS) 2020 [55–57].

3.2.8. Scratch Test

According to the previous methods [58], A549 cells of the logarithmic growth stage were cultured in 6-well plates with two lines on the back, with each well containing a density of 1×10^6 cells/mL. After the cells adhered to the wall, three uniform thin lines were drawn in each well by a sterile pipette tip. Afterward, the medium containing 1% fetal bovine serum (FBS) and various concentrations of compounds were added in wells. After the different time of incubation, the cells were performed by an inverted fluorescence microscope. Then the scratch healing rate was calculated: Migration distance (n h) = edge distance (0 h) − edge distance (12 h or 24 h).

3.2.9. Hoechst Apoptosis Experiment

The nuclear morphological modifications were exhibited by fluorescence pattern. In this assay [58–61], HepG2 cells were seeded in 6-well plates, after 24 h of incubation, different compounds were added at various concentrations, and cells were incubated again for 24 h. Afterward, the medium was removed and cells were washed twice with phosphate-buffered saline (PBS, 10 mM, pH 7.3), and fixed with 4% paraformaldehyde for 15 min at room temperature, then stained with 10 mg/mL Hoechst 33258 in PBS for 20 min at 37 °C in the dark. After incubation, the cells were incubated with an anti-fluorescent attenuator and imaged with an inverted fluorescence microscope.

3.2.10. In Silico Pharmacokinetics

The structures of compound $\mathbf{I_{23}}$ were drawn using ChemDraw (version Ultra 12.0, PerkinElmer Informatics, Waltham, MA, USA) and transformed as SMILES format. In silico drug-likeness predictions were conducted using http://www.swissadme.ch/index.php accessed on 1 May 2022 [62,63].

4. Conclusions

To develop novel microtubule-binding agents for cancer therapy, an array of diacylhydrazine-functionalized cinnamic acid derivatives were facilely synthesized through a two-step process. Antiproliferative bioassays showed that compound $\mathbf{I_{23}}$ exhibited the best antiproliferative activity against three cancer lines with IC_{50} values ranging from 3.36 to 5.99 μM and yielded a lower cytotoxicity towards the normal cell line. Anticancer mechanism investigations suggested that the highly bioactive compound $\mathbf{I_{23}}$ could potentially promote the protofilament assembly of tubulin through fluorescence imaging, tubulin polymerization assay and TEM imaging, thus eventually leading to the stagnation of G2/M phase cell cycle of HepG2 cells. Meanwhile, molecular docking studies revealed that compound $\mathbf{I_{23}}$ could interact with Ser374 residue, Ala233, Leu217, Leu230, Phe272, Pro360, and Arg320 residues, and His229 residue of tubulin directed by the corresponding hydrogen bond interactions, π-alkyl interactions, and π-π stacked interactions, which was similar with that of Taxol located at the S9–S10 stabilizing loop of tubulin. Furthermore, compound $\mathbf{I_{23}}$ could also reduce the cancer cell migration rate and induce HepG2 cells apoptosis. Additionally, the in silico analysis indicated that compound $\mathbf{I_{23}}$ exhibited an acceptable pharmacokinetic profile. Based on the present findings, these simple hydrazide derivatives could be considered as potential microtubule-stabilizer lead structures for future anticancer drug discovery.

Supplementary Materials: The following supporting information can be downloaded at: https://www.mdpi.com/article/10.3390/ijms232012365/s1, References [35,63] are cited in the supplementary materials.

Author Contributions: Conceptualization, X.Z. and Y.-H.F.; methodology, X.Z. and Y.-H.F.; software, Y.-Y.Z. and J.M.; formal analysis, Q.-S.G.; data curation, Y.-Y.Z. and J.M.; writing—original draft preparation, X.Z. and Z.-C.W.; writing—review and editing, X.Z. and Z.-C.W.; visualization, Z.-C.W.; supervision, X.Z. and Z.-C.W.; project administration, X.Z., Z.-C.W. and G.-P.O.-Y.; funding acquisition, X.Z. and Z.-C.W. All authors have read and agreed to the published version of the manuscript.

Funding: We acknowledge the financial supports of National Natural Science Foundation of China (22007022, 32160661), Science and Technology Foundation of Guizhou Province [ZK[2021]034], National Innovation and Entrepreneurship Training Program for College Students (2020033), Frontiers Science Center for Asymmetric Synthesis and Medicinal Molecules, Department of Education, Guizhou Province [Qianjiaohe KY number (2020)004], Program of Introducing Talents of Discipline to Universities of China (111 Program, D20023) and GZU (Guizhou University) Found for Newly Enrolled Talent (No. 202229).

Institutional Review Board Statement: Not applicable.

Informed Consent Statement: Not applicable.

Data Availability Statement: Not applicable.

Conflicts of Interest: The authors declare no conflict of interest.

References

1. Virag, K.; Nyari, T.A. Seasonal variation of cancer mortality in Hungary between 1984 and 2013. *Scand J. Public Health* **2019**, *3*, 746–757. [CrossRef]
2. Nagai, H.; Kim, Y.H. Cancer prevention from the perspective of global cancer burden patterns. *J. Thorac. Dis.* **2017**, *9*, 448–451. [CrossRef] [PubMed]
3. Bray, F.; Ferlay, J.; Soerjomataram, I.; Siegel, R.L.; Torre, L.A.; Jemal, A. Global cancer statistics 2018: GLOBOCAN estimates of incidence and mortality worldwide for 36 cancers in 185 countries. *CA A Cancer J. Clin.* **2017**, *68*, 394–424. [CrossRef]
4. Fu, B.W.; Wang, N.; Tan, H.Y.; Li, S.; Cheung, F.; Feng, Y.B. Multi-component herbal products in the prevention and treatment of chemotherapy-associated toxicity and side effects: A review on experimental and clinical evidences. *Front. Pharmacol.* **2018**, *9*, 1394. [CrossRef] [PubMed]
5. Schirrmacher, V. From chemotherapy to biological therapy: A review of novel concepts to reduce the side effects of systemic cancer treatment. *Int. J Oncol.* **2019**, *54*, 407–419.
6. Seligmann, J.; Twelves, C. Tubulin: An example of targeted chemotherapy. *Future Med. Chem.* **2013**, *5*, 339–352. [CrossRef]
7. Tagliamento, M.; Genova, C.; Rossi, G.; Coco, S.; Rijavec, E.; Dal Bello, M.G.; Boccardo, S.; Grossi, F.; Alama, A. Microtubule-targeting agents in the treatment of non-small cell lung cancer: Insights on new combination strategies and investigational compounds. *Expert Opin. Inv. Drugs* **2019**, *28*, 513–523. [CrossRef]
8. Ferlini, C.; Raspaglio, G.; Cicchillitti, L.; Mozzetti, S.; Prislei, S.; Bartollino, S.; Scambia, G. Looking at drug resistance mechanisms for microtubule interacting drugs: Does TUBB3 work? *Curr. Cancer Drug Targets* **2007**, *7*, 704–712. [CrossRef]
9. Jordan, M.A.; Wilson, L. Microtubules as a target for antiproliferative drugs. *Nat. Rev. Cancer* **2014**, *4*, 253–265. [CrossRef]
10. Kavallaris, M. Microtubules and resistance to tubulin-binding agents. *Nat. Rev. Cancer* **2014**, *10*, 194–204. [CrossRef]
11. Čermák, V.; Dostál, V.; Jelínek, M.; Libusová, L.; Kovář, J.; Rösel, D.; Brábek, J. Microtubule-targeting agents and their impact on cancer treatment. *Eur. J. Cell Biol.* **2020**, *99*, 151075. [CrossRef]

12. Chen, S.M.; Meng, L.H.; Ding, J. New microtubule-inhibiting antiproliferative agents. *Expert Opin. Inv. Drug* **2010**, *19*, 329–343. [CrossRef]
13. Zhao, Y.; Fang, W.; Pors, K. Microtubule stabilising agents for cancer chemotherapy. *Expert Opin. Ther. Pat.* **2009**, *19*, 607–622. [CrossRef] [PubMed]
14. Haider, K.; Rahaman, S.; Yar, M.S.; Kamal, A. Tubulin inhibitors as novel antiproliferative agents: An overview on patents (2013–2018). *Expert Opin. Ther. Pat.* **2019**, *29*, 623–641. [CrossRef]
15. Barreca, M.; Stathis, A.; Barraja, P.; Bertoni, F. An overview on anti-tubulin agents for the treatment of lymphoma patients. *Pharmacol. Therapeut.* **2020**, *211*, 107552. [CrossRef] [PubMed]
16. Tangutur, A.D.; Kumar, D.; Krishna, K.V.; Krishna, S. Microtubule targeting agents as cancer chemotherapeutics: An overview of molecular hybrids as stabilizing and destabilizing agents. *Curr. Top. Med. Chem.* **2017**, *17*, 2523–2537. [CrossRef] [PubMed]
17. Zefirova, O.N.; Diikov, A.G.; Zyk, N.V.; Zefirov, N.S. Ligands of the colchicine site of tubulin: A common pharmacophore and new structural classes. *Russ. Chem. B* **2007**, *56*, 680–688. [CrossRef]
18. Borisy, G.G.; Heald, R.; Howard, J.; Janke, C.; Musacchion, A.; Nogals, E. Microtubules: 50 years on from the discovery of tubulin. *Nat. Rev. Mol. Cell Biol.* **2016**, *17*, 322–328. [CrossRef]
19. Li, W. Drugs Targeting Tubulin Polymerization. *Pharm. Res.-Dordr.* **2012**, *29*, 2939–2942. [CrossRef]
20. Lu, Y.; Chen, J.J.; Xiao, M.; Li, W.; Miller, D.D. An overview of tubulin inhibitors that interact with the colchicine binding site. *Pharm. Res.-Dordr.* **2012**, *29*, 2943–2971. [CrossRef]
21. Naaz, F.; Haider, M.R.; Shafi, S.; Yar, M.S. Anti-tubulin agents of natural origin: Targeting taxol, vinca, and colchicine binding domains. *Eur. J. Med. Chem.* **2019**, *171*, 310–331. [CrossRef]
22. Cao, Y.N.; Zheng, L.L.; Wang, D.; Liang, X.X.; Gao, F.; Zhou, X.L. Recent advances in microtubule-stabilizing agents. *Eur. J. Med. Chem.* **2018**, *143*, 806–828. [CrossRef]
23. Zhao, Y.; Mu, X.; Du, G.H. Microtubule-stabilizing agents: New drug discovery and cancer therapy. *Pharmacol. Therapeut.* **2016**, *162*, 134–143. [CrossRef] [PubMed]
24. Ettinger, D.S. Overview of paclitaxel (Taxol) in advanced lung cancer. *Semin. Oncol.* **1993**, *20*, 46–49. [PubMed]
25. Ojima, I.; Slater, J.C.; Michaud, E.; Kuduk, S.D.; Bounaud, P.Y.; Vrignaud, P.; Vrignaud, P.; Bissery, M.; Veith, J.M.; Pera, P.; et al. Syntheses and structure−activity relationships of the second-generation antiproliferative taxoids: Exceptional activity against drug-resistant cancer cells. *J. Med. Chem.* **1996**, *39*, 3889–3896. [CrossRef]
26. Mooberry, S.L.; Tien, G.; Hernandez, A.H.; Plubrukarn, A.; Davidson, B.S. Laulimalide and isolaulimalide, new paclitaxel-like microtubule-stabilizing agents. *Cancer Res.* **1999**, *59*, 653–660.
27. Shintani, Y.; Tanaka, T.; Nozaki, Y. GS-164, a small synthetic compound, stimulates tubulin polymerization by a similar mechanism to that of Taxol. *Cancer Chemother. Pharmacol.* **1997**, *40*, 513–520. [CrossRef]
28. Haggarty, S.J.; Mayer, T.U.; Miyamoto, D.T.; Fathi, R.; King, R.W.; Mitchison, T.J.; Schreiber, S.L. Dissecting cellular processes using small molecules: Identification of colchicine-like, taxol-like and other small molecules that perturb mitosis. *Chem. Biol.* **2000**, *7*, 275–286. [CrossRef]
29. Marinho, J.; Pedro, M.; Pinto, D.C.G.A.; Silva, A.M.S.; Cavaleiro, J.A.S.; Sunkel, C.E.; Nascimento, M.S.J. 4′-Methoxy-2-styrylchromone a novel microtubule-stabilizing antimitotic agent. *Biochem. Pharmacol.* **2008**, *75*, 826–835. [CrossRef]
30. Reddy, M.V.R.; Akula, B.; Cosenza, S.C.; Lee, C.M.; Mallireddigari, M.R.; Pallela, V.R.; Subbaiah, D.R.C.V.; Udofa, A.; Reddy, E.P. (Z)-1-Aryl-3-arylamino-2-propen-1-ones, highly active stimulators of tubulin polymerization: Synthesis, structure-activity relationship (SAR), tubulin polymerization, and cell growth inhibition studies. *J. Med. Chem.* **2012**, *55*, 5174–5187. [CrossRef]
31. Chakraborti, S.; Das, L.; Kapoor, N.; Das, A.; Dwivedi, V.; Poddar, A.; Chakraborti, G.; Janik, M.E.; Basu, G.; Panda, D.; et al. Curcumin recognizes a unique binding site of tubulin. *J. Med. Chem.* **2011**, *54*, 6183–6196. [CrossRef] [PubMed]
32. Gupta, K.K.; Bharne, S.S.; Rathinasamy, K.; Naik, N.R.; Panda, D. Dietary antioxidant curcumin inhibits microtubule assembly through tubulin binding. *FEBS J.* **2006**, *273*, 5320–5332. [CrossRef]
33. Luo, Y.; Qiu, K.M.; Lu, X.; Liu, K.; Fu, J.; Zhu, H.L. Synthesis, biological evaluation, and molecular modeling of cinnamic acyl sulfonamide derivatives as novel antitubulin agents. *Bioorg. Med. Chem.* **2011**, *19*, 4730–4738. [CrossRef]
34. Sinigersky, V.; Wegener, G.; Schopoy, I. Synthesis and properties of a poly(phenylenevinylene) containing 1, 3, 4-oxadiazole rings. *Eur. Polym. J.* **1993**, *29*, 617–620.
35. Zhang, X.N.; Breslav, M.; Grimm, J.; Guan, K.L.; Huang, A.H.; Liu, F.Q.; Maryanoff, C.A.; Palmer, D.; Patel, M.; Qian, Y.; et al. A new procedure for preparation of carboxylic acid hydrazides. *J. Org. Chem.* **2002**, *67*, 9471–9474. [CrossRef] [PubMed]
36. Kim, D.Y.; Kim, K.H.; Kim, N.D.; Lee, K.Y.; Han, C.K.; Yoon, J.H.; Moon, S.K.; Lee, S.S.; Seong, B.L. Design and biological evaluation of novel tubulin inhibitors as antimitotic agents using a pharmacophore binding model with tubulin. *J. Med. Chem.* **2006**, *49*, 5664–5670. [CrossRef]
37. Greene, T.F.; Wang, S.; Greene, L.M.; Nathwani, S.M.; Pollock, J.K.; Malebari, A.M.; McCabe, T.; Twamly, B.; O'Boyle, N.M.; Zisterer, D.M.; et al. Synthesis and biochemical evaluation of 3-phenoxy-1, 4-diarylazetidin-2-ones as tubulin-targeting antiproliferative agents. *J. Med. Chem.* **2016**, *59*, 90–113. [CrossRef] [PubMed]
38. Chopra, A.; Anderson, A.; Giardina, C. Novel piperazine-based compounds inhibit microtubule dynamics and sensitize colon cancer cells to tumor necrosis factor-induced apoptosis. *J. Biol. Chem.* **2014**, *289*, 2978–2991. [CrossRef] [PubMed]
39. Risinger, A.L.; Li, J.; Bennett, M.J.; Rohena, C.C.; Peng, J.; Schriemer, D.C.; Mooberry, S.L. Taccalonolide Binding to Tubulin Imparts Microtubule Stability and Potent In Vivo Activity. *Cancer Res.* **2013**, *73*, 6780–6792. [CrossRef]

40. Cao, D.; Han, X.L.; Wang, G.C.; Yang, Z.; Peng, F.; Ma, L.; Zhang, R.H.; Ye, H.Y.; Tang, M.H.; Wu, W.S.; et al. Synthesis and biological evaluation of novel pyranochalcone derivatives as a new class of microtubule stabilizing agents. *Eur. J. Med. Chem.* **2013**, *62*, 579–589. [CrossRef]
41. Lu, Y.X.; Wang, Y.; Zhu, W.L. Nonbonding interactions of organic halogens in biological systems: Implications for drug discovery and biomolecular design. *Phys. Chem. Chem. Phys.* **2010**, *12*, 4543–4551. [CrossRef] [PubMed]
42. Johnson, E.R.; Keinan, S.; Mori-Sánchez, P.; Contreras-García, J.; Cohen, A.J.; Yang, W.T. Revealing Noncovalent Interactions. *J. Am. Chem. Soc.* **2010**, *132*, 6498–6506. [CrossRef] [PubMed]
43. Mahadevi, A.S.; Sastry, G.N. Cooperativity in Noncovalent Interactions. *Chem. Rev* **2016**, *116*, 2775–2825. [CrossRef] [PubMed]
44. Nabeshima, K.; Inoue, T.; Shimao, Y.; Sameshima, T. Matrix metalloproteinases in tumor invasion: Role for cell migration. *Pathol. Int.* **2002**, *52*, 255–264. [CrossRef] [PubMed]
45. Ganguly, A.; Yang, H.L.; Sharma, R.; Patel, K.D.; Cabral, F. The role of microtubules and their dynamics in cell migration. *J. Biol. Chem.* **2012**, *287*, 43359–43369. [CrossRef]
46. Wang, G.; Liu, W.; Gong, Z.; Huang, Y.; Li, Y.; Peng, Z. Design, synthesis, biological evaluation and molecular docking studies of new chalcone derivatives containing diaryl ether moiety as potential anticancer agents and tubulin polymerization inhibitors. *Bioorg. Chem.* **2020**, *95*, 103565. [CrossRef] [PubMed]
47. Li, W.L.; Yin, Y.; Shuai, W.; Xu, F.J.; Yao, H.; Liu, J.; Cheng, K.G.; Xu, J.Y.; Zhu, Z.Y.; Xu, S.T. Discovery of novel quinazolines as potential anti-tubulin agents occupying three zones of colchicine domain. *Bioorg. Chem.* **2019**, *83*, 380–390. [CrossRef]
48. Le, Y.; Gan, Y.Y.; Fu, Y.H.; Liu, J.M.; Li, W.; Zou, W.; Zhou, Z.X.; Wang, Z.C.; Ouyang, G.P. Design, synthesis and in vitro biological evaluation of quinazolinone derivatives as EGFR inhibitors for antiproliferative treatment. *J. Enzym. Inhib. Med. Chem.* **2020**, *35*, 555–564. [CrossRef] [PubMed]
49. Gao, F.; Liang, Y.; Zhou, P.F.; Cheng, J.Y.; Ding, K.L.; Wang, Y. Design, synthesis, antiproliferative activities and biological studies of novel diaryl substituted fused heterocycles as dual ligands targeting tubulin and katanin. *Eur. J. Med. Chem.* **2019**, *178*, 177–194. [CrossRef] [PubMed]
50. Yin, Y.; Lian, B.P.; Xia, Y.Z.; Shao, Y.Y.; Kong, L.Y. Design, synthesis and biological evaluation of resveratrol-cinnamoyl derivates as tubulin polymerization inhibitors targeting the colchicine binding site. *Bioorg. Chem.* **2019**, *93*, 103319. [CrossRef]
51. Bai, H.H.; Jin, H.; Yang, F.; Zhu, H.Y.; Cai, J.Y. Apigenin induced MCF-7 cell apoptosis-associated reactive oxygen species. *Scanning* **2014**, *36*, 622–631. [CrossRef] [PubMed]
52. Zhang, Y.W.; Zhao, H.B.; Di, Y.C.; Li, Q.; Shao, D.Y.; Shi, J.L.; Huang, Q.S. Antiproliferative activity of Pinoresinol in vitro: Inducing apoptosis and inhibiting HepG2 invasion. *J. Funct. Foods* **2018**, *45*, 206–214. [CrossRef]
53. Tong, L.Y.; Sun, W.C.; Wu, S.Y.; Han, Y. Characterization of Caerulomycin A as a dual-targeting anticancer agent. *Eur. J. Pharmacol.* **2022**, *922*, 174914. [CrossRef] [PubMed]
54. Yang, L.M.; Ma, X.; Guo, K.R.; Li, J.; Zhang, C.; Wu, L.Q. Dual-functional antitumor conjugates improving the anti-metastasis effect of combretastatin A4 by targeting tubulin polymerization and matrix metalloproteinases. *Eur. J. Med. Chem.* **2022**, *238*, 114439. [CrossRef] [PubMed]
55. Can, S.N.; Lacey, S.; Gur, M.; Carter, A.P.; Yildiz, A. Directionality of dynein is controlled by the angle and length of its stalk. *Nature* **2019**, *566*, 407–410. [CrossRef]
56. Song, Y.L.; Liu, S.S.; Yang, J.; Xie, J.; Zhou, X.; Wu, Z.B.; Liu, L.W.; Wang, P.Y.; Yang, S. Discovery of Epipodophyllotoxin-Derived B_2 as Promising *Xoo*FtsZ Inhibitor for Controlling Bacterial Cell Division: Structure-Based Virtual Screening, Synthesis, and SAR Study. *Int. J. Mol. Sci.* **2022**, *23*, 9119. [CrossRef] [PubMed]
57. Zhou, X.; Ye, H.J.; Gao, X.H.; Feng, Y.M.; Shao, W.B.; Qi, P.Y.; Wu, Z.B.; Liu, L.W.; Wang, P.Y.; Yang, S. The discovery of natural 4′-demethylepipodophyllotoxin from renewable Dysosma versipellis species as a novel bacterial cell division inhibitor for controlling intractable diseases in rice. *Ind. Crop Prod.* **2021**, *174*, 114182. [CrossRef]
58. Hu, X.; Li, L.; Zhang, Q.S.; Wang, Q.Q.; Feng, Z.Z.; Xu, Y.; Xia, Y.; Yu, L.T. Design, synthesis and biological evaluation of a novel tubulin inhibitor SKLB0565 targeting the colchicine binding site. *Bioorg. Chem.* **2020**, *97*, 103695. [CrossRef] [PubMed]
59. Sayeed, I.B.; Vishnuvardhan, M.V.P.S.; Nagarajan, A.; Kantevari, S.; Kamal, A. Imidazopyridine linked triazoles as tubulin inhibitors, effectively triggering apoptosis in lung cancer cell line. *Bioorg. Chem.* **2018**, *80*, 714–720. [CrossRef] [PubMed]
60. Baig, M.F.; Nayak, V.L.; Budaganaboyina, P.; Mullagiri, K.; Sunkari, S.; Gour, J.; Kamal, A. Synthesis and biological evaluation of imidazo [2,1-b]thiazole-benzimidazole conjugates as microtubule-targeting agents. *Bioorg. Chem.* **2018**, *77*, 515–526. [CrossRef] [PubMed]
61. Zhou, X.; Liu, J.M.; Meng, J.; Fu, Y.H.; Wu, Z.B.; Ouyang, G.P.; Wang., Z.C. Discovery of facile amides-functionalized rhodanine-3-acetic acid derivatives as potential anticancer agents by disrupting microtubule dynamics. *J. Enzym. Inhib. Med. Chem.* **2021**, *36*, 1996–2009. [CrossRef] [PubMed]
62. Xiong, G.L.; Wu, Z.X.; Yi, J.C.; Fu, L.; Yang, Z.J.; Hsieh, C.; Yin, M.Z.; Zeng, X.X.; Wu, C.K.; Lu, A.P.; et al. ADMETlab 2.0: An integrated online platform for accurate and comprehensive predictions of ADMET properties. *Nucleic Acids Res.* **2021**, *49*, W5–W14. [CrossRef]
63. Zhou, X.; Feng, Y.M.; Qi, P.Y.; Shao, W.B.; Wu, Z.B.; Liu, L.W.; Wang, Y.; Ma, H.D.; Wang, P.Y.; Yang, S. Synthesis and docking study of *N*-(Cinnamoyl)-*N′*-(substituted)acryloyl hydrazide derivatives containing pyridinium moieties as a novel class of filamentous temperature-sensitive protein Z inhibitors against the intractable *Xanthomonas oryzae* pv. *oryzae infections in rice. J. Agric. Food Chem.* **2020**, *68*, 8132–8142. [CrossRef] [PubMed]

International Journal of *Molecular Sciences*

Review

Recent Advances in PROTACs for Drug Targeted Protein Research

Tingting Yao [1,†], Heng Xiao [2,†], Hong Wang [1,*] and Xiaowei Xu [1,*]

1 State Key Laboratory of Natural Medicines, Key Lab of Drug Metabolism and Pharmacokinetics, China Pharmaceutical University, Nanjing 210009, China
2 Guangdong Provincial Key Laboratory of Virology, Institute of Medical Microbiology, Jinan University, Guangzhou 510632, China
* Correspondence: wanghong991@163.com (H.W.); xw@cpu.edu.cn (X.X.); Tel.: +86-15-895-840-060 (X.X.)
† These authors contributed equally to this work.

Abstract: Proteolysis-targeting chimera (PROTAC) is a heterobifunctional molecule. Typically, PROTAC consists of two terminals which are the ligand of the protein of interest (POI) and the specific ligand of E3 ubiquitin ligase, respectively, via a suitable linker. PROTAC degradation of the target protein is performed through the ubiquitin–proteasome system (UPS). The general process is that PROTAC binds to the target protein and E3 ligase to form a ternary complex and label the target protein with ubiquitination. The ubiquitinated protein is recognized and degraded by the proteasome in the cell. At present, PROTAC, as a new type of drug, has been developed to degrade a variety of cancer target proteins and other disease target proteins, and has shown good curative effects on a variety of diseases. For example, PROTACs targeting AR, BR, BTK, Tau, IRAK4, and other proteins have shown unprecedented clinical efficacy in cancers, neurodegenerative diseases, inflammations, and other fields. Recently, PROTAC has entered a phase of rapid development, opening a new field for biomedical research and development. This paper reviews the various fields of targeted protein degradation by PROTAC in recent years and summarizes and prospects the hot targets and indications of PROTAC.

Keywords: PROTAC; target protein; protein degradation; indication

Citation: Yao, T.; Xiao, H.; Wang, H.; Xu, X. Recent Advances in PROTACs for Drug Targeted Protein Research. *Int. J. Mol. Sci.* **2022**, *23*, 10328. https://doi.org/10.3390/ijms231810328

Academic Editor: Bernhard Biersack

Received: 1 August 2022
Accepted: 5 September 2022
Published: 7 September 2022

1. Introduction

Targeted protein degradation (TPD) is an emerging therapeutic modality that has the potential to solve the dilemma faced by traditional small molecule targeted therapy. Targeted protein degradation currently mainly degrades target proteins through ubiquitin–proteasome and lysosome. At present, molecular glue and PROTAC technology are the fastest growing in the field of targeted protein degradation [1].

Crews et al. introduced PROTAC for the first time in 2001, and PROTAC works by reducing protein levels rather than inhibiting protein function [2–4]. As a bifunctional small molecule compound, typically PROTAC consists of two terminals which are the ligand of the target protein and the specific ligand of E3 ubiquitin ligase, respectively, via a suitable linker [5–10]. PROTAC degrades target proteins through the ubiquitin–proteasome system (UPS). The general process is that PROTAC binds the target protein (POI) and E3 ligase to form a ternary complex, marking the target protein with the label of ubiquitination. The ubiquitinated proteins are recognized and degraded by the intracellular 26S proteasome [11–19] (Figure 1). The E3 ubiquitin ligase has approximately more than 600 members and is the most diverse component of the ubiquitin–proteasome system. The E3 ligases reported in the literature currently used in PROTAC mainly include Cereblon E3 ubiquitin ligase complex (CRBN), Von Hippel–Lindau-containing complex (VHL), inhibitor of apoptosis protein (IAP), and mouse double minute 2 (MDM2). The E3 ligases with the best effect and the highest frequency are mainly CRBN and VHL. Among them, the ligands of CRBN are mainly lenalidomide (Figure 2), thalidomide (Figure 2), and their analogs, while the ligands of VHL are mainly VHL-L (Figure 2) and 3-fluoro-VHL ligand [20–24] (Figure 2). PROTAC structurally connects two ligands through the linkers. The composition and length of the linker play an important role in PROTAC. Generally speaking, the composition and length of the linkers have different effects on degradation activity according to different targets. In addition, linked sites of the linkers also affect degradation activity. The binding sites of POI ligands and E3 ligase ligands are generally in the regions where the ligands are exposed to solvents. The connecting sites are generally connected by amide bonds, carbon atoms, or heteroatoms [25].

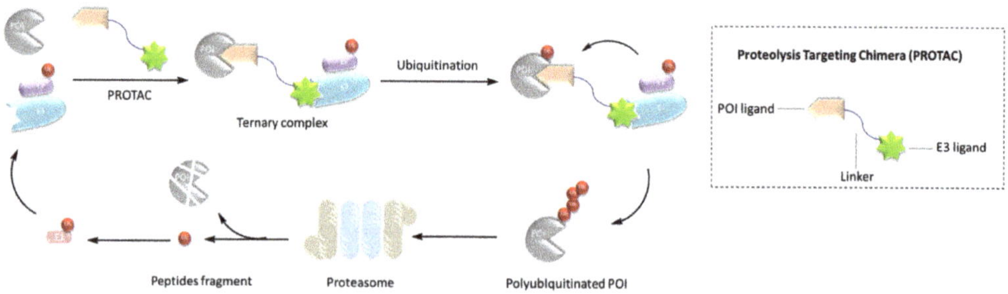

Figure 1. Mechanism of PROTAC.

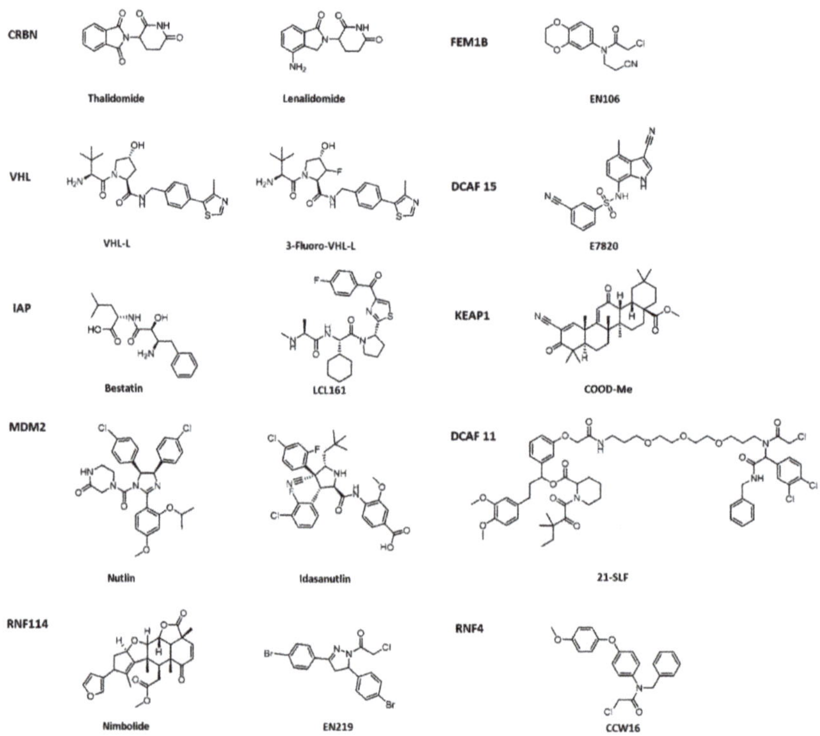

Figure 2. Existing E3 ligands used for targeted protein degradation applications.

Compared with traditional therapies, PROTAC technology has advantages such as wider scope of action, higher activity, and targeting "undruggable" targets. First, PROTAC can degrade the entire target protein to affect protein function, which is expected to solve the potential drug resistance problem faced by current traditional therapies; second, in theory,

PROTAC can grasp the target protein through any corner and gap; therefore, PROTAC can target "undruggable target"; third, PROTAC can also affect non-enzymatic functions and expand the drug space of the target. So far, PROTACs have been successfully used to degrade several distinct target proteins associated with all kinds of illnesses, such as cancer, immune disorders, neurodegenerative conditions, cardiovascular diseases as well as viral infections [26–28]. In particular, 60 successful cases have demonstrated the effectiveness of PROTACs in degrading target proteins, two of which are currently in clinical trials for prostate and breast cancer treatment [29,30]. PROTAC has emerged as a fresh approach to medication development, providing a fresh method of treating disease. This article reviews the five applications of PROTAC in cancer, immune diseases, neurodegenerative diseases, cardiovascular diseases, and viral infections, summarizes and prospects its potential targets and indications.

2. Application of PROTAC in Anticancer

Cancer is one of the worst illnesses threatening human health. In recent years, the treatment of cancer is no longer confined to traditional surgery and radiotherapy and chemotherapy. Targeted therapy and immunotherapy play an important role in anti-cancer treatment. However, there are still no effective targeted drugs for "undruggable targets" such as KRAS and TP53 [31]. The ability to shift the target from "no drug" to "drug" is the most main benefit of PROTAC technology. Traditional targeted drugs need to be firmly bound to the target protein. Since PROTAC protein degrading agent can specifically "label" the target protein only by weakly binding with it, PROTAC degradation agent may solve about 80% of the current "undruggable" proteome. It is a timely help for patients who cannot carry out traditional targeted therapy [32].

2.1. Breast Cancer

Breast cancer is the most frequent malignant tumor in women worldwide, and the research and development of new anti-breast cancer drugs has always attracted much attention. The E3 ubiquitin ligase MDM2 ligand nutlin-3 derivative and poly (ADP) ribose polymerase 1 (PARP1) ligand niraparib derivative PROTACs were used to target PARP1 degradation, thus causing human breast cancer cells MDA-MB-231 to undergo apoptosis [33]. Naito Mikihiko et al. reported that an IAP ligand derivative and estrogen receptor α (ERα) ligand 4-hydroxytamoxifen consists of PROTACs linked by alkyl or PEG between the two [34,35]. Their study showed that PROTAC 1 (Table 1) induces ERα degradation, reactive oxygen species (ROS) production, and necrotic cell death in estrogen-dependent breast cancer MCF-7 cells, with therapeutic potential for ERα-positive breast cancer.

Table 1. Representative PROTACs for cancer.

Indication	PROTAC	Target	Structure	Activity		Ref.
				DC_{50}	D_{max}%	
Breast cancer	1	ERα	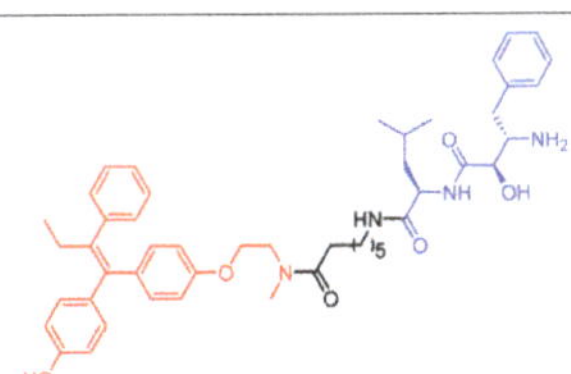	-	-	[34]
	2	ERRα	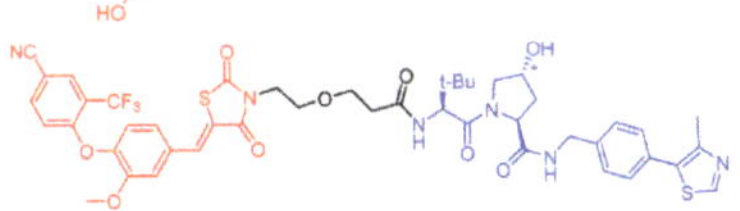	100 nM	86	[36]

Table 1. *Cont.*

Indication	PROTAC	Target	Structure	Activity		Ref.
				DC_{50}	D_{max}%	
	3	FAK	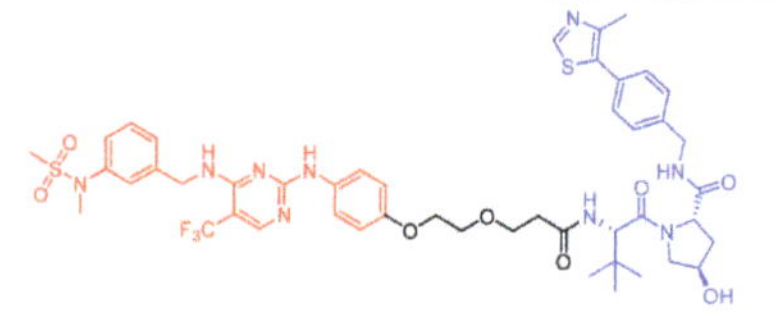	3 nM	99	[37]
	4	FAK	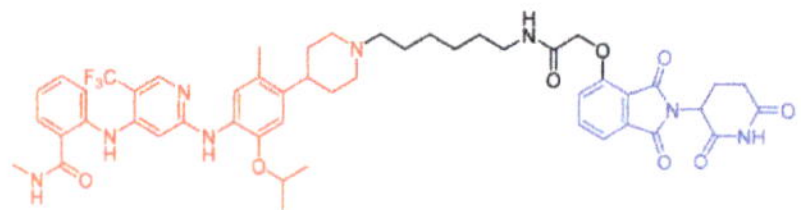	-	>90	[38]
	5	p38α	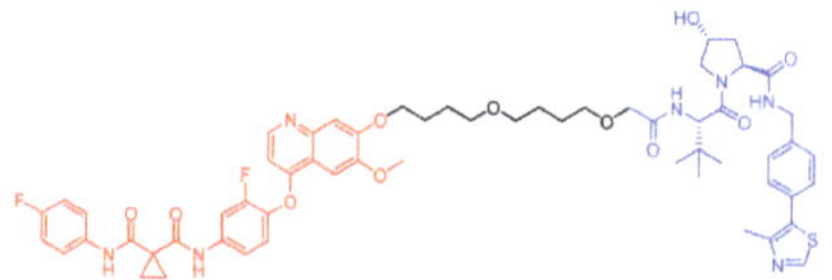	7.16 nM	97.4	[39]
	6	p38δ	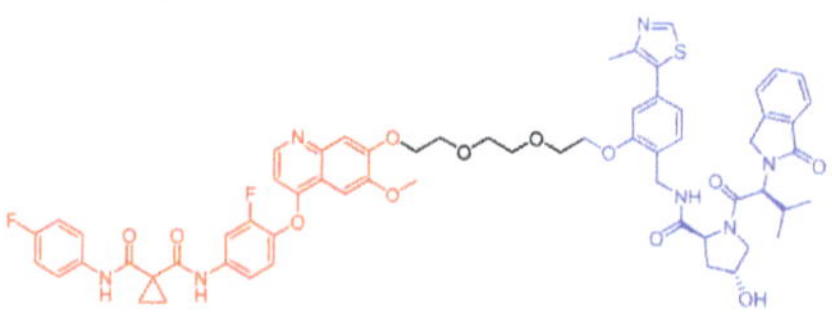	46 nM	99.4	[39]
	7	BCL-X_L	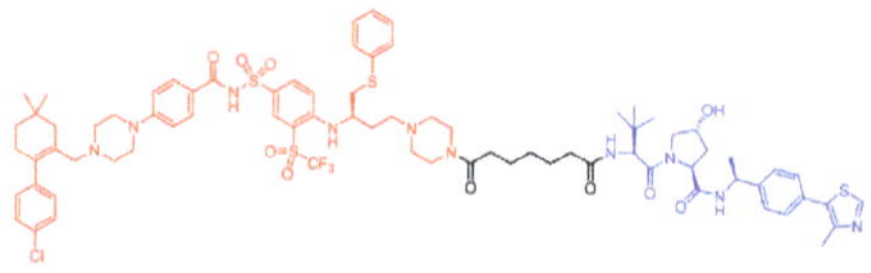	63 nM	90.8	[40]
	8	BRD4	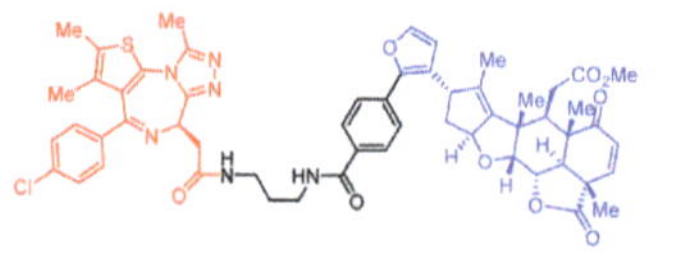	-	-	[41]
	9	PTK6	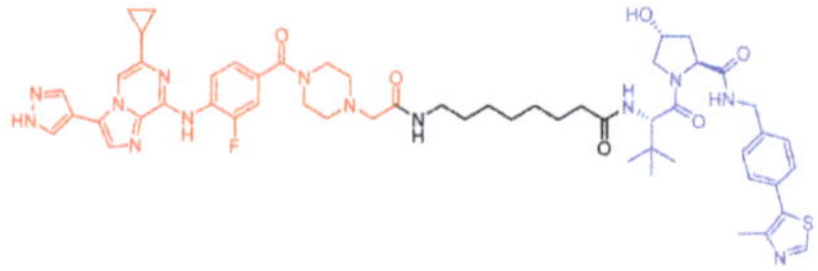	-	-	[26]
	10	RIPK2	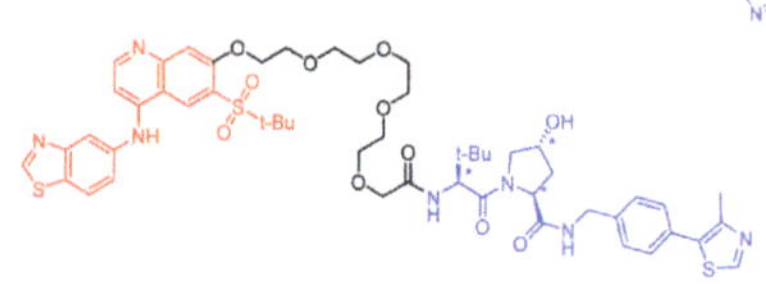	1.4 nM	>95	[36]
AML	11	TRIM24	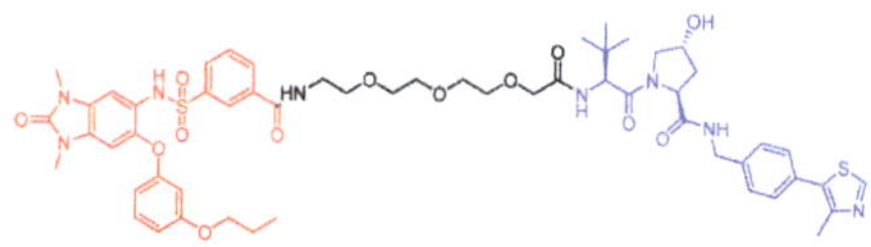	-	-	[42]

Table 1. *Cont.*

Indication	PROTAC	Target	Structure	Activity		Ref.
				DC_{50}	D_{max}%	
	12	MDM2	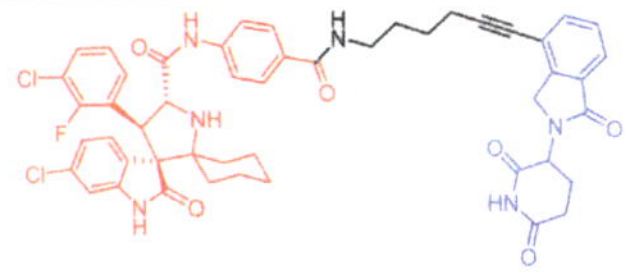	1.5 nM	-	[43]
	13	FLT-3	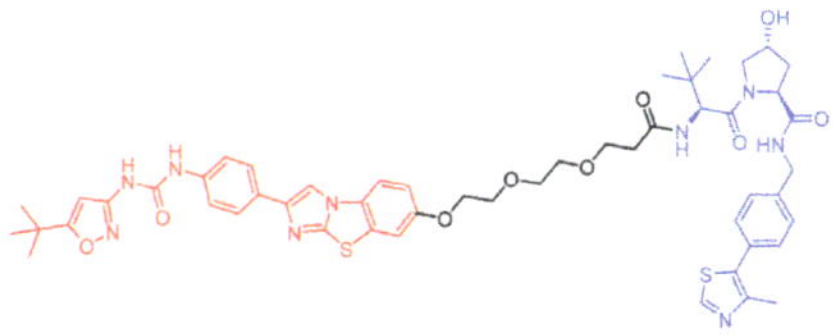	-	-	[44]
	14	STAT3	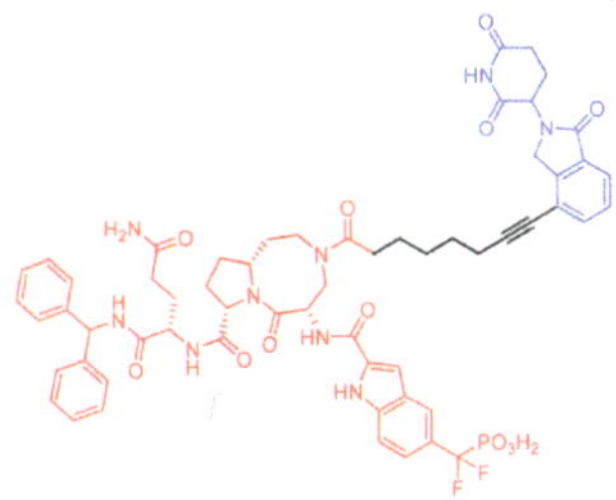	-	>90	[45]
	15	CDK6	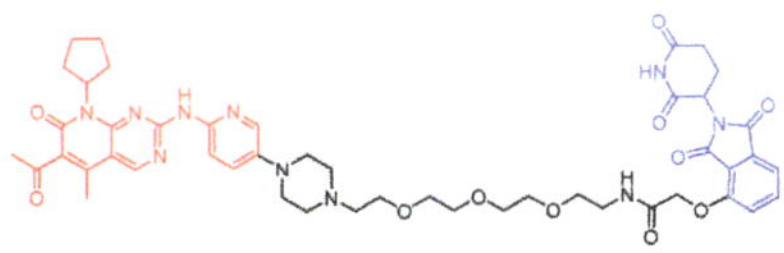	-	-	[46]
	16	BRD4	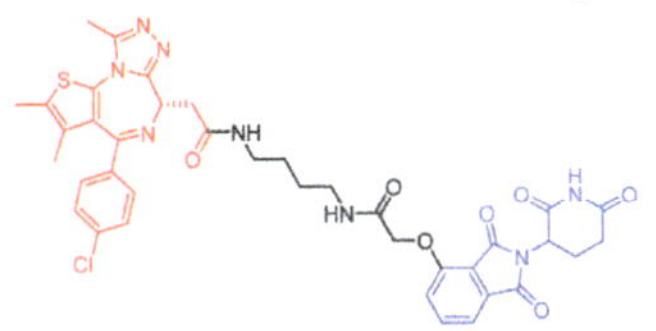	430 nM	-	[47]
	17	BRD4	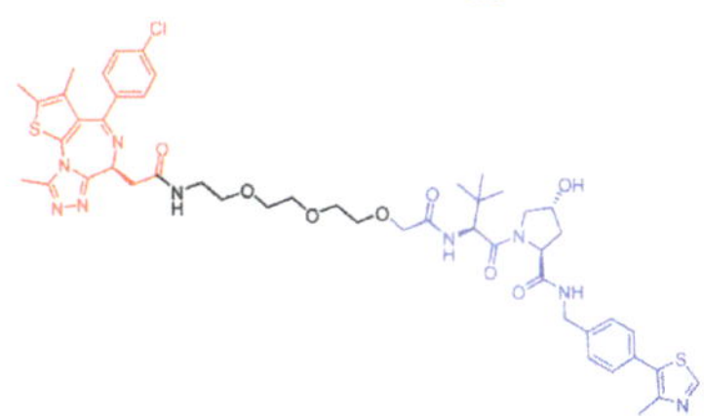	-	-	[15]
	18	BRD4	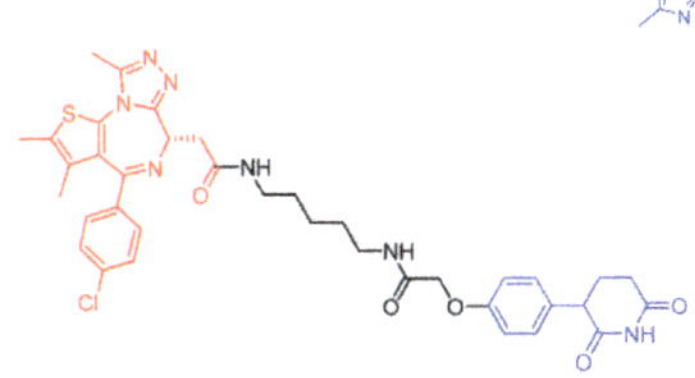	<1 nM	99	[48]

Table 1. *Cont.*

Indication	PROTAC	Target	Structure	Activity		Ref.
				DC_{50}	D_{max}%	
	19	BRD4	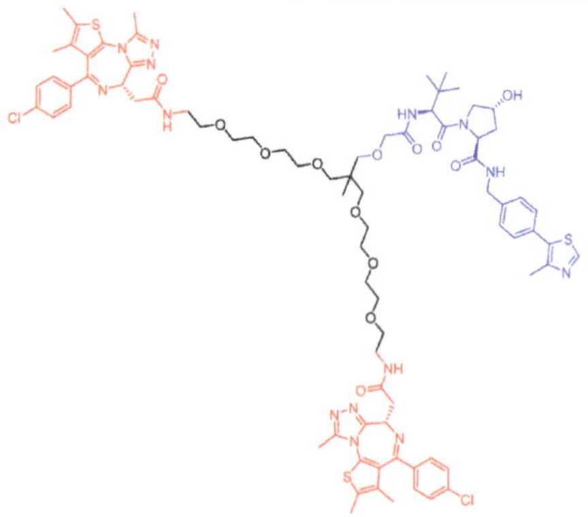	-	-	[49]
T-ALL	20	BCL-X_L	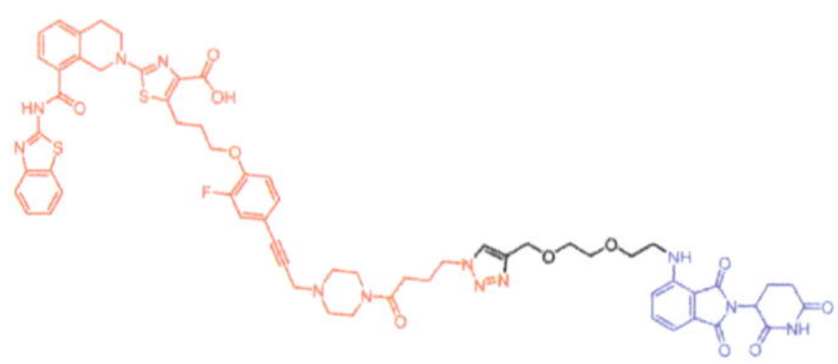	50 nM	>85	[50]
	21	BCL-X_L	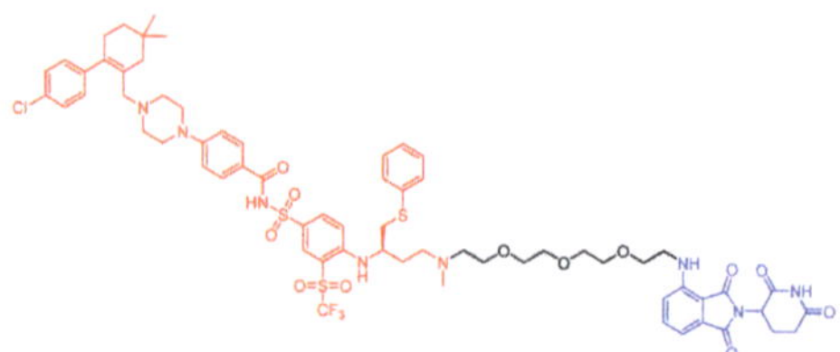	2.5 nM	-	[51]
CML	22	BCR-ABL	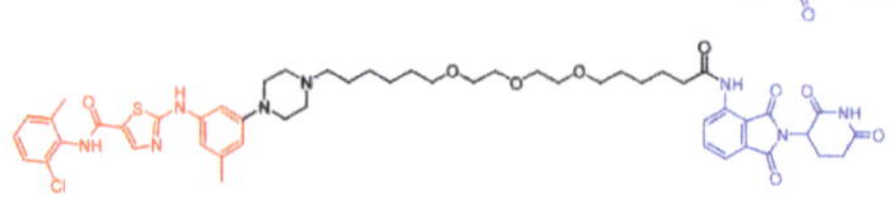	-	>80	[52]
	23	BCR-ABL	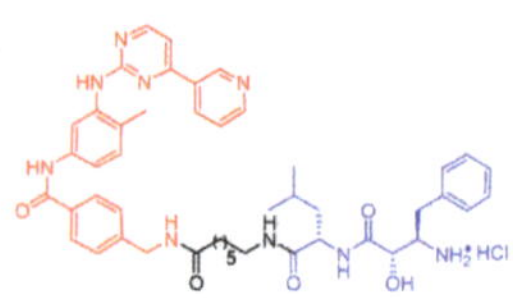	-	-	[53]
	24	BCR-ABL	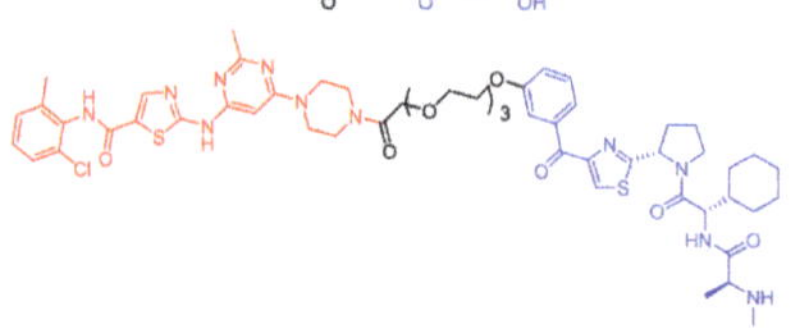	-	-	[54]
TCL	8	BCL-X_L	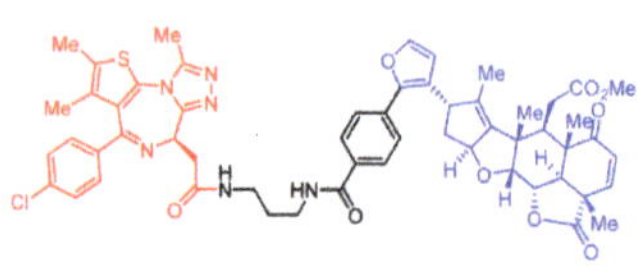	-	-	[55]

Table 1. *Cont.*

Indication	PROTAC	Target	Structure	Activity		Ref.
				DC_{50}	D_{max}%	
	25	BTK	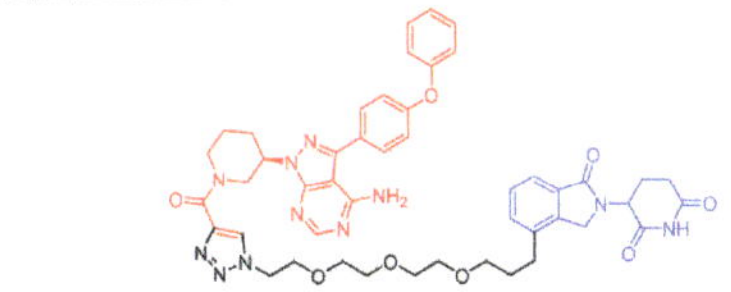	29 nM	-	[56]
	26	BTK	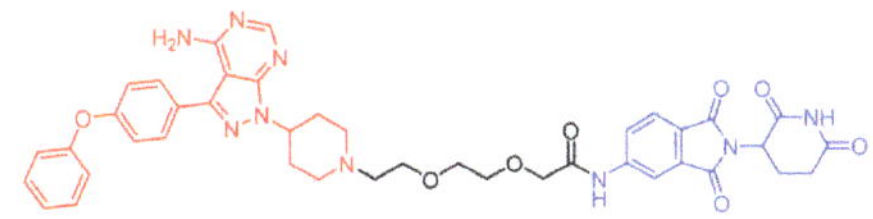	6.2 nM	99	[57]
BCL	27	BTK	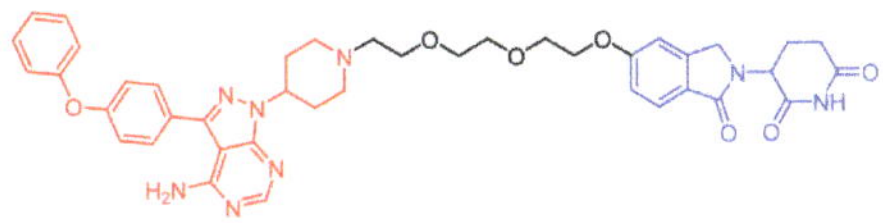	7.9 nM	95	[58]
	28	BTK	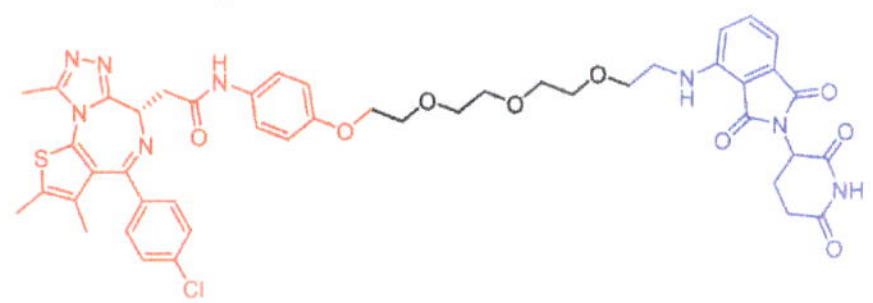	<1 nM	-	[59]
	29	BTK	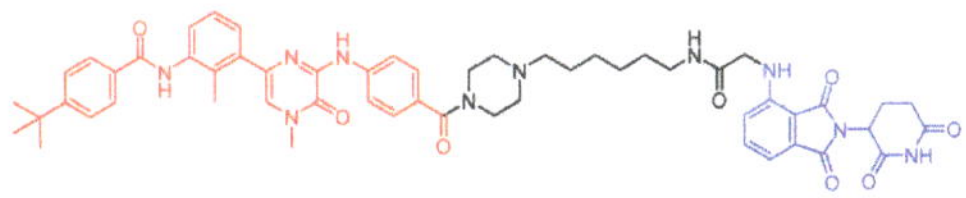	-	-	[60]
	30	BTK	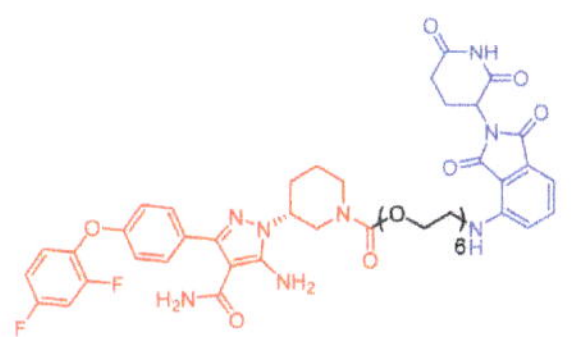	5.9 nM	-	[61]
MM	31	IKZF1-3	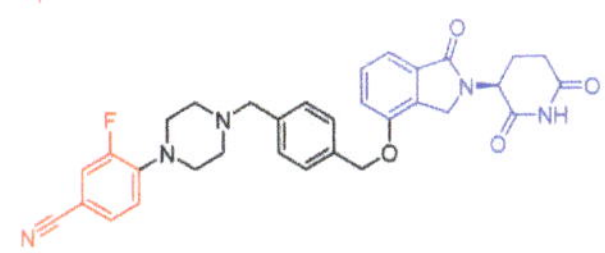	-	-	[62]
	32	CDK6	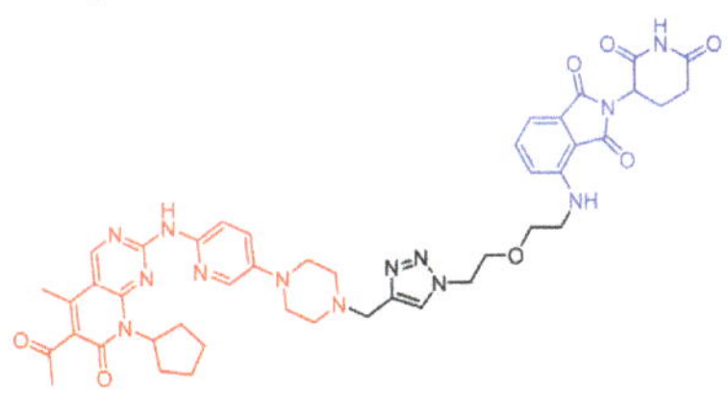	8.6 nM	-	[63]

Table 1. *Cont.*

Indication	PROTAC	Target	Structure	Activity		Ref.
				DC_{50}	D_{max}%	
Colon cancer	33	ERK1-2	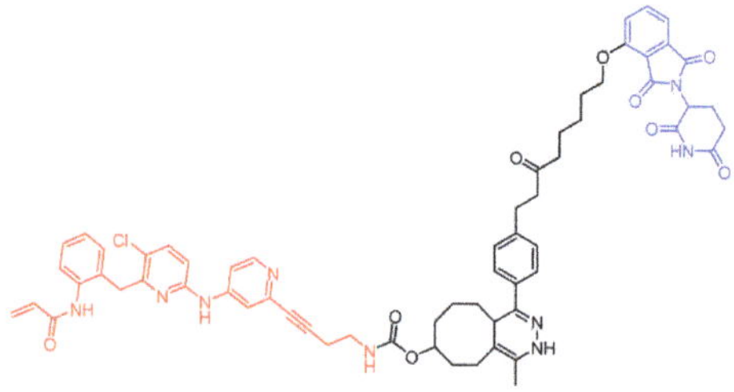	-	-	[64]
	34	CDK9	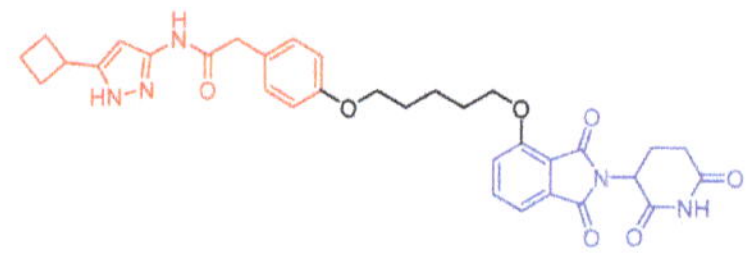	-	-	[65]
	35	BRD4	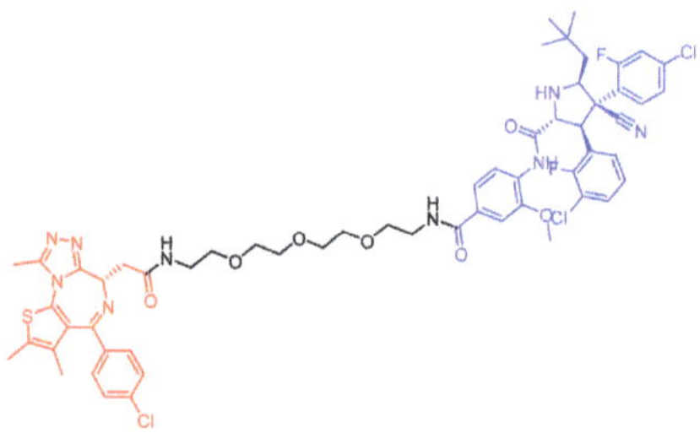	32 nM	98	[66]
	36	TRK	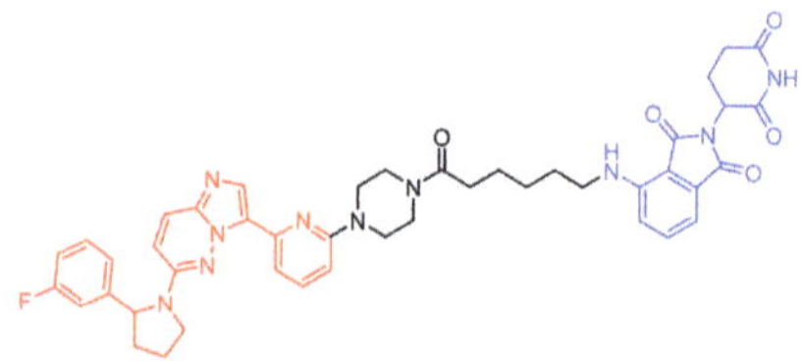	0.48 nM	-	[67]
	37	TRK	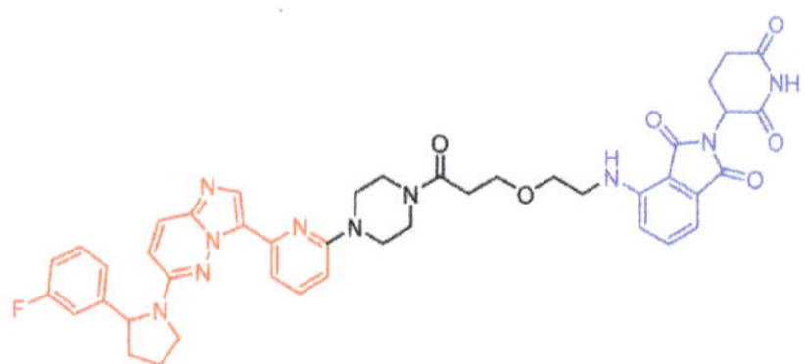	0.36 nM	-	[67]
NSCLC	38	ALK	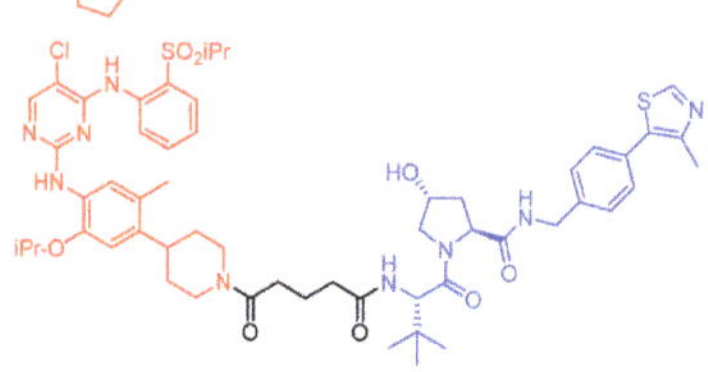	-	-	[68]
	39	ALK	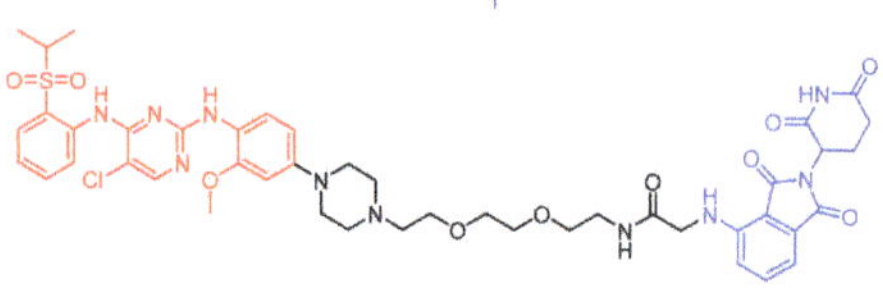	50 nM	-	[69]

Table 1. *Cont.*

Indication	PROTAC	Target	Structure	Activity		Ref.
				DC_{50}	D_{max}%	
	40	ALK	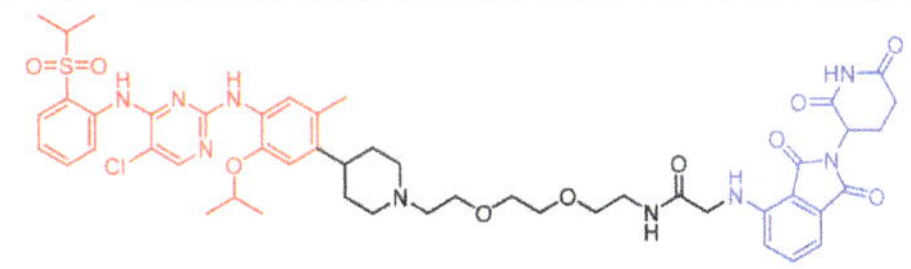	50 nM	-	[69]
	41	EGFR; PARP	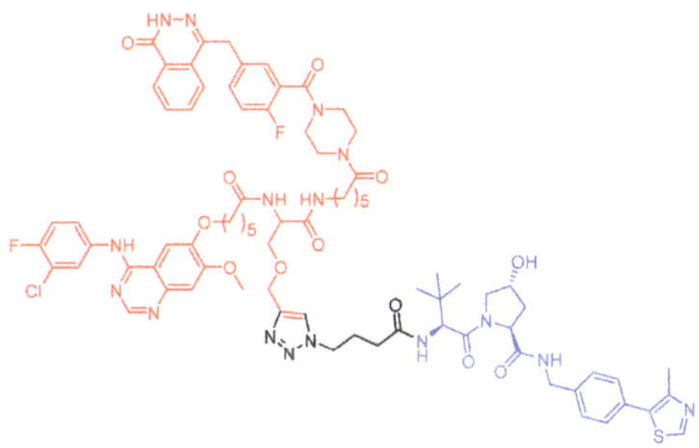	0.47 μM	-	[70]
	42	MIF	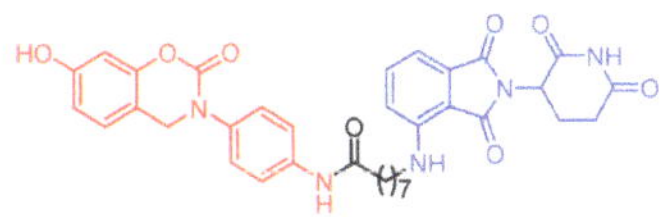	100 nM	>90	[71]
Prostatic cancer	43	AR	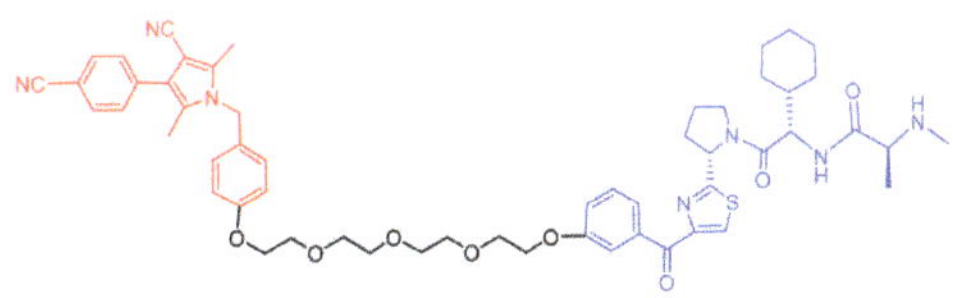	-	-	[72]
Pancreatic cancer	44	TBK1	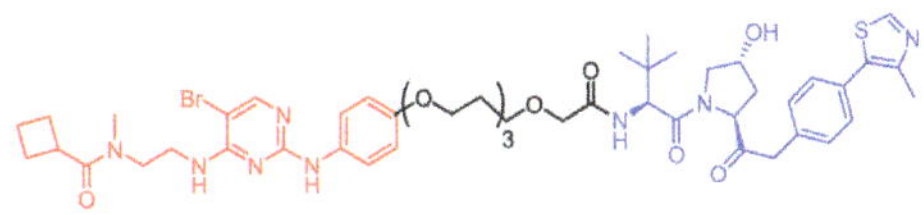	32 nM	96	[73]

Note: red: molecule to bind to POI, black: linker, blue: ligand of E3 ligase.

The Crews group reported the first VHL-based small molecule PROTAC 2 (Table 1) [36] targeting estrogen receptor-related receptor α (ERRα), which was able to specifically reduce ERRα protein levels in McF-7 cells and MDA-MB-231 tumors at nanomolar concentrations. In addition, they also reported another PROTAC molecule, PROTAC 10 (Table 1), which targets the receptor-interacting serine-threonine kinase 2 (RIPK2). PROTAC 10 can significantly degrade RIPK2 in the acute monocytic leukemia cell line THP-1 cells, which may be an effective option for treating acute monocytic leukemia.

Focal adhesion kinase (Fak) is essential for tumor invasion and metastasis [74,75] and acts as a scaffold for kinases and various signaling proteins [76]. Previously, modulation of Fak activity was limited to kinase inhibitors with a low success rate limited success in clinical studies, so the strategy of PROTAC for FAK degradation was investigated. The Crews group synthesized PROTAC 3, a Fak degrader (Table 1), which significantly exceeded the FAK inhibitor defactinib [37] in Fak signaling and cell migration and invasion in human triple-negative breast cancer (TNBC) cells. Schlaepfer David D et al. synthesized a series of PROTACs [38] based on CRBN ligand and FAK inhibitor PND-1186 through a series of CRBN and PEG connections, among which compound PROTAC 4 (Table 1) observed significant degradation of FAK in human pancreatic cancer cell PA-TU-8988 T cells and PA-TU-8988 T xenograft model in nude mice.

The components of p38 Mitogen-activated protein kinase (p38 MAPK), which includes p38, p381, p382, are tissue-specific and react to cytokines and environmental stress [77,78]. Currently, inhibitors targeting p38α have not shown good efficacy and safety [79–81]. The development of selective inhibitors for p38δ is made difficult by the restricted ATP-binding pocket [82–85]. To cause the degradation of p38 or p38 in MDA-MB-231, the Crews group synthesized PROTAC 5 (Table 1) and PROTAC 6 (Table 1) based on foretinib and VHL, which are p38 and p38 selective degraders [39].

The study by Zhou Daohong et al. [40] showed that the B-cell lymphoma extra-large (BCL-X_L)-targeting PROTAC molecule PROTAC 7 (Table 1) combined with docetaxel can more effectively inhibit the growth of tumors in MDA-MB-231 breast cancer xenotransplantation model than docetaxel alone, without causing significant changes in body weight. In addition, it has recently been found that the combination of nimbolide, a covalent ligand of RNF114, with BRD4 inhibitor JQ1 to produce PROTAC 8 (Table 1), which targets BRD4, reduced BRD4 expression levels in 231MFP breast cancer cells [41]. Another study presents PROTACs that target the degradation of protein tyrosine kinase 6 (PTK6), consisting of CRBN and VHL E3 ubiquitin ligase ligands and a published PTK6 inhibitor, as a potential treatment for breast cancer [86]. Among them, PTK6 was degraded by PROTAC 9 (Table 1) in MDA-MB231 triple-negative breast cancer cells. Compared with a single PTK6 inhibitor, the compounds showed a better inhibitory effect on ER+ breast cancer cells and platinum-resistant ovarian cancer cells [26].

2.2. Hematological Tumors

Hematologic malignancy (HM) is a malignant tumor originating from the hematopoietic system. Clinically, there are three common types of leukemia, multiple myeloma, and malignant lymphoma [87]. Some target proteins in hematological malignancies, such as BCL-X_L, MDM2, STAT3, and MALT1, are also difficult to be drugged. The emergence of PROTAC technology brings hope for the treatment of such diseases. Herein, PROTAC molecules targeting hematological malignancies in recent years are introduced.

2.2.1. Leukemia

Leukemia is the most prevalent cancerous blood malignancy, and inhibitors targeting leukemia-related proteins are not effective in the treatment of leukemia. PROTAC technology may solve this dilemma.

The most prevalent acute leukemia in adults is acute myeloid leukemia (AML) [88], which is distinguished by aberrant proliferation and weakened differentiation ability of hematopoietic precursor cells, resulting in a large number of immature leukocytes in the bone marrow, peripheral blood, and even other tissues aggregate, while the number of other normal blood cells decreases dramatically [89]. Tripartite motif-containing protein 24 (TRIM24), also known as transcriptional intermediary factor 1α (TIF1α), is a multi-domain protein involved in the transcriptional regulation of the androgen receptor (AR) and other nuclear receptors [90]. A potential prostate cancer therapy target is TRIM24, which has been implicated in various tumorigenesis and disease progression [91–96]. Bromodomain inhibitors are far from enough as an anticancer strategy. Bradner's team synthesized PROTAC 11 (Table 1) based on the TRIM24 bromodomain inhibitor IACS-9571 and the ligand of VHL. PROTAC 11 recruits VHL to cause the effective and selective degradation of TRIM24 in human acute myeloid leukemia MOLM-13 cells [42]. This study proposes TRIM24 as a fresh target for acute myeloid leukemia creation of drugs, providing a new avenue for "undruggable targets".

MDM2 is an E3 ubiquitin ligase that promotes the degradation of the p53 tumor suppressor gene [97,98]. Wild-type p53 protein is inactivated by overexpressed MDM2 due to its reverse regulation [99]. Therefore, MDM2 is an attractive target for a new therapy for AML specifically for the treatment of MDM2-overexpressing but TP53 wild-type AML. Because p53 stabilization upregulates MDM2 protein levels, which limits the clinical efficacy of MDM2 inhibitors [100], Talpaz Moshe et al. developed an MDM2 PROTAC degrader that binds to and targets MDM2 for degradation, eliminating the inhibition of p53, thereby inducing apoptosis in leukemia cells [101]. Unfortunately, the structure of this compound has not been disclosed. Another team also reported PROTAC 12 (Table 1), a small-molecule degrader targeting MDM2, which effectively induces the rapid degradation of MDM2 in human leukemia cells at a concentration of <1 nm, which can inhibit the growth of B-cell acute lymphoblastic leukemia (B-ALL) RS4-11 cells and human acute myeloid leukemia cell MV4-11 at low nanomolar concentration and induce tumor regression in the RS4-11 xenotransplantation model [43].

FMS-like tyrosine kinase 3 (FLT-3) is a therapeutic target in AML, FLT-3 frequently occurring mutation of an internal tandem duplication (ITD) in the juxta-membrane domain [102,103]. Regarding the limited clinical efficacy of FLT-3 inhibitors [104,105], the Crews group converted the FLT-3 inhibitor quizartinib into the PROTAC molecule PROTAC 13 (Table 1), which induced MV4-11 cells at low nanomolar concentrations and degradation of FLT-3 ITD mutants in MOLM-14 cells. The study showed that PROTAC was able to inhibit cell growth more effectively than the inhibitor alone, and they also demonstrated that the compound was able to induce FLT-3 ITD degradation in vivo. This suggests that the degradation of FLT-3 ITD may provide a useful approach for therapeutic intervention in AML [44].

Signal transducer and activator of transcription 3 (STAT3) is a member of the STAT family, which responds to a variety of cytokines, growth factors, and other signals and activates the expression of downstream genes [106]. Dysregulation of STAT3 contributes to a variety of human cancers as well as many human illnesses. Therefore, STAT3 has long been considered a therapeutic target for diseases such as cancer. Unfortunately, small-molecule drugs targeting STAT3 are difficult to find due to poor specificity and other reasons [107]. In order to solve this problem, Shaomeng Wang's research group used PROTAC technology to design a small molecule PROTAC 14 (Table 1) that can specifically degrade STAT3 in cancer cells. Apoptosis slows the development of a fraction of acute myeloid leukemia and anaplastic large cell lymphoma cell lines, and at well-tolerated dosages, the chemical completely and permanently eradicates tumors in a number of xenotransplantation mice models. Therefore, a possible cancer treatment method is to degrade the STAT3 protein [45].

The cyclin-dependent kinase (CDK) family are serine/threonine kinases [108] that function in cell cycle regulation and transcription. Excessive activation of CDK proteins results in dysregulated cell proliferation that promotes tumor progression, and inhibition of some CDK family members has been shown to be a viable approach to cancer therapy [109]. But the design of selective small-molecule inhibitors is often hindered by similar ligand-binding pockets. Furthermore, current inhibitors cannot disrupt scaffold function [110]. To solve this problem, Gray's team used the PROTAC strategy to describe a phthalimide-based degrader, PROTAC 15 (Table 1), since this degrader forms a different ternary complex with the E3 ligase CRBN; so, this degrader is specific and proteome-wide selective for CDK6. PROTAC 15 exploits the selective dependence of AML cells on CDK6 to rapidly degrade CDK6, enabling dynamic mapping of its direct role in coordinating signaling and gene control in AML [46].

BRD2, BRD3, BRD4, and BRDT are the key members of the bromodomain and extraterminal domain (BET) family of proteins, which are crucial for epigenetic control [111–113]. A large number of small molecule degraders targeting BRD4 have demonstrated therapeutic activity in preclinical models of AML and inflammatory diseases. In 2015, Bradner's lab independently reported CRBN-dependent BET protein degradation of PROTAC molecules—thalidomide-based PROTAC 16 (Table 1) [47]. In 2017, Ciulli et al. described a PROTAC molecule PROTAC 17 (Table 1) related to VHL binders [15], the degrader can bind to BRD4 and VHL to mediate the degradation of BRD4. Due to the structural instability of CRBN binder thalidomide, the Rankovic lab discovered new CRBN binders with higher chemical stability and ligand efficiency in 2021 and designed a novel BET PROTAC molecule PROTAC 18 (Table 1) with higher efficiency. Their data showed that PROTAC 18 can inhibit the viability of human acute myeloid leukemia MV4-11 cells at picomolar concentrations

(IC_{50} = 3 pM) [48]. Subsequently, Ciulli's group designed a trivalent PROTAC molecule PROTAC 19 (Table 1) based on MZ1 and the BET inhibitor MT1 that can bind to two BRD4 proteins, which carries two binding domains for BET proteins and one for the E3 binding domain to which ubiquitin ligase binds. They found that PROTAC 19 was 300-fold more active in degrading BRD2 protein than the existing bivalent PROTAC molecules ARV-771 and MZ1. And PROTAC 19 can more effectively inhibit the viability of MV4-11 cells and induce apoptosis of prostate cancer cell line 22RV1 [49].

T-cell acute lymphoblastic leukemia (T-ALL) is a hematological malignancy originating from immature T-cell precursors. A previous study found that T-ALL was dependent on BCL-X_L [114]. The BCL-2 protein family member BCL-X_L is essential for the survival of cancer cells [115–117]. However, BCL-X_L-specific inhibitors targeted, dose-limiting platelet toxicity, leading to thrombocytopenia, which limits their application in acute leukemia [118]. Konopleva Marina Y investigated the preclinical efficacy of PROTAC7 in T-ALL cell lines in vitro and in living T-ALL patient-derived xenotransplantation (PDX) models. This study shows that T-ALL cells are highly sensitive to PROTAC 7 in vitro. In a living T-ALL PDX model, the use of PROTAC 7 in combination with chemotherapy can alleviate leukemia and prolong patient survival. In conclusion, PROTAC 7 targeting BCL-X_L is an efficient and safe adjuvant therapy in T-ALL [119]. In another study, Zheng Guangrong et al. reported two PROTAC BCL-X_L degraders, PROTAC 20 (Table 1) [50] and PROTAC 21 (Table 1) [51], which can act in a dose- and time-dependent manner. Degrades BCL-X_L in MOLT-4 T-ALL cells with unique selectivity for MOLT-4 cells compared to traditional BCL-X_L inhibitors, suggesting that T-ALL can be improved by converting the inhibitor to PROTAC treatment window.

Chronic myelogenous leukemia (CML) is most often caused by the loss of auto-inhibitory constraint of the c-ABL kinase domain in the oncogenic fusion protein BCR-ABL [120,121]. With the advent of BCR-ABL-targeted tyrosine kinase inhibitors (TKIs), CML has become a chronic but manageable disease, but due to the presence of persistent leukemia stem cells (LSCs), all CML patients must receive lifelong treatment [122–124]. Therefore, targeting BCR-ABL degradation may bring new benefits to CML patients. In 2017, the Crews group demonstrated PROTACs that bind to VHL and CRBN E3 ligases, respectively, to known kinase inhibitors (such as imatinib, bosutinib, and dasatinib). Western blot showed that the CRBN-based PROTAC molecule PROTAC 22 (Table 1) could degrade the BCR-ABL fusion protein and c-ABL-expressed receptors in K562 CML cells. When using the VHL binder, only dasatinib-based PROTACs observed degradation of c-ABL, and no degradation of BCL-ABL was observed for all VHL-based compounds. No degradation of c-ABL and BCL-ABL was observed with imatinib-based PROTACs [52]. In addition, Naito's group developed BCR-ABL degraders PROTAC 23 (Table 1) [54] and PROTAC 24 (Table 1) [53] called SNIPER (ABL), which induced the degradation of BCR-ABL protein in BCR-ABL positive CML cells and inhibited the growth of chronic myeloid leukemia K562 cells. Taken together, the above-mentioned multiple studies suggest that the degradation of BCR-ABL is a potential strategy for the treatment of BCR-ABL-driven chronic myelogenous leukemia.

2.2.2. Malignant Lymphoma

New treatment medicines are urgently required to treat T-cell lymphoma (TCL). It was discovered that BCL-X_L was essential for the survival of the vast majority of T-cell lymphoma cell lines, patient-derived xenografts, and significant patient samples [125]. Therefore, targeted inhibition of BCL-X_L has therapeutic value for some TCL patients. However, targeting BCL-X_L small-molecule inhibitors failed due to targeted toxicity leading to thrombocytopenia. To overcome this toxicity, a study explored the therapeutic effect of PROTAC 8 on several TCL cell lines in vitro and in a mouse model with a TCL xenograft. The results show that PROTAC 8's targeting of BCL-X_L preferentially kills BCL-X_L-dependent TCL cells without causing any obvious platelet damage. In addition, the combination of this degrader and other inhibitors targeting BCL-2 family proteins has broad therapeutic effects on multiple TCL types and other cancer types dependent on BCL-X_L [55].

Human mucosa-associated lymphoid tissue protein 1 (MALT1) is a protease and scaffold protein involved in NF-κB signal transduction, which is essential for cell proliferation and survival [126]. Recent studies have found that MALT1 has therapeutic targeting in ABC-type diffuse large B-cell lymphoma (ABC-DLBCL) [127]. Studies have shown that inhibiting the MALT1 protein causes autoimmune disease and death in mice. On the other hand, the degradation of MALT1 protein also has an effective antitumor effect without causing autoimmunity in mice. These findings prompted Ari Melnick's team to investigate alternative MALT1-targeted treatments for its scaffolding activity. They synthesized many MALT1-targeting PROTACs. The data demonstrated that MALT1 PROTACs induce MALT1 degradation in the human diffuse large B-cell lymphoma cell line OCI-Ly3, suggesting that MALT1 PROTACs may be excellent drugs for the treatment of ABC-DLBCL and other lymphomas [128].

B cell receptor signaling pathway is significantly regulated by Bruton's tyrosine kinase (BTK). It plays a role in the proliferation, differentiation, and death of B cells and is frequently expressed in several hematological malignancies. Therefore, it is regarded as a key target for the therapy of hematological malignancies. Ibrutinib, its irreversible inhibitor, has transformed the way that patients with chronic lymphocytic leukemia (CLL) and other B-cell cancers are treated, but many patients have developed drug resistance [129,130]. BTK degradation can be used as a strategy to solve the resistance of BTK inhibition [57]. Based on this, Rao Yu's research group, Zhu Jun's research group, and Liu Wan Li's research group worked together to successfully and efficiently degrade a variety of clinically relevant mutant BTK proteins by constructing a novel high solubility BTK protein degradation agent PROTAC 25 (Table 1) and overcomes the clinical resistance of B-cell lymphoma (BCL) to the clinical first-line drug ibrutinib caused by BTK protein mutation. More importantly, the effectiveness of the new strategy to overcome tumor resistance has been validated by in vivo experiments [56]. In addition, Crews' group also proposed some BTK-degrading PROTAC molecules [57] as potential treatments for patients with ibrutinib-resistant CLL. This article presents VHL-based and CRBN-based PROTACs, compound PROTAC 26 (Table 1) is a PROTAC based on ibrutinib with CRBN-binder. This degrader degrades BTK with a DC_{50} of 6.2 nM and a D_{max} greater than 99% in human Burkitt lymphocyte Namalwa cells. It has better efficacy than ibrutinib in primary cell samples from C481S patients. Subsequently, since the pharmacokinetic properties of PROTAC 26 were not suitable for further in vivo development, they made a series of structural modifications to the linker and E3 recruiting ligands to synthesize the equally effective PROTAC 27 (Table 1), PROTAC 27 was more potent than PROTAC 26, a better pharmacokinetic profile is expected to further explore BTK degradation in vivo [58]. In addition, the research group also designed a BRD4-targeting compound PROTAC 28 (Table 1) in 2015, which demonstrated rapid, efficient, and prolonged BRD4 degradation in all tested Burkitt's lymphoma (BL) cell lines [59]. Gray's team developed many compounds based on BTK binders with VHL as well as CRBN binders as recruiting groups for E3 ligases. Among them, the compound PROTAC 29 (Table 1) showed BTK degradation ability in the human AML cell line MOLM-14 and the B-cell lymphoma cell line Ramos B cells, and this substance was reported effective when the QD was 50 mg/kg in a patient-derived xenograft mice model of mantle cell lymphoma (MCL) [60]. Calabrese Matthew F et al. also reported 11 BTK PROTACs linked by PEG chains of different lengths, among which PROTAC 30 (Table 1) was able to degrade BTK [61] with high specificity in both Ramos cells and rats.

The second most prevalent hematological malignancy of multiple myeloma (MM) is a frequent malignant tumor brought on by aberrant clonal plasma cell proliferation. It accounts for 1% of all tumors and 10% of hematological malignancies. CRBN ligand immunomodulatory drugs and proteasome inhibitors are commonly used in the treatment of MM. However, due to the refractory and easy recurrence of MM, the search for new therapeutic strategies is still imminent [131]. Lopez-Girona Antonia et al. described many

PROTACs for the treatment of relapsed or refractory multiple myeloma (RRMM), and finally screened PROTAC 31 (Table 1), PROTAC 31 is an Ikaros/Aiolos (IKZF1-3) degrader, which has enhanced antiproliferative and tumoricidal activity in multiple myeloma cell lines, including lenalidomide and pomalidomide resistant cells in vitro, which has strong immune-stimulating activity, currently in phase 2 clinical trials [62]. In addition, Rao Yu's group used a PROTAC strategy to design and synthesize CDK6-targeted degraders. These PROTACs can effectively and specifically degrade CDK6 at low concentrations. A representative palbociclib-derived PROTAC 32 (Table 1) can strongly inhibit multiple proliferation of myeloma, leukemia, and MCL cells. These findings highlight the value and promise of creating therapeutic compounds based on PROTAC [63].

2.3. Colon Cancer

In the United States, colon cancer ranks third among the causes of cancer-related death, and it is considered one of the major cancers that seriously threatens human health, together with lung cancer, prostate cancer, and breast cancer [132]. In recent years, very important progress has occurred in the field of treatment of this common disease, targeting relevant pathogenic proteins for degradation, such as ERK, CDK9, BRD4, TRK, etc., and has been shown to be effective. A recent study demonstrated that a covalent inhibitor-based PROTAC 33 (Table 1) targeting extracellular-regulated kinase 1-2 (ERK1-2) could target malignant melanoma cells ERK1-2 degrades and inhibits the phosphorylated ERK1-2 signaling pathway in A375 and human colon cancer cells HCT116, and it is experimentally confirmed that the binding to CRBN is the key to the degradation of ERK1-2 [64].

The first example of a PROTAC that selectively degrades CDK9 was reported by Rana Sandeep et al., who found that PROTAC 34 (Table 1) degraded CDK9 in a dose-dependent manner in human colorectal cancer cell line HCT116 cells. While the levels of CDK2 and CDK5 in cells remained unchanged, indicating that the compound selectively degrades CDK9, this study suggests that strategies targeting CDK9 degradation may be useful in the treatment of colon cancer [65]. In 2019, the Crews team published a study targeting BRD4 degradation, which showed that a nutin-based PROTAC, PROTAC 35 (Table 1), was able to degrade the human colon cancer cell line HCT116 cells at nanomolar concentrations. They also found that PROTAC 35 inhibited the proliferation of many wild-type p53 cancer cell lines more effectively than PROTACs that degraded BRD4 using VHL [66].

Liu Jing et al. reported two degraders of tropomyosin receptor kinase (TRK), PROTAC 36 and 37 (Table 1), the compound prepared in this paper is composed of CRBN E3 ligase binding agent, which is connected with TRK inhibitor [133] as the target part of TRK. In KM12 colon cancer cells, it was discovered that these two substances decreased the levels of the TPM3-TRKA fusion protein. PROTAC 37 also decreased the levels of the AGBL4-TRKB and ETV6-TRKC fusion proteins. Furthermore, both CRBN-based PROTACs showed favorable plasma exposure levels in mice, thus, these two compounds are valuable chemical tools to study the in vivo function of TRK fusions during colon carcinogenesis [67].

The most cancer-related fatalities occur from lung cancer worldwide, 85% of which are caused by non-small-cell lung cancer (NSCLC). In the past two decades, significant progress has been made in the treatment of NSCLC [134] and targeting pathogenic targets for degradation through the PROTAC strategy has created new opportunities for the treatment of NSCLC as a new therapeutic approach. Hwang's team designed and synthesized several ALK-PROTAC molecules to degrade anaplastic lymphoma kinase (ALK) fusion proteins. One of the compounds, PROTAC 38 (Table 1), effectively induced ALK degradation and inhibited the growth of the ALK fusion-positive cell line SU-DHL-1 and the human NSCLC cell line H3122, and also inhibited tumor growth in the H3122 xenograft model [68]. Furthermore, in 2018, Gray Nathanael S et al. disclosed the development of the first PROTACs targeting ALK degradation. These PROTACs were based on the binding of ALK inhibitors to E3 ubiquitin ligase ligands, and they screened two compounds, PROTAC 39 (Table 1) and 40 (Table 1). These two compounds efficiently induced the degradation of ALK in NSCLC, anaplastic large cell lymphoma (ALCL), and neuroblastoma (NB) cell lines. As potential chemotherapeutics for NSCLC, their data also suggest that compounds targeting ALK degradation also promote the degradation of other kinases, such as PTK2, Aurora A, FER, and RSK1 for further investigation [69]. In the same year, a successful example of targeting EGFR and PARP dual PROTACs was introduced for the first time in an article by Li Hua et al., in which novel dual PROTACs were synthesized using gefitinib, olaparib, CRBN, or VHL E3 ligands as substrates. Among them, compound PROTAC 41 (Table 1) successfully degraded both EGFR and PARP in H1299 human NSCLC cells. This research will substantially expand the PROTAC method's application possibilities and create a new avenue for the development of human non-small cell lung cancer drugs [70].

Macrophage migration inhibitory factor (MIF) is involved in diseases through protein-protein interactions in cancers and inflammations such as melanoma [135], neuroblastoma [136], and lung cancer [137]. Studies have found that MIF is down-regulated by protein degradation. Expressing [138] or reducing the level of MIF [139,140] can reduce tumor metastasis and induce antitumor responses [141]. Therefore, MIF may be a new target for cancer therapy. Dekker Frank J's team reported the first MIF-targeting PROTAC molecule PROTAC 42 (Table 1) in 2021. Their research data showed that this degradation agent can inhibit the proliferation of A549 cells by downregulating the level of MIF in non-small cell lung cancer A549 cells and inhibiting MIF-related signal transduction, proving the potential of PROTAC technology in the treatment of NSCLC [71].

2.4. Prostatic Cancer

Prostate cancer is the second most prevalent cancer to cause mortality in males and has a significant impact on men's health globally [142]. Scientists have never stopped exploring treatment strategies for prostate cancer. Naito's team developed PROTAC 43 (Table 1), which showed potent protein-degrading activity against AR. In addition, PROTAC 43 potently induced caspase activation and apoptosis in prostate cancer cells compared to AR inhibitors. These findings imply results suggest that targeting AR degradation may be one of the treatment approaches for AR-dependent proliferation in prostate cancer [72].

2.5. Pancreatic Cancer

Pancreatic cancer is one of the deadliest malignancies to us. Pancreatic cancer survival rates have remained largely unchanged since the 1960s [143]. The research on pancreatic cancer has never stopped, and by causing the degradation of associated kinase proteins, it may be a possible method for the therapy of pancreatic cancer. The IKK protein kinase family includes TANK-binding kinase 1 (TBK1), a key player in innate immunity. Some studies have found that abnormal TBK1 can cause a variety of diseases and inhibiting the activity of TBK1 can slow or prevent the growth of cancer cells. Therefore, TBK1 is a crucial target for research in the process of tumorigenesis, but there is still no approved TBK1 inhibitor [144]. However, the Crews group reported a PROTAC that induces TBK1 degradation, and PROTAC 44 (Table 1) [73] may serve as an alternative strategy. Interestingly, the Kd of this compound binding to TBK1 in mouse pancreatic cancer Pan02.13 cells was 4 nM. After 16 h of treatment, DC_{50} was 32 nM and D_{max} was 96%. In addition, this compound could not degrade the TBK1 homologous protein IKKε with a binding affinity of 70 nM Kd.

3. Application of PROTAC in Immune Diseases

Immune inflammatory diseases are very common diseases in life, such as rheumatoid arthritis, systemic lupus erythematosus, ulcerative colitis, and so on. These diseases are threatening people's health all the time. Scientists' research on the treatment strategies for these diseases has never stopped. As an emerging strategy, PROTAC has penetrated into the treatment field of immune-inflammatory diseases. The following introduces the use of PROTAC in several applications in immune inflammatory disease targets.

3.1. IRAK3

Interleukin-1 receptor-associated kinase 3 (IRAK3) is a member of the IRAK family [145], and related studies have shown that IRAK3 can inhibit pro-inflammatory signaling in congenital leukocytes [146]. In addition, knockout of the IRAK3 gene in mouse bone marrow cells promotes the proliferation of effector T cells, which is conducive to enhancing the host response to checkpoint inhibition and overcoming immunosuppression [147,148]. Therefore, IRAK3 is a potential target for immune diseases. Edmondson's team first released the IRAK3 PROTAC molecule PROTAC 45 (Table 2) [149] based on the linkage of the IRAK3 ligand and CRBN ligand. Their data showed that more than 98% of IRAK3 was degraded in human monocytic leukemia THP1 cells and primary macrophages. This study provides a good tool for IRAK3 degradation.

Table 2. Representative PROTACs for immune diseases.

Protac	Target	Structure	Activity		Ref.
			DC_{50}	D_{max}%	
45	IRAK3	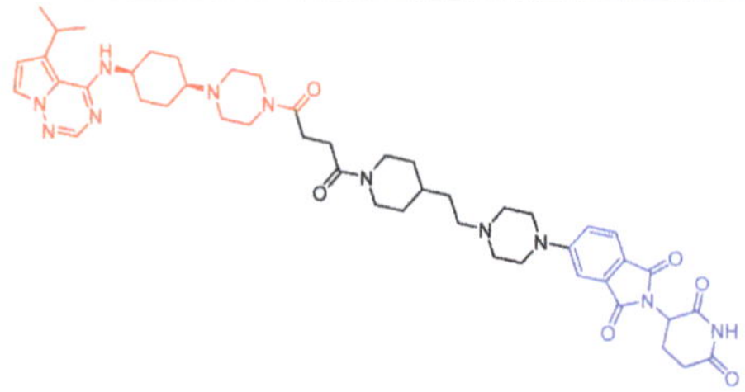	2 nM	98	[149]
46	IRAK4	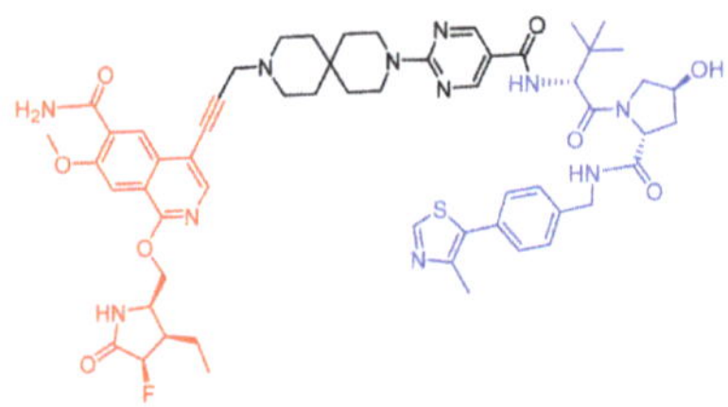	151 nM	-	[150]
47	IRAK4	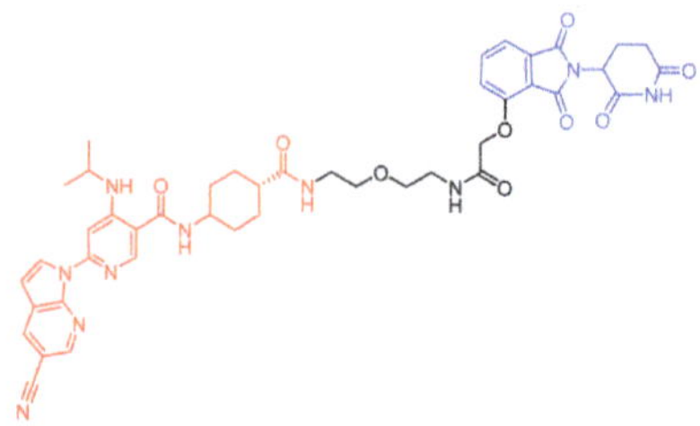	405 nM	90	[151]
48	IRAK4	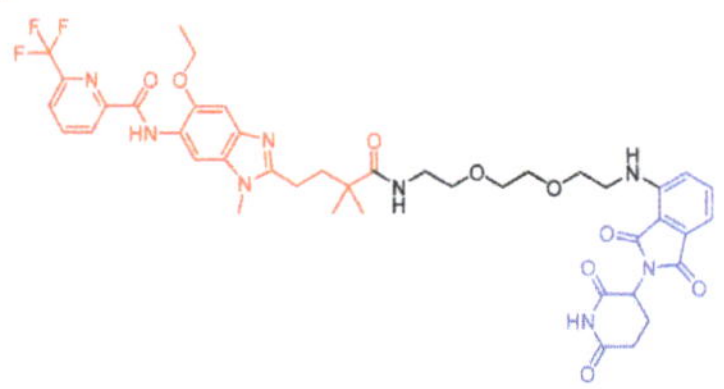	-	-	[152]

Table 2. *Cont.*

Protac	Target	Structure	Activity		Ref.
			DC_{50}	D_{max}%	
49	HDAC3	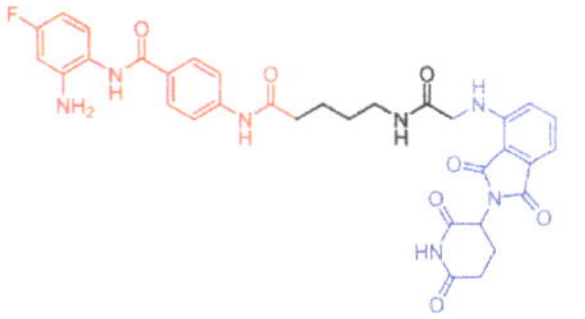	0.32µM	-	[153]
50	HDAC3	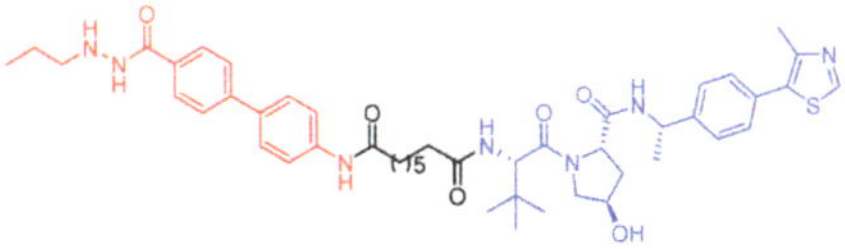	42 nM	-	[154]
51	HDAC6	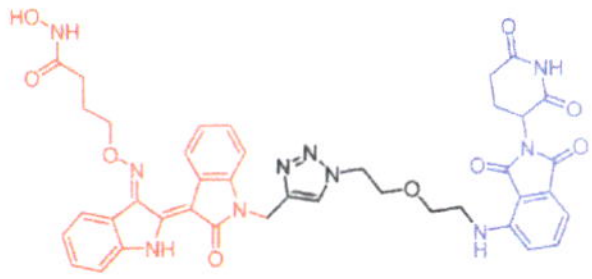	108.9 nM	88	[155]
52	H-PGDs	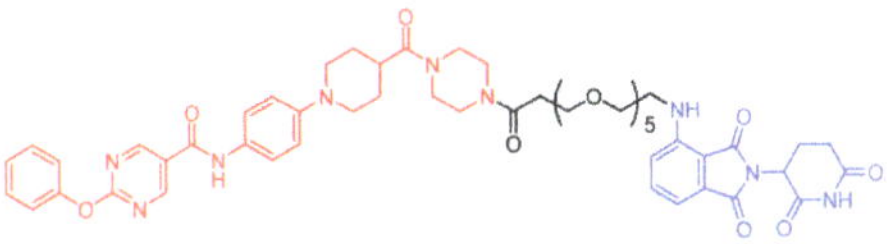	-	-	[156]
53	IDO1	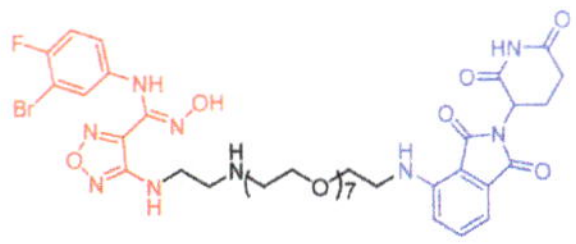	2.84 µM	93	[157]
54	Sirt2	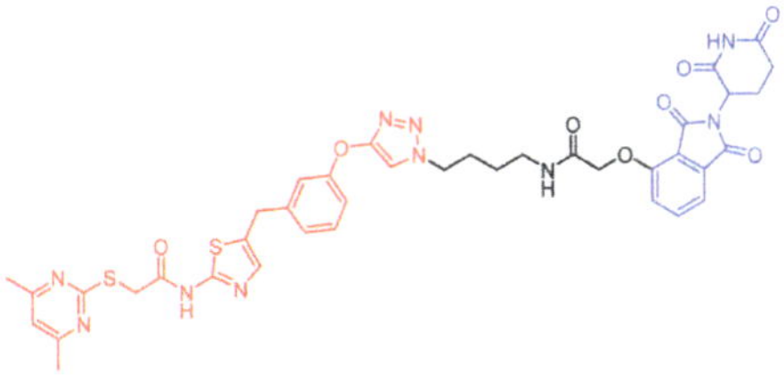	-	-	[158]
55	Sirt2	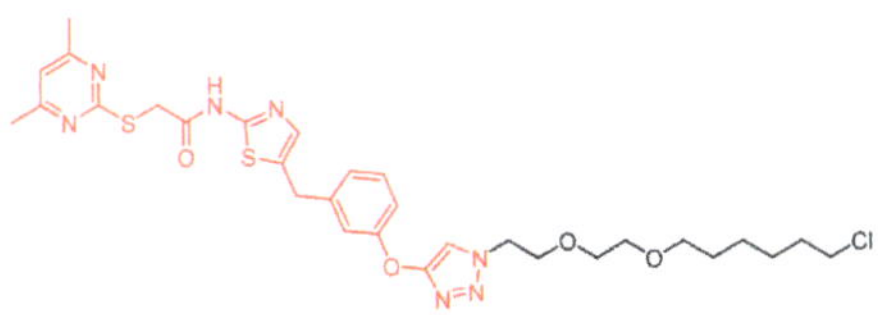	-	-	[159]
56	PCAF-GCN5	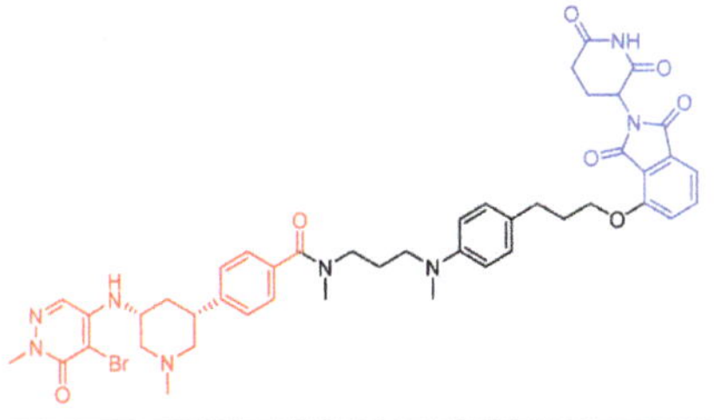	1.5–3 nM	>90	[160]

Table 2. *Cont.*

Protac	Target	Structure	Activity		Ref.
			DC_{50}	D_{max}%	
57	RIPK2	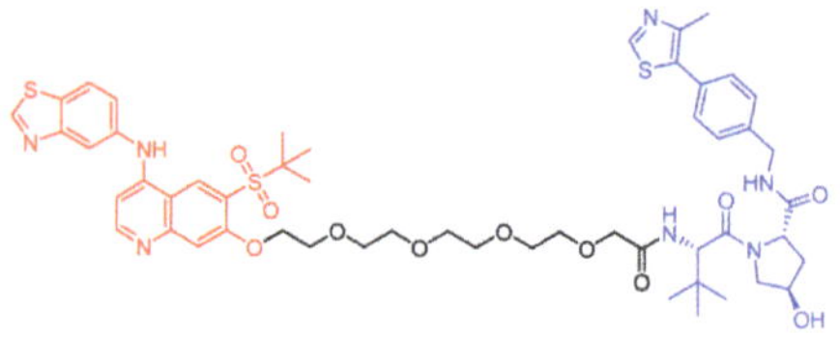	1.4 nM	>95	[36]
58	RIPK2	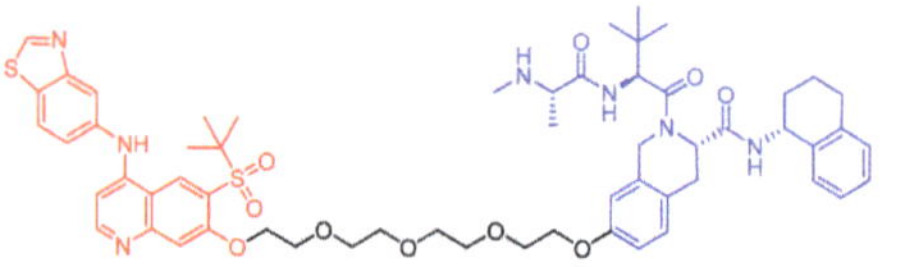	4 nM	-	[161]
59	RIPK2	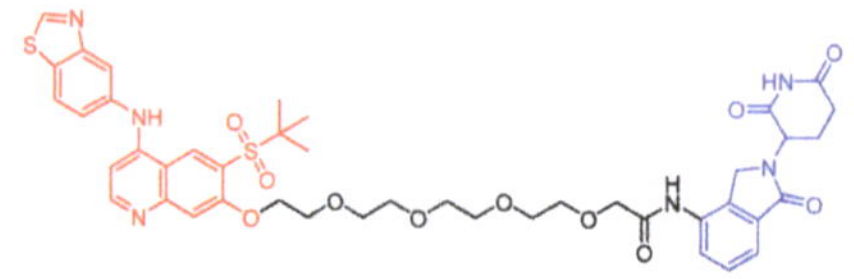	2.5 nM	-	[161]

Note: red: molecule to bind to POI, black: linker, blue: ligand of E3 ligase.

3.2. IRAK4

Interleukin-1 receptor-associated kinase 4 (IRAK4) is a key protein in the immune response mediated by toll-like receptors and interleukin receptors and has been identified as an autoimmune disease and cancer dual target [133,150,162,163]. At present, the inhibitor [164] against IRAK4 has not yet been approved. Therefore, using PROTAC technology to knock out IRAK4 may be an alternative strategy for the treatment of IRAK4-related diseases. Anderson Niall A. et al. developed and designed some PROTACs [150] that induce the degradation of IRAK4, among which the compound PROTAC 46 (Table 2) successfully induced the degradation of IRAK4 in peripheral blood mononuclear cells (PBMC) and human dermal fibroblasts. Dai Xuedong et al. also reported an IRAK4-targeting degrader, PROTAC 47 (Table 2) [151], and their data showed that more than 90% of IRAK4 was rapidly degraded after treating HEK293T cells with this compound for 24 h. In another study, Duan Wenhu et al. [152] also described and screened a compound targeting IRAK4, PROTAC 48 (Table 2), which induces IRAK4 degradation in diffuse large B-cell lymphoma OCILY10 and TMD8 cells. Meanwhile, PROTAC 48 suppressed the proliferation of cell lines expressing the B-cell lymphoma MYD88 L265P mutant and hindered the IRAK4-NF-B signaling pathway. In addition, KT-474, a potential "first-in-class" IRAK4 oral protein degrader developed by Kymera, has entered phase 1 clinical trials. The trial results showed that, in healthy volunteers, a single dose of KT-474 dose-dependently reduced the levels of IRAK4 and various pro-inflammatory cytokines with good safety and tolerability. These studies illustrate the potential application of IRAK4 degraders for the management of oncology and inflammatory indications.

3.3. HDAc3

A class of proteases known as histone deacetylases (HDAcs) are responsible for altering chromosomal shape and controlling gene expression. Inflammatory illnesses such as asthma and chronic obstructive pulmonary disease are affected by HDAc3, which is crucial [165]. Dekker's group [153] synthesized PROTAC 49 (Table 2) for the degradation of HDAc3. PROTAC49 is a linker between the HDAc inhibitor anthranilide derivative and the CRBN ligand pomalidomide. Their results showed that PROTAC49 had a small impact on gene expression in RAW 264.7 macrophages activated by lipopolysaccharide/interferon, and was able to selectively downregulate HDAc3 levels compared to biochemical evidence using siRNA. That same year, Liao's team [154] developed PROTAC 50 (Table 2), the compound PROTAC 50 induces selective and efficient degradation of HDAC3. The dose-limiting toxicity of traditional HDAC inhibitors may be overcome by PROTAC 50's catalytic mechanism of action and isoenzyme selectivity.

3.4. HDAc6

Studies have shown that HDAC6 is indispensable for the assembly and activation of the NLRP3 inflammasome. Furthermore, activation of the NLRP3 inflammasome is primarily dependent on the zinc-finger ubiquitin-binding domain of HDAC6 rather than its deacetylation function [166,167]. Therefore, traditional small-molecule inhibitors targeting the deacetylation domain of HDAC6 are not suitable for inhibiting the activation of the NLRP3 inflammasome. In 2021, He's group reported an HDAC6 degrader PROTAC 51 (Table 2) based on the natural product indirubin derivatives and the CRBN ligand pomalidomide. Their data showed that compound PROTAC 51 can downregulate NLRP3 levels in THP-1 cells that construct the NLRP3 inflammasome activation model, accompanied by downregulation of related cytokines such as IL-1β. This study sheds light on the value of targeted protein degradation strategies in the treatment of inflammatory disorders to some extent [155].

3.5. HPGDs

Excessive hematopoietic prostaglandin synthase 2 (PGD2) causes many diseases such as allergic disease, physiologic sleep, and Duchenne muscular dystrophy [168]. PGD2 synthesis requires the participation of hematopoietic prostaglandin D synthases (H-PGDs); therefore, H-PGDs are possible treatment targets for such diseases, and in vivo research has also shown that inhibition of H-PGDs is available in the management of allergic inflammation [169]. It is vitally necessary to develop novel therapeutic methods that specifically target H-PGDs as the inhibitors for inhibiting H-PGDs that have been produced so far are not therapeutically effective. In 2021, Demizu et al. [156] reported PROTAC 52 (Table 2) for targeting H-PGDs for degradation and sustained inhibition of PGD2 production. Their findings suggest that knockdown of H-PGDs protein through the PROTAC strategy, thereby inhibiting the production of PGD2, has the potential to become a new therapeutic modality.

3.6. IDo1

Indoleamine 2,3-dioxygenase 1 (IDO1) is an enzyme that causes immunological tolerance by preventing T cells from proliferating [170]. Recent studies have found that IDO1 plays an important role in cancer immune escape [171]. In a paper, the Xie team published the first PROTAC 53 (Table 2) that induces downregulation of IDO1 protein levels, a degrader that induced more than 93% of IDO1 protein to be degraded by UPS in HeLa cells [157]. The study not only demonstrates the feasibility of degrading IDO1 but also provides a method to investigate the role of IDO1 protein in tumor immune evasion.

3.7. Sirt2

The sirtuins family includes sirtuin 2 (Sirt2), by deacetylating a variety of substrates, it participates in various biological processes including gene silencing, cell cycle regulation, metabolism, and apoptosis. Sirt2 dysregulation has been linked to cancer development, type II diabetes, bacterial infections, and neurological disorders [172–177]. This suggests that Sirt2 may be a viable drug intervention target. Jung Manfred et al. reported for the first time the compound PROTAC 54 (Table 2) [158] used to induce Sirt2 degradation, which is composed of Sirt2 inhibitor and CRBN ligand. The results showed that PROTAC 54 can selectively and dose-dependently degrade sirt2 compared with Sirt2 inhibition. Subsequently, they designed and developed a Sirt2 degrader PROTAC 55 (Table 2) [159] with a more optimized performance compared with PROTAC 54. PROTAC 55 can degrade Sirt2 at a 10-fold lower concentration than PROTAC 54.

3.8. PCAF/GCN5

A set of epigenetic proteins called P300/CBP-associated factor (PCAF) and general control nonderepressible 5 (GCN5) are essential for a number of cellular processes, such as DNA damage repair, metabolic control, and cell proliferation and differentiation [178]. However, chemical inhibition targeting the PCAF/GCN5 bromodomain was not sufficient to attenuate the inflammatory response of PCAF-deficient immune cells. In 2018, Tough David F et al. first reported a compound [160] targeting PCAF/GCN5 degradation, PROTAC 56 (Table 2), which induced PCAF/GCN5 degradation in THP1 cells at nanomolar concentrations. Meanwhile, the decline in PCAF/GCN5 inhibited the production of many inflammatory factors, which further illustrates the importance of constructing a new PROTAC strategy to prevent inflammation.

3.9. RIPk2

The receptor-interacting protein kinase (RIPK) family consists of RIPK1, RIPK2, and RIPK3, which play important roles in inflammation and innate immunity [179]. Numerous inflammatory cytokines are released when RIPK2 is activated, and dysregulation of this pathway is closely related to autoimmune disorders including inflammatory bowel disease. Despite both RIPK1 and RIPK2 inhibitors having entered clinical studies, RIPK inhibitor medications have not yet received approval. The Crews team announced a RIPK2 degrader PROTAC 57 (Table 2) [36] based on the RIPK2 binder and VHL binder in 2015. Their research data showed that PROTAC 57 can degrade more than 95% of RIPK2 at nanomolar concentrations. In 2020, Harling's team synthesized RIPK2 degraders PROTAC 58 (Table 2) and 59 (Table 2) [161] based on the IAP binder and CRBN binder. They found that compared with PROTAC 57, The effect of PROTAC 58 and 59 in degrading RIPK2 in THP-1 cells is not satisfactory. Although PROTAC 57 has a strong degradation ability, the study found that its binding ability to RIPK2 is much weaker than that of PROTAC 58 and 59.

3.10. ASK1

Apoptosis signal-regulating kinase 1 (ASK1) is a widely expressed protein kinase that is extremely redox-sensitive and is directly involved in the regulation of apoptosis and signaling pathways such as inflammation and fibrosis under oxidative stress [180]. Previous animal-level studies have shown that inhibition of ASK1 can effectively reduce liver and kidney injury and fibrosis and is expected to be a possible target for the therapy of nonalcoholic steatohepatitis (NASH) [181] and diabetic kidney disease (DKD) [182] and other diseases. At present, the development of ASK1 inhibitor GS4997 has entered phase III clinical trials [183], but its two key clinical trials for the treatment of NASH have failed one after another, casting a shadow over the development of drugs for this target. PROTAC technology can achieve low-dose degradation of target proteins. Therefore, targeting ASK1 degradation through PROTAC molecules may develop into a new approach to the treatment of diseases such as NASH and DKD.

4. Application of PROTAC in Neurodegenerative Diseases

Neurodegenerative diseases are an area in need of new therapies and molecular insights. The aggregation of misfolded proteins such as tau protein and α-synuclein protein is the main cause of such diseases, and they cannot be modulated by traditional small molecule drugs; therefore, the treatment of neurodegenerative diseases has always been a challenge. In recent years, the use of PROTAC technology to degrade target proteins has become a new treatment method. Therefore, PROTAC technology is expected to play a potential role in neurodegenerative diseases caused by protein aggregation.

4.1. Alzheimer's Disease

Tau protein is a microtubule-associated protein abundant in neurons and plays a role in axonal transport and microtubule stabilization [184]. Abnormal regulation of tau protein is the cause of numerous neurodegenerative disorders such as Alzheimer's disease (AD).

Therefore, tau is a possible therapeutic target for neurodegenerative disorders. In 2016 and 2018, Li Yan-Mei and Jiang Zhengyu et al. revealed peptide-based induction of tau degradation in PROTACs respectively [185,186]. In 2019, Kargbo's group reported the first small-molecule tau degrader [187]. In this paper, six tau-targeting PROTACs were developed based on CRBN and VHL binders. The results showed that tau-targeting PROTACs successfully degraded tau in human tau-p301L and tau-a152T neurons, and several favorable pharmacokinetic parameters were shown. In the same year, Haggarty's team also announced the synthesis of a series of novel targeting tau PROTACs [188]. The representative compound PROTAC 60 (Table 3) could effectively degrade wild-type and mutant tau. Furthermore, PROTAC 60 preferentially degrades tau in FTD neurons of frontotemporal dementia compared to normal cells. In 2021, Wang Jian-Zhi et al. synthesized a PROTAC based on the binding of tau ligand and VHL ligand, named PROTAC 61 (Table 3) [189]. They used PROTAC 61 in both in vitro and in vivo investigations. The experimental results indicated that PROTAC 61 effectively induced tau degradation under both physiological and pathological conditions. At the same time, they demonstrated that knocking out tau did not cause significant abnormalities in mice. These data suggest that induction of tau protein degradation using PROTAC technology is a possible method for treating neurodegenerative illnesses including AD.

Table 3. Representative PROTACs for neurodegenerative diseases.

Indication	PROTAC	Target	Structure	Activity		Ref.
				DC_{50}	D_{max}%	
AD	60	Tau	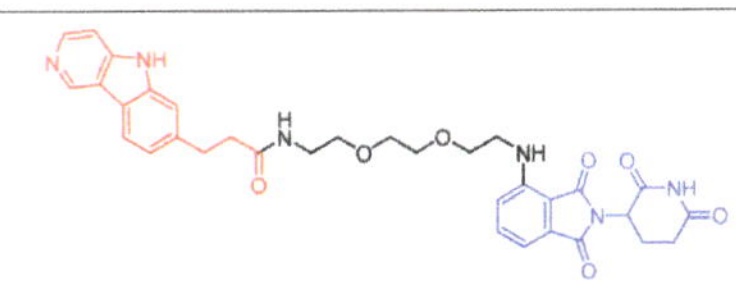	-	75	[188]
	61	Tau	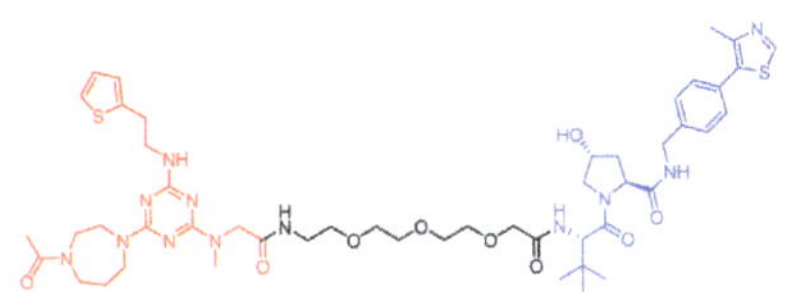	-	-	[189]
	62	GSK-3β	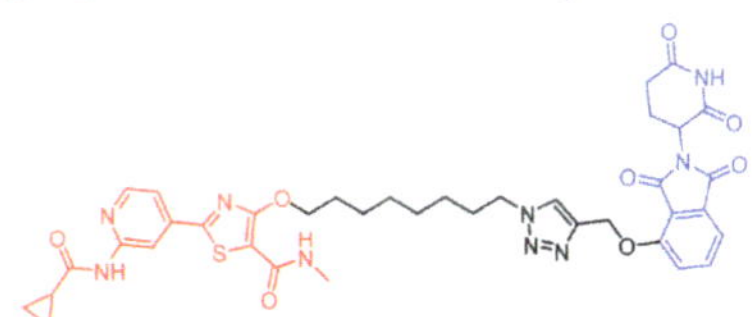	-	-	[190]
HD	63	mHtt	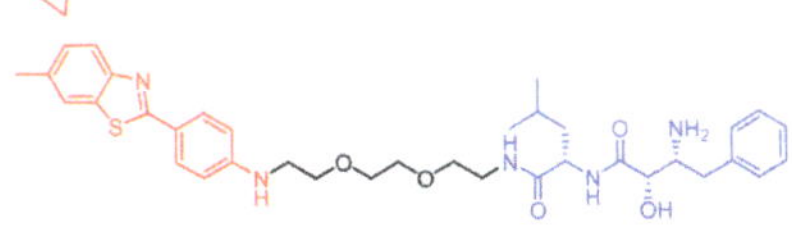	-	-	[191]
	64	mHtt	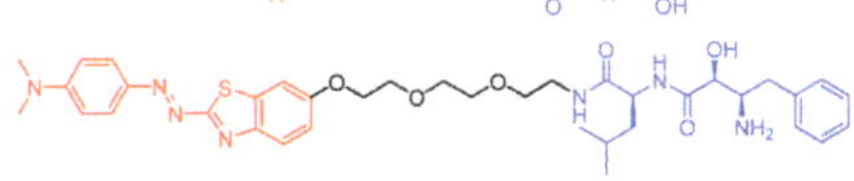	-	-	[191]
	65	mHtt	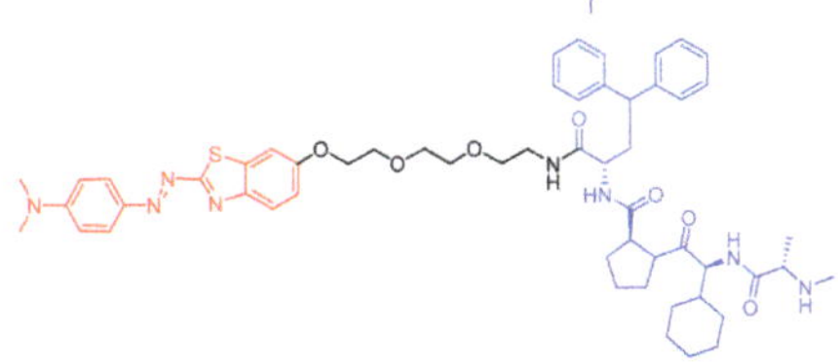	-	-	[192]

Table 3. *Cont.*

Indication	PROTAC	Target	Structure	Activity		Ref.
				DC_{50}	D_{max}%	
PD	66	α-synuclein	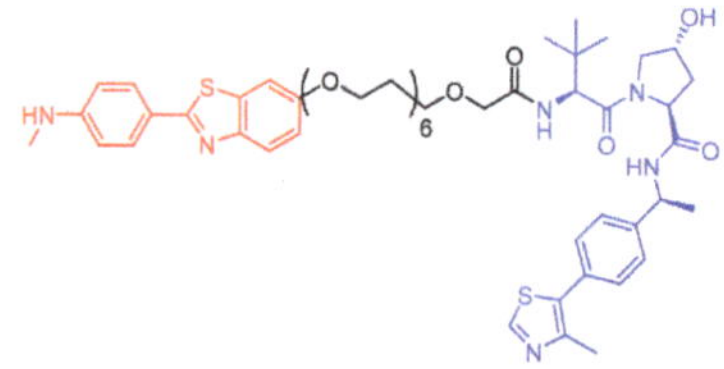	-	65	[193]

Note: red: molecule to bind to POI, black: linker, blue: ligand of E3 ligase.

Glycogen synthase kinase 3 (GSK-3) is a class of serine/threonine protein kinases [194]. Relevant studies have demonstrated that GSK-3β can boost the phosphorylation of tau protein and the production of the amyloid-β peptide to induce AD. And GSK-3β has a strong pro-inflammatory effect leading to neuronal loss [195–197]. Therefore, GSK-3β is thought to be a potential target for neurodegenerative disorders. Sun Haopeng et al. developed many bifunctional PROTACs targeting GSK-3β for the first time [190]. They can achieve nanomolar degradation of GSK-3β. Among them, the representative compound PROTAC 62 (Table 3) can degrade more than 44% of GSK-3β; additionally, it is worth noting that PROTAC 62 can also prevent the death of mouse hippocampal neurons HT-22 cells induced by glutamate. Their study is of great significance, which presents a fresh approach to the creation of GSK-3β degraders.

4.2. Huntington's Disease

An inherited neurological disorder called Huntington's disease (HD) is brought on by the HTT gene's exon 1 developing more than 35 CAG repeats, and the resulting mutant huntingtin (mHtt) accumulates in nerve cells [198]. These aggregates can cause nerve cells to die, which can lead to numerous symptoms, including motor impairment and cognitive deficits. In 2017, Ishikawa's group [191] designed PROTACs 63 (Table 3) and 64 (Table 3) against this target. PROTAC 63 and PROTAC 64 successfully induced downregulation of mHtt protein levels in HD patient primary cells and mHtt-transfected HeLa cells. Subsequently, they synthesized a new PROTAC 65 (Table 3) [192]. The newly synthesized PROTAC was synthesized with IAP inhibitor MV1 and mHtt ligand through PEG linkage, and the newly synthesized compound showed stronger affinity compared with PROTAC 63 and PROTAC 64. Furthermore, the newly synthesized PROTAC was able to degrade mHtt in fibroblasts from HD patients in a time- and dose-dependent manner.

4.3. Parkinson's Disease

Parkinson's disease is a motor system-affecting, progressive neurodegenerative condition, and its main feature is the neuronal cytoplasmic aggregation of Lewy bodies composed of aggregates of α-synuclein protein, leading to neuronal degeneration [199]. In 2020, a PROTAC that induces α-synuclein protein degradation was reported by Kargbo's group [193]. Six key PROTACs were screened in the paper. These six compounds can target α-synuclein protein degradation, and the representative PROTAC 66 (Table 3) can significantly reduce the protein level of α-synuclein in HEK293 TREX u-syn A53T cells, the D_{max} value is 65%. It can be seen that this compound can be utilized as a possible Parkinson's disease medication and has broad prospects.

5. Application of PROTAC in Cardiovascular Diseases

3-hydroxy-3-methylglutaryl coenzyme A reductase (HMGCR) catalyzes 3-hydroxy-3-methylglutaryl coenzyme A in the cholesterol synthesis pathway. HMGCR is a target of statins for the prevention and treatment of cardiovascular diseases [200,201]. In 2020, Luo's team reported a series of PROTAC molecules [202], among which PROTAC 67 (Table 4) has the greatest impact on the HMGCR protein's ability to degrade in Chinese hamster ovary SRD15 cells. Additionally, PROTAC 67 activates the sterol regulatory element-binding protein pathway (SREBP) and blocks cholesterol synthesis. That same year, Xiang's group [203] reported two kinds of lovastatin acid and VHL ligand-conjugated HMGCR targeting PROTAC 68 (Table 4) and 69 (Table 4), and PROTAC 68 could effectively degrade HMGCR in HepG2 cells (DC_{50} = 120 nM). In vivo studies have shown that PROTAC 69 induces HMGCR breakdown and cholesterol reduction in mice with diet-induced hypercholesterolemia.

Table 4. Representative PROTACs for cardiovascular diseases.

Protac	Target	Structure	Activity		Ref.
			DC_{50}	D_{max}%	
67		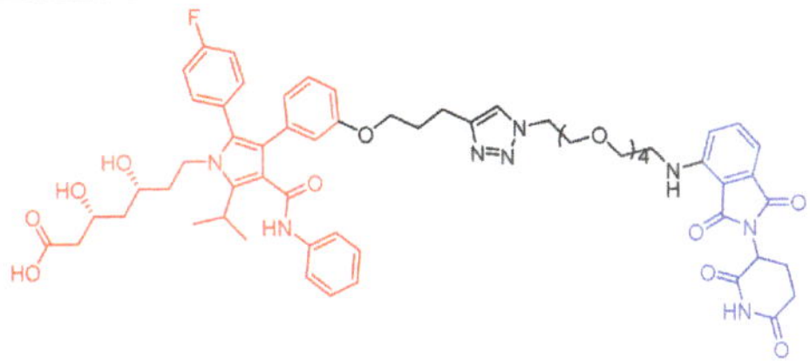	0.1 μM	-	[202]
68	HMGCR	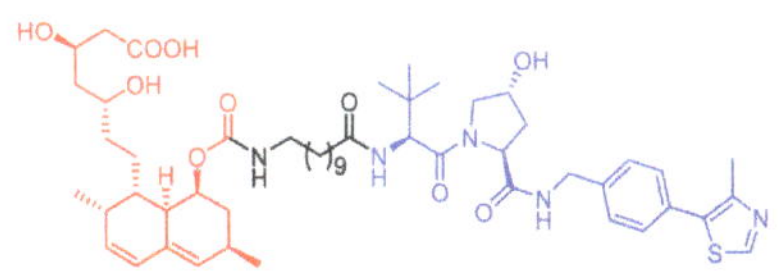	120 nM	76	[203]
69		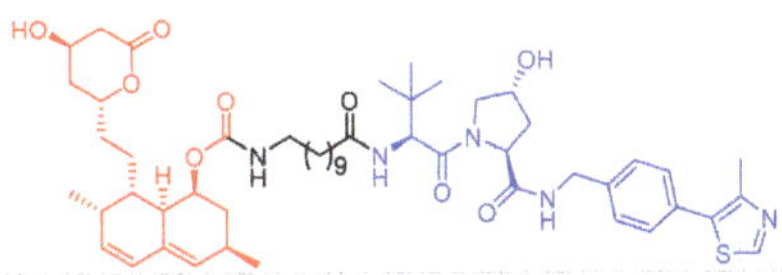	-	56	[203]

Note: red: molecule to bind to POI, black: linker, blue: ligand of E3 ligase.

6. Application of PROTAC in Antiviral

Infection with hepatitis C virus (HCV) is the main cause of chronic liver disease, in which the hepatitis C virus (HCV) NS3 protein plays an important role [204,205]. Although VX-950, an inhibitor of the NS3/4A protease, has been authorized for the treatment of HCV, patients are prone to develop drug resistance, so a new treatment method is urgently needed to solve this problem. Yang Priscilla L. et al. synthesized many NS3-targeting PROTACs [206] by linking VX-950 and CRBN Binder based on the PROTAC strategy. Among them, the representative compound PROTAC 70 (Table 5) can effectively degrade NS3 in human hepatoma adherent Huh7.5 cells. In addition, PROTAC 70 can also degrade V55A and A156S mutant NS3. Therefore, the successful discovery of this degradation agent is a boon for HCV-infected patients. This study also suggests that PROTACs may also be potential antiviral drugs, a strategy that has also been used to target SARS-CoV-2 as it emerges. The major proteases (Mpro and PLpro) [207] and RNA-dependent RNA polymerase (RdRP) [208] of SARS-CoV-2 are currently targeted by small molecule inhibitors [209]. They could be potential targets for PROTAC molecules.

Table 5. Representative PROTACs for antiviral.

Indication	PROTAC	Target	Structure	Activity		Ref.
				DC_{50}	D_{max}%	
HCV	70	NS3	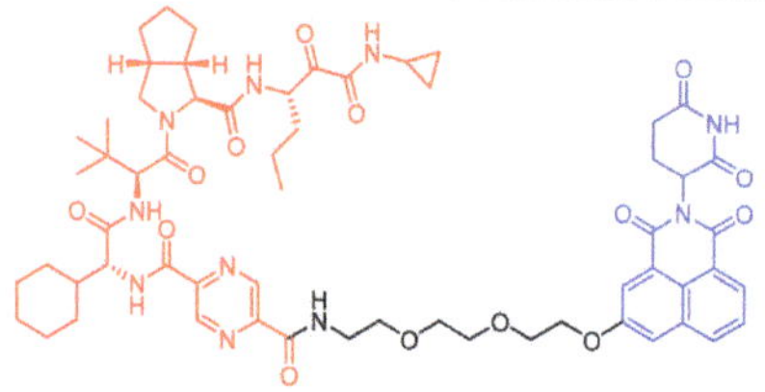	50 nM	-	[206]

Note: red: molecule to bind to POI, black: linker, blue: ligand of E3 ligase.

7. Conclusions and Prospects

PROTAC technology has been developed for 20 years, and some molecules have entered the clinical stage; this sheds light on the huge therapeutic potential of PROTAC in tumors, immune diseases, neurodegenerative diseases, cardiovascular diseases, and viral infections. First, targets for drug resistance were particularly sensitive to PROTACs. The mainstay of cancer treatment in the past has been chemotherapy, but acquired resistance to chemotherapeutic drugs hinders clinical application, resulting in disease recurrence. The subsequent development of kinase inhibitors and immunotherapy have also exposed the problem of drug resistance. Since PROTACs affect protein function, including enzymatic and non-enzymatic functions, by clearing the entire target protein, this technology is expected to address potential drug resistance faced by current treatments. The second is that PROTACs may aim for "undruggable targets." The majority of small-molecule medications or large-molecule antibodies need the active site of a binding enzyme or receptor to work; however, it is thought that over 80% of proteins in human cells don't have these sites. PROTACs, on the other hand, can grab target proteins through any nooks and crannies. Third, PROTACs can affect non-enzymatic functions. Conventional small-molecule medications often work by stopping their targets' enzyme activity. Accumulating studies have shown that PROTACs have the potential to increase the "druggable space" of targets and to regulate protease as well as non-enzyme functions. Some of the difficulties posed by conventional small molecule inhibitors can be overcome. Of course, in terms of clinical practice, PROTAC drugs are still in a relatively early stage, and there are challenges such as slow development of PROTACs and a slow success rate, poor membrane permeability and oral bioavailability, and insufficient evidence from human clinical studies. However, with the accumulation of time and in-depth research, these problems will basically be solved, and once a clinical breakthrough is formed, it will open a new era of drug innovation. PROTAC has a wide range of targets and a huge market. It is believed that with the continuous progress and improvement of this technology, PROTAC can become as successful as small molecule inhibitors, monoclonal antibodies, and immunotherapy, so that more patients with diseases can benefit from it.

Author Contributions: Writing—original draft preparation, T.Y.; review and editing, H.X.; investigation, H.W. and X.X. All authors have read and agreed to the published version of the manuscript.

Funding: This work was supported by Open Research Fund Program of Guangdong Provincial Key Laboratory of Virology (X.X. and H.X.) and the National Natural Science Foundation of China (82073926 to H.W.).

Institutional Review Board Statement: Not applicable.

Informed Consent Statement: Not applicable.

Data Availability Statement: Not applicable.

Conflicts of Interest: The authors declare no conflict of interest.

References

1. Alabi, S.B.; Crews, C.M. Major advances in targeted protein degradation: PROTACs, LYTACs, and MADTACs. *J. Biol. Chem.* **2021**, *296*, 100647. [CrossRef]
2. Sakamoto, K.M.; Kim, K.B.; Kumagai, A.; Mercurio, F.; Crews, C.M.; Deshaies, R.J. Protacs: Chimeric molecules that target proteins to the Skp1-Cullin-F box complex for ubiquitination and degradation. *Proc. Natl. Acad. Sci. USA* **2001**, *98*, 8554–8559. [CrossRef]
3. Gu, S.; Cui, D.; Chen, X.; Xiong, X.; Zhao, Y. PROTACs: An Emerging Targeting Technique for Protein Degradation in Drug Discovery. *BioEssays* **2018**, *40*, e1700247. [CrossRef] [PubMed]
4. Yang, J.; Li, Y.; Aguilar, A.; Liu, Z.; Yang, C.Y.; Wang, S. Simple Structural Modifications Converting a Bona fide MDM2 PROTAC Degrader into a Molecular Glue Molecule: A Cautionary Tale in the Design of PROTAC Degraders. *J. Med. Chem.* **2019**, *62*, 9471–9487. [CrossRef]
5. Tan, L.; Gray, N.S. When Kinases Meet PROTACs. *Chin. J. Chem.* **2018**, *36*, 971–977. [CrossRef]
6. Scheepstra, M.; Hekking, K.F.W.; van Hijfte, L.; Folmer, R.H.A. Bivalent Ligands for Protein Degradation in Drug Discovery. *Comput. Struct. Biotechnol. J.* **2019**, *17*, 160–176. [CrossRef]
7. Sakamoto, K.M. Protacs for treatment of cancer. *Pediatr. Res.* **2010**, *67*, 505–508. [CrossRef] [PubMed]
8. Ottis, P.; Crews, C.M. Proteolysis-Targeting Chimeras: Induced Protein Degradation as a Therapeutic Strategy. *ACS Chem. Biol.* **2017**, *12*, 892–898. [CrossRef]
9. Nguyen, C.; West, G.M.; Geoghegan, K.F. Emerging Methods in Chemoproteomics with Relevance to Drug Discovery. *Methods Mol. Biol.* **2017**, *1513*, 11–22.
10. Raina, K.; Crews, C.M. Chemical inducers of targeted protein degradation. *J. Biol. Chem.* **2010**, *285*, 11057–11060.
11. Nowak, R.P.; DeAngelo, S.L.; Buckley, D.; He, Z.; Donovan, K.A.; An, J.; Safaee, N.; Jedrychowski, M.P.; Ponthier, C.M.; Ishoey, M.; et al. Plasticity in binding confers selectivity in ligand-induced protein degradation. *Nat. Chem. Biol.* **2018**, *14*, 706–714. [CrossRef]
12. An, S.; Fu, L. Small-molecule PROTACs: An emerging and promising approach for the development of targeted therapy drugs. *EBioMedicine* **2018**, *36*, 553–562. [CrossRef]
13. Farnaby, W.; Koegl, M.; Roy, M.J.; Whitworth, C.; Diers, E.; Trainor, N.; Zollman, D.; Steurer, S.; Karolyi-Oezguer, J.; Riedmueller, C.; et al. Publisher Correction: BAF complex vulnerabilities in cancer demonstrated via structure-based PROTAC design. *Nat. Chem. Biol.* **2019**, *15*, 846. [CrossRef]
14. Drummond, M.L.; Williams, C.I. In Silico Modeling of PROTAC-Mediated Ternary Complexes: Validation and Application. *J. Chem. Inf. Modeling* **2019**, *59*, 1634–1644. [CrossRef]
15. Gadd, M.S.; Testa, A.; Lucas, X.; Chan, K.H.; Chen, W.; Lamont, D.J.; Zengerle, M.; Ciulli, A. Structural basis of PROTAC cooperative recognition for selective protein degradation. *Nat. Chem. Biol.* **2017**, *13*, 514–521. [CrossRef]
16. Roy, M.J.; Winkler, S.; Hughes, S.J.; Whitworth, C.; Galant, M.; Farnaby, W.; Rumpel, K.; Ciulli, A. SPR-Measured Dissociation Kinetics of PROTAC Ternary Complexes Influence Target Degradation Rate. *ACS Chem. Biol.* **2019**, *14*, 361–368. [CrossRef]
17. Hughes, S.J.; Ciulli, A. Molecular recognition of ternary complexes: A new dimension in the structure-guided design of chemical degraders. *Essays Biochem.* **2017**, *61*, 505–516.
18. Riching, K.M.; Mahan, S.; Corona, C.R.; McDougall, M.; Vasta, J.D.; Robers, M.B.; Urh, M.; Daniels, D.L. Quantitative Live-Cell Kinetic Degradation and Mechanistic Profiling of PROTAC Mode of Action. *ACS Chem. Biol.* **2018**, *13*, 2758–2770. [CrossRef]
19. Toure, M.; Crews, C.M. Small-Molecule PROTACS: New Approaches to Protein Degradation. *Angew. Chem. Int. Ed. Engl.* **2016**, *55*, 1966–1973. [CrossRef]
20. Komander, D.; Rape, M. The ubiquitin code. *Annu. Rev. Biochem.* **2012**, *81*, 203–229. [CrossRef]
21. Chen, Z.J.; Sun, L.J. Nonproteolytic functions of ubiquitin in cell signaling. *Mol. Cell* **2009**, *33*, 275–286. [CrossRef]
22. Jevtić, P.; Haakonsen, D.L.; Rapé, M. An E3 ligase guide to the galaxy of small-molecule-induced protein degradation. *Cell Chem. Biol.* **2021**, *28*, 1000–1013. [CrossRef]
23. Kannt, A.; Đikić, I. Expanding the arsenal of E3 ubiquitin ligases for proximity-induced protein degradation. *Cell Chem. Biol.* **2021**, *28*, 1014–1031. [CrossRef]
24. Bond, M.J.; Crews, C.M. Proteolysis targeting chimeras (PROTACs) come of age: Entering the third decade of targeted protein degradation. *RSC Chem. Biol.* **2021**, *2*, 725–742. [CrossRef] [PubMed]
25. Bemis, T.A.; La Clair, J.J.; Burkart, M.D. Unraveling the Role of Linker Design in Proteolysis Targeting Chimeras. *J. Med. Chem.* **2021**, *64*, 8042–8052. [CrossRef]
26. Benowitz, A.B.; Jones, K.L.; Harling, J.D. The therapeutic potential of PROTACs. *Expert Opin. Ther. Pat.* **2021**, *31*, 1–24. [CrossRef] [PubMed]
27. Sun, X.; Gao, H.; Yang, Y.; He, M.; Wu, Y.; Song, Y.; Tong, Y.; Rao, Y. PROTACs: Great opportunities for academia and industry. *Signal Transduct. Target. Ther.* **2019**, *4*, 64. [CrossRef]
28. Zeng, S.; Huang, W.; Zheng, X.; Liyan, C.; Zhang, Z.; Wang, J.; Shen, Z. Proteolysis targeting chimera (PROTAC) in drug discovery paradigm: Recent progress and future challenges. *Eur. J. Med. Chem.* **2021**, *210*, 112981. [CrossRef] [PubMed]
29. Neklesa, T.; Snyder, L.B.; Willard, R.R.; Vitale, N.; Raina, K.; Pizzano, J.; Gordon, D.A.; Bookbinder, M.; Macaluso, J.; Dong, H.; et al. An oral androgen receptor PROTAC degrader for prostate cancer. *J. Clin. Oncol.* **2018**, *36*, 381. [CrossRef]

30. Flanagan, J.J.; Qian, Y.; Gough, S.M.; Andreoli, M.; Bookbinder, M.; Cadelina, G.; Bradley, J.; Rousseau, E.; Willard, R.; Pizzano, J.; et al. Abstract P5-04-18: ARV-471, an oral estrogen receptor PROTAC degrader for breast cancer. *Cancer Res.* **2019**, *79*, P5–P04. [CrossRef]
31. Kim, J.; Kim, H.; Park, S.B. Privileged structures: Efficient chemical "navigators" toward unexplored biologically relevant chemical spaces. *J. Am. Chem. Soc.* **2014**, *136*, 14629–14638. [CrossRef] [PubMed]
32. Xi, M.; Chen, Y.; Yang, H.; Xu, H.; Du, K.; Wu, C.; Xu, Y.; Deng, L.; Luo, X.; Yu, L.; et al. Small molecule PROTACs in targeted therapy: An emerging strategy to induce protein degradation. *Eur. J. Med. Chem.* **2019**, *174*, 159–180. [CrossRef] [PubMed]
33. Zhao, Q.; Lan, T.; Su, S.; Rao, Y. Induction of apoptosis in MDA-MB-231 breast cancer cells by a PARP1-targeting PROTAC small molecule. *Chem. Commun.* **2019**, *55*, 369–372. [CrossRef]
34. Okuhira, K.; Demizu, Y.; Hattori, T.; Ohoka, N.; Shibata, N.; Nishimaki-Mogami, T.; Okuda, H.; Kurihara, M.; Naito, M. Development of hybrid small molecules that induce degradation of estrogen receptor-alpha and necrotic cell death in breast cancer cells. *Cancer Sci.* **2013**, *104*, 1492–1498. [CrossRef]
35. Ohoka, N.; Morita, Y.; Nagai, K.; Shimokawa, K.; Ujikawa, O.; Fujimori, I.; Ito, M.; Hayase, Y.; Okuhira, K.; Shibata, N.; et al. Derivatization of inhibitor of apoptosis protein (IAP) ligands yields improved inducers of estrogen receptor α degradation. *J. Biol. Chem.* **2018**, *293*, 6776–6790. [CrossRef]
36. Bondeson, D.P.; Mares, A.; Smith, I.E.; Ko, E.; Campos, S.; Miah, A.H.; Mulholland, K.E.; Routly, N.; Buckley, D.L.; Gustafson, J.L.; et al. Catalytic in vivo protein knockdown by small-molecule PROTACs. *Nat. Chem. Biol.* **2015**, *11*, 611–617. [CrossRef]
37. Cromm, P.M.; Samarasinghe, K.T.G.; Hines, J.; Crews, C.M. Addressing Kinase-Independent Functions of Fak via PROTAC-Mediated Degradation. *J. Am. Chem. Soc.* **2018**, *140*, 17019–17026. [CrossRef]
38. Tanjoni, I.; Walsh, C.; Uryu, S.; Tomar, A.; Nam, J.O.; Mielgo, A.; Lim, S.T.; Liang, C.; Koenig, M.; Sun, C.; et al. PND-1186 FAK inhibitor selectively promotes tumor cell apoptosis in three-dimensional environments. *Cancer Biol. Ther.* **2010**, *9*, 764–777. [CrossRef] [PubMed]
39. Smith, B.E.; Wang, S.L.; Jaime-Figueroa, S.; Harbin, A.; Wang, J.; Hamman, B.D.; Crews, C.M. Differential PROTAC substrate specificity dictated by orientation of recruited E3 ligase. *Nat. Commun.* **2019**, *10*, 131. [CrossRef]
40. Khan, S.; Zhang, X.; Lv, D.; Zhang, Q.; He, Y.; Zhang, P.; Liu, X.; Thummuri, D.; Yuan, Y.; Wiegand, J.S.; et al. A selective BCL-$X_{(L)}$ PROTAC degrader achieves safe and potent antitumor activity. *Nat. Med.* **2019**, *25*, 1938–1947. [CrossRef]
41. Spradlin, J.N.; Hu, X.; Ward, C.C.; Brittain, S.M.; Jones, M.D.; Ou, L.; To, M.; Proudfoot, A.; Ornelas, E.; Woldegiorgis, M.; et al. Harnessing the anti-cancer natural product nimbolide for targeted protein degradation. *Nat. Chem. Biol.* **2019**, *15*, 747–755. [CrossRef] [PubMed]
42. Gechijian, L.N.; Buckley, D.L.; Lawlor, M.A.; Reyes, J.M.; Paulk, J.; Ott, C.J.; Winter, G.E.; Erb, M.A.; Scott, T.G.; Xu, M.; et al. Functional TRIM24 degrader via conjugation of ineffectual bromodomain and VHL ligands. *Nat. Chem. Biol.* **2018**, *14*, 405–412. [CrossRef] [PubMed]
43. Li, Y.; Yang, J.; Aguilar, A.; McEachern, D.; Przybranowski, S.; Liu, L.; Yang, C.Y.; Wang, M.; Han, X.; Wang, S. Discovery of MD-224 as a First-in-Class, Highly Potent, and Efficacious Proteolysis Targeting Chimera Murine Double Minute 2 Degrader Capable of Achieving Complete and Durable Tumor Regression. *J. Med. Chem.* **2019**, *62*, 448–466. [CrossRef]
44. Burslem, G.M.; Song, J.; Chen, X.; Hines, J.; Crews, C.M. Enhancing Antiproliferative Activity and Selectivity of a FLT-3 Inhibitor by Proteolysis Targeting Chimera Conversion. *J. Am. Chem. Soc.* **2018**, *140*, 16428–16432. [CrossRef]
45. Bai, L.; Zhou, H.; Xu, R.; Zhao, Y.; Chinnaswamy, K.; McEachern, D.; Chen, J.; Yang, C.Y.; Liu, Z.; Wang, M.; et al. A Potent and Selective Small-Molecule Degrader of STAT3 Achieves Complete Tumor Regression *In Vivo*. *Cancer Cell* **2019**, *36*, 498–511.e417. [CrossRef] [PubMed]
46. Brand, M.; Jiang, B.; Bauer, S.; Donovan, K.A.; Liang, Y.; Wang, E.S.; Nowak, R.P.; Yuan, J.C.; Zhang, T.; Kwiatkowski, N.; et al. Homolog-Selective Degradation as a Strategy to Probe the Function of CDK6 in AML. *Cell Chem. Biol.* **2019**, *26*, 300–306.e9. [CrossRef]
47. Winter, G.E.; Buckley, D.L.; Paulk, J.; Roberts, J.M.; Souza, A.; Dhe-Paganon, S.; Bradner, J.E. Drug Development. Phthalimide conjugation as a strategy for in vivo target protein degradation. *Science* **2015**, *348*, 1376–1381.
48. Min, J.; Mayasundari, A.; Keramatnia, F.; Jonchere, B.; Yang, S.W.; Jarusiewicz, J.; Actis, M.; Das, S.; Young, B.; Slavish, J.; et al. Phenyl-Glutarimides: Alternative Cereblon Binders for the Design of PROTACs. *Angew. Chem. Int. Ed.* **2021**, *60*, 26663–26670. [CrossRef] [PubMed]
49. Imaide, S.; Riching, K.M.; Makukhin, N.; Vetma, V.; Whitworth, C.; Hughes, S.J.; Trainor, N.; Mahan, S.D.; Murphy, N.; Cowan, A.D.; et al. Trivalent PROTACs enhance protein degradation via combined avidity and cooperativity. *Nat. Chem. Biol.* **2021**, *17*, 1157–1167. [CrossRef] [PubMed]
50. Zhang, X.; Thummuri, D.; He, Y.; Liu, X.; Zhang, P.; Zhou, D.; Zheng, G. Utilizing PROTAC technology to address the on-target platelet toxicity associated with inhibition of BCL-$X_{(L)}$. *Chem. Commun.* **2019**, *55*, 14765–14768. [CrossRef]
51. Zhang, X.; Thummuri, D.; Liu, X.; Hu, W.; Zhang, P.; Khan, S.; Yuan, Y.; Zhou, D.; Zheng, G. Discovery of PROTAC BCL-$X_{(L)}$ degraders as potent anticancer agents with low on-target platelet toxicity. *Eur. J. Med. Chem.* **2020**, *192*, 112186. [CrossRef] [PubMed]
52. Lai, A.C.; Toure, M.; Hellerschmied, D.; Salami, J.; Jaime-Figueroa, S.; Ko, E.; Hines, J.; Crews, C.M. Modular PROTAC Design for the Degradation of Oncogenic BCR-ABL. *Angew. Chem. Int. Ed. Engl.* **2016**, *55*, 807–810. [CrossRef]
53. Demizu, Y.; Shibata, N.; Hattori, T.; Ohoka, N.; Motoi, H.; Misawa, T.; Shoda, T.; Naito, M.; Kurihara, M. Development of BCR-ABL degradation inducers via the conjugation of an imatinib derivative and a cIAP1 ligand. *Bioorg. Med. Chem. Lett.* **2016**, *26*, 4865–4869. [CrossRef] [PubMed]

54. Shibata, N.; Miyamoto, N.; Nagai, K.; Shimokawa, K.; Sameshima, T.; Ohoka, N.; Hattori, T.; Imaeda, Y.; Nara, H.; Cho, N.; et al. Development of protein degradation inducers of oncogenic BCR-ABL protein by conjugation of ABL kinase inhibitors and IAP ligands. *Cancer Sci.* **2017**, *108*, 1657–1666. [CrossRef]
55. He, Y.; Koch, R.; Budamagunta, V.; Lv, D.; Khan, S.; Zhang, X.; Wiegand, J.S.; Zheng, G.; Weinstock, D.M.; Zhou, D. DT2216, a BCL-X_L Proteolysis Targeting Chimera (PROTAC), Is a Potent Anti T-Cell Lymphoma Agent That Does Not Induce Significant Thrombocytopenia. *Blood* **2019**, *134*, 303. [CrossRef]
56. Sun, Y.; Ding, N.; Song, Y.; Yang, Z.; Liu, W.; Zhu, J.; Rao, Y. Degradation of Bruton's tyrosine kinase mutants by PROTACs for potential treatment of ibrutinib-resistant non-Hodgkin lymphomas. *Leukemia* **2019**, *33*, 2105–2110. [CrossRef]
57. Buhimschi, A.D.; Armstrong, H.A.; Toure, M.; Jaime-Figueroa, S.; Chen, T.L.; Lehman, A.M.; Woyach, J.A.; Johnson, A.J.; Byrd, J.C.; Crews, C.M. Targeting the C481S Ibrutinib-Resistance Mutation in Bruton's Tyrosine Kinase Using PROTAC-Mediated Degradation. *Biochemistry* **2018**, *57*, 3564–3575. [CrossRef] [PubMed]
58. Jaime-Figueroa, S.; Buhimschi, A.D.; Toure, M.; Hines, J.; Crews, C.M. Design, synthesis and biological evaluation of Proteolysis Targeting Chimeras (PROTACs) as a BTK degraders with improved pharmacokinetic properties. *Bioorg. Med. Chem. Lett.* **2020**, *30*, 126877. [CrossRef]
59. Lu, J.; Qian, Y.; Altieri, M.; Dong, H.; Wang, J.; Raina, K.; Hines, J.; Winkler, J.D.; Crew, A.P.; Coleman, K.; et al. Hijacking the E3 Ubiquitin Ligase Cereblon to Efficiently Target BRD4. *Chem. Biol.* **2015**, *22*, 755–763. [CrossRef]
60. Dobrovolsky, D.; Wang, E.S.; Morrow, S.; Leahy, C.; Faust, T.; Nowak, R.P.; Donovan, K.A.; Yang, G.; Li, Z.; Fischer, E.S.; et al. Bruton tyrosine kinase degradation as a therapeutic strategy for cancer. *Blood* **2019**, *133*, 952–961. [CrossRef]
61. Zorba, A.; Nguyen, C.; Xu, Y.; Starr, J.; Borzilleri, K.; Smith, J.; Zhu, H.; Farley, K.A.; Ding, W.; Schiemer, J.; et al. Delineating the role of cooperativity in the design of potent PROTACs for BTK. *Proc. Natl. Acad. Sci. USA* **2018**, *115*, e7285–e7292. [CrossRef] [PubMed]
62. Hansen, J.D.; Correa, M.; Nagy, M.A.; Alexander, M.; Plantevin, V.; Grant, V.; Whitefield, B.; Huang, D.; Kercher, T.; Harris, R.; et al. Discovery of CRBN E3 Ligase Modulator CC-92480 for the Treatment of Relapsed and Refractory Multiple Myeloma. *J. Med. Chem.* **2020**, *63*, 6648–6676. [CrossRef]
63. Su, S.; Yang, Z.; Gao, H.; Yang, H.; Zhu, S.; An, Z.; Wang, J.; Li, Q.; Chandarlapaty, S.; Deng, H.; et al. Potent and Preferential Degradation of CDK6 via Proteolysis Targeting Chimera Degraders. *J. Med. Chem.* **2019**, *62*, 7575–7582. [CrossRef]
64. Lebraud, H.; Wright, D.J.; Johnson, C.N.; Heightman, T.D. Protein Degradation by In-Cell Self-Assembly of Proteolysis Targeting Chimeras. *ACS Cent. Sci.* **2016**, *2*, 927–934. [CrossRef]
65. Robb, C.M.; Contreras, J.I.; Kour, S.; Taylor, M.A.; Abid, M.; Sonawane, Y.A.; Zahid, M.; Murry, D.J.; Natarajan, A.; Rana, S. Chemically induced degradation of CDK9 by a proteolysis targeting chimera (PROTAC). *Chem. Commun.* **2017**, *53*, 7577–7580. [CrossRef]
66. Hines, J.; Lartigue, S.; Dong, H.; Qian, Y.; Crews, C.M. MDM2-Recruiting PROTAC Offers Superior, Synergistic Antiproliferative Activity via Simultaneous Degradation of BRD4 and Stabilization of p53. *Cancer Res.* **2019**, *79*, 251–262.
67. Chen, L.; Chen, Y.; Zhang, C.; Jiao, B.; Liang, S.; Tan, Q.; Chai, H.; Yu, W.; Qian, Y.; Yang, H.; et al. Discovery of First-In-Class Potent and Selective Tropomyosin Receptor Kinase Degraders. *J. Med. Chem.* **2020**, *63*, 14562–14575. [CrossRef] [PubMed]
68. Kang, C.H.; Lee, D.H.; Lee, C.O.; Du Ha, J.; Park, C.H.; Hwang, J.Y. Induced protein degradation of anaplastic lymphoma kinase (ALK) by proteolysis targeting chimera (PROTAC). *Biochem. Biophys. Res. Commun.* **2018**, *505*, 542–547.
69. Powell, C.E.; Gao, Y.; Tan, L.; Donovan, K.A.; Nowak, R.P.; Loehr, A.; Bahcall, M.; Fischer, E.S.; Jänne, P.A.; George, R.E.; et al. Chemically Induced Degradation of Anaplastic Lymphoma Kinase (ALK). *J. Med. Chem.* **2018**, *61*, 4249–4255. [CrossRef] [PubMed]
70. Zheng, M.; Huo, J.; Gu, X.; Wang, Y.; Wu, C.; Zhang, Q.; Wang, W.; Liu, Y.; Liu, Y.; Zhou, X.; et al. Rational Design and Synthesis of Novel Dual PROTACs for Simultaneous Degradation of EGFR and PARP. *J. Med. Chem.* **2021**, *64*, 7839–7852.
71. Xiao, Z.; Song, S.; Chen, D.; van Merkerk, R.; van der Wouden, P.E.; Cool, R.H.; Quax, W.J.; Poelarends, G.J.; Melgert, B.N.; Dekker, F.J. Proteolysis Targeting Chimera (PROTAC) for Macrophage Migration Inhibitory Factor (MIF) Has Anti-Proliferative Activity in Lung Cancer Cells. *Angew. Chem. Int. Ed. Engl.* **2021**, *60*, 17514–17521. [CrossRef] [PubMed]
72. Shibata, N.; Nagai, K.; Morita, Y.; Ujikawa, O.; Ohoka, N.; Hattori, T.; Koyama, R.; Sano, O.; Imaeda, Y.; Nara, H.; et al. Development of Protein Degradation Inducers of Androgen Receptor by Conjugation of Androgen Receptor Ligands and Inhibitor of Apoptosis Protein Ligands. *J. Med. Chem.* **2018**, *61*, 543–575. [CrossRef] [PubMed]
73. Crew, A.P.; Raina, K.; Dong, H.; Qian, Y.; Wang, J.; Vigil, D.; Serebrenik, Y.V.; Hamman, B.D.; Morgan, A.; Ferraro, C.; et al. Identification and Characterization of Von Hippel-Lindau-Recruiting Proteolysis Targeting Chimeras (PROTACs) of TANK-Binding Kinase 1. *J. Med. Chem.* **2018**, *61*, 583–598. [CrossRef] [PubMed]
74. Yoon, H.; Dehart, J.P.; Murphy, J.M.; Lim, S.T. Understanding the roles of FAK in cancer: Inhibitors, genetic models, and new insights. *J. Histochem. Cytochem.* **2015**, *63*, 114–128. [CrossRef]
75. Sulzmaier, F.J.; Jean, C.; Schlaepfer, D.D. FAK in cancer: Mechanistic findings and clinical applications. *Nat. Rev. Cancer* **2014**, *14*, 598–610. [CrossRef]
76. Lee, B.Y.; Timpson, P.; Horvath, L.G.; Daly, R.J. FAK signaling in human cancer as a target for therapeutics. *Pharmacol. Ther.* **2015**, *146*, 132–149. [CrossRef]
77. Cuadrado, A.; Nebreda, A.R. Mechanisms and functions of p38 MAPK signalling. *Biochem. J.* **2010**, *429*, 403–417. [CrossRef]
78. Cargnello, M.; Roux, P.P. Activation and function of the MAPKs and their substrates, the MAPK-activated protein kinases. *Microbiol. Mol. Biol. Rev.* **2011**, *75*, 50–83. [CrossRef]
79. Han, J.; Lee, J.D.; Bibbs, L.; Ulevitch, R.J. A MAP kinase targeted by endotoxin and hyperosmolarity in mammalian cells. *Science* **1994**, *265*, 808–811. [CrossRef] [PubMed]

80. Genovese, M.C. Inhibition of p38: Has the fat lady sung? *Arthritis Rheum.* **2009**, *60*, 317–320. [CrossRef] [PubMed]
81. Astolfi, A.; Manfroni, G.; Cecchetti, V.; Barreca, M.L. A Comprehensive Structural Overview of p38α Mitogen-Activated Protein Kinase in Complex with ATP-Site and Non-ATP-Site Binders. *ChemMedChem* **2018**, *13*, 7–14. [CrossRef] [PubMed]
82. Escós, A.; Risco, A.; Alsina-Beauchamp, D.; Cuenda, A. p38γ and p38δ Mitogen Activated Protein Kinases (MAPKs), New Stars in the MAPK Galaxy. *Front. Cell Dev. Biol.* **2016**, *4*, 31. [CrossRef] [PubMed]
83. Alevy, Y.G.; Patel, A.C.; Romero, A.G.; Patel, D.A.; Tucker, J.; Roswit, W.T.; Miller, C.A.; Heier, R.F.; Byers, D.E.; Brett, T.J.; et al. IL-13-induced airway mucus production is attenuated by MAPK13 inhibition. *J. Clin. Investig.* **2012**, *122*, 4555–4568. [CrossRef] [PubMed]
84. Yurtsever, Z.; Scheaffer, S.M.; Romero, A.G.; Holtzman, M.J.; Brett, T.J. The crystal structure of phosphorylated MAPK13 reveals common structural features and differences in p38 MAPK family activation. *Acta Crystallogr. Sect. D Biol. Crystallogr.* **2015**, *71*, 790–799. [CrossRef]
85. Yurtsever, Z.; Patel, D.A.; Kober, D.L.; Su, A.; Miller, C.A.; Romero, A.G.; Holtzman, M.J.; Brett, T.J. First comprehensive structural and biophysical analysis of MAPK13 inhibitors targeting DFG-in and DFG-out binding modes. *Biochim. Biophys. Acta* **2016**, *1860*, 2335–2344. [CrossRef]
86. Zeng, H.; Belanger, D.B.; Curran, P.J.; Shipps, G.W., Jr.; Miao, H.; Bracken, J.B.; Siddiqui, M.A.; Malkowski, M.; Wang, Y. Discovery of novel imidazo[1,2-a]pyrazin-8-amines as Brk/PTK6 inhibitors. *Bioorg. Med. Chem. Lett.* **2011**, *21*, 5870–5875. [CrossRef] [PubMed]
87. Hu, D.; Shilatifard, A. Epigenetics of hematopoiesis and hematological malignancies. *Genes Dev.* **2016**, *30*, 2021–2041. [CrossRef]
88. Deschler, B.; Lübbert, M. Acute myeloid leukemia: Epidemiology and etiology. *Cancer* **2006**, *107*, 2099–2107. [CrossRef]
89. Döhner, H.; Weisdorf, D.J.; Bloomfield, C.D. Acute Myeloid Leukemia. *N. Engl. J. Med.* **2015**, *373*, 1136–1152. [CrossRef] [PubMed]
90. Le Douarin, B.; Zechel, C.; Garnier, J.M.; Lutz, Y.; Tora, L.; Pierrat, P.; Heery, D.; Gronemeyer, H.; Chambon, P.; Losson, R. The N-terminal part of TIF1, a putative mediator of the ligand-dependent activation function (AF-2) of nuclear receptors, is fused to B-raf in the oncogenic protein T18. *Embo J.* **1995**, *14*, 2020–2033. [CrossRef]
91. Tsai, W.W.; Wang, Z.; Yiu, T.T.; Akdemir, K.C.; Xia, W.; Winter, S.; Tsai, C.Y.; Shi, X.; Schwarzer, D.; Plunkett, W.; et al. TRIM24 links a non-canonical histone signature to breast cancer. *Nature* **2010**, *468*, 927–932. [CrossRef]
92. Cui, Z.; Cao, W.; Li, J.; Song, X.; Mao, L.; Chen, W. TRIM24 overexpression is common in locally advanced head and neck squamous cell carcinoma and correlates with aggressive malignant phenotypes. *PLoS ONE* **2013**, *8*, e63887. [CrossRef] [PubMed]
93. Groner, A.C.; Cato, L.; de Tribolet-Hardy, J.; Bernasocchi, T.; Janouskova, H.; Melchers, D.; Houtman, R.; Cato, A.C.B.; Tschopp, P.; Gu, L.; et al. TRIM24 Is an Oncogenic Transcriptional Activator in Prostate Cancer. *Cancer Cell* **2016**, *29*, 846–858. [CrossRef]
94. Li, H.; Sun, L.; Tang, Z.; Fu, L.; Xu, Y.; Li, Z.; Luo, W.; Qiu, X.; Wang, E. Overexpression of TRIM24 correlates with tumor progression in non-small cell lung cancer. *PLoS ONE* **2012**, *7*, e37657. [CrossRef]
95. Liu, X.; Huang, Y.; Yang, D.; Li, X.; Liang, J.; Lin, L.; Zhang, M.; Zhong, K.; Liang, B.; Li, J. Overexpression of TRIM24 is associated with the onset and progress of human hepatocellular carcinoma. *PLoS ONE* **2014**, *9*, e85462. [CrossRef] [PubMed]
96. Wang, J.; Zhu, J.; Dong, M.; Yu, H.; Dai, X.; Li, K. Knockdown of tripartite motif containing 24 by lentivirus suppresses cell growth and induces apoptosis in human colorectal cancer cells. *Oncol. Res.* **2014**, *22*, 39–45. [CrossRef] [PubMed]
97. Freedman, D.A.; Wu, L.; Levine, A.J. Functions of the MDM2 oncoprotein. *Cell. Mol. Life Sci.* **1999**, *55*, 96–107. [CrossRef]
98. Wu, X.; Bayle, J.H.; Olson, D.; Levine, A.J. The p53-mdm-2 autoregulatory feedback loop. *Genes Dev.* **1993**, *7*, 1126–1132. [CrossRef]
99. Momand, J.; Zambetti, G.P.; Olson, D.C.; George, D.; Levine, A.J. The mdm-2 oncogene product forms a complex with the p53 protein and inhibits p53-mediated transactivation. *Cell* **1992**, *69*, 1237–1245. [CrossRef]
100. Wang, S.; Sun, W.; Zhao, Y.; McEachern, D.; Meaux, I.; Barrière, C.; Stuckey, J.A.; Meagher, J.L.; Bai, L.; Liu, L.; et al. SAR405838: An optimized inhibitor of MDM2-p53 interaction that induces complete and durable tumor regression. *Cancer Res.* **2014**, *74*, 5855–5865. [CrossRef]
101. Kandarpa, M.; Peterson, L.F.; Potu, H.; Ramappan, M.; Liu, Y.; Polk, A.; Wang, S.; Talpaz, M. Improved Anti-Leukemic Pre-Clinical Efficacy of a Protac Based MDM2 Degrader in a Large AML Cohort. *Blood* **2019**, *134*, 2670. [CrossRef]
102. Thiede, C.; Steudel, C.; Mohr, B.; Schaich, M.; Schäkel, U.; Platzbecker, U.; Wermke, M.; Bornhäuser, M.; Ritter, M.; Neubauer, A.; et al. Analysis of FLT3-activating mutations in 979 patients with acute myelogenous leukemia: Association with FAB subtypes and identification of subgroups with poor prognosis. *Blood* **2002**, *99*, 4326–4335. [CrossRef] [PubMed]
103. Smith, C.C.; Wang, Q.; Chin, C.S.; Salerno, S.; Damon, L.E.; Levis, M.J.; Perl, A.E.; Travers, K.J.; Wang, S.; Hunt, J.P.; et al. Validation of ITD mutations in FLT3 as a therapeutic target in human acute myeloid leukaemia. *Nature* **2012**, *485*, 260–263. [CrossRef]
104. Grunwald, M.R.; Levis, M.J. FLT3 inhibitors for acute myeloid leukemia: A review of their efficacy and mechanisms of resistance. *Int. J. Hematol.* **2013**, *97*, 683–694. [CrossRef] [PubMed]
105. Pratz, K.W.; Cortes, J.; Roboz, G.J.; Rao, N.; Arowojolu, O.; Stine, A.; Shiotsu, Y.; Shudo, A.; Akinaga, S.; Small, D.; et al. A pharmacodynamic study of the FLT3 inhibitor KW-2449 yields insight into the basis for clinical response. *Blood* **2009**, *113*, 3938–3946. [CrossRef]
106. Johnson, D.E.; O'Keefe, R.A.; Grandis, J.R. Targeting the IL-6/JAK/STAT3 signalling axis in cancer. *Nat. Rev. Clin. Oncol.* **2018**, *15*, 234–248. [CrossRef]
107. Yang, J.; Stark, G.R. Roles of unphosphorylated STATs in signaling. *Cell Res.* **2008**, *18*, 443–451. [CrossRef]
108. Malumbres, M.; Barbacid, M. Cell cycle, CDKs and cancer: A changing paradigm. *Nat. Rev. Cancer.* **2009**, *9*, 153–166. [CrossRef]
109. Sánchez-Martínez, C.; Lallena, M.J.; Sanfeliciano, S.G.; de Dios, A. Cyclin dependent kinase (CDK) inhibitors as anticancer drugs: Recent advances (2015–2019). *Bioorg. Med. Chem. Lett.* **2019**, *29*, 126637. [CrossRef] [PubMed]

110. Kollmann, K.; Heller, G.; Schneckenleithner, C.; Warsch, W.; Scheicher, R.; Ott, R.G.; Schäfer, M.; Fajmann, S.; Schlederer, M.; Schiefer, A.I.; et al. A kinase-independent function of CDK6 links the cell cycle to tumor angiogenesis. *Cancer Cell* **2013**, *24*, 167–181. [CrossRef]
111. Filippakopoulos, P.; Knapp, S. Targeting bromodomains: Epigenetic readers of lysine acetylation. *Nat. Rev. Drug Discov.* **2014**, *13*, 337–356. [CrossRef]
112. Shi, J.; Vakoc, C.R. The mechanisms behind the therapeutic activity of BET bromodomain inhibition. *Mol. Cell.* **2014**, *54*, 728–736. [CrossRef]
113. Ghoshal, A.; Yugandhar, D.; Srivastava, A.K. BET inhibitors in cancer therapeutics: A patent review. *Expert Opin. Ther. Pat.* **2016**, *26*, 505–522. [CrossRef]
114. Chonghaile, T.N.; Roderick, J.E.; Glenfield, C.; Ryan, J.; Sallan, S.E.; Silverman, L.B.; Loh, M.L.; Hunger, S.P.; Wood, B.; DeAngelo, D.J.; et al. Maturation stage of T-cell acute lymphoblastic leukemia determines BCL-2 versus BCL-X_L dependence and sensitivity to ABT-199. *Cancer Discov.* **2014**, *4*, 1074–1087. [CrossRef] [PubMed]
115. Czabotar, P.E.; Lessene, G.; Strasser, A.; Adams, J.M. Control of apoptosis by the BCL-2 protein family: Implications for physiology and therapy. *Nat. Rev. Mol. Cell Biol.* **2014**, *15*, 49–63. [CrossRef]
116. Igney, F.H.; Krammer, P.H. Death and anti-death: Tumour resistance to apoptosis. *Nat. Rev. Cancer* **2002**, *2*, 277–288. [CrossRef]
117. Adams, J.M.; Cory, S. The Bcl-2 apoptotic switch in cancer development and therapy. *Oncogene* **2007**, *26*, 1324–1337. [CrossRef]
118. Will, B.; Zhou, L.; Vogler, T.O.; Ben-Neriah, S.; Schinke, C.; Tamari, R.; Yu, Y.; Bhagat, T.D.; Bhattacharyya, S.; Barreyro, L.; et al. Stem and progenitor cells in myelodysplastic syndromes show aberrant stage-specific expansion and harbor genetic and epigenetic alterations. *Blood* **2012**, *120*, 2076–2086. [CrossRef]
119. Zhang, Q.; Khan, S.; Zhang, X.; Kuruvilla, V.M.; Ghotbaldini, S.; Wells, J.; Baran, N.; Cai, T.; Han, L.; Ferrando, A.; et al. Targeting BCL-X_L By Protac DT2216 Effectively Eliminates Leukemia Cells in T-ALL Pre-Clinical Models. *Blood* **2019**, *134*, 3870. [CrossRef]
120. Hantschel, O.; Superti-Furga, G. Regulation of the c-Abl and Bcr-Abl tyrosine kinases. *Nat. Rev. Mol. Cell Biol.* **2004**, *5*, 33–44. [CrossRef] [PubMed]
121. Hantschel, O.; Warsch, W.; Eckelhart, E.; Kaupe, I.; Grebien, F.; Wagner, K.U.; Superti-Furga, G.; Sexl, V. BCR-ABL uncouples canonical JAK2-STAT5 signaling in chronic myeloid leukemia. *Nat. Chem. Biol.* **2012**, *8*, 285–293. [CrossRef] [PubMed]
122. Wertheim, J.A.; Forsythe, K.; Druker, B.J.; Hammer, D.; Boettiger, D.; Pear, W.S. BCR-ABL-induced adhesion defects are tyrosine kinase-independent. *Blood* **2002**, *99*, 4122–4130. [CrossRef]
123. Ichim, C.V. Kinase-independent mechanisms of resistance of leukemia stem cells to tyrosine kinase inhibitors. *Stem Cells Transl. Med.* **2014**, *3*, 405–415. [CrossRef] [PubMed]
124. Hamilton, A.; Helgason, G.V.; Schemionek, M.; Zhang, B.; Myssina, S.; Allan, E.K.; Nicolini, F.E.; Müller-Tidow, C.; Bhatia, R.; Brunton, V.G.; et al. Chronic myeloid leukemia stem cells are not dependent on Bcr-Abl kinase activity for their survival. *Blood* **2012**, *119*, 1501–1510. [CrossRef] [PubMed]
125. Koch, R.; Christie, A.L.; Crombie, J.L.; Palmer, A.C.; Plana, D.; Shigemori, K.; Morrow, S.N.; Van Scoyk, A.; Wu, W.; Brem, E.A.; et al. Biomarker-driven strategy for MCL1 inhibition in T-cell lymphomas. *Blood* **2019**, *133*, 566–575. [CrossRef] [PubMed]
126. Hamp, I.; O'Neill, T.J.; Plettenburg, O.; Krappmann, D. A patent review of MALT1 inhibitors (2013-present). *Expert Opin. Ther. Pat.* **2021**, *31*, 1079–1096. [CrossRef]
127. Fontan, L.; Goldstein, R.; Casalena, G.; Durant, M.; Teater, M.R.; Wilson, J.; Phillip, J.; Xia, M.; Shah, S.; Us, I.; et al. Identification of MALT1 feedback mechanisms enables rational design of potent antilymphoma regimens for ABC-DLBCL. *Blood* **2021**, *137*, 788–800. [CrossRef]
128. Fontan, L.; Hatcher, J.; Scott, D.; Qiao, Q.; Us, I.; Du, G.; Durant, M.; Wilson, J.; Wu, H.; Gray, N.; et al. Chemically Induced Degradation of MALT1 to Treat B-Cell Lymphomas. *Blood* **2019**, *134*, 2073. [CrossRef]
129. Zhang, H.; Qiu, L. Chapter 8—Bruton's Tyrosine Kinase (BTK) Inhibitors as Sensitizing Agents for Cancer Chemotherapy. In *Protein Kinase Inhibitors as Sensitizing Agents for Chemotherapy*; Chen, Z.-S., Yang, D.-H., Eds.; Academic Press: Cambridge, MA, USA, 2019; pp. 109–124.
130. Woyach, J.A.; Furman, R.R.; Liu, T.M.; Ozer, H.G.; Zapatka, M.; Ruppert, A.S.; Xue, L.; Li, D.H.; Steggerda, S.M.; Versele, M.; et al. Resistance mechanisms for the Bruton's tyrosine kinase inhibitor ibrutinib. *N. Engl. J. Med.* **2014**, *370*, 2286–2294. [CrossRef]
131. Nijhof, I.S.; van de Donk, N.; Zweegman, S.; Lokhorst, H.M. Current and New Therapeutic Strategies for Relapsed and Refractory Multiple Myeloma: An Update. *Drugs* **2018**, *78*, 19–37. [CrossRef]
132. Labianca, R.; Beretta, G.D.; Kildani, B.; Milesi, L.; Merlin, F.; Mosconi, S.; Pessi, M.A.; Prochilo, T.; Quadri, A.; Gatta, G.; et al. Colon cancer. *Crit. Rev. Oncol. Hematol.* **2010**, *74*, 106–133. [CrossRef] [PubMed]
133. Choi, H.S.; Rucker, P.V.; Wang, Z.; Fan, Y.; Albaugh, P.; Chopiuk, G.; Gessier, F.; Sun, F.; Adrian, F.; Liu, G.; et al. (R)-2-Phenylpyrrolidine Substituted Imidazopyridazines: A New Class of Potent and Selective Pan-TRK Inhibitors. *ACS Med. Chem. Lett.* **2015**, *6*, 562–567. [CrossRef] [PubMed]
134. Nagano, T.; Tachihara, M.; Nishimura, Y. Molecular Mechanisms and Targeted Therapies Including Immunotherapy for Non-Small Cell Lung Cancer. *Curr. Cancer Drug Targets* **2019**, *19*, 595–630. [CrossRef]
135. Soumoy, L.; Kindt, N.; Ghanem, G.; Saussez, S.; Journe, F. Role of Macrophage Migration Inhibitory Factor (MIF) in Melanoma. *Cancers* **2019**, *11*, 529. [CrossRef]
136. Cavalli, E.; Ciurleo, R.; Petralia, M.C.; Fagone, P.; Bella, R.; Mangano, K.; Nicoletti, F.; Bramanti, P.; Basile, M.S. Emerging Role of the Macrophage Migration Inhibitory Factor Family of Cytokines in Neuroblastoma. Pathogenic Effectors and Novel Therapeutic Targets? *Molecules* **2020**, *25*, 1194. [CrossRef]

137. Coleman, A.M.; Rendon, B.E.; Zhao, M.; Qian, M.W.; Bucala, R.; Xin, D.; Mitchell, R.A. Cooperative regulation of non-small cell lung carcinoma angiogenic potential by macrophage migration inhibitory factor and its homolog, D-dopachrome tautomerase. *J. Immunol.* **2008**, *181*, 2330–2337. [CrossRef]
138. Charan, M.; Das, S.; Mishra, S.; Chatterjee, N.; Varikuti, S.; Kaul, K.; Misri, S.; Ahirwar, D.K.; Satoskar, A.R.; Ganju, R.K. Macrophage migration inhibitory factor inhibition as a novel therapeutic approach against triple-negative breast cancer. *Cell Death Dis.* **2020**, *11*, 774. [CrossRef]
139. Zhang, M.; Yan, L.; Kim, J.A. Modulating mammary tumor growth, metastasis and immunosuppression by siRNA-induced MIF reduction in tumor microenvironment. *Cancer Gene Ther.* **2015**, *22*, 463–474. [CrossRef]
140. Oliveira, C.S.; de Bock, C.E.; Molloy, T.J.; Sadeqzadeh, E.; Geng, X.Y.; Hersey, P.; Zhang, X.D.; Thorne, R.F. Macrophage migration inhibitory factor engages PI3K/Akt signalling and is a prognostic factor in metastatic melanoma. *BMC Cancer* **2014**, *14*, 630. [CrossRef] [PubMed]
141. Balogh, K.N.; Templeton, D.J.; Cross, J.V. Macrophage Migration Inhibitory Factor protects cancer cells from immunogenic cell death and impairs anti-tumor immune responses. *PLoS ONE* **2018**, *13*, e0197702. [CrossRef]
142. Nguyen-Nielsen, M.; Borre, M. Diagnostic and Therapeutic Strategies for Prostate Cancer. *Semin. Nucl. Med.* **2016**, *46*, 484–490. [CrossRef] [PubMed]
143. Ansari, D.; Tingstedt, B.; Andersson, B.; Holmquist, F.; Sturesson, C.; Williamsson, C.; Sasor, A.; Borg, D.; Bauden, M.; Andersson, R. Pancreatic cancer: Yesterday, today and tomorrow. *Future Oncol.* **2016**, *12*, 1929–1946. [CrossRef]
144. Muvaffak, A.; Pan, Q.; Yan, H.; Fernandez, R.; Lim, J.; Dolinski, B.; Nguyen, T.T.; Strack, P.; Wu, S.; Chung, R.; et al. Evaluating TBK1 as a therapeutic target in cancers with activated IRF3. *Mol. Cancer Res.* **2014**, *12*, 1055–1066. [CrossRef]
145. Su, L.C.; Xu, W.D.; Huang, A.F. IRAK family in inflammatory autoimmune diseases. *Autoimmun. Rev.* **2020**, *19*, 102461. [CrossRef] [PubMed]
146. Rhyasen, G.W.; Starczynowski, D.T. IRAK signalling in cancer. *Br. J. Cancer* **2015**, *112*, 232–237. [CrossRef]
147. Zhang, Y.; Diao, N.; Lee, C.K.; Chu, H.W.; Bai, L.; Li, L. Neutrophils Deficient in Innate Suppressor IRAK-M Enhances Anti-tumor Immune Responses. *Mol. Ther.* **2020**, *28*, 89–99. [CrossRef] [PubMed]
148. Kesselring, R.; Glaesner, J.; Hiergeist, A.; Naschberger, E.; Neumann, H.; Brunner, S.M.; Wege, A.K.; Seebauer, C.; Köhl, G.; Merkl, S.; et al. IRAK-M Expression in Tumor Cells Supports Colorectal Cancer Progression through Reduction of Antimicrobial Defense and Stabilization of STAT3. *Cancer Cell* **2016**, *29*, 684–696. [CrossRef]
149. Degorce, S.L.; Tavana, O.; Banks, E.; Crafter, C.; Gingipalli, L.; Kouvchinov, D.; Mao, Y.; Pachl, F.; Solanki, A.; Valge-Archer, V.; et al. Discovery of Proteolysis-Targeting Chimera Molecules that Selectively Degrade the IRAK3 Pseudokinase. *J. Med. Chem.* **2020**, *63*, 10460–10473. [CrossRef]
150. Nunes, J.; McGonagle, G.A.; Eden, J.; Kiritharan, G.; Touzet, M.; Lewell, X.; Emery, J.; Eidam, H.; Harling, J.D.; Anderson, N.A. Targeting IRAK4 for Degradation with PROTACs. *ACS Med. Chem. Lett.* **2019**, *10*, 1081–1085. [CrossRef] [PubMed]
151. Zhang, J.; Fu, L.; Shen, B.; Liu, Y.; Wang, W.; Cai, X.; Kong, L.; Yan, Y.; Meng, R.; Zhang, Z.; et al. Assessing IRAK4 Functions in ABC DLBCL by IRAK4 Kinase Inhibition and Protein Degradation. *Cell Chem. Biol.* **2020**, *27*, 1500–1509.e13. [CrossRef]
152. Chen, Y.; Ning, Y.; Bai, G.; Tong, L.; Zhang, T.; Zhou, J.; Zhang, H.; Xie, H.; Ding, J.; Duan, W. Design, Synthesis, and Biological Evaluation of IRAK4-Targeting PROTACs. *ACS Med. Chem. Lett.* **2021**, *12*, 82–87. [CrossRef] [PubMed]
153. Cao, F.; de Weerd, S.; Chen, D.; Zwinderman, M.R.H.; van der Wouden, P.E.; Dekker, F.J. Induced protein degradation of histone deacetylases 3 (HDAC3) by proteolysis targeting chimera (PROTAC). *Eur. J. Med. Chem.* **2020**, *208*, 112800. [CrossRef] [PubMed]
154. Xiao, Y.; Wang, J.; Zhao, L.Y.; Chen, X.; Zheng, G.; Zhang, X.; Liao, D. Discovery of histone deacetylase 3 (HDAC3)-specific PROTACs. *Chem. Commun.* **2020**, *56*, 9866–9869. [CrossRef] [PubMed]
155. Cao, Z.; Gu, Z.; Lin, S.; Chen, D.; Wang, J.; Zhao, Y.; Li, Y.; Liu, T.; Li, Y.; Wang, Y.; et al. Attenuation of NLRP3 Inflammasome Activation by Indirubin-Derived PROTAC Targeting HDAC6. *ACS Chem. Biol.* **2021**, *16*, 2746–2751. [CrossRef]
156. Yokoo, H.; Shibata, N.; Naganuma, M.; Murakami, Y.; Fujii, K.; Ito, T.; Aritake, K.; Naito, M.; Demizu, Y. Development of a Hematopoietic Prostaglandin D Synthase-Degradation Inducer. *ACS Med. Chem. Lett.* **2021**, *12*, 236–241. [CrossRef]
157. Hu, M.; Zhou, W.; Wang, Y.; Yao, D.; Ye, T.; Yao, Y.; Chen, B.; Liu, G.; Yang, X.; Wang, W.; et al. Discovery of the first potent proteolysis targeting chimera (PROTAC) degrader of indoleamine 2,3-dioxygenase 1. *Acta Pharm. Sin. B.* **2020**, *10*, 1943–1953. [CrossRef] [PubMed]
158. Schiedel, M.; Herp, D.; Hammelmann, S.; Swyter, S.; Lehotzky, A.; Robaa, D.; Oláh, J.; Ovádi, J.; Sippl, W.; Jung, M. Chemically Induced Degradation of Sirtuin 2 (Sirt2) by a Proteolysis Targeting Chimera (PROTAC) Based on Sirtuin Rearranging Ligands (SirReals). *J. Med. Chem.* **2018**, *61*, 482–491. [CrossRef] [PubMed]
159. Schiedel, M.; Lehotzky, A.; Szunyogh, S.; Oláh, J.; Hammelmann, S.; Wössner, N.; Robaa, D.; Einsle, O.; Sippl, W.; Ovádi, J.; et al. HaloTag-Targeted Sirtuin-Rearranging Ligand (SirReal) for the Development of Proteolysis-Targeting Chimeras (PROTACs) against the Lysine Deacetylase Sirtuin 2 (Sirt2)*. *Chembiochem* **2020**, *21*, 3371–3376. [CrossRef]
160. Bassi, Z.I.; Fillmore, M.C.; Miah, A.H.; Chapman, T.D.; Maller, C.; Roberts, E.J.; Davis, L.C.; Lewis, D.E.; Galwey, N.W.; Waddington, K.E.; et al. Modulating PCAF/GCN5 Immune Cell Function through a PROTAC Approach. *ACS Chem. Biol.* **2018**, *13*, 2862–2867. [CrossRef]
161. Mares, A.; Miah, A.H.; Smith, I.E.D.; Rackham, M.; Thawani, A.R.; Cryan, J.; Haile, P.A.; Votta, B.J.; Beal, A.M.; Capriotti, C.; et al. Extended pharmacodynamic responses observed upon PROTAC-mediated degradation of RIPK2. *Commun. Biol.* **2020**, *3*, 140. [CrossRef]

162. Chaudhary, D.; Robinson, S.; Romero, D.L. Recent advances in the discovery of small molecule inhibitors of interleukin-1 receptor-associated kinase 4 (IRAK4) as a therapeutic target for inflammation and oncology disorders. *J. Med. Chem.* **2015**, *58*, 96–110. [CrossRef]
163. Akira, S.; Takeda, K. Toll-like receptor signalling. *Nat. Rev. Immunol.* **2004**, *4*, 499–511. [CrossRef] [PubMed]
164. McElroy, W.T. Interleukin-1 receptor-associated kinase 4 (IRAK4) inhibitors: An updated patent review (2016–2018). *Expert Opin. Ther. Pat.* **2019**, *29*, 243–259. [CrossRef] [PubMed]
165. Dokmanovic, M.; Clarke, C.; Marks, P.A. Histone deacetylase inhibitors: Overview and perspectives. *Mol. Cancer Res.* **2007**, *5*, 981–989. [CrossRef]
166. Boyault, C.; Sadoul, K.; Pabion, M.; Khochbin, S. HDAC6, at the crossroads between cytoskeleton and cell signaling by acetylation and ubiquitination. *Oncogene* **2007**, *26*, 5468–5476. [CrossRef]
167. Magupalli, V.G.; Negro, R.; Tian, Y.; Hauenstein, A.V.; Di Caprio, G.; Skillern, W.; Deng, Q.; Orning, P.; Alam, H.B.; Maliga, Z.; et al. HDAC6 mediates an aggresome-like mechanism for NLRP3 and pyrin inflammasome activation. *Science* **2020**, *369*, eaas8995. [CrossRef]
168. Lewis, R.A.; Soter, N.A.; Diamond, P.T.; Austen, K.F.; Oates, J.A.; Roberts, L.J., 2nd. Prostaglandin D2 generation after activation of rat and human mast cells with anti-IgE. *J. Immunol.* **1982**, *129*, 1627–1631. [PubMed]
169. Aritake, K.; Kado, Y.; Inoue, T.; Miyano, M.; Urade, Y. Structural and functional characterization of HQL-79, an orally selective inhibitor of human hematopoietic prostaglandin D synthase. *J. Biol. Chem.* **2006**, *281*, 15277–15286. [CrossRef]
170. Platten, M.; Wick, W.; Van den Eynde, B.J. Tryptophan catabolism in cancer: Beyond IDO and tryptophan depletion. *Cancer Res.* **2012**, *72*, 5435–5440. [CrossRef]
171. Godin-Ethier, J.; Hanafi, L.A.; Piccirillo, C.A.; Lapointe, R. Indoleamine 2,3-dioxygenase expression in human cancers: Clinical and immunologic perspectives. *Clin. Cancer Res.* **2011**, *17*, 6985–6991. [CrossRef] [PubMed]
172. Zhao, T.; Alam, H.B.; Liu, B.; Bronson, R.T.; Nikolian, V.C.; Wu, E.; Chong, W.; Li, Y. Selective Inhibition of SIRT2 Improves Outcomes in a Lethal Septic Model. *Curr. Mol. Med.* **2015**, *15*, 634–641. [CrossRef]
173. Eskandarian, H.A.; Impens, F.; Nahori, M.A.; Soubigou, G.; Coppée, J.Y.; Cossart, P.; Hamon, M.A. A role for SIRT2-dependent histone H3K18 deacetylation in bacterial infection. *Science* **2013**, *341*, 1238858. [CrossRef]
174. Park, S.H.; Zhu, Y.; Ozden, O.; Kim, H.S.; Jiang, H.; Deng, C.X.; Gius, D.; Vassilopoulos, A. SIRT2 is a tumor suppressor that connects aging, acetylome, cell cycle signaling, and carcinogenesis. *Transl. Cancer Res.* **2012**, *1*, 15–21. [PubMed]
175. Donmez, G.; Outeiro, T.F. SIRT1 and SIRT2: Emerging targets in neurodegeneration. *EMBO Mol. Med.* **2013**, *5*, 344–352. [CrossRef]
176. Kim, H.S.; Vassilopoulos, A.; Wang, R.H.; Lahusen, T.; Xiao, Z.; Xu, X.; Li, C.; Veenstra, T.D.; Li, B.; Yu, H.; et al. SIRT2 maintains genome integrity and suppresses tumorigenesis through regulating APC/C activity. *Cancer Cell* **2011**, *20*, 487–499. [CrossRef] [PubMed]
177. Yang, M.H.; Laurent, G.; Bause, A.S.; Spang, R.; German, N.; Haigis, M.C.; Haigis, K.M. HDAC6 and SIRT2 regulate the acetylation state and oncogenic activity of mutant K-RAS. *Mol. Cancer Res.* **2013**, *11*, 1072–1077. [CrossRef]
178. Humphreys, P.G.; Bamborough, P.; Chung, C.W.; Craggs, P.D.; Gordon, L.; Grandi, P.; Hayhow, T.G.; Hussain, J.; Jones, K.L.; Lindon, M.; et al. Discovery of a Potent, Cell Penetrant, and Selective p300/CBP-Associated Factor (PCAF)/General Control Nonderepressible 5 (GCN5) Bromodomain Chemical Probe. *J. Med. Chem.* **2017**, *60*, 695–709. [CrossRef]
179. Humphries, F.; Yang, S.; Wang, B.; Moynagh, P.N. RIP kinases: Key decision makers in cell death and innate immunity. *Cell Death Differ.* **2015**, *22*, 225–236. [CrossRef] [PubMed]
180. Schuster, S.; Feldstein, A.E. NASH: Novel therapeutic strategies targeting ASK1 in NASH. *Nat. Rev. Gastroenterol. Hepatol.* **2017**, *14*, 329–330. [CrossRef]
181. Schuster-Gaul, S.; Geisler, L.J.; McGeough, M.D.; Johnson, C.D.; Zagorska, A.; Li, L.; Wree, A.; Barry, V.; Mikaelian, I.; Jih, L.J.; et al. ASK1 inhibition reduces cell death and hepatic fibrosis in an Nlrp3 mutant liver injury model. *JCI Insight* **2020**, *5*, e123294. [CrossRef] [PubMed]
182. Tesch, G.H.; Ma, F.Y.; Nikolic-Paterson, D.J. ASK1: A new therapeutic target for kidney disease. *Am. J. Physiol. Renal Physiol.* **2016**, *311*, F373–F381. [CrossRef] [PubMed]
183. Nelson, C.H.; Etchevers, K.; Yi, S.; Breckenridge, D.; Hepner, M.; Patel, U.; Ling, J.; Mathias, A. Pharmacokinetics, Safety, and Tolerability of Selonsertib, an Apoptosis Signal-Regulating Kinase 1 (ASK1) Inhibitor, Following First-in-Human Single and Multiple Ascending Doses in Healthy Subjects. *Clin. Pharmacokinet.* **2020**, *59*, 1109–1117. [CrossRef]
184. Hof, P.R.; Morrison, J.H. The aging brain: Morphomolecular senescence of cortical circuits. *Trends Neurosci.* **2004**, *27*, 607–613. [CrossRef]
185. Chu, T.T.; Gao, N.; Li, Q.Q.; Chen, P.G.; Yang, X.F.; Chen, Y.X.; Zhao, Y.F.; Li, Y.M. Specific Knockdown of Endogenous Tau Protein by Peptide-Directed Ubiquitin-Proteasome Degradation. *Cell Chem. Biol.* **2016**, *23*, 453–461. [CrossRef] [PubMed]
186. Lu, M.; Liu, T.; Jiao, Q.; Ji, J.; Tao, M.; Liu, Y.; You, Q.; Jiang, Z. Discovery of a Keap1-dependent peptide PROTAC to knockdown Tau by ubiquitination-proteasome degradation pathway. *Eur. J. Med. Chem.* **2018**, *146*, 251–259. [CrossRef]
187. Kargbo, R.B. Treatment of Alzheimer's by PROTAC-Tau Protein Degradation. *ACS Med. Chem. Lett.* **2019**, *10*, 699–700. [CrossRef] [PubMed]
188. Silva, M.C.; Ferguson, F.M.; Cai, Q.; Donovan, K.A.; Nandi, G.; Patnaik, D.; Zhang, T.; Huang, H.T.; Lucente, D.E.; Dickerson, B.C.; et al. Targeted degradation of aberrant tau in frontotemporal dementia patient-derived neuronal cell models. *Elife* **2019**, *8*, e45457. [CrossRef] [PubMed]
189. Wang, W.; Zhou, Q.; Jiang, T.; Li, S.; Ye, J.; Zheng, J.; Wang, X.; Liu, Y.; Deng, M.; Ke, D.; et al. A novel small-molecule PROTAC selectively promotes tau clearance to improve cognitive functions in Alzheimer-like models. *Theranostics* **2021**, *11*, 5279–5295. [CrossRef]

190. Jiang, X.; Zhou, J.; Wang, Y.; Liu, X.; Xu, K.; Xu, J.; Feng, F.; Sun, H. PROTACs suppression of GSK-3β, a crucial kinase in neurodegenerative diseases. *Eur. J. Med. Chem.* **2021**, *210*, 112949. [CrossRef]
191. Tomoshige, S.; Nomura, S.; Ohgane, K.; Hashimoto, Y.; Ishikawa, M. Discovery of Small Molecules that Induce the Degradation of Huntingtin. *Angew. Chem. Int. Ed. Engl.* **2017**, *56*, 11530–11533. [CrossRef]
192. Tomoshige, S.; Nomura, S.; Ohgane, K.; Hashimoto, Y.; Ishikawa, M. Degradation of huntingtin mediated by a hybrid molecule composed of IAP antagonist linked to phenyldiazenyl benzothiazole derivative. *Bioorg. Med. Chem. Lett.* **2018**, *28*, 707–710. [CrossRef] [PubMed]
193. Kargbo, R.B. PROTAC Compounds Targeting α-Synuclein Protein for Treating Neurogenerative Disorders: Alzheimer's and Parkinson's Diseases. *ACS Med. Chem. Lett.* **2020**, *11*, 1086–1087. [CrossRef] [PubMed]
194. Embi, N.; Rylatt, D.B.; Cohen, P. Glycogen synthase kinase-3 from rabbit skeletal muscle. Separation from cyclic-AMP-dependent protein kinase and phosphorylase kinase. *Eur. J. Biochem.* **1980**, *107*, 519–527. [CrossRef]
195. L'Episcopo, F.; Drouin-Ouellet, J.; Tirolo, C.; Pulvirenti, A.; Giugno, R.; Testa, N.; Caniglia, S.; Serapide, M.F.; Cisbani, G.; Barker, R.A.; et al. GSK-3β-induced Tau pathology drives hippocampal neuronal cell death in Huntington's disease: Involvement of astrocyte-neuron interactions. *Cell Death Dis.* **2016**, *7*, e2206. [CrossRef]
196. Phiel, C.J.; Wilson, C.A.; Lee, V.M.; Klein, P.S. GSK-3alpha regulates production of Alzheimer's disease amyloid-beta peptides. *Nature* **2003**, *423*, 435–439. [CrossRef]
197. Sirerol-Piquer, M.; Gomez-Ramos, P.; Hernández, F.; Perez, M.; Morán, M.A.; Fuster-Matanzo, A.; Lucas, J.J.; Avila, J.; García-Verdugo, J.M. GSK3β overexpression induces neuronal death and a depletion of the neurogenic niches in the dentate gyrus. *Hippocampus* **2011**, *21*, 910–922. [CrossRef]
198. Reiner, A.; Dragatsis, I.; Dietrich, P. Genetics and neuropathology of Huntington's disease. *Int. Rev. Neurobiol.* **2011**, *98*, 325–372. [PubMed]
199. Ozansoy, M.; Başak, A.N. The central theme of Parkinson's disease: α-synuclein. *Mol. Neurobiol.* **2013**, *47*, 460–465. [CrossRef]
200. Baigent, C.; Keech, A.; Kearney, P.M.; Blackwell, L.; Buck, G.; Pollicino, C.; Kirby, A.; Sourjina, T.; Peto, R.; Collins, R.; et al. Efficacy and safety of cholesterol-lowering treatment: Prospective meta-analysis of data from 90,056 participants in 14 randomised trials of statins. *Lancet* **2005**, *366*, 1267–1278.
201. Stone, N.J.; Robinson, J.G.; Lichtenstein, A.H.; Bairey Merz, C.N.; Blum, C.B.; Eckel, R.H.; Goldberg, A.C.; Gordon, D.; Levy, D.; Lloyd-Jones, D.M.; et al. 2013 ACC/AHA guideline on the treatment of blood cholesterol to reduce atherosclerotic cardiovascular risk in adults: A report of the American College of Cardiology/American Heart Association Task Force on Practice Guidelines. *J. Am. Coll. Cardiol.* **2014**, *63*, 2889–2934. [CrossRef] [PubMed]
202. Li, M.X.; Yang, Y.; Zhao, Q.; Wu, Y.; Song, L.; Yang, H.; He, M.; Gao, H.; Song, B.L.; Luo, J.; et al. Degradation versus Inhibition: Development of Proteolysis-Targeting Chimeras for Overcoming Statin-Induced Compensatory Upregulation of 3-Hydroxy-3-methylglutaryl Coenzyme A Reductase. *J. Med. Chem.* **2020**, *63*, 4908–4928. [CrossRef]
203. Luo, G.; Li, Z.; Lin, X.; Li, X.; Chen, Y.; Xi, K.; Xiao, M.; Wei, H.; Zhu, L.; Xiang, H. Discovery of an orally active VHL-recruiting PROTAC that achieves robust HMGCR degradation and potent hypolipidemic activity *in vivo*. *Acta Pharm. Sin. B* **2021**, *11*, 1300–1314. [CrossRef]
204. Choo, Q.L.; Kuo, G.; Weiner, A.J.; Overby, L.R.; Bradley, D.W.; Houghton, M. Isolation of a cDNA clone derived from a blood-borne non-A, non-B viral hepatitis genome. *Science* **1989**, *244*, 359–362. [CrossRef] [PubMed]
205. Schulze zur Wiesch, J.; Schmitz, H.; Borowski, E.; Borowski, P. The proteins of the Hepatitis C virus: Their features and interactions with intracellular protein phosphorylation. *Arch. Virol.* **2003**, *148*, 1247–1267. [CrossRef]
206. De Wispelaere, M.; Du, G.; Donovan, K.A.; Zhang, T.; Eleuteri, N.A.; Yuan, J.C.; Kalabathula, J.; Nowak, R.P.; Fischer, E.S.; Gray, N.S.; et al. Small molecule degraders of the hepatitis C virus protease reduce susceptibility to resistance mutations. *Nat. Commun.* **2019**, *10*, 3468. [CrossRef] [PubMed]
207. Ghosh, A.K.; Brindisi, M.; Shahabi, D.; Chapman, M.E.; Mesecar, A.D. Drug Development and Medicinal Chemistry Efforts toward SARS-Coronavirus and COVID-19 Therapeutics. *ChemMedChem* **2020**, *15*, 907–932. [CrossRef]
208. Yin, W.; Mao, C.; Luan, X.; Shen, D.D.; Shen, Q.; Su, H.; Wang, X.; Zhou, F.; Zhao, W.; Gao, M.; et al. Structural basis for inhibition of the RNA-dependent RNA polymerase from SARS-CoV-2 by remdesivir. *Science* **2020**, *368*, 1499–1504. [CrossRef]
209. Boras, B.; Jones, R.M.; Anson, B.J.; Arenson, D.; Aschenbrenner, L.; Bakowski, M.A.; Beutler, N.; Binder, J.; Chen, E.; Eng, H.; et al. Discovery of a Novel Inhibitor of Coronavirus 3CL Protease for the Potential Treatment of COVID-19. *bioRxiv* **2021**. [CrossRef]

International Journal of *Molecular Sciences*

MDPI

Article

Antitumor Effects of a New Retinoate of the Fungal Cytotoxin Illudin M in Brain Tumor Models

Benedikt Linder [1,†], Miroslava Zoldakova [2,†], Zsuzsanna Kornyei [3], Leonhard H. F. Köhler [2], Sebastian Seibt [2], Dominic Menger [1], André Wetzel [2], Emília Madarász [3], Rainer Schobert [2], Donat Kögel [1] and Bernhard Biersack [2,*]

1 Experimental Neurosurgery, Frankfurt University Hospital, Theodor-Stern-Kai 7, 60590 Frankfurt am Main, Germany
2 Organic Chemistry 1, University of Bayreuth, Universitätsstrasse 30, 95440 Bayreuth, Germany
3 Laboratory of Cellular and Developmental Neurobiology, Institute of Experimental Medicine of the Hungarian Academy of Sciences, Szigony utca 43, HU-1083 Budapest, Hungary
* Correspondence: bernhard.biersack@yahoo.com or bernhard.biersack@uni-bayreuth.de; Tel.: +49-921-552673
† These authors contributed equally to this work.

Abstract: While the fungal metabolite illudin M (**1**) is indiscriminately cytotoxic in cancer and non-malignant cells, its retinoate **2** showed a greater selectivity for the former, especially in a cerebral context. Illudin M killed malignant glioma cells as well as primary neurons and astrocytes at similarly low concentrations and destroyed their microtubule and glial fibrillary acidic protein (GFAP) networks. In contrast, the ester **2** was distinctly more cytotoxic in highly dedifferentiated U87 glioma cells than in neurons, which were even stimulated to enhanced growth. This was also observed in co-cultures of neurons with U87 cells where conjugate **2** eventually killed them by induction of differentiation based on the activation of nuclear receptors, which bind to retinoid-responsive elements (RARE). Hence, illudin M retinoate **2** appears to be a promising drug candidate.

Keywords: illudin M; retinoic acid; neuronal cells; brain tumors; anticancer agents

Citation: Linder, B.; Zoldakova, M.; Kornyei, Z.; Köhler, L.H.F.; Seibt, S.; Menger, D.; Wetzel, A.; Madarász, E.; Schobert, R.; Kögel, D.; et al. Antitumor Effects of a New Retinoate of the Fungal Cytotoxin Illudin M in Brain Tumor Models. *Int. J. Mol. Sci.* **2022**, *23*, 9056. https://doi.org/10.3390/ijms23169056

Academic Editor: Tullio Florio

Received: 20 July 2022
Accepted: 11 August 2022
Published: 13 August 2022

1. Introduction

The sesquiterpene illudin M (**1**) was first isolated from the culture broth of *Omphalotus olearius* mushrooms [1]. After a pre-activating reduction of its enone by NADPH-dependent oxido-reductases or glutathione (Nu^1), **1** can alkylate DNA, RNA and proteins (Nu^2) via opening of the spirocyclopropane and thereby induce apoptotic cell death [2]. Although illudin M, similar to its congener illudin S and the related semisynthetic irofulven, is highly efficacious against cancer cells, its indiscriminate toxicity has prevented clinical applications [3,4]. Occasionally, illudin derivatives with reduced toxicity and improved therapeutic indices were reported [5]. Irofulven even underwent several phase II clinical trials but proved largely ineffective, except for some prostate and pancreatic cancers [6].

Retinoic acid (RA) induces differentiation of various types of stem cells, including cancer and neural stem cells, and a retardation of cancer cell proliferation [7,8]. These effects are mediated by nuclear retinoic acid receptors, usually consisting of heterodimers between RAR and RXR proteins, and sometimes by interfering with estrogen signaling [9,10]. Retinoids were particularly efficacious against various human carcinomas when applied as part of combination regimens with standard drugs such as cisplatin or with HDAC and DNA methyltransferase inhibitors [11,12]. Here, we report on the selective impact of a new illudin M retinoate **2** on glioma and stem-cell rich astrocytoma cells, when compared with normal neurons and astrocytes.

2. Results

Retinoate **2** was prepared by esterification of all-*trans* retinoic acid with **1** under Yamaguchi conditions (Scheme 1) [13]. The antiproliferative activities of compounds **1** and

2 were first evaluated by cell viability (MTT) assays against a panel of seven cancer and three non-malignant cell types. These comprised human 518A2 melanoma, HL-60 leukemia, the multi-drug resistant carcinomas KBv1^{+Vbl} cervix and MCF-7^{+Topo} breast, the HT-29 colon carcinoma, the cancer stem cell-rich rat C6 astrocytoma, human U87 glioma as well as non-malignant primary mouse astrocytes and neurons, and the NE-4C neuroectodermal stem cells isolated from the fore- and midbrain vesicles of p53-deficient 9-day-old mouse embryos [14]. Generally, the retinoate **2** was 100-fold less efficacious than **1** against the non-neural cancer cell lines, with the exception of MCF-7^{+Topo}, which responded equally and well to both compounds (Table 1). This cell line is estrogen-dependent and also overexpresses ABC-transporters of the BCRP (breast cancer resistance protein) type, which are assumed to contribute to cancer stem cell resistance and inefficient trespassing of some drugs through the blood brain barrier [15,16]. Against the brain-derived cells, e.g., neuronal progenitors (NE-4C), neurons, astrocytes (Table 2), and glioma lines C6, U87, U251, and MZ-54 (Table 1), **1** again showed similar activities with IC_{50} (48 h) values ranging from 50 to 400 nM. In contrast, the retinoate **2**, while being generally less active than **1**, displayed a significantly greater cytotoxicity against the human U87, U251 and MZ-54 glioblastoma cells than against normal neurons and astrocytes (Tables 1 and 2). The NE-4C neuronal progenitor cells were also sensitive to treatment with compound **2** (Table 1).

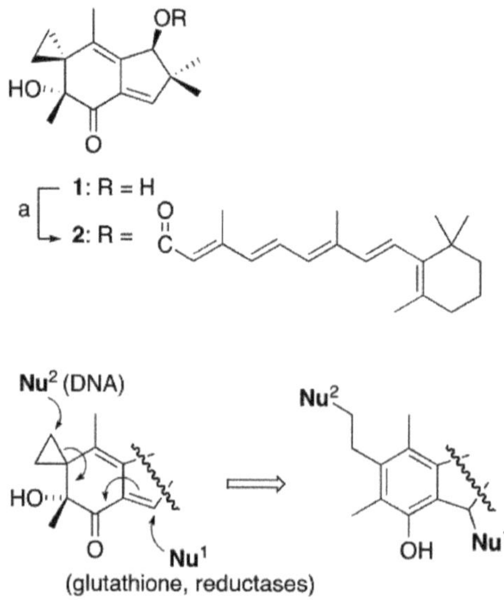

Scheme 1. Synthesis of the retinoate **2** of illudin M (**1**) and mechanism of action. Reagents and conditions: (a) retinoic acid, $C_6H_2Cl_3COCl$, DMF, NEt_3, then **1**, DMAP, toluene, 16 h, r.t., 41%.

Compound **1** is a natural prodrug and undergoes activation by reaction with bionucleophiles such as glutathione forming an activated cyclopropane, which readily reacts with other bionucleophiles, for example, nucleobases of DNA (Scheme 1). The reactivity of the illudin M scaffold was significantly reduced by esterification, explaining the higher antiproliferative activity and toxicity of **1** when compared with ester conjugates [13]. In order to investigate whether retinoic conjugate **2** is also less reactive than **1**, the fading of the enone band of compounds **1** and **2** was monitored at 330 nm in the presence of glutathione, showing a stabilization of conjugate **2** compared with **1** (Figure 1). The stability of conjugate **2** in cell medium was also investigated using HPLC techniques, and **2** showed a high stability under these conditions (cf. Table S1, Supporting Information). A cleavage of the ester bond over the time of observation can be ruled out as neither free **1** nor free all-*trans* retinoic acid were detectable by HPLC. Thus, the anticancer effects of **2** are not based on any **1** formed upon degradation of **2**, but on the intact molecule.

Table 1. Inhibitory concentrations IC_{50} (µM) [1] of **1** and **2** in cancer cells including several GBM cells.

Compd./Cell Line	1		2	
	24 h	48 h	24 h	48 h
518A2	0.04 ± 0.01	0.02 ± 0.01	5.0 ± 0.3	4.0 ± 0.7
HL60	0.007 ± 0.002	0.001 ± 0.000	6.8 ± 2.8	4.5 ± 1.1
$KBv1^{+Vbl}$	0.003 ± 0.001	0.002 ± 0.000	1.6 ± 0.4	1.2 ± 0.1
$MCF\text{-}7^{+Topo}$	0.35 ± 0.13	0.05 ± 0.01	2.5 ± 0.6	0.08 ± 0.01
HT-29	1.5 ± 0.6	0.06 ± 0.01	38 ± 4	4.0 ± 1.3
C6	1.0 ± 0.3	0.40 ± 0.06	40 ± 4	12 ± 1.3
U87	0.63 ± 0.04	0.13 ± 0.03	37 ± 3	2.7 ± 0.7
MZ-54	0.55 ± 0.16	0.06 ± 0.01	n.d.	5.48 ± 1.2
U251	0.3 ± 0.05	0.12 ± 0.02	n.d.	4.51 ± 1.3

[1] Values are derived from dose–response curves obtained by measuring the percentage of viable cells relative to untreated controls after 24/48 h exposure of test compounds using an MTT assay. Values represent means of four independent experiments. n.d., not determined.

Table 2. Inhibitory concentrations IC_{50} (µM) [1] of **1** and **2** in non-malignant brain-derived cells.

Compd./Cell Line	1		2	
	24 h	48 h	24 h	48 h
neurons	0.60 ± 0.12	0.19 ± 0.04	26 ± 1	14 ± 2
astrocytes	1.5 ± 0.4	0.20 ± 0.03	45 ± 2	33 ± 4
NE-4C	0.40 ± 0.10	0.05 ± 0.02	7.0 ± 3.2	2.0 ± 1.2

[1] Values are derived from dose–response curves obtained by measuring the percentage of viable cells relative to untreated controls after 24/48 h exposure of test compounds using an MTT assay. Values represent means of four independent experiments.

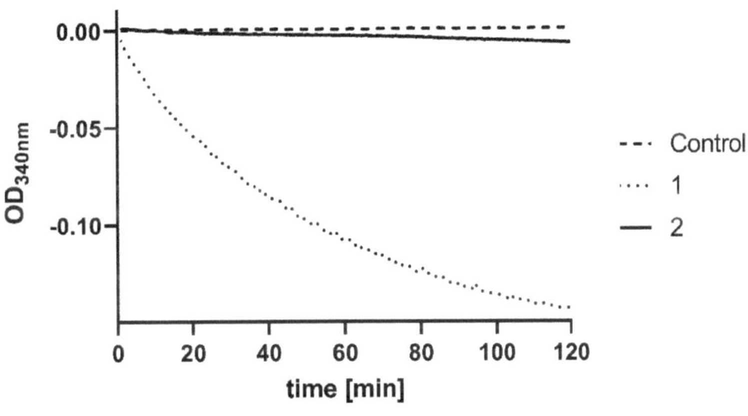

Figure 1. Time-resolved measurement of the absorption at 340 nm (enone group) of 200 µM **1**, **2** or solvent (DMSO) mixed with 6.5 mM glutathione in phosphate-buffered saline (PBS). Data are means ± SD of at least four independent measurements every 30 s.

More typical of retinoic acid is its contribution to cell differentiation during vertebrate embryonic development by interacting with nuclear receptors, which bind to specific retinoid-responsive elements (RARE) within the promoters of subordinate genes, activating their transcription [17–20]. To verify that retinoate **2** also binds to RARE, we used a reporter assay with F9 teratocarcinoma cells featuring a RARE located within the *cis*-acting regulatory sequences of the human retinoic acid receptor alpha gene and immediately upstream of the *E. coli lacZ* genes, which thus acquire retinoid responsivity [21]. *LacZ* expression in cells treated with various concentrations of **2** was then quantitated by densitometric analysis of β-galactosidase activity visualized by X-Gal staining [22]. Figure 2 shows that compound **2** bound to the F9 RARE-*lacZ* reporter cells in a concentration-dependent manner.

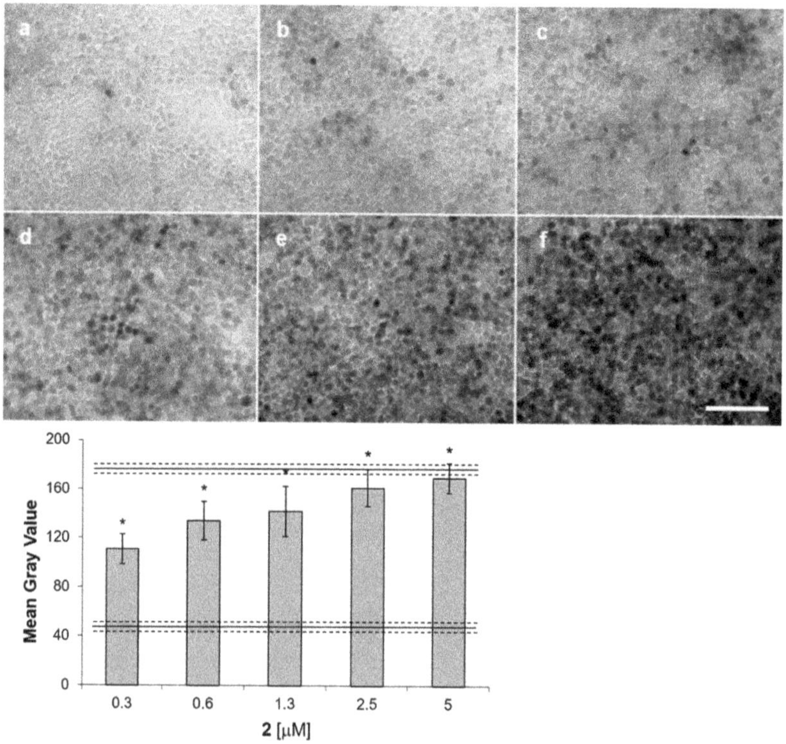

Figure 2. F9 RARE-*lacZ* reporter cells show concentration-dependent activation upon treatment with **2** for 20 h. Top: bright-field images of cells treated with **2** ((**a**): 0 μM, (**b**): 0.3 μM, (**c**): 0.6 μM, (**d**): 1.3 μM, (**e**): 2.5 μM, (**f**): 5.0 μM), and β-galactosidase activity was visualized by X-Gal staining (scale bar: 100 μm). Bottom: densitometric image analyses (mean ± SD); a greater gray value means stronger RARE binding. The threshold lines at 45 and 175 represent the control and the effect of 10 nM all-*trans* retinoic acid. * $p < 0.001$, Student's t test, groups treated with **2** were compared to control, $n = 5$.

While binding to the RARE of these reporter cells, retinoate **2**, unlike all-*trans* retinoic acid itself, did not induce neuronal differentiation of NE-4C neural stem cells even at concentrations as high as 1 μM. NE-4C cells and their differentiated progeny neurons can be immunohistochemically distinguished by the presence of β-tubulin III only in the latter (Figure 3). While short term (48 h) **1** application (1 μM) did not completely disorganize neuronal networks, it affected astrocytes within neuronal cultures, resulting in altered expression/localization of glial fibrillary acidic protein (GFAP), a major cytoskeletal element of glial cells responsible for maintaining mechanical cell stability [23]. As illustrated by fluorescent immunocytochemistry, GFAP filaments were markedly pitted after 48 h exposure to **1**, while retinoate **2** did not cause similar alterations to astrocytes (Figure 4). Moreover, neuronal cells developed normally when exposed to 1 μM retinoate **2** for as long as 3, 6 or 8 days (Figure 5).

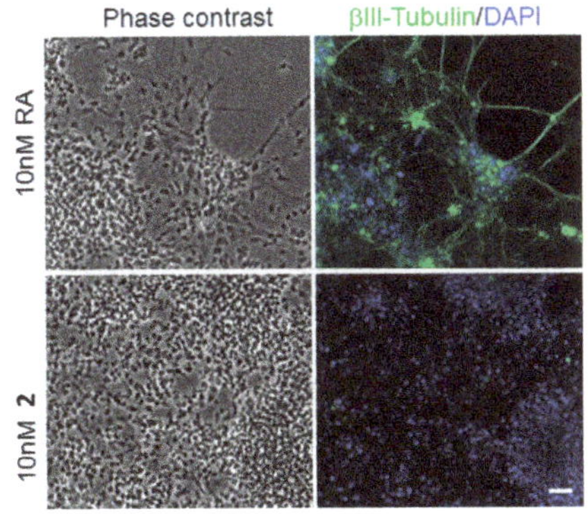

Figure 3. Upper row: NE-4C neural stem cells differentiate to neurons upon treatment with 10 nM of all-*trans* retinoic acid (RA). Bottom row: retinoate **2** (10 nM) did not elicit differentiation of NE-4C stem cells. Right column shows immunohistographs after staining for neuron specific ß-tubulin III (green) and for nuclei with DAPI (blue). Scale bar: 50 μm.

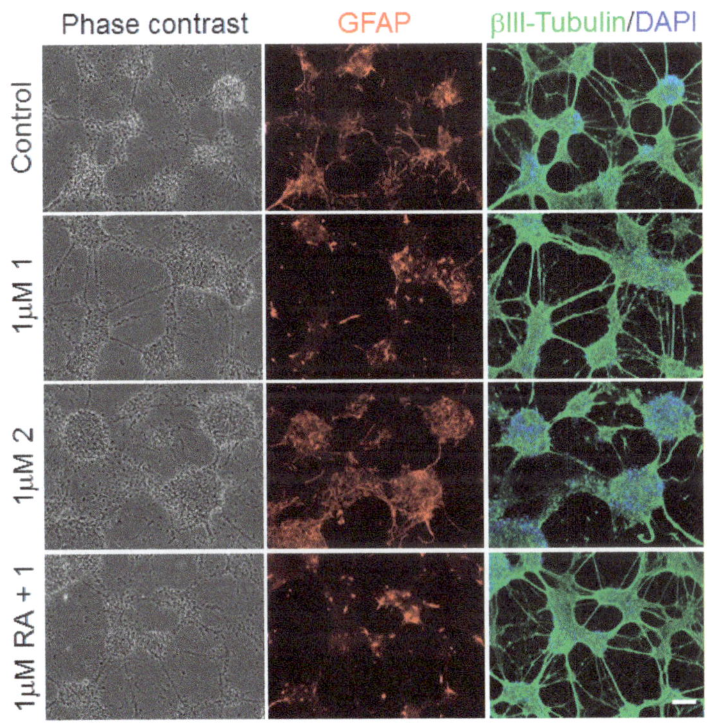

Figure 4. Immunostaining of primary neurons derived from E14, 5 forebrains treated with 1 μM of **1**, **2** or all-*trans* retinoic acid (RA) plus **1**. Blue: nuclei stained with DAPI; green: tubulin stained with βIII-tubulin specific antibody; red: GFAP stained with α-GFAP antibody. Fluorescence microscope Zeiss Axiovert 200 M, 63× objective, 1024 × 1024 res. Scale bar: 50 μm.

In contrast to astrocytes and neuronal cells, cultures of U87 glioma cells were far more sensitive to retinoate **2** at concentrations between 1 and 10 μM, undergoing rapid apoptosis (Figure 6). Further studies on the pro-apoptotic effects of **1** and **2** were carried out with

MZ-54 glioblastoma cells. Annexin V/propidium iodide (PI) flow cytometry of MZ-54 cells treated with **1** or **2** revealed a dose- and time-dependent increase in cell death (Figure 7).

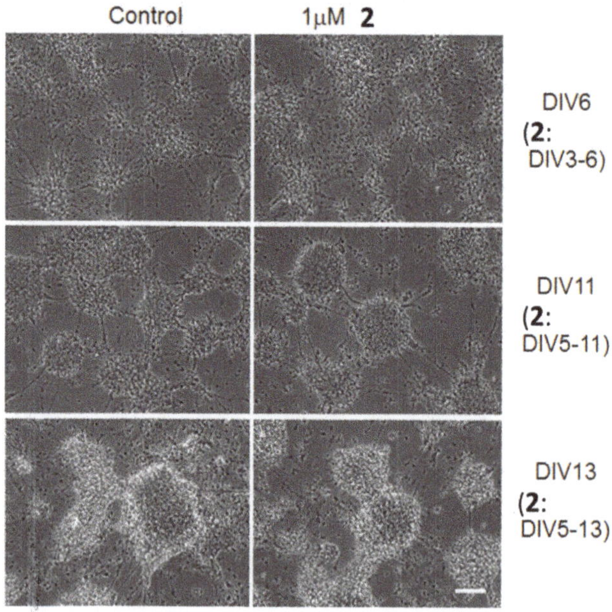

Figure 5. Micrographs of cultures of mouse embryonal neuronal cells untreated (control) or treated with 1 μM of **2**; DIV, days in vitro. Scale bar: 100 μm.

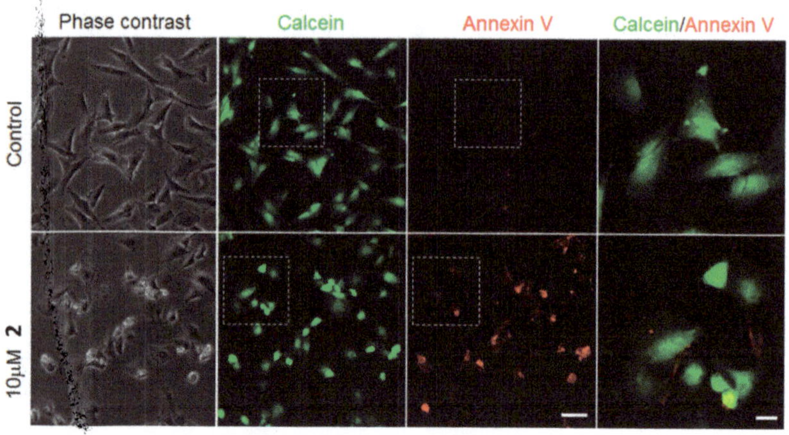

Figure 6. U87 glioma cells undergo apoptosis upon 48 h treatment with 10 μM of **2**. Cells in the early stages of apoptosis show calcein (green)/Annexin V-Cy3 (red) double labeling. Scale bars: 50 μm; 20 μm for inserts.

The selectivity of illudins **1** and **2** for U87 glioblastoma cells versus normal neuronal cells was evaluated by applying them to co-cultures of these cells serving as a surrogate brain tumor model. To this end, isolated murine primary neurons were grown on IBIDI dishes and after 5 days and the establishment of neuronal aggregates, the glioma U87 cells (~ 25,000 cells) were added to the neuronal culture [24]. After 12 h, **1** and **2** were applied in 1 μM concentrations and the response of the different cells was monitored over a period between 15 and 48 h post incubation (Figure 8). In the control samples, the number of proliferating U87 cells increased with time and overgrew the diminishing number of differentiating neurons. In contrast, in the samples treated with **1**, both the neurons and the U87 cells had undergone considerable cellular fragmentation as early as 15 h after treatment, leaving merely cell debris after 48 h. Co-cultures incubated with retinoate **2** featured healthy and compact neuronal aggregates (not quantified numerically) after 15 h compared with control cells. After 48 h exposure to **2**, many neurons were still present in the culture.

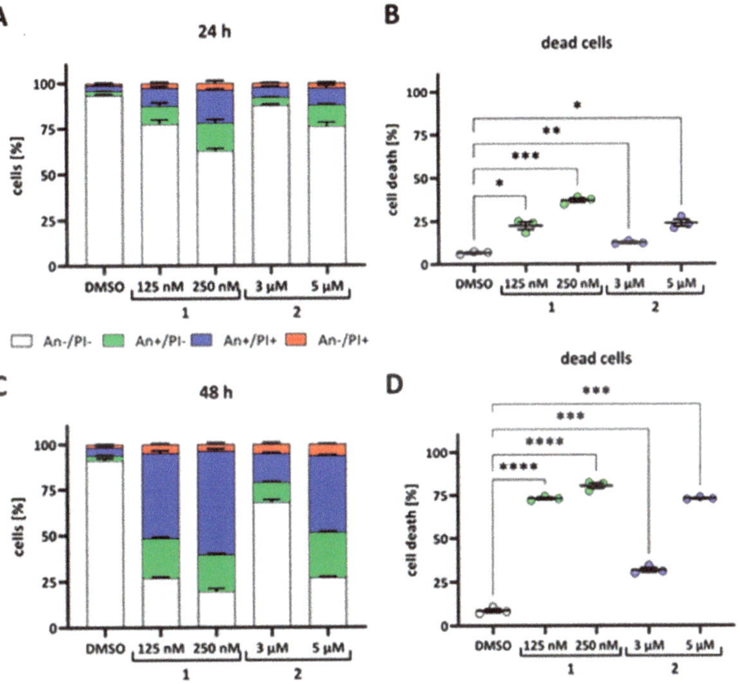

Figure 7. (**A–D**). Annexin V/PI flow cytometry of MZ–54 cells upon treatment with **1** or **2**, mean + SEM; Brown–Forsythe ANOVA test with Dunnett's T3 multiple comparison test (compared with DMSO control); * < 0.05; ** < 0.01; *** < 0.001; **** < 0.0001.

For further investigation of drug efficacy and selectivity, an **ex vivo** tumor growth assay employing organotypic brain slice cultures of adult murine brains were used. For this purpose, GFP-expressing NCH644 (NCH644^{GFP+}) glioma stem-like cells were transplanted onto these slices, and after 1 day, the treatment was started and renewed three times per week, similar to our previous studies [25,26]. These cultures are suitable ex vivo models, which simulate original brain architecture and the presence of vessels, including angiogenesis and microvascular proliferation as well as specific homing of GBM cells toward those vessels, while requiring much fewer laboratory animals when compared with common in vivo animal drug tests [27–29]. Both **1** and **2** showed antitumor activity in the organotypic brain cancer model using NCH644^{GFP+} at a dose of 5 µM (Figure 9). The activity of **2** was significantly reduced at doses of 0.1 and 0.5 µM, which is in line with results from the in vitro MTT experiments, hinting at a considerable stability of the ester conjugate **2** under these conditions (Figure 10).

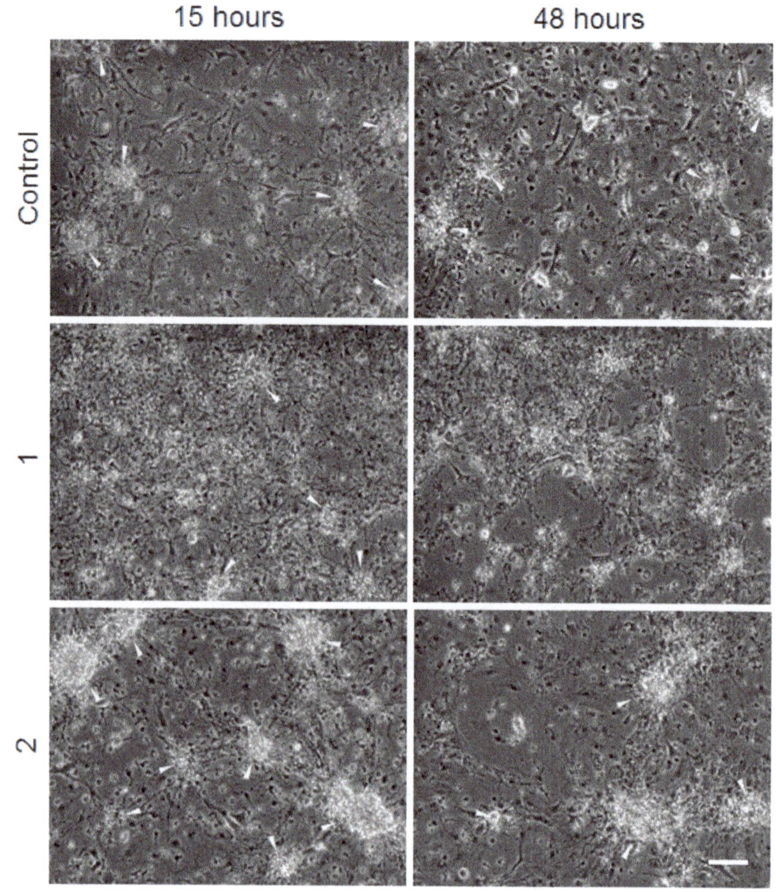

Figure 8. Co-culture of neuronal cells and U87 glioma cells in the absence or presence of 1 μM of **1** or **2**. The neuronal aggregates are marked by arrowheads. Scale bar: 100 μm.

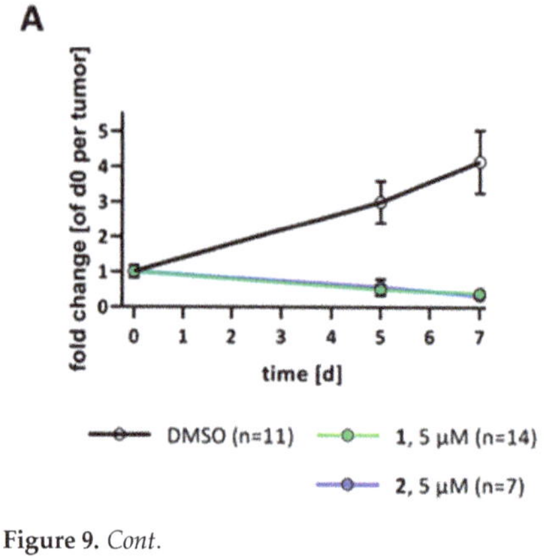

Figure 9. *Cont.*

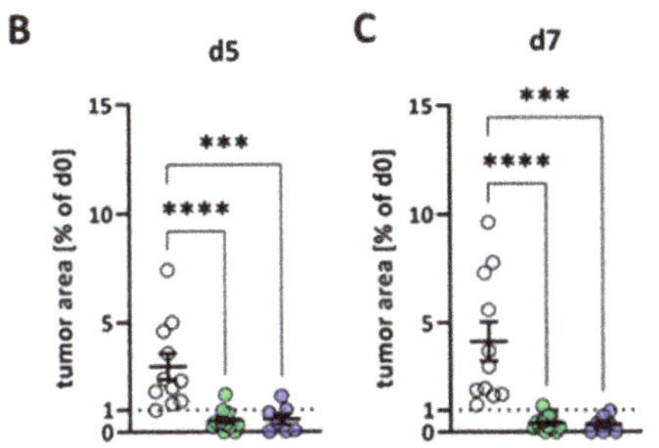

Figure 9. Ex vivo tumor growth assay using organotypic brain slice cultures in the absence (DMSO) or presence of 5 μM of **1** or **2**. (**A**) Growth curves of the NCH644^{GFP+} tumors over time after treatment with solvent (DMSO, black line) or 5 μM **1** (green line) or **2** (blue line) normalized to the tumor size one day after transplantation (d0) depicted as mean +/− SEM. (**B**,**C**) Point plots of the data summarized in (**A**) after treatment for (**B**) 5 days or (**C**) 7 days). Dashed line on y = 1 display the original tumor size. ***: $p < 0.001$; ****: $p < 0.0001$; two-Way ANOVA with Tukey's multiple comparisons test (GraphPad Prism 7).

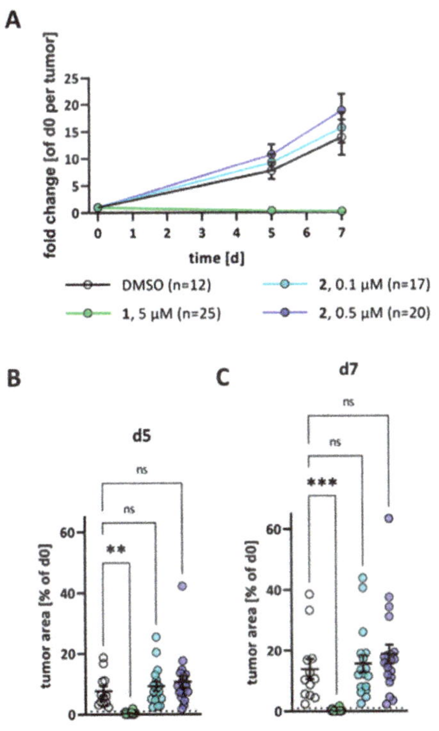

Figure 10. Ex vivo tumor growth assay using organotypic brain slice cultures in the absence (DMSO) or presence of 5 μM of **1** or 0.1 μM or 0.5 μM **2**. (**A**) Growth curves of the NCH644^{GFP+} tumors over time after treatment with solvent (DMSO, black line) or 5 μM **1** (green line), 0.1 μM (light blue line) or 0.5 μM **2** (blue line) normalized to the tumor size one day after transplantation (d0) depicted as mean +/− SEM. (**B**,**C**) Point plots of the data summarized in (**A**) after treatment for (**B**) 5 days or (**C**) 7 days). Dashed line on y = 1 display the original tumor size. **: $p < 0.001$; ***: $p < 0.0001$; two-way ANOVA with Tukey's multiple comparisons test (GraphPad Prism 7).

3. Discussion

Illudin M (**1**) is a fungal product with a unique DNA-damaging mechanism. However, its high toxicity toward laboratory animals stopped its development as an anticancer agent at an early stage [5]. Esterification of the secondary alcohol of **1** led to compounds with reduced toxicity and increased selectivity [13]. Conjugation of **1** to *all*-trans retinoic acid via ester bonding led to remarkable results. In vitro MTT experiments revealed that MCF-7^{+Topo} breast carcinoma cells responded well to ester **2**. These cells are multidrug-resistant cells generated by the exposure to topotecan. In addition to their hormone-sensitive character, they overexpress the BCRP (breast cancer resistance protein) transporter, which plays a role for cancer stem cell resistance, and impaired crossing of drugs through the blood brain barrier [15,16]. The strong time dependence of the activity of ester **2** in the MCF-7^{+Topo} breast carcinoma cells can be a hint at a metabolism of **2** in these cells leading to active illudin M. Similarly, a strong time dependence of compound **2** activity was observed in HT-29 colorectal cancer cells, which are also ABC-transporter expressing multidrug-resistant cells [30]. The multidrug-resistant phenotype is characterized by an increased pH value and alkalinity in the cytoplasm of cancer cells, which might facilitate the hydrolysis of retinoate **2** and the release of **1** from **2** in these cells [31,32]. In addition, U87 glioblastoma cells exhibited strong time-dependent sensitivity to compound **2**. Although intrinsically not multidrug-resistant, ABCG2/BCRP expression can be induced in U87 cells by treatment with anticancer active drugs such as perphenazine and prochlorperazine [33]. If this is also the case for **2**, it remains to be elucidated.

Ester **2** unlike **1** is distinctly more cytotoxic in the cancer stem cell-rich rat C6 astrocytoma cell line than in normal astrocytes. This is somewhat paradoxical since retinoic acid is a known inhibitor of c-Jun N-terminal kinase (JNK), and illudin derivatives such as irofulven were shown to induce apoptosis via activation of JNK and ERK [34–36]. However, C6 astrocytoma cells, unlike normal astrocytes, naturally express cannabinoid CB_1 receptors, which enhance JNK activity upon binding of certain unsaturated fatty acids such as arachidonates, and maybe also of retinoic acid [37,38]. The latter was found to induce apoptosis in C6 glioma cells at relatively high concentrations [39].

The new illudin M retinoate **2** exhibited distinct antiproliferative and apoptosis-inducing effects on various glioblastoma cells. The high selectivity of **2** was confirmed in the co-culture assay using U87 glioma cells and neuronal cells and is corroborated by the high stability of this ester conjugate in cell medium under the conditions applied for the in vitro experiments. Finally, our application of the compound in an organotypic model further underscored the selective targeting of tumor cells within a complex non-transformed environment, since no signs of toxicity could be observed in the tumor-free regions of the brain slices. OTCs are a widely used, 3R-compatible and valuable ex vivo method to recapitulate the physiological brain environment of GBM over an extended time period. This demonstrates vividly the selective antitumor activity of compound **2**. In addition, compound **2** was able to exert its effects via RARE in F9 embryonal carcinoma cells, while it kept its illudin-type properties to a certain extent. This observation indicates an excellent interplay of the retinoate and illudin moieties of the conjugate molecule **2** and warrants a more detailed study of its mechanism of action in tumor cells vs. healthy cells, including models allowing to assess/compare drug effects in a more authentic brain environment. Further studies are planned in order to substantiate the drug-like properties of **2** and the potential of **2** as a promising drug candidate against glioma and multidrug-resistant tumors. This approach will also uncover if **2** can effectively cross the blood–brain barrier (BBB). Other illudin derivatives such as irofulven have been shown to enhance survival of intracranial glioma xenografts in mice, suggesting active crossing of the BBB [40].

4. Materials and Methods

4.1. General

Starting materials and reagents were purchased from Sigma-Aldrich (Taufkirchen, Germany). Illudin M was isolated according to published procedures [13]. The following instruments were used: melting points (uncorrected), Gallenkamp (Cambridge, UK); IR spectra, Perkin-Elmer Spectrum One FT-IR spectrophotometer (Rodgau, Germany) with ATR sampling unit; nuclear magnetic resonance spectra, BRUKER Avance 300 spectrometer (Billerica, MA, USA); chemical shifts are given in parts per million (δ) downfield from tetramethylsilane as internal standard; mass spectra, Varian MAT 311A (EI, Palo Alto, CA, USA).

4.2. *Synthesis of Compound* 2

Retinoic acid (100 mg, 0.33 mmol) was suspended in dry DMF (1 mL), and Et_3N (53 μL, 0.36 mmol) and 2,4,6-trichlorobenzoyl chloride (59 μL, 0.36 mmol) were added. The reaction mixture was stirred at room temperature for 20 min. Illudin M (83 mg, 0.33 mmol) and DMAP (81 mg, 0.66 mmol) dissolved in dry toluene (5 mL) were added and the reaction mixture was stirred at room temperature for 16 h. After dilution with ethyl acetate and washing with water, the organic phase was dried over Na_2SO_4 and concentrated in vacuum. The residue was purified by column chromatography (silica gel 60). Yield: 64 mg (0.12 mmol, 36%); yellow oil; R_f = 0.61 (ethyl acetate/*n*-hexane 1:4); ν_{max} (ATR)/cm^{-1}: 3492, 2962, 2928, 2866, 1696, 1606, 1580, 1447, 1360, 1255, 1235, 1139, 1105, 966, 945, 821, 730; ^{1}H-NMR (300 MHz, $CDCl_3$): δ 0.3–0.5 (1 H, m, 9-H^a), 0.8–1.0 (2 H, m, 8-H^a, 9-H^b), 1.00 (6 H, s, 2 × Me), 1.1–1.6 (17 H, m, 4 × Me, 2 × CH_2, 8-H^b), 1.69 (3 H, s, Me), 1.9–2.1 (5 H, m, Me, CH_2), 2.35 (3 H, s, Me), 3.55 (1 H, s, OH), 5.68 (1 H, s, 3-H), 5.77 (1 H, s, 2′-H), 6.0–6.3 (4 H, m, 4′-H, 6′-H, 8′-H, 9′-H), 6.50 (1 H, s, 1-H), 6.9–7.1 (1 H, m, 5′-H); ^{13}C-NMR (75.5 MHz, $CDCl_3$): δ 6.0 (C-8), 8.8 (C-9), 12.9 (7′-Me), 14.0 (3′-Me), 14.5 (4-Me), 19.2 (cyclohexyl-CH_2), 20.8 (2-Me), 21.7 (cyclohexyl-C*Me*), 24.8 (6-Me), 26.8 (2-Me), 29.0 (cyclohexyl-C*Me*$_2$), 31.5 (C-5), 33.1 (cyclohexyl-CH_2), 34.3 (cyclohexyl-CMe_2), 39.6 (cyclohexyl-CH_2), 49.0 (C-2), 76.1 (C-6), 78.1 (C-3), 117.8 (C-2′), 128.9 (C-9′), 129.4 (C-6′), 130.1 (cyclohexyl-CMe), 131.4 (C-5′), 133.6 (C-7a), 134.9, 135.1 (C-3a, C-4), 137.2 (C-4′), 137.7 (cyclohexyl-C=CMe), 140.0 (C-7′), 146.5 (C-1), 153.7 (C-3′), 166.9 (CO_2), 200.3 (CO); *m/z* (EI) 293 (13), 282 (13), 231 (11), 209 (20), 177 (21), 84 (100).

4.3. *Cells and Cell Culture*

518A2 (Department of Radiotherapy, Medical University of Vienna, Vienna, Austria) melanoma, KB-V1^{Vbl} (ACC-149) cervix carcinoma, MCF-7^{Topo} (ACC-115) breast carcinoma, HT29 (ACC-299) colon carcinoma were cultivated in Dulbecco's Modified Eagle Medium (DMEM) supplemented with 10% fetal bovine serum and 1% antibiotic-antimycotic at 37 °C, 5% CO_2, and 95% humidity. To keep MCF-7^{Topo} and KB-V1^{Vbl} cells resistant, the maximum-tolerated doses of topotecan or vinblastine were added to the cell culture medium 24 h after every passage. The human HL60 leukemia cells were obtained from the German National Resource Center for Biological Material (DSMZ), Braunschweig. Neuroectodermal stem cells NE-4C (ATCC No. CRL-2925), primary mouse astrocytes, rat glioma C6 (ATCC No. CCL-107) and human glioma U87 (ATCC No. HTB-14) were used [41].

NE-4C cells derived from the anterior brain vesicles of p53-deficient 9-day-old mouse embryos were maintained in Minimum Essential Medium (MEM; Sigma Aldrich, Taufkirchen, Germany) supplemented with 5% fetal calf serum (FCS; Invitrogen-Gibco, Carlsbad, CA, USA), 4 mM glutamine (Sigma-Aldrich, Taufkirchen, Germany) and 40 ug/mL gentamycin (Chinoin, Budapest, Hungary). For maintenance, subconfluent cultures were regularly split by trypsinization (0.05 *w*/*v*% trypsin in PBS) into poly-L-lysine-coated Petri dishes. Primary astrocytes were isolated from whole brains of neonatal (P0-P3) mice, as described earlier [24]. In brief, meninges were removed, and the tissue pieces were subjected to enzymatic dissociation, using 0.05% *w*/*v* trypsin and 0.05% *w*/*v* DNase for 10 min at room temperature. The cells were plated onto poly-L-lysine coated plastic surfaces and grown in MEM supplemented with 10% fetal calf serum, 4 mM glutamine and 40 ug/mL gentamycin. Primary neuronal cultures were prepared from embryonic (E15–E16) mouse (CD1) forebrains, as described earlier [42]. In brief, cell suspensions were prepared by mechanical dissociation over a nylon mesh with a pore diameter of 40–42 μm. The cells were seeded at 10^5 cells/cm^2 density onto PLL-coated surfaces. The cultures were maintained in MEM supplemented with 5% FCS, 4 mM glutamine, 40 µg/mL gentamicin and 2.5 µg/mL amphotericin B (Sigma). In co-culture experiments, U87 glioma cells were seeded onto established neuronal cultures at DIV3 in a 25,000 cell/cm^2 density.

U251 (formerly known as U-373 MG; ECACC 09063001), Mz54 (CVCL_M406) human glioblastoma cell lines were cultured in DMEM+GlutaMax supplemented with 10% fetal bovine serum and 1% penicillin/streptomycin mixture. The human glioma cell line Mz54 was obtained from the Dept. of Neurosurgery, University Medical Centre, Johannes Gutenberg University Mainz, Germany, where this line was isolated from a recurrent grade IV glioblastoma. U251 and Mz54 cells were used in passages between 10 and 25 after re-authentication and 40 to 55 after culture establishment, respectively. Only mycoplasma-free cell cultures were used.

NCH644^{GFP+} have been described previously and are derived from NCH644, which were kindly provided by Christel Herold-Mende (University Hospital Heidelberg, Germany) and were cultured as free-floating spheres in serum-free medium [34,35,43]. Specifically, the cells were cultured in Neurobasal-A medium (Gibco, Darmstadt, Germany) supplemented with 1× B27 (Gibco), 100 U/mL penicillin, 100 µg/mL streptomycin (P/S, Gibco), 1 × GlutaMAX (Gibco), 20 ng/mL epidermal growth factor (EGF, Peprotech, Hamburg, Germany), 20 ng/mL fibroblast growth factor (FGF, Peprotech) and 2 µg/mL puromycin for GFP-expressing cells.

4.4. Cell-Based Assay

Non-GBM cells (5 × 10^4 cells/mL, 100 µL/well) were grown in 96-well plates for 24 h. Then, they were treated with various concentrations of the test compounds or vehicle (DMSO, or EtOH) for 72 h at 37 °C. After the addition of 12.5 µL of a 0.5% MTT solution in PBS, the cells were incubated for 3 h at 37 °C so that the water-soluble MTT could be converted to formazan crystals. Then, the plates were centrifuged (300× *g*, 5 min, 4 °C), the medium withdrawn, and the formazan dissolved in 25 µL of DMSO containing 10% sodium dodecylsulfate (SDS) and 0.6% acetic acid for at least 2 h at 37 °C. In the case of the HL60 cells, after 2 h, the precipitate of formazan crystals was redissolved in a 10% solution of SDS in DMSO containing 0.6% acetic acid.

Adherently grown GBM cells were seeded at 5 × 10^4 cells/mL and floating spheres at 8 × 10^4 cells/mL in 100 µL/well and incubated as above for 48 or 72 h at 37 °C. Afterward, 10 µL of a 5 mg/mL MTT solution in PBS was added to the cells for 3 h at 37 °C. Then, the spheres were centrifuged shortly to collect the cells on the bottom of the plate, and the medium was removed by careful pipetting for both adherently grown and floating sphere GBM cultures. Afterward, formazan was dissolved in a mixture (24:1 *v:v*) of isopropanol and 1 M HCl using 100 µL for adherent cells by shaking the plates for at least 20 min. The absorbance of formazan (λ = 570 nm), and background (λ = 630 nm) was measured with a microplate reader (Tecan Spark, Tecan Deutschland GmbH, Crailsheim, Germany).

4.5. RA Reporter Assay

The F9 embryonal carcinoma cell line, stably transfected with the 1.8 kb promoter sequence of RARb2 coupled to the *lacZ* gene, was used to measure active retinoids [44]. The assay is appropriate for detection of all-*trans* RA but is able to detect other retinoid isomers as well. The F9 reporter cells were maintained in 10% FCS containing DMEM in the presence of 400 ug/mL G418. A day before the assay, F9 cell were seeded onto 24-well plates (100,000 cell/well). The cells were than treated with various concentrations of **2**. After 20 h, the β-galactosidase activity was visualized by X-Gal staining, which was followed by densitometric image analyses from n = 3 fields of view (10× magnification) of n = 4 independent cultures.

4.6. Reactivity Test with Glutathione

In a cell-free activity assay, the reaction of **1** and **2** with glutathione was measured by the decreasing absorption at 340 nm (enone group). Then, 2 µL of the substances (10 mM in DMSO) was mixed with 98 µL glutathione solution (6.5 µM in PBS) in a 96-well plate. The absorption was measured at 340 nm every 30 s for 2 h at rt (Tecan F200). All experiments were performed at least four times.

4.7. Cell Death and Apoptosis Induction

U251 und Mz54 glioblastoma cells (6×10^5 cells/mL, 0.5 mL/well) were seeded in 24-well plates. The next day, the cells were treated with **1**, **2** or vehicle (DMSO) for 48 h. Prior to the measurement, the cells were harvested and stained using Annexin-V-APC (BD Biosciences, Heidelberg, Germany) and propidium iodide (PI, 0.05 mg/mL, Sigma-Aldrich, Taufkirchen, Germany). Cells were analyzed with BD Accuri C6 flow-cytometer (BD Biosciences), and data processing was performed using BD Accuri C6 software (BD Biosciences).

4.8. Immunohistochemistry

Astroglia cells grown on poly-L-lysine-coated glass cover slips were fixed with 4% paraformaldehyde in PBS for 20 min at room temperature and then permeabilized with Triton X-100 (0.1% *v*/*v* in PBS; 5 min). Nonspecific antibodies were blocked by incubation with 3% FBS in PBS (room temperature, 1 h). α-GFAP antibody (rabbit, DAKO) was used in a dilution of 1:2000 and was visualized by anti-rabbit IgG Alexa 594 (1:1000). DAPI (4′,6-diamidino-2-phenylindole) was used for nuclei staining. Fluorescence images were captured manually on a Zeiss Axiovert 200 M microscope fitted with 20- to 60-fold zoom and a Zeiss AxioCam MRm digital camera. Further applied antibodies were tubulin III (Sigma T5076), used at a dilution of 1:1000, and Annexin V. Cy3.18 (Sigma A4963).

4.9. Adult Organotypic Brain Slice Cultures and Ex Vivo Tumor Growth Assay

Adult organotypic brain slice culture was carried out as described previously [26,45,46]. Briefly, the brains from adult C57BL/6 J (Envigo, Horst, The Netherlands) were dissected, and dura mater was removed after the mice were euthanized. Subsequently, mouse brains were placed in warm (35–40 °C) 2% low-melting agarose (Carl Roth, Karlsruhe, Germany) and cut on a Vibratome VT1000 (Leica, Wetzlar, Germany) in 150 µm thick sections. These sections were placed on Milli-cell culture inserts (Merck KGaA, Darmstadt, Germany) and cultured in 6-well plates using FCS-free medium consisting of DMEM/F12 supplied with 1× B27 and 1× N2 supplement and 100 U/mL penicillin and 100 µg/mL streptomycin (all from Gibco, Darmstadt, Germany). One day later, multiple spheres were placed on the mouse brain slices. Adequate spheres were prepared by seeding 275.000 $NCH644^{GFP+}$ cells in a total volume of 5 mL in a T25 flask. One day after sphere transplantation, pictures were taken (day 0), and the treatment was started, which was refreshed 3 times per week. Tumor growth was evaluated using FIJI, after which pictures were taken regularly with a Nikon SMZ25 stereomicroscope equipped with a P2-SHR Plan Apo 2 × objective operated by NIS elements software. As the tumor size was normalized to the size on day 0, growth curves were created.

Supplementary Materials: The following supporting information (stability tests) can be downloaded at: https://www.mdpi.com/article/10.3390/ijms23169056/s1.

Author Contributions: Conceptualization, B.L., M.Z., E.M., R.S., D.K. and B.B.; methodology, B.L., M.Z., Z.K., L.H.F.K., S.S., D.M., A.W. and B.B.; validation, B.L., M.Z. and Z.K.; formal analysis, B.L. and M.Z.; investigation, B.L., M.Z., Z.K., L.H.F.K., S.S., D.M. and B.B.; resources, B.L., E.M., R.S. and D.K.; data curation, B.L., M.Z. and Z.K.; writing—original draft preparation, B.L., M.Z., Z.K. and B.B.; writing—review and editing, B.L., M.Z., Z.K., R.S. and D.K.; supervision, E.M., R.S., D.K. and B.B.; project administration, R.S. and B.B.; funding acquisition, B.L., E.M., R.S. and D.K. All authors have read and agreed to the published version of the manuscript.

Funding: R.S. and B.L. are grateful to the Deutsche Forschungsgemeinschaft (grants Scho 402/12-2 and LI 3687/2-1).

Institutional Review Board Statement: The necessary mouse sacrifices were a killing for scientific purposes according to the German animal protection law (Tierschutzgesetz) and only require reporting to the local authority without prior approval.

Informed Consent Statement: Not applicable.

Data Availability Statement: Data supporting reported results can be obtained from the corresponding author upon request.

Acknowledgments: B.L. and D.K. would like to thank Hildegard König for technical assistance. Additionally, B.L. and D.K. thank Stefan Liebner (Edinger Institute, University Hospital Frankfurt) for access to the SMZ25 stereomicroscope. The authors are grateful to Jonathan Corcoran and Malcolm Maden for providing the F9 RA reporter cells.

Conflicts of Interest: The authors declare no conflict of interest.

References

1. Anchel, M.; Hervey, A.; Robbins, W.J. Antibiotic substances from Basidiomycetes. VII. *Clitocybe illudens*. *Proc. Natl. Acad. Sci. USA* **1950**, *36*, 300–305. [CrossRef] [PubMed]
2. Dick, R.A.; Yu, X.; Kensler, T.W. NADPH alkenal/one oxidoreductase activity determines sensitivity of cancer cells to the chemotherapeutic alkylating agent irofulven. *Clin. Cancer Res.* **2004**, *10*, 1492–1499. [CrossRef]
3. McMorris, T.C.; Yu, J.; Lira, R.; Dawe, R.; MacDonald, J.R.; Waters, S.J.; Estes, L.A.; Kelner, M.J. Structure-activity studies of antitumor agent irofulven (hydroxy-methylacylfulvene) and analogues. *J. Org. Chem.* **2001**, *66*, 6158–6163. [CrossRef] [PubMed]
4. Kelner, M.J.; McMorris, T.C.; Taetle, R. Preclinical evaluation of illudins as anticancer agents: Basis for selective cytotoxicity. *J. Natl. Cancer Instit.* **1990**, *82*, 1562–1656. [CrossRef] [PubMed]
5. Schobert, R.; Knauer, S.; Seibt, S.; Biersack, B. Anticancer active illudins: Recent developments of a potent alkylating compound class. *Curr. Med. Chem.* **2011**, *18*, 790–807. [CrossRef]
6. Senzer, N.; Arsenau, J.; Richards, D.; Berman, B.; MacDonald, J.R.; Smith, S. Irofulven demonstrates clinical activity against metastatic hormone-refractory prostate cancer in a phase 2 single-agent trial. *Am. J. Clin. Oncol.* **2005**, *28*, 36–42. [CrossRef]
7. Schlett, K.; Madarász, E. Retinoic acid induced neural differentiation in a neuroectodermal cell line immortalized by p53 deficiency. *J. Neurosci. Res.* **1997**, *47*, 405–415. [CrossRef]
8. Tang, X.-H.; Gudas, L.J. Retinoids, retinoic acid, and cancer. *Annu. Rev. Pathol. Mech. Dis.* **2011**, *6*, 345–364. [CrossRef]
9. Chambon, P. A decade of molecular biology of retinoic acid receptors. *FASEB J.* **1996**, *10*, 940–954. [CrossRef]
10. Hua, S.; Kittler, R.; White, K.P. Genomic antagonism between retinoic acid and estrogen signaling in breast cancer. *Cell* **2009**, *137*, 1259–1271. [CrossRef]
11. Aebi, S.; Kroning, R.; Cenni, B.; Sharma, A.; Fink, D.; Los, G.; Weisman, R.; Howell, S.B.; Christen, R.D. All-trans retinoic acid enhances cisplatin induced apoptosis in human ovarian adenocarcinoma and in squamous head and neck cancer cells. *Clin. Cancer Res.* **1997**, *3*, 2033–2038. [PubMed]
12. Tavares, T.S.; Nanus, D.; Yang, X.J.; Gudas, L.J. Gene microarray analysis of human renal cell carcinoma: The effects of HDAC inhibition and retinoid treatment. *Cancer Biol. Ther.* **2008**, *7*, 1607–1618. [CrossRef] [PubMed]
13. Schobert, R.; Biersack, B.; Knauer, S.; Ocker, M. Conjugates of the fungal cytotoxin illudin M with improved tumour specificity. *Bioorg. Med. Chem.* **2008**, *16*, 8592–8597. [CrossRef] [PubMed]
14. Benda, P.; Lightbody, J.; Sato, G.; Sweet, W. Differentiated rat glial cell strain in tissue culture. *Science* **1968**, *61*, 370–371. [CrossRef] [PubMed]
15. Levenson, A.S.; Jordan, V.C. MCF-7: The first hormone-responsive breast cancer cell line. *Cancer Res.* **1997**, *57*, 3071–3078.
16. Robey, R.W.; Polgar, O.; Deeken, J.; To, K.W.; Bates, S.E. ABCG2: Determining its relevance in clinical drug resistance. *Cancer Metastasis Rev.* **2007**, *26*, 39–57. [CrossRef]
17. Umesono, K.; Murakami, K.K.; Thompson, C.C.; Evans, R.M. Direct repeats as selective response elements for the thyroid hormone, retinoic acid and vitamin D3 receptors. *Cell* **1991**, *64*, 1255–1266. [CrossRef]
18. Leid, M.; Kastner, P.; Lyons, R.; Nakshatri, H.; Saunders, M.; Zacharewski, T.; Chen, J.-Y.; Staub, A.; Garnier, J.-M.; Mader, S.; et al. Purification, cloning, and RXR identity of the HeLa cell factor with which RAR or TR heterodimerizes to bind target sequences efficiently. *Cell* **1992**, *68*, 377–395. [CrossRef]
19. De Luca, L. Retinoids and their receptors in differentiation, embryogenesis, and neoplasia. *FASEB J.* **1991**, *5*, 2924–2933. [CrossRef] [PubMed]
20. Maden, M. Retinoic acid in the development, regeneration and maintenance of the nervous system. *Nat. Rev. Neurosci.* **2007**, *8*, 755–765. [CrossRef] [PubMed]
21. Wagner, M.; Han, B.; Jessell, T.M. Regional differences in retinoid release from embryonic neural tissue detected by an in vitro reporter assay. *Development* **1992**, *116*, 55–66. [CrossRef] [PubMed]

22. Reynolds, A.; Lundblad, V. Yeast vectors and assays for expression of cloned genes. In *Current Protocols in Molecular Biology*; Ausubel, F.A., Brent, R., Kingston, R.E., Moore, D.D., Seidman, J.G., Smith, J.A., Struhl, K., Eds.; Greene Publishing and Wiley-Interscience: New York, NY, USA, 1989; pp. 13.6.1–13.6.4.
23. Yang, Z.; Wang, K.K. Glial fibrillary acidic protein: From intermediate filament assembly and gliosis to neurobiomarker. *Trends Neurosci.* **2015**, *38*, 364–374. [CrossRef]
24. Környei, Z.; Slávik, V.; Szabó, B.; Gócza, E.; Czirók, A.; Madarász, E. Humoral and contact interactions in astroglia/stem cell co-cultures in the course of glia-induced neurogenesis. *Glia* **2005**, *49*, 430–444. [CrossRef]
25. Campos, B.; Wan, F.; Farhadi, M.; Ernst, A.; Zeppernick, F.; Tagscherer, K.E.; Ahmadi, R.; Lohr, J.; Dictus, C.; Gdynia, G.; et al. Differentiation therapy exerts antitumor effects on stem-like glioma cells. *Clin. Cancer Res.* **2010**, *16*, 2715–2728. [CrossRef] [PubMed]
26. Gerstmeier, J.; Possmayer, A.-L.; Bozkurt, S.; Hoffmann, M.E.; Dikic, I.; Herold-Mende, C.; Burger, M.C.; Münch, C.; Kögel, D.; Linder, B. Calcitriol promotes differentiation of glioma stem-like cells and increases their susceptibility to temozolomide. *Cancers* **2021**, *13*, 3577. [CrossRef] [PubMed]
27. Minami, N.; Maeda, Y.; Shibao, S.; Arima, Y.; Ohka, F.; Kondo, Y.; Maruyama, K.; Kusuhara, M.; Sasayama, T.; Kohmura, E.; et al. Organotypic brain explant culture as a drug evaluation system for malignant brain tumors. *Cancer Med.* **2017**, *6*, 2635–2645. [CrossRef]
28. De Boüard, S.; Herlin, P.; Christensen, J.G.; Lemoisson, E.; Gauduchon, P.; Raymond, E.; Guillamo, J.-S. Antiangiogenic and inti-invasive effects of sunitinib on experimental human glioblastoma. *Neuro- Oncol.* **2007**, *9*, 412–423. [CrossRef]
29. Soubéran, A.; Tchoghandjian, A. Practical review on preclinical human 3D glioblastoma models: Advances and challenges for clinical translation. *Cancers* **2020**, *12*, 2347. [CrossRef]
30. Abdalla, A.N.; Malki, W.H.; Qattan, A.; Shahid, I.; Hossain, M.A.; Ahmed, M. Chemosensitization of HT29 and HT29-5FU cell lines by a combination of a multi-tyrosine kinase inhibitor and 5FU downregulates ABCC1 and inhibits PIK3CA in light of their importance in Saudi colorectal cancer. *Molecules* **2021**, *26*, 334. [CrossRef]
31. Reshkin, S.J.; Bellizzi, A.; Caldeira, S.; Albarani, V.; Poignee, M.; Alunni-Fabbroni, M.; Casavola, V.; Tommasino, I. Na^+/H^+ exchanger-dependent intracellular alkalinization is an early event in malignant transformation and plays an essential role in the development of subsequent transformation-associated phenotypes. *FASEB J.* **2000**, *14*, 2185–2197. [CrossRef]
32. Boscoboinik, D.; Gupta, R.S.; Epand, R.M. Investigation of the relationship between altered intracellular pH and multidrug resistance in mammalian cells. *Br. J. Cancer* **1990**, *61*, 568–572. [CrossRef] [PubMed]
33. Otreba, M.; Stojko, J.; Kabala-Dzik, A.; Rzepecka-Stojko, A. Perphenazine and prochlorperazine decrease glioblastoma U-87 MG cell migration and invasion: Analysis of the ABCB1 and ABCG2 transporters, E-cadherin, α-tubulin and integrins (α3, α5, and β1) levels. *Oncol. Lett.* **2022**, *23*, 182. [CrossRef] [PubMed]
34. Zheng, X.; Shen, G.; Yang, X.; Liu, W. Most C6 cells are cancer stem cells: Evidence from clonal and population analyses. *Cancer Res.* **2007**, *67*, 3691–3697. [CrossRef] [PubMed]
35. Lee, H.Y.; Sueoka, N.; Hong, W.-K.; Mangelsdorf, D.J.; Claret, F.X.; Kurie, J.M. All-trans-retinoic acid inhibits Jun N-terminal kinase by increasing dual-specificity phosphatase activity. *Mol. Cell. Biol.* **1999**, *19*, 1973–1980. [CrossRef]
36. Wang, W.; Waters, S.J.; MacDonald, J.R.; Roth, C.; Shentu, S.; Freeman, J.; Von Hoff, D.D.; Miller, A.R. Irofulven (6-hydroxymethylacylfulvene, MGI 114)-induced apoptosis in human pancreatic cancer cells is mediated by ERK and JNK kinases. *Anticancer Res.* **2002**, *22*, 559–564.
37. Sanchez, C.; Galve-Roperh, I.; Canova, C.; Brachet, P.; Guzman, M. D9-Tetrahydrocannabinol induces apoptosis in C6 glioma cells. *FEBS Lett.* **1998**, *436*, 6–10. [CrossRef]
38. Rueda, D.; Galve-Roperh, I.; Haro, A.; Guzman, M. The CB1 cannabinoid receptor is coupled to the activation of c-Jun N-terminal kinase. *Mol. Pharmacol.* **2000**, *58*, 814–820. [CrossRef]
39. Tang, K.; Cao, L.; Fan, S.Q.; Wu, M.H.; Huang, H.; Zhou, Y.H.; Zhou, M.; Tang, Y.L.; Wang, R.; Zeng, F.; et al. Effect of all-trans retinoic acid on C6 glioma cell proliferation and differentiation. *Zhong Nan Da Xue Xue Bao Yi Xue Ban* **2008**, *33*, 892–897.
40. Friedman, H.S.; Keir, S.T.; Houghton, P.J.; Lawless, A.A.; Bigner, D.D.; Waters, S.J. Activity of irofulven (6-hydroxymethylacylfulvene) in the treatment of glioblastoma multiforme-derived xenografts in mice. *Cancer Chemother. Pharmacol.* **2001**, *48*, 413–416. [CrossRef]
41. Zoldakova, M.; Kornyei, Z.; Brown, A.; Biersack, B.; Madarász, E.; Schobert, R. Effects of a combretastatin A4 analogous chalcone and its Pt-complex on cancer cells: A comparative study of uptake, cell cycle and damage to cellular compartments. *Biochem. Pharmacol.* **2010**, *80*, 1487–1496. [CrossRef]
42. Jády, A.G.; Nagy, Á.M.; Köhidi, T.; Ferenczi, S.; Tretter, L.; Madarász, E. Differentiation-dependent energy production and metabolite utilization: A comparative study on neural stem cells, neurons, and astrocytes. *Stem Cells Dev.* **2016**, *25*, 995–1005. [CrossRef]
43. Campos, B.; Gal, Z.; Baader, A.; Schneider, T.; Sliwinski, C.; Gassel, K.; Bageritz, J.; Grabe, N.; von Deimling, A.; Beckhove, P.; et al. Aberrant self-renewal and quiescence contribute to the aggressiveness of glioblastoma. *J. Pathol.* **2014**, *234*, 23–33. [CrossRef] [PubMed]
44. Sonneveld, E.; van den Brink, C.E.; van der Leede, B.J.; Maden, M.; and Van Der Saag, P.T. Embryonal carcinoma cell lines stably transfected with mRARbeta2-lacZ: Sensitive system for measuring levels of active retinoids. *Exp. Cell Res.* **1999**, *250*, 284–297. [CrossRef]

45. Linder, B.; Wehle, A.; Hehlgans, S.; Bonn, F.; Dikic, I.; Rödel, F.; Seifert, V.; Kögel, D. Arsenic trioxide and (-)-gossypol synergistically target glioma stem-like cells via inhibition of Hedgehog and Notch signaling. *Cancers* **2019**, *11*, 350. [CrossRef] [PubMed]
46. Remy, J.; Linder, B.; Weirauch, U.; Konovalova, J.; Marschalek, R.; Aigner, A.; Kögel, D. Inhibition of PIM1 blocks the autophagic flux to sensitize glioblastoma cells to ABT-737-induced apoptosis. *Biochim. Biophys. Acta Mol. Cell. Res.* **2019**, *1866*, 175–189. [CrossRef] [PubMed]

www.ingramcontent.com/pod-product-compliance
Lightning Source LLC
Chambersburg PA
CBHW061839210926
51850CB00006B/930
9783036583112

MDPI
St. Alban-Anlage 66
4052 Basel
Switzerland
Tel. +41 61 683 77 34
Fax +41 61 302 89 18
www.mdpi.com

International Journal of Molecular Sciences Editorial Office
E-mail: ijms@mdpi.com
www.mdpi.com/journal/ijms